ITT Technical Institute - 166
5905 Stewart Parkway
Douglasville, GA 30135

FUNDAMENTALS OF MICROCONTROLLERS

AND APPLICATIONS IN EMBEDDED SYSTEMS

(With the PIC18 Microcontroller Family)

Ramesh S. Gaonkar

State University of New York
O.C.C. Campus at Syracuse

THOMSON

DELMAR LEARNING™

Australia   Canada   Mexico   Singapore   Spain   United Kingdom   United States

Fundamentals of Microcontrollers and Applications in Embedded Systems (with the PIC18 Microcontroller Family)

Ramesh S. Gaonkar

Vice President, Technology and Trades ABU:
David Garza

Director of Learning Solutions:
Sandy Clark

Executive Editor:
Stephen Helba

Managing Editor:
Larry Main

Senior Product Manager:
Michelle Ruelos Cannistraci

Marketing Director:
Deborah S. Yarnell

Marketing Manager:
Guy Baskaran

Marketing Coordinator:
Shanna Gibbs

Director of Production:
Patty Stephan

Content Project Manager:
Christopher Chien

Technology Project Manager:
Kevin Smith

Senior Editorial Assistant:
Dawn Daugherty

Library of Congress Cataloging-in-Publication Data

Gaonkar, Ramesh S.
  Fundamentals of microcontrollers and applications in embedded systems (with the PIC18 microcontroller family) / Ramesh Gaonkar.
    p. cm.
  Includes bibliographical references.
  ISBN 1-4018-7914-4
  1. Programmable controllers. 2. Microelectronics. 3. Embedded computer systems. I. Title.
  TJ223.P76G36 2006
  629.8'95––dc22
                    2006019204

NOTICE TO THE READER

Dedication:

To my wife: Shaila

Without her unwavering support,
understanding, and encouragement,
the completion of this textbook
would not have been possible.

BRIEF CONTENTS

TABLE OF CONTENTS

ix

This text is intended primarily for undergraduate students in technology and engineering curricula and also for practicing engineers who are working in embedded systems. It is a comprehensive treatment of the micro-controller fundamentals that includes architecture, assembly language programming, and applications of microcontrollers in embedded systems. Microchip PIC18F microcontroller families are used in the illustrations. The text assumes a course in digital logic as a prerequisite; however, it does not assume any background in programming.

The purpose of this text is threefold: one is to teach the basic principles underlying the microprocessor/microcontroller and its applications; the second is to teach assembly language; and the third is to expose students to the latest technology that is used in industry. These three objectives can be accomplished by using the PIC18F microcontroller as an illustration. The text is aimed at two levels: an introductory course in microcontrollers and an advanced course in designing microcontroller-based systems.

CHANGING TECHNOLOGY AND 8-BIT MICROCONTROLLERS

The rationale for 8-bit microcontroller:

1.   In the last 30+ years, microprocessor technology has gone from 4-bit processors to 32-, 64-, and 128-bit processors. Initially, we focused on the details of the microprocessor as a device. As we gained considerable experience, the basic principles underlying the device began to evolve. Now it is time to focus on the principles underlying the microprocessor/microcontroller rather than on the device.

2.   From the teaching point of view, the basic concepts are easier to teach with the 8-bit than with the high-end microprocessors.

3.   The 8-bit microcontrollers have already established their market in embedded systems, which remains a growth area.

Why Assembly Language

To understand the underlying hardware of the microcontroller, and troubleshoot the system, a thorough knowledge of assembly language is essential. In industry, even though the majority of the embedded system programming is done in C, some critical segments of the software code are written in assembly language. This text is intended for the entry-level course in the microcontroller area, and it assumes no background in programming, a situation likely to emerge in many engineering technology programs. In my opinion, it is very difficult to teach assembly and C languages and microcontroller concepts in one course.

The 18F series of microcontrollers are at the high end of the spectrum of devices available from Microchip; however, medium-level and low-end series of microcontrollers are commonly used in many products. In the systems where available memory is limited, assembly language is likely to be the language of choice for program development.

Features of PIC18F

If new product development in the area of embedded systems is based on a microcontroller, industries tend to select either the PIC series, the Motorola 68HC family, 8051 family variants, or the Atmel AVR series of microcontrollers. However, this year, Microchip (PIC devices) has ranked number one among the suppliers of 8-bit devices.

1. The PIC microcontroller family of devices is designed with the Harvard architecture, and students in engineering and technology should be exposed to the concepts of the Harvard architecture.

2. The PIC microcontroller family range includes 8-pin to 100-pin devices and can be used in various embedded systems—from simple to complex.

3. PIC18F trainers are easily available for the educational market at a reasonable cost, and they are supported by the manufacturer (Microchip).

4. Product development work (writing, assembling, executing, and troubleshooting programs and hardware) for the PIC devices is facilitated by free IDE (Integrated Development Environment) tools available from Microchip.

Approach and Rationale

The microprocessor is a general-purpose programmable logic device. A thorough understanding of the microprocessor demands the concepts and skills from two different disciplines: hardware concepts from digital electronics and programming skills from computer science. Hardware is the physical structure of the microprocessor, and the programming makes it alive; one without the other is meaningless. The integration of a microprocessor, memory, and I/O on a single chip leads to the concept of the microcontroller.

Some of my assumptions and observations are as follows:

1. Software (instructions) is an integral part of the microprocessor and demands as much attention as the hardware.

2. In industry, for the development of microcontroller-based projects, a much larger percentage of time and effort is devoted to software than to hardware.

3. In embedded systems, hardware and software are closely interlinked; hardware and software designs begin almost simultaneously. Therefore, the partitioning tasks between hardware and software and integrating these tasks at a later stage is an essential step in designing embedded systems.

4. Technology and engineering students tend to be hardware oriented and have considerable difficulty in programming.

5. Students have difficulty in understanding mnemonics and realizing the critical importance of flags.

Therefore, in this text, the contents are presented with an integrated approach to hardware and software in the context of the PIC18F microcontroller. The text begins with basic concepts of a microcontroller using PIC18F family of devices, introduces its assembly language, illustrates various data transfer processes through interfacing peripherals, and finally integrates concepts discussed in the text by designing a simple embedded system.

Manuscript Organization

1. **Chapters 1 through 4** focus on basic hardware concepts underlying a microprocessor-based product and associated programming in assembly language. Chapter 1 presents an overview of the primary components (microprocessor, memory, and I/O) of a microprocessor-based product, and Chapter 2 introduces the PIC18F microcontroller. Chapter 3 discusses assembly language programming in the context of the PIC18F, and Chapter 4 outlines the processes of writing, assembling, and executing the assembly language programs using IDE.

2. **Chapters 5 through 8** deal with the assembly language programming of the PIC18F microcontroller. Chapters 5 and 6 introduce assembly language mnemonics of the PIC18F microcontroller with simple illustrative programs. Chapters 7 and 8 examine the concepts of stacks and subroutines and illustrate several programs as subroutines. The contents are presented in a step-by-step format. Each instruction is reviewed briefly, and a group of instructions are illustrated through simple programs. Each illustrative program begins with a problem statement, provides the analysis of the problem, illustrates the program, explains the programming steps, and demonstrates steps in troubleshooting by using the MPLAB Simulator.

3. **Chapters 9 through 13** illustrate various concepts and processes of data transfer such as interrupts, parallel and serial I/O by interfacing peripherals with the PIC18F452/4520 microcontroller. Each illustration explains the hardware and describes how hardware and software work together in transferring data between the microcontroller and its peripherals. Chapter 9 introduces the interfacing concepts and illustrates the interfacing of peripheral devices such as switches, LEDs, a keyboard, and an LCD, and Chapter 10 deals with the concepts related to the interrupt, and its applications. Chapter 11 describes the operations of timers and how they are used in conjunction with CCP (Capture, Compare, and PWM) modules in PIC18F microcontrollers. Chapter 12 introduces the concepts of data converters and illustrates the interfacing of sensors using the A/D converter module. Chapter 13 explains the basic concepts in serial communication and various protocols such as EIA-232, SPI, and I^{2}C. Applications of these protocols are illustrated using USART and MSSP modules in the PIC microcontroller.

4. **Chapter 14** integrates hardware concepts and software techniques by illustrating the design processes of an embedded system—a Time and Temperature Monitoring System (TTMS). This project brings together many concepts discussed in the text.

5. **Appendix A:** This appendix includes the complete set of PIC18F instructions explained with illustrative examples, listed in alphabetical order. This is an extremely useful feature: students need easy access to the instruction set with complete explanations.

Complementary Instructional Materials on CD and Web Sites

- The accompanying CD includes: 1) assembly language programs discussed in the text that can be assembled and executed either on the Microchip IDE® Simulator or an available trainer; 2) a PIC18 IDE Simulator by Oshonsoft, valid for a semester after installation, that can be used for some I/O interfacing exercises; 3) data sheets of the PIC18F4520 microcontroller and various other devices.

- The Microchip IDE software can be downloaded from http://www.microchip.com

- Additional programs in C language and updates related to the PIC microcontroller are available on the textbook's Online Companion at www.electronictech.com.

Features of the Text:

1. Comprehensive Integrated Approach to Hardware and Software.

   The text is divided in three sections and the content is comprehensive: basic concepts in hardware, assembly language, and integration of hardware and software in interfacing peripherals and designing an embedded system.

2. Organization Appropriate to Teaching

   Most microprocessor courses are laboratory oriented. Initially, this textbook covers enough hardware architecture essentials to prepare students to write programs; students will be able to write simple programs beginning with Chapter 2. When students are able to write programs, the textbook returns to hardware topics.

3. Step-By-Step Approach for Software

   Students are exposed to programming in a systematic way. This is the area where students need assistance. Often students enter this type of course with no programming background. The text has been designed to supply that needed support.

   The programming section includes numerous illustrative examples. These examples begin with a problem statement, provide the analysis of the problem, illustrate the program, and explain the programming steps. Chapters 5 through 8 differ from any available texts, and are the major strength of the book. Instructions are explained in detail through the use of examples, not just listed from the Microchip manual.

   Examples and explanations of instructions enhance students' understanding of programming. An instruction set with full explanation is essential in writing programs.

4. Text is laboratory oriented

   This feature also reinforces the integral aspect of software with hardware. Chapters 5 through 8 can be used for laboratory experiences. The section includes illustrative programs that can be used to verify programming concepts on a simulator or single-board demo board.

5. In-Depth Treatment of I/O Interfacing.

   Chapters 9 through 13 are devoted to I/O interfacing and various peripherals. Illustrative examples in these chapters can be used in laboratory to perform hardware experiments.

6. Synthesis of the Concepts through Project Design

   Chapter 14 is concerned with the design of a temperature control and display project that synthesizes all the concepts discussed in earlier chapters.

7. Appendices include details of software for laboratory experiments.

   MPLAB IDE, an integrated software package that includes an assembler, a debugger, and a simulation program, is available from Microchip. Students can use the source code included on the CD in laboratory experiments.

A WORD WITH FACULTY

This text is an attempt to share my classroom experiences, my course development efforts, and my observations in industrial practices.

This text can be used flexibly to meet the objectives of various courses at the undergraduate level. If used for a one-semester introductory course with 50 percent hardware and 50 percent software emphasis, the following chapters are recommended: Chapters 1 through 4 for hardware lectures, Chapters 5 through 8 for software laboratory sessions, and Chapter 9 for interfacing. If the course is heavily oriented towards hardware, and project oriented, Chapters 1 through 4 and Chapters 9 through 14 are recommended, and necessary programs can be selected from Chapters 5 through 8. If the course is heavily oriented toward software, Chapters 1 through 8 and selected portions of Chapter 9 are recommended. For a two-semester course, it is best to use the entire text.

Technology and engineering students tend to be hardware oriented and have considerable difficulty in programming. I have made a few suggestions in the following section: A Word with Students.

An Instructor's Manual with answers to end-of-chapter questions is available (Delmar order number 1401879152).

A WORD WITH STUDENTS

The designing of microcontroller-based embedded systems is an exciting, challenging, and growing field; it will pervade industry for decades to come. To meet the challenges of this growing technology, you will have to be conversant with the programmable aspect of the microcontroller. Assembly language programming is a process of problem solving and communication in a strange language of mnemonics. Most often, hardware-oriented students find this communication process very difficult. One of the questions frequently asked by students is "How do I get started in a given programming assignment?" How does an architect approach building a house? The first step is to break down the entire problem into small tasks and organize these tasks in an architectural drawing. Similarly, the programming tasks can be split into small tasks and organized by drawing a flowchart. You should never begin just writing instructions without a flowchart. Once the flowchart has been drawn, label the nodes where arrows come together, and translate each block into instructions. Many of you encounter difficulties in translating decision-making (jump) blocks. You should ask two questions: 1) what is the flag (condition) we are looking for? and 2) do we want jump instruction based on the condition that is true or false? To get some help in translating flowcharts into instructions, examine various types of programs and imitate them. You can begin by studying the illustrative program relevant to an assignment, its flowchart, its analysis, program description, and particularly the comments. Read the instructions from Appendix A as necessary and pay attention to the flags. This text is written in such a way that simple programming aspects of the microcontroller can be self-taught. Once you master the elementary programming techniques, then interfacing and design becomes exciting and fun.

ACKNOWLEDGMENTS

The author is recognized for the publication of a book, but there are many helping hands behind the scene: the family, friends, colleagues, reviewers, and editors, and without whose contributions this textbook would not have seen the light of the day.

My sincere thanks to my family members: my wife, Shaila, for her unwavering support for my writing activities and creating a conducive home environment that enables me to write, and quietly stepping in to taking over many tasks for which I am responsible. Also thanks to my daughters, Nelima and Vanita, for their continued interest in my publishing ventures.

Several persons have made valuable contributions to this text. I would like to extend my sincere appreciation to Tab Cox, my former student, my colleagues in our evening division, and an engineer at a local company, who introduced me to PIC controllers and was always there to answer any of my questions. Similarly, I cannot give thanks enough to my reviewers—their contributions were invaluable, and they deserve credit for shaping this book for its technical contents. They read every word of my rough draft, gave invaluable technical help in pointing out errors and sometimes in editing my awkward phrasing and articles. I appreciate the painstaking efforts and numerous suggestions of my reviewers. Many thanks to

Norman Ahlhelm, Central Texas College, Killeen, TX
Walt Bankes, Rochester Institute of Technology, Rochester, NY
David Barth, Edison Community College, Piqua, OH
James Britton, Southeastern Oklahoma State University, Durant, OK
Ron Krahe, Penn State Behrend, Erie, PA
Frank Lanzer, Anne Arundel Community College
Charles Osborn, DeVry University, Fremont, CA
Lloyd Stallkamp, Montana State University, Havre, MT

If this text reads well, the credit goes to Dave Franke, who edited the rough draft, as well as to Eileen Troy and her colleagues at GEX Publishing Services who saw through the entire process of copyediting and the production. I thank Dave Garza and Steve Helba for initiating the project, and appreciate all the support and efforts of Dawn Daugherty, Michelle Ruelos Cannistraci, and Christopher Chien from Thomson Delmar Learning for seeing the project through to completion.

I would appreciate any communication about the text from the reader, so please feel free to send me an e-mail message or contribute to Web site information.

Ramesh S. Gaonkar
State University of New York
O.C.C. Campus at Syracuse
Syracuse, New York 13215
E-mail: gaonkarr@sunyocc.edu

CHAPTER 1

MICROPROCESSOR AND MICROCONTROLLER FUNDAMENTALS

OVERVIEW

The microcontroller has become integral to many industrial and consumer products, such as laboratory instruments, medical equipment, automobile dash boards, wireless phones, and home appliances. The operations of these products or systems are controlled or managed behind the scenes by the microcontroller; thus known as embedded systems. In some of these systems, the microcontroller is used as a stand-alone device; in others, it is part of a larger system. The microcontroller consists primarily of three components: microprocessor, memory, and input/output (I/O) ports on a single semiconductor chip. If the system is designed using three separate (discrete) components rather than integrated on the same chip, it is called a "microprocessor-based" system. The physical components of these systems are referred to as the hardware, and the instructions that control the operations of these systems are called the software.

In these embedded systems, the microprocessor is the central player; it processes data based on the instructions stored in memory, communicates with memory and I/Os, and controls the timing of all its operations. The instructions (programs) that are stored in memory are in binary 0s and 1s referred to as the "machine language." However, the programs are written either in an assembly language or a high-level language.

This chapter describes hardware components of an embedded system and programming languages used to operate or manage that system. It also explains how various types of numbers and characters we use in our daily lives are represented in binary numbers. Finally, based on these concepts, a simple embedded system—a time and temperature display—is described.

OBJECTIVES

- List the components of an embedded microcontroller-based system.
- Explain the difference between a microcontroller and microprocessor, and MCU and MPU.
- Explain the difference between a microcontroller-based and a microprocessor-based system.
- Draw a block diagram of a microprocessor-based system showing the microprocessor, memory, I/Os, and buses.
- Describe the functions of the address bus, data bus, and control signals.
- Explain the relationship between the address lines on a memory chip and number of registers inside the chip.
- Explain the functions of control signals Read (RD) and Write (WR).
- Explain differences between machine, assembly, and high-level languages, and list the advantages and disadvantages of an assembly language and a high-level language.
- Explain how the following numbers and characters are represented in an 8-bit microprocessor: signed and unsigned numbers, Binary Coded Decimal (BCD) numbers, and *American Standard Code for Information Interchange* (ASCII) characters.
- List the components of a time-temperature embedded system and draw its block diagram.

1.1 EMBEDDED SYSTEMS AND MICROCONTROLLERS

In our daily lives, we are surrounded by appliances and products such as microwave ovens, digital clocks, copying machines, mobile phones, and music systems that are run or controlled by electronic devices, called microcontrollers. These microcontrollers are not visible to the user; rather they are embedded inside the system. Those who use a Personal Computer (PC) or microcomputer are familiar with three major components of the PC system: microprocessor, memory, and I/O as discrete or individual components. A **microcontroller** is an electronic device that includes these three components: microprocessor, memory, and I/O (Figure 1-1) on a single semiconductor unit, called an integrated circuit. The terms microcontroller and microcontroller unit (MCU) are used synonymously. In addition to these three components, the microcontroller includes many support devices (such as timers, A/D converters, and serial I/O) as shown in Figure 1-1, and the functions of these devices will be explained later. In a very limited way, a microcontroller can be seen as a human head that includes brain, memory, eyes, ears, and mouth. The microprocessor processes information similar to the brain, memory stores information, and I/O devices either receive information or send out information similar to eyes, ears, and mouth. The human brain is very complex compared to the microprocessor; however, the processes of elementary functions, such as adding two numbers, can be similar in the brain and the microprocessor.

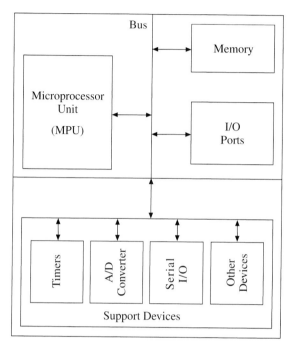

FIGURE 1-1   Block Diagram—Microcontroller or Microcontroller Unit (MCU)

In an embedded system, the microcontroller can be a part of a larger system such as a copying machine, which is responsible for carrying out a specific function, or it may be the sole component of a stand-alone system such as a digital clock. Our focus in this text is on fundamental concepts underlying the microcontroller. Therefore, we will discuss embedded systems first in terms of the following three components: microprocessor, memory, and I/O.

1.1.1 Microprocessor (Processor, CPU, and MPU)

In the world of digital electronics, the term **processor** is generally understood as an electronic computing device that processes information composed of **bits—binary digits 0 and 1**—that make up the instructions written in memory (described later) by the user. The embedded system has two major components: hardware and software. The hardware is the system's physical component, such as its electronic circuit, either assembled on a printed circuit board with discrete components or fabricated on a semiconductor wafer called a processor. The software is a set of instructions the user writes in memory so that the hardware can perform an intended task, such as adding two binary numbers. As an analogy, the processor can be compared to an old-fashioned player piano that runs on a drum and a music sheet. The physical structure of the piano is hardware and music sheet is its software. Or in this digital world, it can be compared to a CD player—the physical structure of the player is the hardware, the CD is similar to memory, and the music written on the CD is software.

The processor consists primarily of three components: the Arithmetic Logic Unit (ALU), registers, and the control circuit (Figure 1-2). The ALU performs the computing task specified by the instruction, registers are used for temporary storage, and the control unit is used for timing and enabling various other operations. The processor is called by different names, including Central Processing Unit (CPU), microprocessor, and Microprocessor Unit (MPU) based on the process used in fabricating the circuit and its application, as described in the following section.

Central Processing Unit (CPU): This term is commonly used in the field of Computer Science and was used in early computers when the processors were built using discrete components. Even today, a processing unit in a computer is referred to as CPU.

Microprocessor is a processor built using the Very-Large-Scale Integration (VLSI) technique (described in Section 1.6) on a single semiconductor surface. The entire circuit is known as an integrated circuit, commonly referred to as a chip.

Microprocessor Unit (MPU): This term is synonymous to the microprocessor or CPU, and it performs all of the same functions as a CPU.

In technical literature as well as in this text, all four terms—processor, microprocessor, CPU, and MPU are used synonymously.

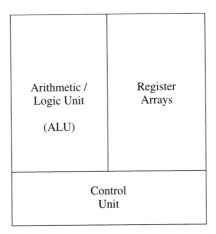

FIGURE 1-2   Block Diagram—Microprocessor or Microprocessor
                  Unit (MPU)

1.1.2 Memory

Memory is an essential component of a microprocessor- or microcontroller-based system. Without memory, a microprocessor cannot function. Memory is a semiconductor chip that stores binary information—instructions and data—and supplies that information to the microprocessor when requested. Memory is made up of registers arranged in a sequence, and each register can store a fixed number of bits. A common size for the register is 8 bits wide.

Memory may be viewed as a page to write on or a printed application form with a fixed number of lines that can be numbered. Each line has space for a fixed number of characters (Figure 1-3). In our analogy, each line on the form is a register that can store a fixed number of bits; the entire page (or form) is memory. Memory is a group of registers that can store a fixed number of bits, and each register can be assigned a number called an address. For example, if we take 16 registers, we can number them in decimal from 0 to $15|_{10}$ or in binary with a 4-bit numbering system from 0000_2 to 1111_2, as shown in Figure 1-3. The lines A_3 to A_0 in Figure 1-3 are called the address lines, and the binary numbers on these lines identify a register.

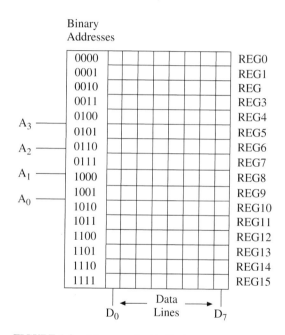

FIGURE 1-3   Memory—Inside Registers

Memory used in microprocessor-based systems is divided into two major groups: Read/Write (R/W) memory and Read-Only Memory (ROM). As the name suggests, in R/W memory, the processor can write into memory, which means that the processor can store information in registers. It can also read (get information from) memory. The R/W memory, popularly called RAM (Random Access Memory), is volatile, which means the information stored in memory disappears when the power is turned off. On the other hand, ROM is non-volatile, meaning that information stored is permanent, and the processor can only read from it. In ROM, information is always available for reading when the power is turned back on. In our analogy,

---

[1] The subscript digit 10 refers to decimal numbers and 2 refers to binary numbers.

R/W memory is like a blackboard in the classroom: it is erased at the end of the class and written on anew for the next class. ROM is like pages in a book, text that cannot be changed.

1.1.3 Input/Output (I/O)

The third major component of a microprocessor-based system is I/O devices, also known as peripherals. As the term I/O suggests, there are two types of devices: input and output.

Input Devices such as switches and keyboards are used to enter binary information, data and instructions, in microprocessor-based systems.

Output Devices such as light-emitting diodes (LEDs), seven-segment LEDs, video monitors, or printers are used to display information; the microprocessor (after processing information), sends results to these devices for display.

1.1.4 Bus

The three components described in the previous section are connected by a group of wires called the **bus**. The bus is a common communication path between processor, memory, and I/O. The concept of a bus is distinct from wires used in connecting analog circuits in two ways: (1) wires used for the bus function as a group, and (2) the same wires are connected to multiple devices. The reason the processor can use the same bus to communicate with multiple devices is that the processor selects only one device at a time with which to communicate. The process is similar to airplanes using the same runway monitored by the air traffic controller. Here we are describing the bus as a single entity, but in reality, there are three buses in microprocessor systems: address bus, data bus, and control bus (described in detail in Section 1.2).

1.1.5 How Does the Microprocessor-Based System Work?

Figure 1-4 shows a block diagram of a microprocessor-based system: it includes a microprocessor, R/W memory, ROM, switches for input, and LEDs for output. All the components are connected by a system bus. Let us assume that a program is written in ROM that tells the processor to read the logic levels (on/off) of the switches and turn on the LEDs that correspond to switches that are on.

When the system is turned on, the MPU goes to the program in ROM at the first memory location and fetches the first instruction using the system bus and decodes it. The instruction tells the processor to enable the input port and read positions of the switches. Next the processor goes to the ROM again to get the next instruction, decodes the instruction, and turns on the LEDs. Thus the processor executes one instruction at a time in a sequence until the end of the program.

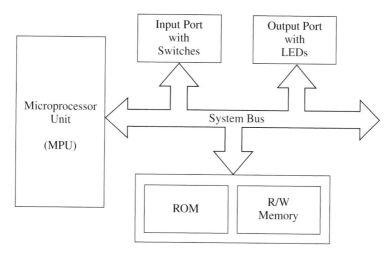

FIGURE 1-4   Microprocessor-Based System with System Bus

1.2 MICROPROCESSOR-BASED SYSTEMS: INTERNAL VIEW WITH SYSTEM BUS

In the last section, we discussed microprocessor-based systems in terms of three components: microprocessor, memory, and I/O at the functional level. Now we will discuss these components again in more depth and focus on how they operate using the system bus.

1.2.1 Microprocessor Architecture

The **microprocessor**, often referred to simply as a **processor**, is a binary computing and logic device (electronic circuit) on a semiconductor chip. It is a clock-driven, general-purpose, register-based programmable device that processes binary data according to the instructions provided in a storage device called memory. It accepts binary data from input devices such as a keyboard and sends the results to output devices such as a video screen. The microprocessor receives its binary instructions to perform its operations from the memory, and it executes those instructions in a sequence based on its clock frequency. Now, let us examine the definition of the microprocessor in more depth.

1. It is a **binary** (digital) **register-based** device that can perform **computing** (arithmetic operations add, subtract, etc.) and **logic** (AND, OR, etc.) functions on data stored in memory, or directly from input devices such as a keyboard. It performs its operations with registers as a unit or a single bit in a register.

2. It is a **general-purpose** device, meaning it can be used for a variety of purposes such as computing, word processing, image processing, and providing timing to turn machines on and off.

3. It is a **programmable** device, meaning it can be made to perform (execute) the functions within its capability based on the user's choice. For example, a radio can be programmed to select a station by changing the value of a tuning capacitor using the

"tuning" knob. In the case of the microprocessor, the programmer selects the instructions for a task to be accomplished, and these instructions are stored in memory for the processor to read and execute. Writing these instructions in an appropriate sequence is called **programming**; a group of instructions is called software (as explained in the previous section).

4. It executes instructions in a **sequence** as stored in memory with the timing of its **clock**. The execution sequence continues from one memory location to the next until it receives an instruction to change the sequence or stop. Therefore, in programming, the sequence of instructions is very important.

The microprocessor is a digital electronic device that includes various circuits, including registers, counters, and adders, on a single chip. To perform any arithmetic/logic operation, the processor needs the bit pattern or pulses to activate or enable these circuits. For example, in our digital electronics lab, to add two binary numbers in an adder circuit, we need to load two numbers in two registers and enable the registers and the adder by sending clock pulses. The adder then provides the sum. Similarly, to perform any function such as add or subtract, the microprocessor needs pulses or bit patterns of 0s and 1s, called **binary instructions**. The designer of the microprocessor implements these bit patterns or instructions in a given processor based on the number of functions it is required to perform. The entire set of these binary instructions is called the **machine language** of the processor. For example, if the engineering design team for a microprocessor uses 8-bit patterns, the team can have $2^8 = 256$ bit patterns or combinations. The team can select certain patterns to perform the necessary functions (such as add, subtract, etc.). This is similar to any other spoken language. For example, there are 26 letters in the English language and a combination of certain letters are selected to represent an operation or meaning. For example, when a two-year old child hears the word *sit*, the child's brain interprets the word and activates the necessary muscles to sit down. Similarly, these binary instructions activate a certain sequence of operations or signals (such as enabling a register or resetting a counter) inside the processor to perform a function. The sequence of operations that takes place internally is directed by the instructions inside the processor, called **microcode**. The user writes a program for the processor to perform a function that is converted into bit pattern (machine language of the processor) and stored in memory. When the processor is asked to execute the program, it reads these bit patterns from memory to activate the circuits and performs the functions specified in the program.

Where does the programmer get these bit patterns? Bit patterns are determined by the engineering team that designs the microprocessor. The engineering team determines the functions the microprocessor should perform and designs the necessary electronic circuits to be on the chip. The bit pattern necessary to activate these circuits is embedded at the design stage. The microprocessor's manufacturer provides the user with a list of instructions based on the design of the bit pattern necessary to perform various functions.

How does the processor read these instructions from memory? The processor communicates with memory using three buses: address, data, and control.

Address Bus is a group of lines (wires) that the processor uses to identify memory registers and I/O devices. It is unidirectional, which means that the information flows in one direction (as shown in Figure 1-5) from the processor to memory and I/Os. The size of the address bus varies from processor to processor depending upon the applications for which the processor is being designed. The address bus is used as a number scheme; the addresses are assigned to memory registers and I/Os using the address bus. As an analogy, to identify or number 1000 houses, we need a 3-digit decimal system ranging from 000 to 999_{10} (assuming zero is the first number). Similarly, a 4-bit binary numbering scheme (4-bit address bus) can identify 16 registers (2^4 = 16) from 0000 to 1111_2. A microprocessor with a 12-bit address bus is capable of addressing or identifying 4096 (2^{12} = 4096) memory registers. At present, we have various microcontrollers that have MPUs with 16-bit address bus; thus capable of addressing 64K (2^{64}=65,536) memory. Typically, microprocessors used in PCs have an address bus 32 to 36 bits wide.

Data Bus is a group of lines that are used to transfer binary data or instructions. It is bidirectional because data flow in both directions—from the microprocessor to memory and output devices and from memory and input devices to microprocessor. As mentioned earlier, the programmer writes or stores instructions in memory. When the processor reads an instruction from memory, data flow from memory to the processor, and when the processor writes into memory, data flow from the processor to memory. Similarly, when the processor displays its results at an output device such as LEDs, data flow from the processor to an I/O device.

Control Bus is a group of individual lines called Read and Write that are used for timing signals to enable memory and I/O devices. We will refer these as control lines rather than a control bus. Now let us examine how these buses are used by the processor for communication.

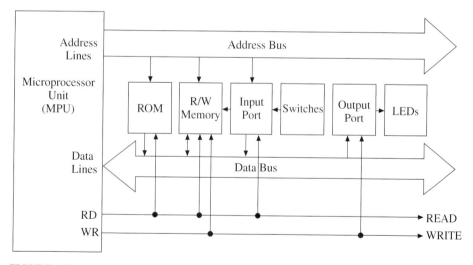

FIGURE 1-5   Microprocessor-Based System with Address Bus, Data Bus, and Control Signals

When the processor reads an instruction from memory, the processor identifies the memory register by sending the address of the register on the address bus, and it asserts the Read timing signal to enable the memory chip. The memory places the contents of the memory register (instruction code) on the data bus, and the processor receives that code in its instruction decoder for interpretation.

Processor Size is the number of bits a microprocessor can operate on at a time. The processor size determines the size of the registers and the design parameters. Intel, a semiconductor company, designed the first microprocessor, the Intel 4004, in the early 1970s. The design was based on 4 bits and it was known as a 4-bit microprocessor. This 4-bit processor was quickly superseded by 8-bit processors, followed by 16-, 32-, 64-, and 128-bit processors in the next three decades. A group of 8 bits is defined as a **byte**, and a group of 4 bits is a **nibble**. When 16-bit processors were designed in the late 1970s, the term **word** came to be known as a group of 16 bits, and that was followed by the terms **double word** for 32 bits and **quad word** for 64 bits. The width of the data bus is generally used to label the size of the processor.

1.2.2 Memory and Its Internal View

As discussed in the previous section, memory is a storage device (semiconductor chip), consists of registers, stores binary information, and it is an integral part of a microprocessor system.

In microprocessor-based systems, the information that is temporary or altered, such as input from a keyboard, is stored in R/W memory. The microprocessor uses this memory to read from and write to when the system is on. On the other hand, the information stored in ROM is permanent; therefore, this type of memory is used to store binary information that is generally not altered, such as programs that run the system.

The R/W memory is divided into two groups: static and dynamic (Figure 1-6). The static memory, made up of transistors, stores binary information (0/1) in the form of voltage. It is high speed, but relatively expensive. Microcontrollers have static memory on their chips. The dynamic memory, made up of Field Effect Transistors (FETs), stores information as capacitive charge. It is slow in response when the MPU attempts to read from it, but less expensive. Dynamic memory is generally used when a large amount of memory is required, such as in a Personal Computer.

Similarly, ROM is divided into two groups: permanent and semi-permanent. The permanent ROM includes two types of memory: masked ROM and Programmable Read-Only Memory (PROM). Manufacturing masked ROM is a specialized and expensive process for small orders, used only for high-volume production of units by memory manufacturers. On the other hand, PROM memory can be programmed in a laboratory using a special unit called a PROM programmer. In PROM memory, once a program is stored, it is permanent and can be programmed only once; it is also known as OTP (one time programmable).

As shown in Figure 1-6, the semi-permanent memory includes three types of memory: Erasable Programmable Read-Only Memory (EPROM), Electrically Erasable Programmable Read-Only Memory (EEPROM), and flash memory. The flash memory is the latest development in memory technology. Before the flash memory, EPROM was commonly used in microcontrollers. As the name suggests, the EPROM memory chip can be entirely erased and reprogrammed with new instructions. However, the erasing process requires a special quartz window on the chip, and exposure to ultraviolet light for more than 10 minutes. On the other hand, information in EEPROM is altered by sending electrical signals to specific registers, and if necessary, the entire chip can be erased in milliseconds. However, this memory is expensive.

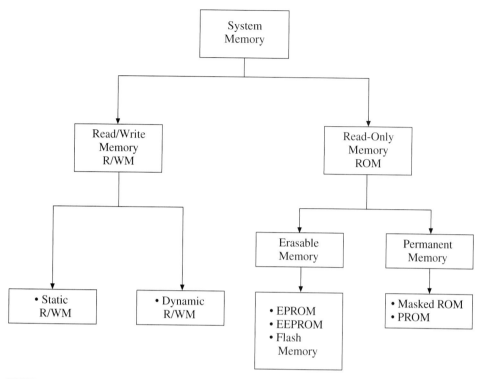

FIGURE 1-6   Memory Classification

Flash memory is similar to EEPROM, but it has many advantages over EEPROM and EPROM. Flash memory is much less expensive than EEPROM and can be erased and programmed in blocks of registers, rather than an individual register as in EEPROM, by sending electrical signals. Flash memory can be reprogrammed on the circuit board without the necessity of removing it from the embedded system. EPROM must be removed from the circuit for reprogramming. The erasing and programming process in the flash memory can be repeated more than one million times without degrading its storage capacity and does not require a special programmer or eraser. It can be programmed with low-voltage signals, making it suitable in low-voltage applications. Flash memory is becoming very common in microcontrollers replacing EEPROM and EPROM. The availability of flash memory in microcontrollers is one of the primary reasons for the expansion of microcontroller applications in many products.

The size of a memory chip is frequently specified in terms of the total number of registers on the chip and the number of bits in each register or total number of bits it stores. For example, a 256×8 memory chip represents 256 registers, with each register having space for 8 bits. The size of this memory chip can be labeled as 256 bytes or 2048 ($256 \times 8 = 2048$) bits. The number of bits a memory register holds is called a **memory word**. The 256×8 memory chip can be viewed as having 256 lines to write instructions or data, each with 8 bits. The number of registers in a memory chip is always a power of 2 in order to maximize the use of a given number of address lines. For example, 10 bits can generate 1024 (2^{10}) numbers; thus, 10 address lines are capable of addressing 1024 registers. In the digital world, the number 1024 is defined as 1 K (upper-case K to differentiate from lower-case k, which represents 1000 in the analog world) because 1024 is the closest approximation to thousand.

How does the microprocessor identify these memory registers to write to or read from? The answer lies in the analogy of how we identify houses on a given road—we number them. In a given neighborhood, let us assume that we have five roads with 100 houses on each road, for a total of 500 houses. We need a 3-digit decimal system with the capacity of defining 1000 numbers from 000 to 999. For each road we can use the groups of numbers such as: 000-099, 100-199, 200-299, or 900-999, leaving some numbers unused or free for future houses. Similarly, the microprocessor has a numbering scheme, called addressing, that assigns binary numbers to each register in a memory chip.

Let us assume in our microprocessor-based system that the microprocessor has a 16-bit numbering system, meaning it uses 16 bits on 16 address lines to assign numbers to memory registers. Therefore, this microprocessor has the capability of assigning $65,536_{10}$ (2^{16} =65536 = 64 K) addresses from 0000 to FFFF in hexadecimal. This entire range is known as memory space or map. In our system, if we have a 1K Byte memory chip with 1024 registers, the memory chip needs 1024 addresses. In decimal numbers, this memory chip needs a block of addresses from 0000 to 1023 that is equivalent to 0000 to 03FF in hexadecimal. The system designer[2] can design an interfacing circuit that can assign any block of 1024 addresses in the entire memory space of 64 K. As an example, this 1K memory chip can be assigned any block of addresses such as 0000_H to $03FF_H$ (at the beginning of the memory space), 4000_H to $43FF_H$, 8000_H to $83FF_H$, or the highest block $FC00_H$ to $FFFF_H$.

In microprocessor-based systems, memory is separate from the microprocessor; thus, the system designer can interface multiple chips and assign any address to a given chip within its memory space. However, in a microcontroller, memory is already on the chip with the processor; therefore, the addresses of memory in the microcontrollers are already assigned by the chip designer[3] and the microcontroller manufacturer provides that information to the users or system designers.

1.2.3 I/O (Input/Output): Peripherals versus Interfacing Devices (I/O Ports)

The microprocessor communicates with the outside world through I/O devices. The input devices bring binary information into the processor and the output devices transfer (copy) the binary information out of the processor. The term I/O device is used with various meanings; therefore, we need to define the term very carefully. For example, a keyboard is connected to a microprocessor through an interfacing device such as a buffer (explained later). The term I/O device can refer to the keyboard, the buffer, or both. In this text, we refer to the keyboard as a peripheral and the buffer as an interfacing I/O device or I/O port.

In microprocessor systems, we have peripherals such as a keyboard, switches, and a mouse that provide binary information as input to the processor. Similarly, we have output peripherals such as a video screen, printer, Liquid Crystal Display (LCD) panel, and LEDs that accept

---

[2] The term system designer refers to the person that designs microprocessor- or microcontroller-based products or systems.

[3] The term chip designer refers to the designer (or design team) of a microprocessor or a microcontroller.

information from the processor and display it. These peripherals need to be interfaced with the microprocessor through I/O interfacing chips, such as buffers and latches (described in the next paragraph) known as I/O ports. In microcontrollers, these buffers and latches are already designed and fabricated on the microcontroller chip and assigned addresses.

Buffer and Latch

A buffer (such as the 74LS244) is a logic device that allows binary information to pass through when enabled, much like a gate. Some buffers are bidirectional and allow information to flow in both directions. In microprocessor-based systems, the buffer is used as an input and output interfacing device, and a system designer can assign a binary address to the buffer, called I/O port address. When the microprocessor asserts its Read/Write control line, it enables the buffer to pass binary bits through from one side of the buffer to the other. The microprocessor can access the input peripheral, such as a keyboard connected to the buffer, only through that address. In microcontrollers, these port addresses are already assigned by the chip designer.

    A **latch** is a storage device that can store one bit. The latch (such as the 74LS373) includes eight latches on a chip that can store 8 bits. When the latch is enabled, it can receive information from the processor, latch it (hold on to it), and display it at output peripherals such as LEDs. In microprocessor-based systems, the latch is used as an output interfacing device, and a system designer can assign a binary address to the latch, called I/O port address, similar to that of a buffer. The microprocessor can output the information to the latch by accessing it through its port address. In microcontrollers, the latches/buffers, called I/O ports, are built in to the chip with the I/O addresses already assigned by the chip designer.

1.2.4 How Does the Microprocessor-based System Execute a Program?

Figure 1-7(a) (which is an expansion of Figure 1-5) shows a microprocessor-based system that includes a microprocessor, ROM, R/W memory, an input port connected to on/off switches, and an output port connected to LEDs. The processor has 16-bit address bus, 8-bit data bus, and two control signals—Read and Write. Memory and I/O ports are connected to the processor using these buses and assigned addresses.

    As shown in Figure 1-7(a), ROM is a 2 KB (2048 × 8) and R/WM is a 1 KB (1024 × 8) memory chip. The address of ROM begins at 0000 and goes up to $07FF_H$, and the address range of R/W memory is 2000_H to $23FF_H$. The addresses of input and the output ports are 8000_H and 8001_H, respectively. The addresses of memory and I/O ports are designed using interfacing circuits and selected arbitrarily for the illustration. ROM has a system monitoring program that checks the initial conditions of the system and directs the program execution to a given memory location in R/W memory.

    A simple program of three 2-byte instructions is written and stored in R/W memory in locations from 2000_H to 2005_H as shown in Figure 1-7(b), which is an expanded version of R/W memory in Figure 1-7(a). The process of writing a program and entering it into memory is discussed in Chapter 4. These three instructions, as illustrated in Figure 1-7(b), are: (1) read the logic levels (on/off) of switches at the input port 8000_H, (2) display the reading at the output port 8001_H, and (3) stop the execution and wait. Figure 1-7(b) shows the instructions in English statements, but the memory will have the binary equivalent of these instructions in its registers. When the system is turned on, the system monitor program will check the initial conditions of

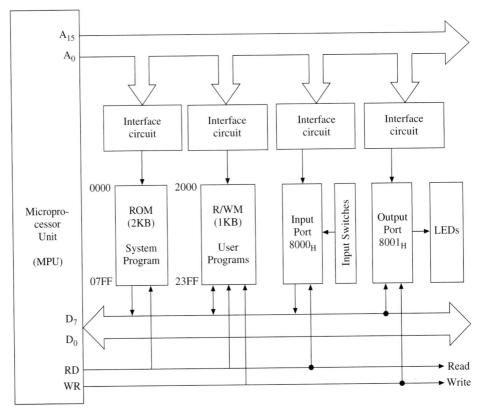

FIGURE 1-7 (a)   Microprocessor-Based System—Expansion of Figure 1-5

2000	READ
2001	PORT 8000$_H$
2002	DISPLAY AT
2003	PORT 8001$_H$
2004	STOP
	R/WM User Programs
23FF	

FIGURE 1-7 (b)

the system, and then direct the processor to location 2000$_H$. The processor places the address 2000$_H$ of the first instruction on the address bus, selects the register 2000$_H$, and enables the memory by asserting the Read signal. The processor gets the binary code on the data bus and places the code in its instruction register, which decodes the instruction based on the micro-code (the set of instructions already designed inside the processor). The processor then determines that it is a two-byte instruction. It places the next address 2001$_H$ on the address bus and fetches the remaining part of the instruction. To execute the instruction, the processor places the address of the input port 8000_H on the address bus, enables the buffer by sending the Read signal, reads the on/off positions of the switches via the data bus, and then saves the reading in one of its internal registers. It then repeats the same process to get the second instruction. To execute the second instruction, the processor places the address of the output port 8001_H on the address bus and the reading of the switches from the internal register on the data bus. It then enables the latch by asserting the Write signal and turns on the corresponding LEDs. The processor repeats the same process called **Fetch**, **Decode**, **and Execute** for the third instruction, then stops the execution and waits for further instructions. The processor reads and executes one instruction at a time in a given sequence.

This process of the program execution can best be compared with the process of putting together a toy that requires assembly or setting up a printer for our computer. We get a set of instructions on a paper written in a sequence. We read one instruction at a time, interpret the instruction, and perform the task. In our analogy, the paper on which these instructions are written is similar to memory and our brain is equivalent to the processor.

1.2.5 Review of Important Concepts

The microprocessor-based system consists of three components: microprocessor, memory, and I/O, and the microprocessor communicates with memory and I/O using the system bus made up of the address bus, the data bus, and the control signals.

1. The microprocessor
   - is a clock-driven binary electronic device that can perform arithmetic and logical operations based on the set of instructions designed in the circuit.
   - reads instructions from memory and performs the specified tasks.
   - reads data from input devices and writes results to output devices.
   - communicates with memory and I/O using the system bus.

2. The memory
   - is an electronic storage device consisting of registers.
   - stores binary data and instructions, called programs.
   - provides the instructions and data to the processor when requested.

3. The input/output ports
   - are buffers and latches that are used in interfacing peripherals, such as keyboards and LEDs.
   - are accessed by the MPU through their port addresses.
   - provide binary information to the MPU (input port) or receive binary information from the MPU (output port).

4. The system bus is a communication path between the microprocessor, memory, and I/O ports consists of three buses: an address bus, data bus, and control signals.

5. The address bus is unidirectional and the processor uses the address bus to select memory registers and I/O ports. The number of address lines determines the addressing capacity of the microprocessor.

6. The data bus is a set of bidirectional lines on which data flow in both directions—between the microprocessor and memory or I/O. The size of the data bus determines the largest number of data bits that can be transferred by the processor at one time.

7. The control signals—Read and Write—are timing signals that are used to enable memory and I/O ports.

8. The microprocessor is a sequential computing device that executes one instruction at a time and goes from one memory register to the next until it is told otherwise. The execution of an instruction is a three-step process: fetch, decode, and execute.

1.3 SOFTWARE: FROM MACHINE TO HIGH-LEVEL LANGUAGES

The processor is a binary computing device capable of understanding binary instructions of 0s and 1s, called machine language. However, it is difficult for human beings to write and understand programs in sets of 0s and 1s. Therefore, manufacturers provide English-like words or abbreviations called **mnemonics** representing each binary instruction. Programs written in mnemonics are called **Assembly language**. The major drawback of assembly language is that every program must be based on the unique design of the processor; therefore, these programs are machine-specific or machine-dependent. An assembly language program written for a system on an Intel processor will not work on a system based on a Motorola processor. To circumvent this problem, **high-level languages** such as Visual BASIC, C, C++, and Java were devised. The programs written in high-level languages are expected to be machine-independent and should work on any machine.

1.3.1 Machine Language

A machine language is made up of binary instructions, and a binary instruction is made up of a group of bits. In a typical 8-bit microprocessor, a group of 8 bits is used by the chip designer to form the machine language. The group of 8 bits can have 256 (2^8) combinations. The chip designer selects certain combinations to implement various operations such as the addition of two registers. Based on these combinations, internal circuits are designed, and the processor can perform those operations when such combinations are read from memory.

As examples, the following combinations represent the addition process in the specified processors.

Machine Instruction	Hex Code	Mnemonics	Description	Manufacturers
1 0 0 0 0 0 0 0	80	ADD B	Add register B to Accumulator	Intel 8085
0 0 1 0 1 0 0 0	28	ADD A, R0	Add register R0 to Accumulator	Intel 8051
0 0 0 1 1 0 1 1	1B	ABA	Add Accumulators A and B	Motorola 6811

If we examine the machine codes, we find that they are all different for the same operation of the addition as if these are three different languages such as English, German, and French. The conversion of binary code into Hex code is for the convenience of the user; representing 8 bits in two Hex digits is easier and less prone to error than writing combinations of eight 0s and

1s. However, it is still difficult to write, understand, and troubleshoot programs in Hex code; therefore, manufacturers provide English-like words (mnemonics) to represent each operation. These mnemonics form the assembly language for a given processor.

The mnemonics are made up by the processor's design team; they can be either very cryptic or easy to read. For example, in Motorola 6811, the letter A represents the addition in the instruction ABA; the instruction ADD A in 8051, R0 specifies the addition and the registers to be added; the instruction ADD B in 8085 specifies only register B, and the accumulator is implied.

Some instructions need more than 8 bits, which is similar to our language. Some verbs or commands are incomplete without a follow-up word. For example, the verb Go must be followed by the word that tells the person "where to go." Similarly, when we want to load a byte in a register, the load command must be followed by a byte to be loaded. For example, to load the number $3F_H$ in the accumulator, the instructions in 8051 and 6811 microcontrollers are as follows:

    0 1 1 1 0 1 0 0   0 0 1 1 1 1 1 1 = 74 3F = MOV A, 3FH: Intel 8051 Microcontroller
    1 0 0 0 0 1 1 0   0 0 1 1 1 1 1 1 = 86 3F = LDAA $3FFH: Motorola 6811 Microcontroller

If we examine these instructions, we find that each manufacturer has chosen different words (such as MOV and LDA) to load a byte and a different style of representing Hex numbers (3FH Vs $3F). Furthermore, some instructions are 3 bytes or 4 bytes long depending on the operation to be performed and the design of the processor.

1.3.2 Assembly Language

As mentioned in the previous section, an assembly language is made up of mnemonics that represent various operations that can be performed by the processor. The primary advantage of assembly language over machine language is that it is easier for the programmer to write, understand, and communicate program logic in mnemonics than in combinations of 0s and 1s. However, these mnemonics must be converted into machine language for execution by the processor. The conversion can be done either manually or by using a computer.

In manual assembly, the programmer can look up Hex codes in the list of instructions provided by the manufacturer and enter the Hex code in the trainers memory using a keyboard. Nowadays this process is obsolete, and the use of computers to write assembly language programs is almost universal. The programmer uses a program called an editor (such as Note-Pad, Wordpad, or Microsoft Word on a PC) to write an assembly language program and uses another program called an **assembler** that translates mnemonics into Hex and binary code. In addition, the assembler can check for syntax errors and assign memory locations for each instruction. Then the binary code is transferred from the computer into the memory of the system that is being designed or a trainer.

Because the mnemonics have one-to-one correspondence with the machine (binary) instructions, assembly language is the most efficient in executing the program, using the least memory space for storage; there is no additional overhead in the program. An assembly language is generally used in small systems for which memory is limited and in real-time systems for which fast execution is essential. The real-time system is defined as a system in which a response is required for an external event (input) within a given time. For example, the anti-lock braking system (ABS) in a car is a real-time system that requires an instantaneous response to slippery road conditions.

As mentioned, the major disadvantage of assembly language is that it is specific to a given processor. It is machine-dependent and not portable from one system to another. In addition, it is somewhat difficult to write and troubleshoot compared to the high-level languages, which are discussed in the following section.

1.3.3 High-Level Languages

The high-level programming languages were developed to be machine-independent. In other words, a program in a high-level language written for a computer with an Intel processor should be able to work on a computer with a Motorola processor. Among high-level languages, C, C++, Visual BASIC, and JAVA, C is widely used in designing embedded systems. The instructions in these languages are written in English-like statements and follow a set of rules and conventions.

But how do the statements in English get translated into the machine code of the processor being used? By using another program called a compiler, which translates high-level language as input and produces the machine code as output. The input to the compiler is called the source code, and the output of the compiler is called the object code that can be executed by the processor being used. Therefore, each computer must have compilers for each high-level language based on its processor.

To translate each high-level language instruction from an English-like statement into object code (binary code) requires multiple machine codes. Therefore, programs written in a high-level language get translated into a large object code, which requires larger memory for storage and longer time for execution than those of assembly language. However, the primary advantages of high-level languages are machine independence and ease in writing and troubleshooting.

1.4 DATA FORMAT

The microprocessor is a binary device that processes a group of data bits based on its register size. However, in our daily life, we use various types of numbers and symbols such as unsigned and signed (positive and negative) integers, decimal, and text symbols. Therefore, we need to represent these numbers and symbols in binary code within the register size of the processor. The representation and processing of these various types of numbers and symbols are discussed in the following sections.

1.4.1 Unsigned and Signed Integers

In an 8-bit register, the largest integer that can be processed is FF_H (255_{10}). Therefore, the range of unsigned integers in an 8-bit register is from 00 to FF_H, a total of 256 integers, including zero.

The next issue is the representation of signed (positive and negative) integers without the symbols of + and - because the processor is incapable of understanding anything other than the binary digits 0 and 1. Therefore, in an 8-bit register we use the most significant bit, D7[4] for signs; the plus symbol is represented by 0 and the minus symbol is represented by 1. The remaining 7 bits D6-D0 are used for the magnitude of an integer. Therefore, the maximum positive integer that can be represented in an 8-bit register is: 0 1 1 1 1 1 1 1 = $7F_H$ ($+127_{10}$). The range of positive integers is 0 to $7F_H$ so the total number of positive integers = 128 including zero, as shown below. Similarly, we have 128 negative integers starting with bit D7 = 1, indicating a negative sign. The range of negative integers is 80_H to FF_H (-128_{10} to -1_{10}). Any number larger than 80 is a negative integer, making the total range of signed integers that can be represented in an 8-bit register -128_{10} to $+127_{10}$.

Positive Integers		Negative Integers	
D7 D6 D5 D4 D3 D2 D1 D0		D7 D6 D5 D4 D3 D2 D1 D0	
0] 0 0 0 0 0 0 0 = 00_H = 00_{10}		1] 0 0 0 0 0 0 0 = 80 = -128_{10}	
0] 1 1 1 1 1 1 1 = $7F_H$ = $+127_{10}$		1] 1 1 1 1 1 1 1 = FF = -1_{10}	
↑		↑	
Sign Bit		Sign Bit	

If we examine the negative integers from 81_H to FF_H, we find that they are equivalent to 2's complement of $7F_H$ to 01_H respectively; therefore, we can represent the negative integers as 2's complement of positive integers (See Appendix D for 2's Complement Operations). The number 80_H is unique; 2's complement of 80_H is 80_H. In computer systems, a negative number is represented by its 2's complement equivalent.

---

[4] In an 8-bit register, when bits represent data they are shown as D7 to D0; D7 being the most significant bit. When bits represent individual bits, they are shown as B7-B0. However, they are synonymous.

EXAMPLE 1.1

Find the decimal values of the Hex number $9F_H$ and 57_H as represented in a microprocessor-based system.

Solution:

The problem statement does not specify whether these are signed numbers or unsigned numbers. Therefore, we need to calculate the decimal values assuming both types of numbers.

1. If the byte $9F_H$ is an unsigned number, its decimal value $= 159_{10}$ as shown:

   $9F_H = 9 \times 16^1 + (15) \times 16^0 = 144 + 15 = 159_{10}$

   (If you are not familiar with the process of number conversion, see Appendix D: Number Systems and Hex Arithmetic.)

   If the byte $9F_H$ is a signed number, it is a negative number in the 2's complement because bit D7 is 1.

   Therefore, the negative signed value of $9F_H$ can be calculated by taking 2's complement of $9F_H$.

   2's complement of $9F_H = 61_H = 6 \times 16^1 + 1 \times 16^0 = -97_{10}$

2. The decimal value of the unsigned or the signed number 57_H is the same because it is a positive number $= +87_{10}$

1.4.2 Binary Coded Decimal (BCD) Numbers

In our daily life, we use decimal numbers to represent magnitudes of various quantities. For example, temperature is displayed as 25 degrees Celsius. Therefore, we need to represent these numbers in Binary Coded Decimal (BCD) numbers. Digits 0 to 9 can be represented by four binary digits from 0000 to 1001. Therefore, we can represent 2-digit BCD numbers in an 8-bit register. For example, 25 decimal can be shown as 0010 0101 in BCD. However, this is just a representation, similar to writing phonetics of Arabic in English alphabets. Otherwise, the actual equivalent of $25_H = 37_{10}$.

The decimal number 25 represented above is called a packed BCD. However, in many applications, we need to separate the number 25 as 02 and 05, called unpacked BCD. For example, to display 25 at 7-segment LEDs, we will have to separate (unpack) 2 and 5 to display them at 2 different LEDs. This process is called unpacking (see Chapter 6 for more details and illustrations of the unpacking).

1.4.3 Alphanumeric (ASCII) Code

The microprocessor is a digital device that uses the language of binary numbers (0s and 1s) exclusively; however, communication in English is based on the letters of the alphabet, decimal numbers, and punctuation symbols. We need a code that can translate these alphanumeric symbols into binary language. The code that is commonly used is called the *American Standard Code for Information Interchange* (ASCII).

ASCII is a 7-bit code with 128 (2^7) combinations. Each combination from 00 to $7F_H$ is assigned either a letter, a decimal digit, a symbol, or a machine command. For example, decimal digits 0 to 9 are represented by 30_H to 39_H, and lower-case a to z are represented by 61_H to $7A_H$ (a complete list of ASCII characters is included in Appendix E). Typical examples of peripherals that use ASCII code include the keyboard of a microcomputer and a printer. When the digit 7 is typed on a keyboard, the processor receives 37_H in binary, and the system programs translate these ASCII characters into appropriate binary values.

There is also an extended ASCII code that is used in microcomputers. This is an 8-bit code with 256 combinations; these additional combinations are used for graphic characters.

1.5 MICROPROCESSOR (MPU)- AND MICROCONTROLLER (MCU)-BASED SYSTEMS

As mentioned at the beginning of this chapter, we are surrounded by products and systems that are controlled by a microprocessor or microcontroller. Whether a product or system uses a microprocessor or microcontroller depends primarily on the requirements of memory, speed of processing, necessity of additional components, and cost effectiveness. For example, an image-processing system requires a large memory and high-speed processing. On the other hand, household appliances, such as dishwashers and microwave ovens, do not require large memory or high-speed processing. To integrate various concepts, we will illustrate a simple embedded system in the following section that displays time and temperature using an MPU. We will reexamine the same system in the context of a microcontroller to illustrate the differences between the MPU-based and MCU-based embedded systems. We will discuss this same system throughout the text using the PIC18 microcontroller and add details as we begin to learn additional hardware and software concepts.

1.5.1 Project Statement

Design a microprocessor-based system that displays time and temperature alternatively every five seconds at an Liquid Crystal Display (LCD) module. The system should monitor the temperature in a room continuously and turn on a fan if the temperature goes higher than a set point and turn on a heater if the temperature falls below a set point.

1.5.2 System Hardware

Figure 1-8 shows the block diagram of the time-temperature display system. It includes MPU, flash and R/W (RAM) memory, two input ports, and two output ports. All components are connected through a system bus.

Microprocessor Unit (MPU)

The MPU is a general-purpose microprocessor with the address bus, the data bus, and control signals shown as the system bus in Figure 1-8. Once the system is turned on, the MPU goes to memory location 0000_H in flash memory and begins executing the program. It reads temperature from the input port PORTA and stores the reading in R/W memory. The program processes the binary reading, checks whether it is within the set limit, calculates its equivalent

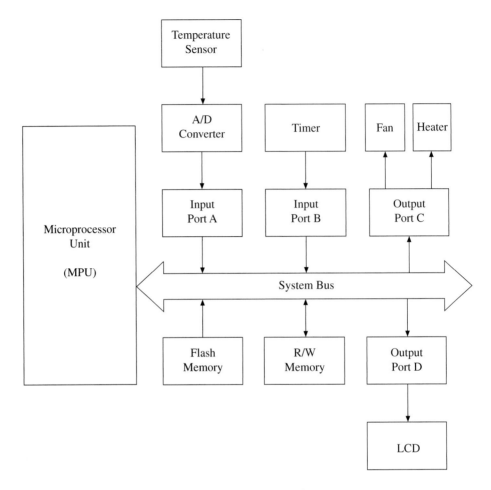

FIGURE 1-8   Microprocessor-Based Time and Temperature System

value in Celsius, converts it into ASCII characters, and sends out the temperature reading to the
LCD output port PORTC for display. If the reading is not within the set limits, the MPU turns
on either the heater or the fan. Next, the MPU reads the timer from the input port PORTB, converts it into time, waits for the specified time interval, and displays the time at the LCD. This
process is repeated continuously and time and temperature are displayed alternatively.

Memory

The system includes two types of memory: flash and R/W memory. The flash memory stores
the permanent program that monitors the system, processes temperature readings, and displays
time and temperature at the LCD. The R/W memory is used for storing temporary data.

Input and Output (I/O)

Figure 1-8 shows two input ports and two output ports: the input ports enable the MPU to read temperature and the timer and display time and temperature on the LCD and turn on/off the fan and the heater.

Input: The system includes two input ports: one to read the temperature and the other to read the timer. Using a transducer (a device that converts one form of energy into another), the temperature is converted into an analog electrical signal. Examples of transducers include devices such as a microphone, speaker, or thermocouple. The microphone converts sound energy into electrical signals, the speaker converts electrical energy into sound, and the thermocouple converts heat energy into electrical signals. Nowadays, a semiconductor device called a temperature sensor, similar to a transistor in appearance, is available as a transducer. The temperature sensor generates voltage proportional to the temperature; however, this is an analog signal and must be converted into a digital signal for the processor to read it. Figure 1-8 shows a device called an Analog-to-Digital (A/D) converter that converts analog voltage from the temperature sensor into the equivalent binary value in 8 bits. The MPU obtains this reading by accessing input port PORTA.

The second input port, PORTB, is connected to the timer chip using a buffer as an input port. The timer is an electronic device with a register that holds a count. This count is decremented (or incremented) at every clock cycle, and the time is calculated by the difference between two counts multiplied by the clock period. The MPU reads the count by accessing this input port, calculating the time, and updating the clock on the LCD. In reality, the buffer shown in Figure 1-8 is unnecessary because timers have their built-in buffers on the timer chips.

Output: The system shown in Figure 1-8 has two latches interfaced as output ports: one for the heater and the fan, and the second for the LCD. The heater and the fan are controlled by individual bits of the output port; therefore, they require only one port. The MPU turns them on and off by sending appropriate bits. To display a character on the LCD, the MPU accesses it as an output port and sends ASCII or special graphic characters on its data bus and writes to the LCD by asserting control signals.

1.5.3 System Software (Programs)

A group of programs that monitors the entire system is called the system software. In small systems this software is called a monitor program, and in large complex systems it is called an operating system. These programs are stored either in ROM, PROM, EPROM, or flash memory; Figure 1-8 shows the flash memory.

1.5.4 Microcontroller-based Systems

Figure 1-9 shows the same time-temperature system discussed in the previous section except this system is based on a microcontroller rather than a microprocessor. In a microcontroller, microprocessor (MPU), memory, and I/O ports plus some additional support devices such as timers and A/D converters are all inside the chip. Therefore, in a microcontroller-based system:

1. The buses are not available to the user (with some exceptions in which external memory can be connected).

2. Memory and I/O addresses are already assigned during the chip design process.
3. I/O pins are available to the user to connect peripherals.
4. I/O ports are generally multiplexed, meaning they can be used for many purposes by programming control registers inside the microcontroller. For example, an I/O port can be set up as an input port or an output port; an individual pin can be set up either as input pin or output pin.
5. Many of the support devices such as A/D converter, timers, and serial I/O devices (discussed in Chapter 2) are available on the chip.

Figure 1-9 shows a microcontroller that includes an MPU, memory, A/D converter, timer, and three I/O ports: A, B, and C. The temperature sensor is connected to one of the pins of PORTA that provides an analog signal to the A/D converter. Two pins of PORTA are used as output pins to control the fan and the heater. I/O ports on a microcontroller are generally multiplexed to perform multiple functions; they need to be set up through software to perform specific functions. PORTB is set up as an output port and interfaced with the LCD. The control signals necessary to write to the LCD are generated by using PORTC.

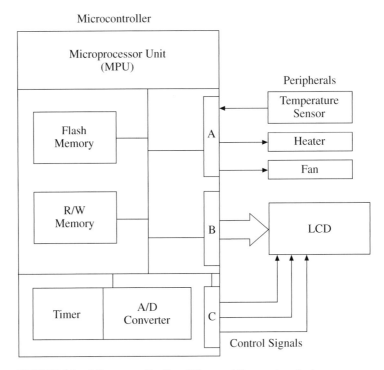

FIGURE 1-9   Microcontroller-Based Time and Temperature System

The system software programs are stored in flash memory using a process defined by the chip designer. Though additional instructions are necessary to set up various devices in a microcontroller-based system, the programs will be similar to those in the microprocessor-based system discussed earlier.

1.6 HISTORICAL PERSPECTIVE AND LOOK AHEAD

Since the invention of the transistor, semiconductor technology has gone through revolutionary changes. Semiconductor fabrication technology is now capable of placing millions of transistors on a single chip. Digital gates are being replaced by application-specific devices, and microprocessors have grown in size from 4 bits to 64 and 128 bits. Here we will take a brief look back at the past, discuss the present state of technology, and peek into future.

1.6.1 Technological Advances

The transistor has become the building block of modern electronics, and is considered the most important invention of the twentieth century. The transistor was invented in 1947 by William Shockley, John Bardeen, and Walter Brattain,[5] three scientists at Bell Laboratories, the research arm of AT&T. Prior to 1947, electronics circuits were designed using vacuum tubes, which were bulky, unreliable, and produced a lot of heat. In contrast, the transistor is small and reliable, has a fast switching speed, and consumes very little power. Finally, one of the transistor's major advantages is that it can be mass produced on silicon semiconductor wafers, which can be cut into individual chips.

In 1955, Dr. Shockley left Bell Labs and formed his own company. The offshoot of that company was Fairchild Semiconductor, formed by Robert Noyce and Gordon Moore. During that period, one of the major problems was how to reduce the number of components (such as transistors and resistors) needed to create high-quality electronic systems. The solution to this problem was the design of an integrated circuit (IC) whereby multiple transistors, diodes, and resistors could be etched on a single chip. In 1960, Jack Kilby at Texas Instruments and Robert Noyce at Fairchild Semiconductor invented the IC independently. Around the same time, Texas Instruments manufactured a series of logic gates (AND, OR, NAND, and NOR) known as the 7400 series using IC technology; this process of placing a few transistors, diodes, and resistors on a silicon chip came to be known as Small-Scale Integration (SSI). As fabrication technology improved, circuits requiring around 100 transistors on a single chip were designed and produced, which came to be known as Medium-Scale Integration (MSI). This was followed by Large-Scale Integration (LSI), which placed tens of thousands of transistors on a chip, and Very-Large-Scale Integration (VLSI), which placed hundreds of thousands of transistors on a chip. In the 1990s, over a million transistors were placed on one chip, a technology called ULSI (Ultra-Large-Scale Integration). Today, we have developed a method of putting an entire system on a chip, as with a microcontroller, known as System-On-Chip (SOC).

1.6.2 Birth of a Microprocessor and Its Revolutionary Impact

In 1968, Robert Noyce and Gordon Moore left Fairchild Semiconductor and formed a new company known as Intel, which is now world renowned for designing and producing microprocessors. Initially, the company was primarily in the business of designing memory. In 1969, it received a contract from a Japanese company, Busicom, to design a programmable calculator. The original specifications by Busicom required twelve different chips. Ted Hoff, an Intel engineer on the project, came up with a bold and innovative approach to design the

---

[5] In 1956, all three scientists received the Nobel prize in physics for their invention of the transistor.

calculator. He suggested a design based primarily on two chips: (1) a general-purpose electronic circuit that could perform various logic functions and (2) a storage medium, called memory, made up of registers that could supply 0s and 1s to activate the necessary circuits in the first chip. Thus the programmable calculator was designed and used for different functions by writing programs into its memory chip. Later, Intel realized its potential as a computing device with many applications, coined the term microprocessor, and released it as the 4004 microprocessor in 1971. This was the beginning of the microprocessor revolution.

The 4004 was a 4-bit processor with 2300 transistors, 640 bytes of memory addressing capacity, and a 108-kHz clock. Thirty years later, there are 64-bit (as well as 128-bit) processors designed with more than forty million transistors; they have 64 gigabytes of memory-addressing capacity and can run at frequencies higher than 1 GHz. Gordon Moore, co-founder of Intel, predicted that the number of transistors per IC will be doubled every eighteen months; this came to be known as "Moore's law."

1.6.3 von Neumann (Princeton) versus Harvard Architecture and CISC versus RISC Processors

In the 1940s two major classes of computer architecture emerged: von Neumann (or Princeton) and Harvard. The von Neumann model uses a single storage structure, called memory, for instructions and data. In contrast, the Harvard architecture uses physically separate memory units and buses for instructions and data.

In a computer based on von Neumann architecture, the CPU can process either instructions or data, one at a time, because the instructions and data use the same buses. In a computer with Harvard architecture, the CPU can access data while completing the execution of an instruction. Thus, computers with Harvard architecture can be faster than that of computers designed with von Neumann architecture. The first microprocessor—Intel 4004—was designed based on the von Neumann model.

Over the years, fabrication technology advanced by leaps and bounds, causing the price of microprocessor chips to drop considerably, and applications began to infiltrate every aspect of our daily lives. The demand for high-speed processing began to increase, and to meet that demand, CPUs were designed with higher and higher frequencies.

In addition, CPU designers added more and complex instructions. The design of the control unit and the instruction decoder became very complex, occupying more than 50 percent of the chip area. This left less space for registers, resulting in the reduction of the number of registers on the chip. Therefore, to execute the code, the MPU needed to access the memory more frequently, which slowed execution. This design approach was classified as Complex Instruction Set Computing or Computers (CISC) and chips were called CISC processors. Meanwhile, research studies discovered that programs used only about 20 percent of the instructions frequently and the remaining 80 percent very rarely. A new design philosophy emerged from these studies called Reduced Instruction Set Computing/Computers (RISC) that simplified the hardware design by including frequently used simple instructions and improved the performance. The design of RISC processors generally include the following features:

- The number of instructions is minimized.
- The number of addressing modes and the instructions that access memory are minimized.
- Most instructions are executed in one cycle.
- The number of registers in the CPU is larger than that of CISC processors.

In comparison, CISC processors are nearly opposite RISC processors.

1.6.4 Microprocessors to Microcontrollers

The microprocessor era began with the Intel 4004 4-bit processor. This pioneer was quickly replaced by 8-bit processors such as Intel 8008/8080/8085, Zilog Z80, Motorola 6800, and MOS Technology 6502. In the 1970s, microcomputers were designed using these 8-bit microprocessors. IBM replaced the term microcomputer with Personal Computer (PC) when it began to market computers designed with the 16-bit Intel 8088 processor. Since the 1980s, two distinct trends emerged in the microprocessor field: (1) the design of larger and faster processors for high-speed processing, and (2) the integration of various devices such as the microprocessor, memory, I/O ports, timers, and data converters on a single chip.

The first trend, increasing the size and speed, was led by the Intel 8088 processor used in PCs, followed by several processors such as the Intel 80XXX and Pentium processors, the Motorola 68000 series, and AIM (alliance of Apple, IBM, and Motorola) POWER PCs. The second trend, toward integration, was led by the 8-bit Intel 8051 microcontroller, followed by the Zilog Z8, the Motorola 68HC series, the Atmel AVR series, the Microchip PIC series, and microcontrollers from many other manufacturers. Currently, microcontrollers are available in sizes ranging from 4-bit to 32-bit; typically, an automobile may have one 32-bit, several 16-bit, and tens of 8-bit microcontrollers. A typical home may have more than fifty microcontrollers built in to its various phones, security systems, fax machines, and computers. These are all considered "embedded control systems," in which the running, controlling, and monitoring go on behind the scenes.

Among all the different types of microcontrollers available in the market, the number of products using 8-bit microcontrollers has experienced explosive growth in the last decade, and the number of applications is growing. The focus of this text is 8-bit microcontrollers and their embedded applications using the PIC18 microcontroller family.

The PIC18 is an 8-bit microcontroller designed with the RISC philosophy and Harvard architecture. It has separate memory units with independent address and data buses for instructions and data. The embedded system described in Section 1.5 will be expanded in subsequent chapters using the PIC18 microcontroller and its assembly language.

SUMMARY

The major concepts discussed in this chapter can be summarized as follows:

1. A microcontroller- or microprocessor-based system is defined as an electronic system that is controlled either by a microcontroller or a microprocessor. The system is divided into two major segments: hardware and software. The term "hardware" is used to describe its physical components and "software" describes the programs that provide instructions to direct its operations.

2. A microcontroller is an integrated electronic computing device that includes three major components on a single chip: the MPU, memory, and I/O ports. These components are connected by common communication lines called the system bus.

3. The MPU is a group of electronic circuits fabricated on a semiconductor chip that can read instructions written in memory and process (compute) binary data according to those instructions.

4. Memory is a semiconductor storage device consisting of registers; these registers store binary bits. Two major categories of memory are the Read-Only Memory (ROM) and the Read/Write Memory (R/WM also known as RAM)

5. I/O ports are latches and buffers that can be interfaced with peripherals, such as keyboards and LEDs.

6. The system bus is divided into three groups: the address bus, data bus, and control signals.

7. The address bus is a group of unidirectional lines that carry the addresses of memory or I/O devices, and these addresses are initiated by the MPU.

8. The data bus is a group of bidirectional lines that carry data between MPU and memory or I/O devices.

9. The control lines are timing signals that initiate various read-write operations of the MPU.

10. The MPU is capable of decoding and executing binary instructions that are designed into its electronic circuits; these binary instructions are known as microcode.

11. In an 8-bit processor, data can be represented in various formats. To represent signed numbers, bit D7 is used to indicate the sign. 0 is for positive numbers and 1 is for negative numbers. Negative numbers are represented in 2's complement format.

12. Binary Coded Decimal (BCD) numbers are represented by a group of 4 bits. A byte can hold two BCD numbers called packed BCD numbers.

13. American Standard Code for Information Interchange (ASCII) is a 7-bit code with 128 combinations used to represent alphanumeric symbols (alphabets and numbers). An extended ASCII is based on the combinations of 8 bits; thus it represents 256 combinations.

14. The assembly language mnemonics have a one-to-one correspondence with the machine language, and thus, the assembly language is very efficient in execution and requires minimum memory storage. However, the major drawback is that it is machine-specific.

15. Programs in high-level languages such as Visual Basic, C, and C++ are written in English-like statements. They are translated into the processor's machine language by compilers, making them machine-independent. However, these programs require large memory space and longer time for execution than that of assembly language programs.

QUESTIONS

1.1   List the components of a microprocessor-based system.

1.2   What is the difference between a microprocessor and a microcontroller?

1.3   Explain the difference between the terms Microprocessor, MPU, and CPU.

1.4   List the two major categories of memory and explain their functions.

1.5   Define the terms Bit, Byte, and Word.

1.6   Explain the functions of input and output devices with examples.

1.7   Is a scanner connected to a PC an input device or an output device?

1.8   Explain the functions of address bus, data bus, and control lines.

1.9   Explain why the address bus is unidirectional and the data bus is bidirectional.

1.10   Calculate the number of bits that can be stored in 1 K-byte (KB) memory.

1.11   Calculate the number of registers in 8 KB memory and the address of the last register in Hex (assuming the address of the first register is 0000).

1.12   Calculate the number of registers in 4 MB memory and the address of the last register in Hex (assuming the address of the first register is 00000).

1.13   If the processor has a 12-bit address bus, calculate its memory addressing capacity.

1.14   If the processor has a 21-bit address bus, calculate its memory addressing capacity.

1.15   If the last memory address in a given memory chip is $07FF_H$, calculate the size of the memory chip.

1.16   If the address range of flash memory in a microcontroller is 00000_H to $1FFFF_H$, calculate the size of the memory.

1.17   In a microcontroller, R/W memory is assigned the address range from 2000_H to $21FF_H$; calculate the size of the R/W memory.

1.18   How are the signed numbers represented in 8-bit MPU?

1.19   Calculate the decimal value of the Hex integer 78_H if it is an unsigned number.

1.20   Calculate the decimal values of the Hex integer 98_H if it is a signed number as well as if it is an unsigned number.

1.21   Calculate the Hex equivalent to represent a negative decimal number -12_{10} in an 8-bit microprocessor.

1.22   Find the Hex equivalent of the decimal number 138_{10} and show its binary representation in an 8-bit processor.

1.23   Find the Hex equivalent of the negative decimal number -138_{10} and show its binary representation in an 8-bit processor.

1.24   Define ASCII code and explain why the total number of codes is limited to 128.

1.25   Find the ASCII codes for upper-case letters A and Z, and lower-case letters a and z from Appendix E.

1.26   Given lower-case ASCII letters, suggest a logical operation to make them upper-case.

1.27   Explain why assembly language programs are efficient in execution.

1.28   What is a major advantage of writing programs in a high-level language?

CHAPTER *2*

MICROCONTROLLER ARCHITECTURE—PIC18F FAMILY

OVERVIEW

In Chapter 1, we discussed embedded systems in two major segments: hardware and software. Hardware, the physical component, is a microcontroller and software is a set of instructions stored in memory for the microcontroller to execute. This chapter focuses on the hardware aspect and Chapter 3 deals with the software aspect of the PIC18F family of microcontrollers.

A microcontroller, as defined in Chapter 1, consists primarily of three components on a single semiconductor chip: a microprocessor unit (MPU), memory, and I/O ports. In this chapter, we will discuss these components in the context of the PIC18FXXX and PIC18FXXXX family of microcontrollers. The Xs represent the actual version of the processor; thus, 4 Xs represent the upgraded version of 3 Xs. In this text, the term PIC18F refers to all versions of this family. We will examine: (1) how these components are connected using internal buses, (2) pinouts of I/O ports available to the user, and (3) how data are transferred to outside peripherals. In addition, this chapter provides an overview of the functions of various support devices such as timers, A/D converters, Capture, Compare, and Pulse width Modulation (CCP) modules, and serial I/O devices. These devices and their applications in embedded systems are discussed in individual chapters later in the textbook. This chapter concludes with an illustration of a simple assembly language program displaying a byte at an I/O port of PIC 18F microcontroller.

The microcontroller is a complex device, and learning its internal architecture demands understanding many concepts from digital electronics. This chapter is intended to provide an overview of the PIC microcontroller family. The reader should focus on understanding the underlying principles of a microcontroller rather than the details of PIC microcontrollers. This chapter provides an overall view of the PIC 18F microcontrollers with few details.

OBJECTIVES:

- Identify major components of the PIC18F microcontroller family.
- List the internal registers of the PIC18F microcontroller and describe its Harvard architecture.
- Identify the parallel and serial processes of data transfer.
- List conditions and processes of data transfers: status check and interrupt.
- List the support devices in 18F microcontrollers and describe their functions.
- Explain a set of simple assembly language instructions and the underlying machine code.
- Explain how a program is stored in memory and executed by MPU.

2.1 PIC18F MICROCONTROLLER FAMILIES

PIC microcontrollers are designed using the Harvard architecture (as explained in Chapter 1) that includes a separate memory for data and for programs (instructions). Figure 2-1 shows a simplified block diagram of a typical PIC microcontroller that includes five blocks: microprocessor unit (MPU), program memory, data memory, I/O ports, and support devices. There are many versions of PIC18F microcontrollers, ranging in package size from 18 pins to 100 pins, and their clock frequencies range from 25 MHz to 48 MHz. The primary differences between these versions are in available memory and I/O ports. The program memory varies from 4 KB to 128 KB, data memory from 256 bytes to 3968 bytes, data EEPROM from 128 bytes to 1 KB (some versions do not include EEPROM), and I/O pins from 16 to 70.

2.1.1 Microprocessor Unit (MPU)

A typical MPU is divided into three segments: Arithmetic Logic Unit (ALU), Registers, and Control Unit, as shown in Figure 1-3 in Chapter 1. Figure 2-2 shows a further expansion of that block diagram, specific to the microprocessor unit (MPU) in the PIC controllers. It also shows various internal busses that connect the MPU to memory.

Arithmetic Logic Unit (ALU)

The ALU includes electronic circuits (such as the adder, comparator, and flags) that are designed to perform arithmetic and logic functions, such as addition, subtraction, logical AND, OR, and Exclusive OR. The register that is used to perform these functions is generally called the accumulator. In this microcontroller family, it is called the Working Register (WREG), and it is 8 bits wide. The ALU also shows the block called Instruction Decoder, which is used to interpret an instruction; this is an internal function to the ALU. When the MPU fetches an instruction from memory, it places the instruction in the Instruction Decoder, which decodes the instruction and directs the processor on how to execute that instruction. This process is similar

Microcontroller Unit (MCU)

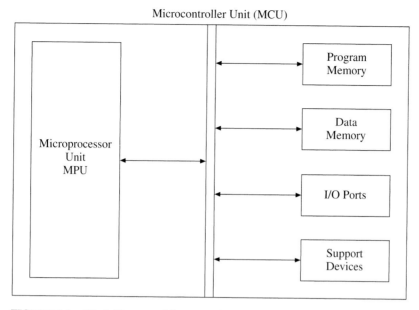

FIGURE 2-1   Block Diagram—Microcontroller Unit (MCU)

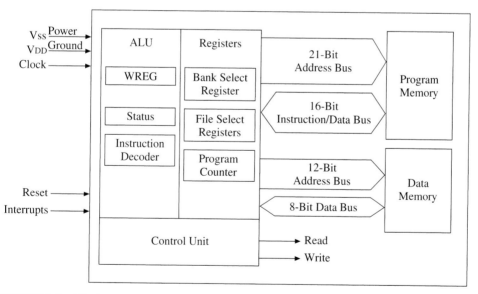

FIGURE 2-2   Block Diagram—MPU and Memory

to how our brains function. When we hear a word or a sentence, it is interpreted by the brain, and an action is initiated based on that interpretation.

Another register in the ALU is a STATUS register, also called the flag register. Five bits of this register B0 to B4 (discussed in Chapter 3) are used to indicate the status of arithmetic and logic operations. For example, bit B0 of the STATUS register is used to indicate the carry status of an arithmetic operation. If an addition of two numbers generates a carry, bit B0 is set to one, indicating that the carry has been generated. Similarly, other bits B1 to B4 are used as flags for other types of data conditions or results. These flags and their critical importance in assembly language programming are discussed in Chapters 3 and 5.

Registers

The register segment of the MPU includes various types of registers that are used to hold memory addresses. Three registers are shown in Figure 2-2—Bank Select Register (BSR), File Select Registers (FSR), and Program Counter (PC). These registers are defined here briefly, but explained in more details in Section 2.1.2.

The Program Counter (PC) is a 21-bit register that holds the program memory address of the instruction that is to be read next by the MPU. The MPU uses the program counter as a memory pointer (locator) to fetch an instruction. When the MPU fetches an instruction from memory, it also increments the address in the Program Counter to point to the next memory location.

The BSR is a four-bit register that is used in direct addressing the data memory, and the FSRs are 16-bit registers that are used in indirect addressing (explained in the Memory Section 2.1.3).

Control Unit

This unit is designed to provide timing and control signals to various Read and Write operations. This unit oversees the binary information flow between the microprocessor, memory, and I/O.

2.1.2 Buses

In Chapter 1, we defined the functions of the address bus, data bus, and control bus (signals). In this chapter, we will examine the details of PIC18F buses.

Address Bus

The PIC18F family of microcontrollers has two different address spaces: one is for the program memory that stores instructions, and the other is for data memory that stores data (this is discussed again in the next section—Memory). The 21-bit address bus is used for program memory and the 12-bit address bus is used for data memory. Therefore, the PIC18F microcontroller has maximum capacity of addressing 2 Meg (2^{21} = 2097152) memory registers for programs and 4 K (2^{12} = 4096) memory registers for data.

Data Bus

The PIC18F microcontroller has two data buses—a 16-bit data bus for program memory and an 8-bit data bus for data memory. The size of the data bus determines the largest number that can be transferred between MPU and other devices in one cycle.

Control Bus

As discussed in Chapter 1, these are single control signals that are initiated by the MPU when it is writing into or reading from a device; these are called Read and Write signals. The Read and Write signals are relevant only when a device such as an LCD or external memory are interfaced with the MCU; otherwise, a discussion of these signals is not necessary for one to understand MCU operations.

EXAMPLE 2.1

Problem:

Calculate the address space available, in digital units, for a microcontroller that has MPU with a 13-bit address bus.

Solution:

1. The 13-bit address bus can generate $2^{13} = 8192$ numbers. Therefore, this MPU can address 8 K (8196/1024 = 8) of memory in digital units.

2. Another way the address space can be calculated is by converting 13 bits of 1s into decimal because the maximum address is all 1s. In Hex, 13 bits of 1s = $1FFF_H = 8195_{10}$ plus initial address 0 = 8196 = 8 K. This is similar to a decimal numbering system. For example, given a two-digit decimal numbering system, we can generate $10^2 = 100$ numbers or the largest number 99 plus initial zero = 100.

2.1.3 Memory

As mentioned in the previous section, PIC controllers include two types of memory: program memory and data memory, as shown in Figure 2-2, each having its own address bus and data bus (Harvard architecture). The MPU with the Harvard architecture has concurrent access to both memories, resulting in faster execution of the programs.

Program Memory

This memory is used to store the instructions of a program. The older versions of PIC MCUs, designated with the letter C (such as 18C), have EPROM and newer versions, designated with letter F (such as 18F), have flash memory. As shown in Figure 2-2, memory has 21-bit address bus and a 16-bit data bus.

With a 21-bit address bus, the entire address range, called the memory map or memory space, spans from all zeros to all 1s—000000 to $1FFFFF_H$—with a total of 2 MB of memory as shown in Figure 2-3a. However, the presently available versions of the PIC18 family have program memory from 4 KB to 128 KB. Figure 2-3a shows the memory range of PIC18F452/4520 from 0000 to 07FFF$_H$; the unused address lines of the 21-bit address bus always read 0.

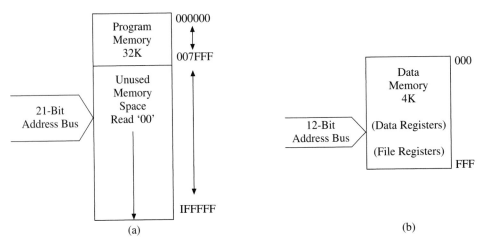

FIGURE 2-3   P1C18F252—(a) Program Memory with Addresses (b) Data Memory With Addresses

The 16-bit data bus carries 16 bits at a time. All but four instructions in the PIC18 family of controllers are 16 bits (two bytes) or one word in length; thus, these instructions are read by the processor in one cycle. The remaining four instructions are of 32-bit size; thus require two processor cycles.

Data Memory

This memory, as shown in Figure 2-2, has a 12-bit address bus and an 8-bit data bus. With the 12-bit address bus, the address space ranges from 000 to FFF$_H$ (Figure 2-3b). To avoid confusion between the program memory and data memory, we will refer to data memory as data registers or file registers (as used in the instruction set provided by the manufacturer—explained in Chapter 3).

Figure 2-4 illustrates the data register map of the 18F452/4520 microcontroller, and these registers with 12-bit addresses can be addressed (or accessed) in three ways: (1) directly, by using the Bank Select Register (BSR), (2) indirectly, by using File Select Registers (FSR), and (3) some special-purpose registers though the Access Bank. At this point in the discussion, it will be difficult to follow these various methods of accessing data registers without an understanding of instructions, which will be described in Chapter 5.

The address map of the data registers ranges from 000 to FFF$_H$ (4096 addresses), and the entire map is divided into 16 banks (0 to F$_H$); each bank has 256 registers (00 to FF$_H$). In direct addressing, the BSR provides a 4-bit address that is combined with 8 bits specified in an instruction resulting in a 12-bit address. The address map begins with Bank0 with the range from 000$_H$ to 0FF$_H$, and ends in BankF$_H$ with the address range from F00$_H$ to FFF$_H$. In the indirect addressing, a FSR is used to hold the address of a data register, and the MPU uses the FSR to access the data register.

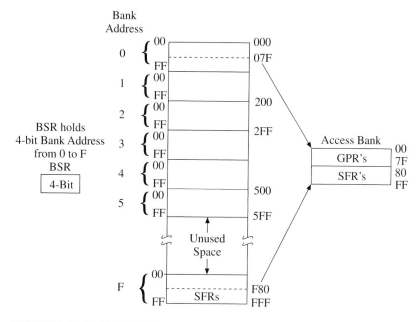

FIGURE 2-4   P1C18F452/4520—Data Memory Showing Access Bank and Address Range

In addition, Figure 2-4 shows one special bank called the Access Bank that is made up of the first 128 data registers (00 to $7F_H$), which are called General Purpose Registers (GPUs) and the last 128 registers ($F80_H$ to FFF_H), which are called Special Function Registers (SFRs). The SFRs are used by the MPU to perform many functions such as defining addresses of I/O ports and operations of various registers (explained in Chapter 3). In an instruction, if the access bit 'a' = 0 (explained later in Chapter 3), the registers in the access bank can be directly addressed without the BSR register, which results in faster execution of the programs.

Data EEPROM

The Electrically Erasable Programmable Read Only Memory (EEPROM) is functionally similar to flash memory, but each register can be individually accessed for read or write operations, rather than accessing a block of memory locations as in flash memory. EEPROM memory is used for frequent read and write operations, depending on the version of PIC controller; some have no EEPROM memory, some have 256 bytes, and some have 1024 bytes of EEPROM. It is important to note that this memory is not a part of the data memory space; it is addressed through special function registers.

2.1.4 I/O Ports

As discussed in Chapter 1, I/O ports are interfacing devices (such as buffers and latches) that enable the user to connect I/O peripherals such as keyboards, switches, printers, and LEDs to MCUs. In the PIC family of MCUs, the number of I/O ports varies from 3 to 10 (16 pin to 70 pins), identified by alphabetic letters such as PORTA and PORTB. Many of I/O ports have eight pins but the number varies depending on the functions an I/O port is designed for, and the majority of pins are "multiplexed" (meaning they can be set up to perform different functions). As an example, PORTB (Figure 2-5) has eight "bidirectional" I/O pins—data can flow in both directions—from the MPU to an I/O such as LED as an output and from a keyboard to the MPU as an input. In addition, all pins RB0 to RB7 (except RB4) are also multiplexed, meaning they can be set up to perform other functions as well.

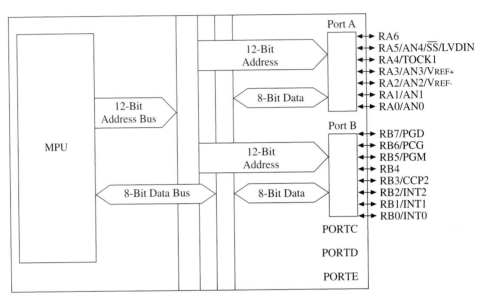

FIGURE 2-5   P1C18F452—I/O Ports A and B (Ports C, D, E are shown without details)

These I/O ports are already assigned addresses in the design stage of the MCU. These addresses are stored in SFRs as discussed in the Data Memory Section. For example, the addresses of PORTA and PORTB are $F80_H$ and $F81_H$, respectively, and these addresses are stored in SFRs labeled PORTA and PORTB.

Because the individual pins of an I/O port are designed to perform multiple functions, these pins must be set up or initialized by writing instructions in associated special function registers. For example, the pins of PORTB can be set up either as an input or an output by writing to the special function register TRISB (Data Direction Register with the address $F93_H$). The use of these I/O ports is illustrated briefly in Section 2.6 and in detail with various applications in Chapter 9.

2.2 PROCESSES OF DATA TRANSFER BETWEEN A MICROCONTROLLER AND OUTSIDE PERIPHERALS

In the previous section, we examined the components of PIC18F MCU and internal buses. Now we will examine the modes and processes of data transfer between the MCU and external peripherals.

The process of data transfer is divided into two modes: parallel and serial. These data transfers occur under different conditions: unconditional, status-check (polling), and interrupt. These topics are discussed in the following sections.

2.2.1 Parallel Data Transfer

In the parallel data transfer mode, a complete word (based on the size of the data bus such as 8-bit or 16-bit) is transferred simultaneously over the entire data bus. The data transfer from I/O ports to peripherals such as a seven-segment LED or a printer (except printers connected through the USB port in a personal computer) is in the parallel mode.

In the PIC18F microcontroller family, if we connect eight I/O pins—RB0 to RB7—of PORTB to eight LEDs and write instructions to output data to PORTB, the MPU uses all eight data lines to output data to the LEDs (see Illustration in Section 2.6).

2.2.2 Serial Data Transfer

In the serial mode, data are transferred one bit at a time over a single line. For example, to transmit information over a telephone line using a modem, a parallel word (8-bit) is converted into a stream of bits and transmitted as one bit at a time. There are many peripherals such as a scanner, keyboard, and mouse that are designed to communicate in the serial mode. The data transmission in the serial mode is slower than that of the parallel mode; however, the serial I/O has many advantages. The serial mode requires only three signal lines: "transmit," "receive," and "Ground" and is suitable for transmission over a fairly long distance.

2.2.3 Interrupts and Status Check: Conditions of Data Transfer

The processes of data transfer or communication in a microcontroller can be divided into two groups: MPU initiated versus externally (external to the MPU) requested. In externally requested, a signal requesting a service (attention) is sent to the MPU. If the MPU acknowledges the request, it interrupts the task it is performing, provides the necessary service, and goes back to the previous task. This process of communication is called interrupt. These processes are summarized in Figure 2-6.

MPU Initiated

There are two types of MPU-initiated data transfer: unconditional and conditional. For unconditional data transfer, the user assumes that a peripheral is ready to accept the data any time and writes the necessary instructions to output data to the peripheral. An example of unconditional data transfer is our LED I/O port discussed in Section 2.6. For conditional data transfer, the user writes instructions (in the program) to check whether a peripheral is ready to accept data, and when the peripheral acknowledges that it is ready to accept data, the processor sends data to that peripheral. This can be done either by the technique of polling or by using handshake signals. An example of conditional data transfer is when data are being sent to a printer that is much slower in its operation than the rate at which the processor can output data. Typically, the MPU

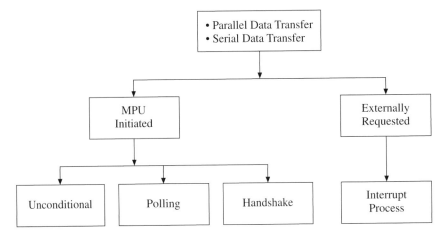

FIGURE 2-6   Processes and Conditions of Data Transfer

checks whether the printer is ready to accept a byte, and when it is ready, a data byte is sent, and the processor waits in a loop until the printer finishes printing that byte, and when the printer is ready again, the processor sends the next byte.

Externally Requested Interrupts

An interrupt is a process whereby a signal is generated, by an external device or a peripheral, to interrupt the processor and request a service. The interrupts are divided into two groups: hardware interrupts and software interrupts. The hardware interrupts are generated by two different sources: external peripherals connected to I/O ports such as a keyboard, and sources inside the controller, such as a timer (described in Section 2.3.1). Software interrupts are generated to handle errors. For example, if a program encounters an operation of dividing a number by zero, this operation will generate an interrupt alerting the processor that the result has an indeterminate value. In some processors software interrupts are called exceptions or traps. There are no software interrupts built in to PIC microcontrollers.

A hardware interrupt is like a telephone call whereby a ring lets you know that someone is requesting to talk with you. You stop whatever you are doing and answer the telephone call. Similarly, when an interrupt is generated and accepted by the processor, it stops whatever program it is executing, goes to the directed memory location, executes the set of instructions (called a service routine) for that interrupt, and goes back to whatever it was doing before the interrupt. In our telephone analogy, if we do not want to be interrupted, we can take the receiver off the hook. Similarly, we can disable the interrupt process in a microcontroller.

The interrupt process is essential to the efficient operation of the microcontroller. For example, if the controller is printing a long file, and we want to enter text using the keyboard without interrupts, we would have to wait until the printing is completed. However, because of the availability of interrupts, the keyboard can interrupt the MPU and request the processor to accept the input from the keyboard. The processor accepts the input from the keyboard, displays it on the screen, and goes back to printing again.

The PIC18 family of microcontrollers has three (some versions have four) pins on PORTB to accept external interrupts. The number of internal interrupts generated by support devices (such as timers) varies with the version of the device. The interrupt process is set up and controlled using SFRs. The details of the interrupt process with applications are discussed in Chapter 10.

Reset

Reset is a special type of external interrupt. The primary function of a reset is to establish or reestablish proper initial conditions and begin processing. In the PIC18 family, a microcontroller is reset in various ways. For example, the microcontroller is reset various ways such as (1) by a manual switch, (2) when the power is turned on, called Power-On Reset, and (3) when the power goes below the specified voltage, called Brown-Out Reset.

2.3 SUPPORT DEVICES

In addition to the microprocessor (MPU), memory, and I/O ports, a microcontroller includes many support devices that enable the MPU to perform the various modes of data transfer described in the previous section. For example, in a personal computer, there is a timer that keeps track of a clock, a serial I/O circuit that connects to a modem and transmits information to telephone lines, and a circuit that connects to a printer. However, in a microcontroller, these various functions are performed by support devices that are on the same chip. This controller includes many support devices such as timers, A/D converter, parallel and serial communications, and Capture/Compare/Pulse Width Modulation (CCP) modules. The support devices in the PIC18 family are introduced in the following sections to provide an overview. The details are discussed in later chapters.

2.3.1 Timers

In digital systems, time is calculated by counting. There are various approaches to counting and calculating time; however, the basic concept is as follows: A count (number) is loaded into a register and the count is incremented (increased by one) or decremented (decreased by one) at every clock cycle. The count is read at a given condition, such as when the register reaches zero or a certain count. The difference between the initial count and the final count multiplied by the clock period provides the time elapsed.

In the PIC18F family, the number of timers varies from three to five, and they are labeled as Timer0 through Timer4. Timers are divided into two categories: timers with 8-bit registers and timers with 16-bit registers. Timers can be used to calculate time or to count events. When a timer reaches its maximum count or a specified limit, it can generate an interrupt signal (described in the previous section) that alerts the processor. The timer operations are controlled by the associated SFRs. For example, the SFR register T0CON (Timer0 Control) determines the operation of TIMER0. The user must load an appropriate byte in the T0CON register, and each bit in the register determines various operational parameters of TIMER0 such as source of clock frequency, 8-bit or 16-bit mode, and counting on rising or falling edge pulse.

Timers and Capture/Compare/PWM (CCP) modules shown in Figure 2-7 can be used to perform functions such as capture, compare, and pulse width modulation (PWM). The following examples illustrate a few applications of these functions, and these applications are discussed in detail in later chapters.

1. The capture function can be used to measure the period of an incoming signal. The capture module can be configured to record the count between two rising (or falling) pulses of an incoming signal, and the resulting count is used to calculate the period of the incoming signal.

2. The compare function can be used to generate time delays, trigger a special event signal, or generate a periodic wave form. A timer value is constantly compared with the value loaded into a capture module, and when the values match, a signal is generated.

3. The PWM function can generate pulse waveforms with a specified duty cycle and period that can be used in applications such as motor control. The SFRs associated with PWM are used to specify the period and the duty cycle.

2.3.2 Serial Data Communication

As mentioned earlier, there are many peripherals that can receive or transmit only one bit at a time. An obvious example is data transmission over telephone lines. In this situation, an 8-bit data byte on the data lines of the MPU must be converted into a stream of single bits using a shift register or a modem card (which also includes a shift register). The device that transmits data is called a transmitter and the device that receives data is called a receiver.

The serial I/O can be synchronous or asynchronous. In the synchronous format, a transmitter and a receiver have the same clock that synchronizes the data transfer. In the asynchronous format, data transfer can occur any time. Therefore, the transmitter must alert the receiver, check its readiness to accept data, and inform the receiver of the end of data transmission. This timing information becomes a part of data transmission. There is an established protocol for the asynchronous serial I/O called RS232 (discussed in Chapter 13).

The PIC18F family of microcontrollers includes modules that support both types of serial data transmission; they are shown as the Master Synchronous Serial Port (MSSP) and the Addressable Universal Synchronous/Asynchronous Receiver/Transmitter (USART) in Figure 2-7. These modules and their applications are described briefly in the following sections.

Master Synchronous Serial Port (MSSP)

This module is used for synchronous serial data transfer and can be configured in one of two modes: (1) Serial Peripheral Interface (SPI) or (2) Inter-integrated Circuit (I²C). These modes were developed by two manufacturers with different protocols (procedures), and at this point, the details of these modes are unimportant. What is important is the understanding of basic concepts in serial I/O. In this mode, we can use three pins from PORTC: a clock pin (SCK) for synchronous data transfer, a serial data in (SDI) pin, and a serial data out (SDO) pin. Examples of serial peripherals that are interfaced with the microcontroller include shift registers and EEPROM. Multiple peripherals can be interfaced with the PIC18 microcontroller to transmit and

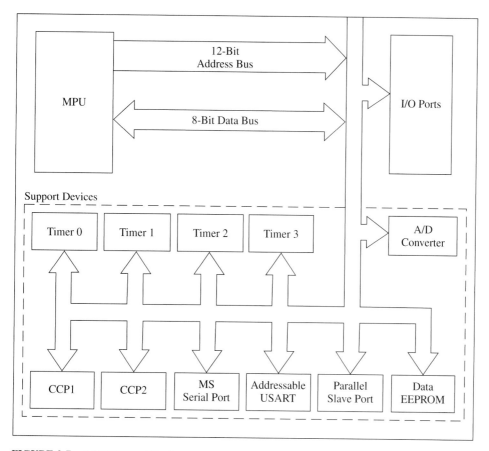

FIGURE 2-7   MCU Support Devices

receive data. The microcontroller, called master, is responsible for data transfer. The peripherals, called slaves, provide or receive data. To implement this process of data transfer, many parameters need to be configured such as the rate of data transfer, the definition of the pulse edge (rising or falling) used for data transfer, and the mode. All these parameters are configured by using associated SFRs. The MSSP module is also used to expand the number of I/O ports by using shift registers.

Addressable USART

This module also deals with serial data transmission in the asynchronous and synchronous modes. It uses RS232 (EIA 232) protocol, which is the standard that was originally designed for data transfer between a computer and a terminal using a modem. The standard defines various electrical and mechanical aspects of this data transmission such as signal voltages, data format, data rates, and number of pins. In addition to modem interfacing, this module is used for data transfer between two microcontrollers to expand the number of I/O ports.

2.3.3 Analog-to-Digital (A/D) Converter

The microcontroller is a digital device that processes only discrete binary electrical signals (0s and 1s). Real-life quantities, such as sound, light, temperature, and pressure, are non-electrical and analog in nature, meaning they have a continuous set of values. Therefore, these quantities must first be converted into analog electrical signals and then into equivalent binary values. For example, in a speech recognition system, a microphone converts our speech into an analog electrical signal, and an A/D converter converts the analog signal into equivalent binary signals. An A/D converter is an electronic chip that accepts an analog signal as input and provides equivalent binary output.

The PIC18F family has a 10-bit A/D converter, meaning that it converts its analog input signal into an equivalent 10-bit binary number. The user can connect multiple analog sources to designated pins divided among various I/O ports. For example, PORTA has four pins—AN0 to AN3—that can be configured as analog inputs. There are several SFRs associated with the A/D conversion that must be programmed to set up different parameters for the conversion process. The concepts underlying the A/D conversion and its applications using PIC controllers are fully described in Chapter 12.

2.3.4 Parallel Slave Port (PSP)

Figure 2-7 shows a module called parallel slave port (PSP). PORTD can be set up as a parallel slave port whereby other microcontrollers or microprocessors can read from or write to this port. Thus, PORTD can be used to setup communication between PIC18F MCU and other MCUs.

2.3.5 Special Features

Microcontrollers are being used for many applications, some of which require special features. For example, applications that run on batteries prefer low power consumption. For medical instruments, safety is a critical issue, and prevention of software piracy is an important issue for manufacturers. Many of these features are provided by the PIC18 family of microcontrollers. They are briefly described in the following section.

1. Sleep Mode: This is a power-down mode that reduces the power when the microcontroller is idle by executing the SLEEP instruction.
2. Watchdog Timer (WDT): The WDT module is used primarily to locate software bugs. A well-written program is expected to be executed in a given amount of time, and instructions are written to reset the timer at the end of this program. If the program is caught in an infinite loop, the WDT will not be reset within the specified time. This will generate a system reset.
3. Code Protection: This protection is provided to prevent the copying of software by competitors. The entire Data EEPROM can be protected from external reads and writes by setting up certain bits in the appropriate SFR.

4. In-Circuit Serial Programming: Two pins—Data and Clock—are provided to enter the program serially after the microcontroller is already on a board. This feature enables manufacturers to enter, modify, and update the software without removing the MCU from the board.

5. In-Circuit Debugger: This feature enables the user to debug software as it is being executed. Two pins on PORTB—RB6 and RB7—can be set up in the serial mode, and internal registers can be observed on a PC screen as instructions are being executed.

Points to Remember

Figure 2-8 shows the entire block diagram of one of the microoocontrollers of PIC18F family architecture. The points to emphasize are as follows:

1. The microcontroller consists of three components: MPU, memory, and I/O.
2. MPU executes all the operations (such as arithmetic and logic) after reading the binary instructions stored in memory.
3. The microcontroller designed with Harvard architecture has two separate memories: a program memory to store instructions, and data memory to store data.
4. Data memory is divided into banks and has two groups of registers: (1) general-purpose registers (GPRs) that can be used for various arithmetic and logic operations, and (2) special function registers (SFRs) that store I/O addresses and parameter information for various operations.
5. The address bus for program memory is 21 bits wide, enabling the MPU to address 2 MB of memory. The address bus for data memory is 12 bits wide providing the MPU with the capability of addressing 4 K of memory.
6. The data communication processes are divided into two modes: parallel and serial. In the parallel mode, multiple bits are transferred simultaneously, and in the serial mode, one bit is transmitted at a time.
7. The processes of data transfer are divided primarily in two categories: MPU initiated and externally requested. The MPU initiated category is divided into two groups: unconditional and conditional data transfer. For the conditional data transfer, the MPU checks the readiness of peripherals for data transfer. External requests are initiated either by peripherals or by internal support devices such as timers. If the MPU accepts a request, it stops the present task (processing) and executes the instructions that are necessary for the request. After the completion of those instructions, it goes back to the previous task. This is called the interrupt process.
8. The Reset is a specialized interrupt that restores the initial conditions for the microcontroller.
9. In addition to the three major components (MPU, memory, and I/O), the microcontroller includes various support devices such as timers, A/D converter, CCP modules, and serial I/O. These support devices, after completion of their tasks, set a flag or generate interrupts to inform the MPU.
10. The PIC18F452/4520 is a microcontroller that includes many special features, such as sleep mode for power saving, watchdog timer to detect software bugs, code protection, serial programming, and debugging.

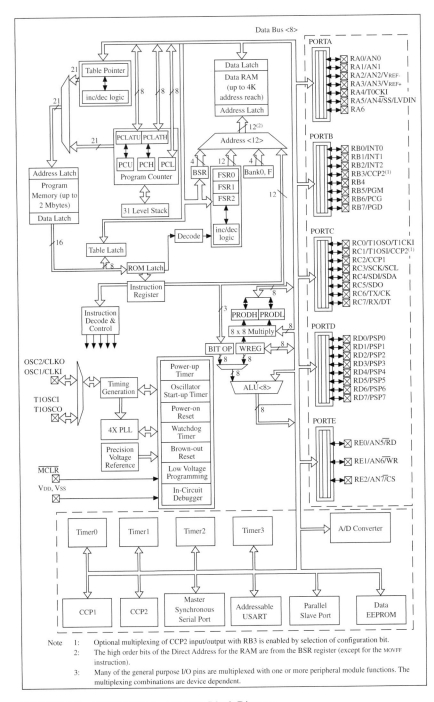

FIGURE 2-8   P1C18F4X2 Architecture Block Diagram

2.4 MICROCHIP PIC FAMILY OF DEVICES

In previous sections, we focussed on one family of microcontrollers, namely the PIC18. This is considered a high-end microcontroller; Microchip also has low-end and middle-range microcontroller families as listed in Table 2-1. The low-end microcontrollers are based on 12-bit wide instructions, the middle range on 14-bit wide instructions, and the high-end on 16-bit wide instructions. The number of pins on a package varies anywhere from 6 pins to 100 pins. A recent addition is a family of 16-bit microcontrollers (see Table 2-1). For comparison, we will discuss the PIC16F family in the next section.

The hardware architecture of the PIC16F family is similar to the PIC18F family. Figure 2-9 shows MPU, program memory, data memory, I/O ports, and support devices of the PIC16F MCU. The support devices shown include similar features to that of the PIC18F family. The primary differences are in the size of the memory capacity and the instruction set.

The program counter (PC) is only 13 bits wide, making it capable of accessing 8 K of program memory with the address range of 0000 to $1FFF_H$. The effective address bus for data memory is 9 bits wide; thus it can have 512 bytes of data memory.

The instruction set includes only 35 instructions as opposed to 77 instructions in the18F family. However, many of the tasks that can be performed with one or two instructions by an 18F microcontroller may require a set of instructions for a 16F microcontroller. For example, the 16FXXX family does not have a multiply instruction, but it can perform multiplication by repeating the addition operation.

2.5 PIC18F INSTRUCTIONS AND ASSEMBLY LANGUAGE

In the previous four sections, we examined the architecture (hardware) of the PIC18F family of microcontrollers. In this section, we will illustrate an elementary process of writing assembly language instructions and their execution.

The PIC18F microcontroller family has total of 77 instruction types. That means the MPU in this microcontroller is capable of understanding and performing 77 different types of operations. These instructions are upward compatible with the instruction sets of other families, which have either 33 or 35 instructions. The term upward compatible means a program written for another PIC microcontroller family will run on a PIC18F microcontroller. In the PIC18F instruction set, all but four instructions are 16-bit word length; the remaining four instructions are 32-bit length. These instructions can be grouped in different categories such as "load (move or copy)," "arithmetic," and "logic." These categories will be discussed in Chapter 3. Here we will simply examine the binary format and assembly language mnemonics of a few instructions.

2.5.1 Instruction Description: Copy (Load) 8-Bit Number into W Register

- Binary Format: 0000 1110 XXXX XXXX (any 8-bit number)

This is a 16-bit instruction. The first 8 bits tell the processor that it should copy the following 8-bit number into the W register. This instruction type is already designed into the processor, and the user needs to write this instruction into its program memory to load an 8-bit number in W register. The major difficulty for the user is to find the code from the binary instruction list,

TABLE 2-1

List of Selected Microcontroller Families from Microchip

Part No.	Program OTP/Flash	EE PROM	RAM	Total Pins	I/O Pins	ADC	Analog Comp.	Digital Timers/WDT	Serial I/O	CCP/ECCP	Max Speed MHz	Instruction Size	Total Instructions
10F200	256x12 Flash		16	8	4			1-8 bit, 1-WDT			4	12-bit	33
10F220	256x12 Flash		16	8	4	2x8-bit	.	1-8 bit, 1-WDT			8	12-bit	33
12F510	1536x12 Flash		38	8	6	3x8-bit	1	1-8 bit, 1-WDT			8	12-bit	33
16F506	1536x12 Flash		67	14	12	3x8-bit	2	1-8 bit, 1-WDT			20	12-bit	33
16C55A	768x12 OTP		24	28	20			1-8 bit, 1-WDT			40	12-bit	33
16CR58B	3072x12 ROM		73	18	12			1-8 bit, 1-WDT			20	12-bit	33
12F683	2048x14 Flash	256	128	8	6	4x10-bit	1	1-16 bit, 2-8 bit, 1-WDT			20	14-bit	35
16F687	2048x14 Flash	256	128	20	18	12x10-bit	2	1-16 bit, 1-8 bit, 1-WDT	EU/I^2C/SPI		20	14-bit	35
18F1230	2048x16 Enh Flash	128	256	18-28	16	4x10-bit	3	2-16 bit, 1-WDT	EU		40	16-bit	77
18F4520	16384x16 Enh Flash	256	1536	40-44	36	13x10-bit	2	1-8 bit, 3-16 bit, 1-WDT	EU/ MI^2C /SPI	1/1	40	16-bit	77
18F6527	24576x16 Enh Flash	1024	3936	64	54	12x10-bit	2	2-8 bit, 3-16 bit, 1-WDT	2EU/ 2-MI^2C /SPI	2/3	40	16-bit	77
18F8622	32768x16 Enh Flash	1024	3936	80	70	16x10-bit	2	2-8 bit, 3-16 bit, 1-WDT	2EU/ 2-MI^2C /SPI	2/3	40	16-bit	77
18F96J60	32768x16 Flash		2048	100	72	16x10-bit	2	2-8 bit, 3-16 bit, 1-WDT	2EU/ 2-MI^2C /SPI	2/3	42	16-bit	77
24FJ128GA-010	65536x16 Flash		8192	100-128	86	16x10-bit	2	5-16 bit, 1-WDT	2-UART 2-I^2C/SPI	5	32	16-bit	77

Abbreviations: 1) ADC: Analog-Digital Converter, 2) AUSART: Addressable USART, 3) CCP: Capture/Compare/PWM, 4) ECCP: Enhanced CCP,

5) EU: Enhanced USART, 6) Enh Flash: Enhanced Flash, 7) I^2C: Inter-integrated Circuit Bus, 8) MI^2C/SPI: Master I^2C /SPI, 9) OTP: One-Time Programmable,

10) SPI: Serial Peripheral Interface, 11) USART: Universal Synchronous/Asynchronous Receiver/Transmitter, 11) WDT: Watchdog Timer

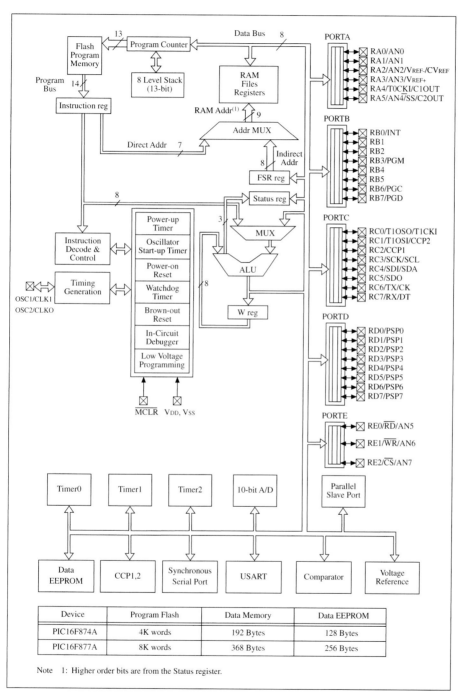

Device	Program Flash	Data Memory	Data EEPROM
PIC16F874A	4K words	192 Bytes	128 Bytes
PIC16F877A	8K words	368 Bytes	256 Bytes

Note   1: Higher order bits are from the Status register.

FIGURE 2-9   P1C16F87 Architecture Block Diagram

interpret, and enter it into the memory without an error. To mitigate this difficulty, the manufacturer or its design team provides mnemonics corresponding to each instruction. The design teams selects letters or abbreviations that can help the user understand the operation of the binary instruction. For example, the binary instruction on page 45 is expressed in mnemonics as follows:

- Mnemonics: MOVLW 8-bit

Interpretation: MOVLW suggests: Move L to W, meaning copy the literal (8-bit number that follows) into W register.

2.5.2 Instruction Description: Copy Contents of W Register into Data (File) Register

- Binary Format: 0110 1110 8-bit address of a file register in MCU

In this instruction, the first 8-bit unit tells the processor that it should copy the contents of the W register into the register, the address of which is given in the next 8 bits. In this MCU, the I/O PORTC is assigned the address 82_H. If we substitute 82_H in the place of 8-bit address, this instruction outputs (copies) the contents of W register to PORTC. The complete mnemonics format of this instruction is as follows:

- Mnemonics: MOVWF PORTC, a

Interpretation: Copy the contents of W register into a file register F that is given next and "a" specifies that the register PORTC resides in the Access bank described earlier. In writing this instruction in a program, the address of PORTC must be defined.

By knowing the mnemonics of the binary instruction set, the user can write programs in assembly language. But the microprocessor does not understand mnemonics. These mnemonics must be translated into binary codes. An assembler, a software program, is used to translate these mnemonics into corresponding binary codes and store the codes in program memory starting at a specified memory location. The processor can read these binary instructions from memory and perform the tasks specified in the instructions.

2.6 ILLUSTRATION: DISPLAYING A BYTE AT AN I/O PORT OF PIC18F452 MICROCONTROLLER

In previous sections, we examined the architecture of the PIC18F microcontroller family and a few copy instructions. We will use those instructions to display a byte at one of the I/O ports.

2.6.1 Problem Statement

Write instructions to light up alternate LEDs starting with LED0 at PORTC shown in Figure 2-10.

2.6.2 Problem Analysis

This problem can be divided into two parts: hardware and software. The hardware part deals with interfacing LEDs at the I/O port PORTC, and the software part deals with writing instructions to set up PORTC and turn on alternate LEDs.

Hardware

PORTC is a "bidirectional" port, meaning it can be set up as an input port or an output port. Therefore, it should be set up, by writing instructions, as an output port that can transfer binary bits to LEDs. Figure 2-10 shows that the cathodes of LEDs are connected to ground through current limiting resistors; therefore, logic 1 is required to turn on LEDs.

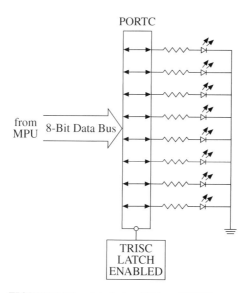

FIGURE 2-10   Interfacing LEDs to PORTC

Program (Software)

To turn on alternate LEDs at PORTC, the following steps are necessary:

- Setup PORTC as an output port by sending 0s to the TRISC register. (TRISC is defined below.)
- Load a byte 55H (01010101) in the W register—alternate 0s and 1s to turn on LEDs.
- Write (copy) the contents of the W register into PORTC.

Each I/O port has a setup register called TRIS, a part of the special function registers (SFRs). To set up all pins of PORTC as outputs, we need to send all logic 0s to the TRISC (address F94$_H$) register. PORTC and TRISC have been assigned the addresses F82$_H$ and F94$_H$ respectively by the designer of the MCU. The instructions to set up PORTC as an output port and turn on the alternate LEDs are listed in the following section.

2.6.3 Assembly Language Program

	Instructions	Comments
MOVLW	00	; Load 0s in W register
MOVWF	TRISC	; Place all 0s in TRIS register to set up PORTC as output

	Instructions	Comments
MOVLW	0 x 55[1], 0	; Byte to turn on alternate LEDs
MOVWF	PORTC	; Output 55H to turn on alternate LEDs
SLEEP		; Go into power down mode

To store these instructions in program memory, we need to look up binary code in the manufacturer's manual. If we are writing these instructions using an assembler (described in Chapter 4), the assembler will look up the codes and another program can copy these codes into the memory of the MCU. Each of these instructions are word (16 bits) length and require two memory locations. Now let us store these instructions in program memory starting from memory locations 000000. The program memory has 21-bit addresses, but we will ignore the first two zeros and use only the remaining four Hex numbers. Our instructions will be stored in memory as follows:

Memory Address	Memory Contents Binary Code	Hex	Mnemonics	Comments
0000	0 0 0 0 0 0 0 0	0 0	MOVLW 00	Load W register with 0s
0001	0 0 0 0 1 1 1 0	0 E		
0002	1 0 0 1 0 1 0 0	9 4	MOVWF TRISC,0	Place all 0s in TRISC
0003	0 1 1 0 1 1 1 0	6 E		TRISC address[2]
0004	0 1 0 1 0 1 0 1	55	MOVLW 0x55	Hex Byte 55H to turn on
0005	0 0 0 0 1 1 1 0	0 E		Alternate LEDs
0006	1 0 0 0 0 0 1 0	82	MOVWF PORTC, 0	Turn on LEDs
0007	0 1 1 0 1 1 1 0	6 E		PORTC address[3]
0008	0 0 0 0 0 0 1 1	0 3	SLEEP	Power down
0009	0 0 0 0 0 0 0 0	0 0		

The important point to remember is that the binary code given in the second column is stored in memory, and that is the only language the processor can execute. The memory addresses shown in Column 1 are not stored in memory; instead addresses are placed on the address bus by the MPU when it begins the execution.

In the MPLAB IDE simulator (explained in Chapter 4), the contents of the program memory are shown as follows. The memory addresses are displayed in the word format beginning with

---

[1] In assembly language programming, Hex byte is written as 0xByte or with the subscript H.

[2] & [3] TRISC and PORTC are in Access memory; therefore, the addresses $F94_H$ and $F82_H$ are coded as 94_H and 82_H.

an even address, and the Hex code is shown with the contents of the odd memory first followed by the contents of the even memory. This format avoids any confusion and improves readability.

Memory Word Address	Hex Code	Mnemonics	Comments
000000	0 E 0 0	MOVLW 00	Load W register with 0s
000002	6 E 9 4	MOVWF TRISC,0	Set up PORTB as output
000004	0 E 5 5	MOVLW 0x55	Byte 55H to turn on LEDs
000006	6 E 82	MOVWF PORTC, 0	Turn on LEDs
000008	0 0 03	SLEEP	Power down

2.6.4 Program Execution

When the processor is reset, its Program Counter is cleared and holds the 21-bit address 000000_H, and the program execution begins. The processor places the 21-bit address on the address bus, and the address decoder decodes the address and selects two memory registers 0000 and 0001, asserts the Read signal, enables the memory chip, and then the memory places the code 0E 00 on the 16-bit instruction data bus. The processor places the code in the instruction decoder, which interprets the instruction, and the processor executes the task specified by the instruction within the clock cycle. The program counter is incremented to the next address, 0002_H, when the previous address is placed on the address bus. At the beginning of the next clock, the processor places the next address on the address bus, gets the next code, and executes the next instruction. This process is repeated until all the instructions are executed. When the instruction MOVWF PORTC is executed, the byte 55_H is sent to PORTC and alternate LEDs are lit up. See Figure 2-11 for the execution of the instruction MOVWF   PORTC, 0 (Hex code: 6E 82).

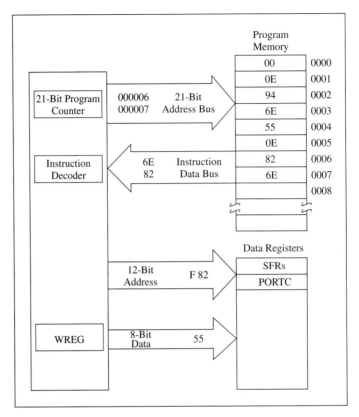

FIGURE 2-11   Execution of Instruction MOVWF PORTC

2.7 EMBEDDED SYSTEM: TEMPERATURE AND TIME DISPLAY

In previous sections, we discussed the architecture of the PIC18F family of microcontrollers and a few instructions to illustrate the process of program execution. Now we can expand the block diagram of our Embedded System: Temperature and Display, discussed in Chapter 1 (Figure 1-8).

Figure 2-12 shows the expanded version of Figure 1-8, including the PIC18F microcontroller with three I/O ports: PORTA, PORTB, and PORTC. A temperature sensor is connected as an input to the line RA0 of PORTA, and a LCD (liquid crystal display) is connected as an output to PORTB. Two signals of PORTA—RA6 and RA5—are used as output lines to turn on the heater and the fan. Eight data lines RB7-RB0 of PORTB are connected to eight data lines of the LCD module. Three lines of PORTC, RC1, RC2, and RC3 are used as control signals. RC1 enables the LCD, RC2 selects the register (RS) inside the LCD, and RC3 is used to read from and write to the LCD module. This system can be designed with only two I/O ports; three signals to control LCD could have been used from PORTA. To minimize the complexity, we separated control functions from I/O functions.

The temperature sensor is an electronic chip that senses the ambient temperature and converts it into an equivalent analog voltage in millivolts. The output of the temperature sensor is connected to RA0; therefore, RA0 should be setup as AN0 to accept an analog signal. AN0 functions as an input to the internal A/D converter that in turn converts the analog input signal into an equivalent 10-bit digital signal. The program initiates the conversion process, and at the end of the conversion an interrupt signal is generated by the A/D converter. The MPU acknowledges the interrupt and reads the digital reading and stores it into data memory.

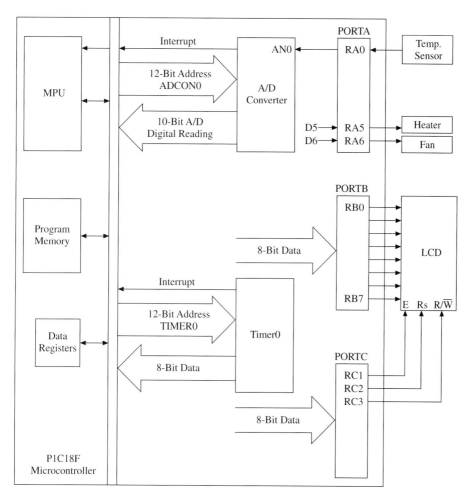

FIGURE 2-12   Embedded System—Microcontroller-Based Time and Temperature System

The second part of the embedded system is to design a clock and display time. The time is calculated by setting up a timer for a given count. At each clock cycle, the count is incremented, and when the counter reaches the maximum count, it is reset to 0. This is similar to the odometer in a car. When the odometer reaches 999,999, it is reset to 000,000 and begins again. When the timer goes from the maximum count to zero, it is called an overflow, and when an overflow occurs, it generates an interrupt and alerts the MPU of the overflow. The program keeps track of these overflows, calculates the time, and updates the clock time.

Based on the discussion in the previous paragraphs, the software (program) stored in program memory is expected to perform the following tasks:

- Set up PORTA as an input port and PORTB as an output port by loading appropriate bytes in the associated TRIS registers.
- Set up the LCD module for proper display.
- Initialize the timer with an appropriate count.
- Set up the interrupt processes.
- Begin the conversion process, and at the end of the conversion, when an interrupt is generated, read and store the digital equivalent reading.
- Convert the digital reading into equivalent temperature reading in degrees Celsius or Fahrenheit and display it on the LCD.
- If the temperature reading is above or below the set points, turn on/off the fan/heater.
- When the timer overflows, calculate the time. When the time reaches one second, update the clock and display the time at the LCD.
- The tasks described above are continuously repeated and displayed on the LCD module at a given interval. The LCD display alternates between displaying temperature and time.

We will continue to discuss this project and add more details throughout the text. In Chapters 5 through 9, when appropriate instructions are introduced, we will write the above tasks to be performed by the PIC18F microcontroller as modules. In Chapters 9 through 13 we will examine the hardware.

SUMMARY

This chapter focused primarily on the hardware architecture of 18F microcontroller families. The features of the 18F microcontroller are summarized at the end of Section 2.3, and the main concepts underlying the operations of the microcontroller are listed as follows:

1. The 18F microcontroller is designed using the Harvard architecture, and it includes MPU, program memory, data memory (registers), I/O ports, and many support devices such as timers, serial I/O, CCP modules, and an A/D converter.

2. The program memory and data memory (registers) have their own address and data buses that enable the MPU to access instructions and data in the same clock cycle.

3. The modes of data transfer are divided into two groups: parallel and serial. In the parallel mode, multiple bits are transferred simultaneously over the entire data bus, and in the serial mode, only one bit is transferred at a time by converting the parallel bits into a stream of serial bits.

4. The conditions of data transfer are divided primarily in two groups: MPU-initiated and Externally Requested. The MPU-initiated data transfer sends data unconditionally or by checking the readiness of the device. The Externally Requested data transfer is initiated by an external peripheral or internal sources, such as support devices. This process of data transfer is known as an interrupt.

5. The software (the set of instructions) is an integral part of microprocessor-based systems. The instructions are stored in memory, and the MPU reads each instruction using its bus structure, decodes it, and executes that instruction.

QUESTIONS, ASSIGNMENTS, AND SIMULATION EXERCISES

2.1   What is an MCU?

2.2   List the major components of an MPU.

2.3   Explain the special features of the Harvard architecture.

2.4   What is the function of an address bus?

2.5   If an MPU has 19 address lines, calculate the addressing capacity of the MPU.

2.6   If an MPU is capable of addressing 128 K-byte of memory, calculate the size of its address bus.

2.7   Calculate the number of 8-bit registers in 128 KB of memory.

2.8   PIC18F MCU has a 12-bit address bus for its data memory. If we increase the size to 14 bits, what is the size of the data memory?

2.9   PIC16F MCU has a 13-bit address bus. Calculate the addressing capacity of its MPU.

2.10   What is the function of the W register in the PIC18 MCU?

2.11   What is the function of a program counter (PC) and its size in the PIC18 MCU?

2.12   What is the size and the function of the BSR register?

2.13   What is the size of the FSR register? Explain the significance of its size.

2.14   What is the size of the address bus for the data memory in the PIC18F MCU and how many data registers can be accessed by this bus?

2.15   What are the special function registers and their address range in the 18F MCU?

2.16   What is an I/O port and what is the size of the address of an I/O port in the PIC18 MCU?

2.17   Explain multiplexing and how I/O ports are set up to perform a given function.

2.18   What is an advantage of EEPROM over flash memory?

2.19   List the two primary modes of data transfer.

2.20   What is the data transfer mode used for transferring information via modem?

2.21   Is a fax machine a serial or parallel device?

2.22   Explain the reason the MPU should check the readiness of a peripheral device such as a printer.

2.23   What is an interrupt process?

2.24   Why is an interrupt process classified as an externally initiated process?

2.25   List sources that can interrupt the PIC18F MPU.

2.26   Can the interrupt process be disabled?

2.27   List typical support devices included on the MCU chip.

2.28   What is a timer and how is it used in MCU-based systems?

2.29   What is the function of an A/D converter and why is it necessary in MCU-based systems?

2.30   Specify the number of instructions in a PIC18 MCU and the bit size of most instructions.

SIMULATION EXERCISES USING PIC18 SIMULATOR IDE

2.1 SE Refer to Appendix F – Section F.1.2. Build the microcontroller system as suggested in the appendix and enter the Hex code of the program discussed in Section 2.6. Execute the program as outlined in the appendix.

2.2 SE Change the code of the first instruction from 0E 00 to 0E 0F and execute the program. Explain your results.

2.3 SE Change the code of the third instruction in relation to 2.1SE from 0E 55 to 0E AA and execute the program. Explain your results.

CHAPTER 3

PIC18F PROGRAMMING MODEL AND ITS INSTRUCTION SET

OVERVIEW

This chapter introduces a programming model of the PIC18F microcontroller and explains the functions of various registers in its arithmetic logic unit (ALU) and data memory (also referred to as data or file registers). The chapter also discusses the flags that represent various data conditions after arithmetic and logic operations and emphasizes their critical importance in writing assembly language programs.

The PIC18 instruction set is divided into seven functional categories: data copy, arithmetic, logic, bit manipulation, program redirection, table pointers, and machine control. Instructions in each category are listed and their primary features are summarized. To show the format of the instructions and how they are designed at the level of machine language, the instruction set is grouped in the following types of operations: byte-oriented, bit-oriented, literal, and program redirection operations.

The chapter includes a simple program that illustrates an application of some these instructions. The program also shows how the binary instructions are stored in memory and how they are executed using the pipelining structure.

OBJECTIVES

- List the registers and their functions in the programming model of the PIC18F microcontroller.
- List the data condition flags stored in the STATUS register.
- Explain how the status of the four data condition flags is used to redirect the program execution.
- List the functional categories of the PIC18F microcontroller instruction set and illustrate each category with at least two instructions.
- Define the terms "opcode" and "operand," and recognize the terms in a given instruction of the PIC18F instruction set.
- Explain how data memory (registers) are divided into banks and how registers are accessed using Bank Select Register (BSR).
- Write a set of instructions to perform simple operations such as copy, add, and subtract.
- Explain how the instructions are executed in a pipeline structure.

3.1 PIC18F PROGRAMMING MODEL

In Chapter 2, we examined the hardware aspects of the PIC18/16F microcontrollers (MCUs) and their microprocessor units (MPUs). In this chapter, we will begin to explore the software aspects of the microprocessor. In Chapter 1, we defined the microprocessor as an electronic computing device that can read and understand binary instructions from memory and execute the tasks specified in the instructions. These binary instructions are known as the machine language of the microprocessor; however, the programmer writes programs either in assembly language or high-level language. To write instructions in assembly language, the programmer needs to know the internal architecture of the microprocessor. This architecture includes accumulators, registers, and flags (discussed later). The representation of the internal architecture, necessary to write assembly language programs, is called a programming model of the microprocessor. Each microprocessor has its own programming model, and its assembly language instructions are directly related to this model. In this text, we focus on PIC18F452/4520 microcontrollers; however, the programming model represents the entire 18F family.

Figure 3-1 shows a programming model of the PIC18F MCU that is divided into two major groups of registers. Registers that hold data for arithmetic and logic functions are from Arithmetic Logic Units (ALUs), and registers called special function registers (SFRs) are from data memory. The programming model includes the working register (WREG), flag register (STATUS), file select registers (FSRs), multiplication register, program counter, stack, and SFRs. All the registers shown in the programming model are defined in SFRs. These registers and their functions are described in the following sections.

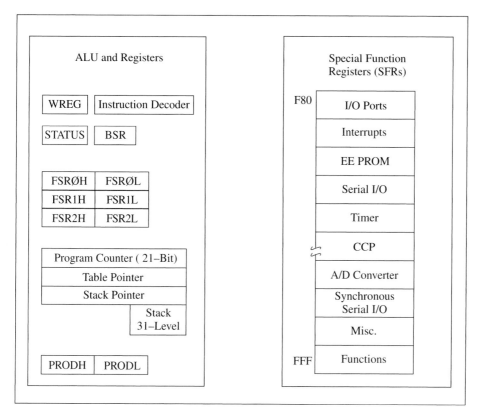

FIGURE 3-1 P1C18F Programming Model

3.1.1 Working Register (WREG)

The Working Register in the PIC18F MCU is similar to an accumulator in other processors. It is an 8-bit register in the ALU that is used in all arithmetic and logic operations. One of the powerful features in this MCU is that the result of an arithmetic/logic operation can be saved in the WREG or in the other operand register (data register).

3.1.2 Bank Select Register (BSR)

The Bank Select Register (BSR) is an 8-bit register but uses only the lower 4 bits to specify the data bank from 0 to F, and the upper 4 bits are always zero. The data memory has 4096 registers and is divided into 16 banks, each with 256 registers as explained in Chapter 2 (Figure 2-4). The BSR provides the higher-order 4 bits of a 12-bit address of a data register.

3.1.3 STATUS Register

The STATUS Register is an 8-bit register that uses 5 individual bits, B0 to B4, called flags, reflecting the data conditions (the status) of an operation; the remaining 3 bits, B5–B7, are unused.

The data condition flags reflect the nature of the result after an operation. For example, as discussed in Chapter 2, when an addition operation generates a sum larger than 8 bits, the operation generates a carry and bit B0, also known as the carry flag (Figure 3-2), in the STATUS register is set to 1. The five flags are known as C (Carry), DC (Digit Carry), Z (Zero), OV (Overflow), and N (Negative). The programmer can use the four flags (C, Z, OV, and N) to make decisions or redirect the program. The remaining flag DC is used internally by the processor for BCD operations.

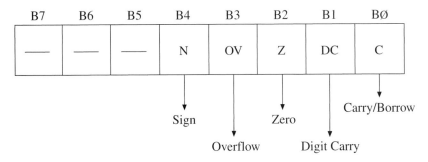

FIGURE 3-2   Flags in STATUS Register

The processor has flip-flops in its ALU segment that are set or reset according to the result of an operation. The outputs of the flip-flops are copied into specific bit positions of the STATUS register to provide information about the status of the flags. The user may examine this register if necessary. The flags affected by an operation are specified in the description of an instruction. All operations do not affect the flags, and some operations affect only selected flags.

- C—Carry/Borrow Flag. The carry flag represents two arithmetic operations: addition and subtraction. The flag is set to 1 when an addition generates a carry and a subtraction generates a borrow from the most significant bit of the result; otherwise, it is reset. For an 8-bit addition (or a subtraction), a carry (or a borrow) from data bit D7[1] sets the carry flag, and the flag is stored in bit position B0 of the STATUS register. The instruction set includes two branch instructions (BC—Branch on Carry and BNC—Branch on No Carry) based on this flag. This flag is relevant in arithmetic operations of unsigned numbers, and it should be ignored in operations of signed numbers.
- DC—Digit Carry Flag. The DC flag is set when a carry is generated by bit B3 (4th low-order bit) to B4 in arithmetic operations. There is no branch instruction associated with this flag; this flag is used internally by the processor for BCD (Binary Coded Decimal, see Section 3.2) operations.
- Z—Zero Flag. The zero flag is set to 1 when the result (all 8 bits) of an operation is zero; otherwise, it is reset to 0. To clarify any potential confusion, let us restate: the zero flag

---

[1]An individual bit is referred to with a letter B. It is also referred to as the letter D when it is a data bit. Both terms are synonymous.

is set to 1 when a result is zero, and it is reset to 0 when the result is not zero. Typically, the zero flag is used in a counting operation. For example, if a register is set up as a down counter to count from 50 to zero for a 50-car parking lot and when the count goes to zero, the zero flag will be set.

The instruction set includes two branch instructions (BZ—Branch on Zero and BNZ—Branch on No Zero) and five skip instructions (INCFSZ, DECFSZ, TTSTFSZ, BTFSC, and BTFSS) based on the Zero flag. See Table 3-7 for additional details).

- OV—Overflow Flag. The OV flag relates to the operations of signed numbers and it occupies the bit position B3 in the STATUS register. For the operations of unsigned numbers, the user can ignore this flag.

  In an 8-bit operation, the largest unsigned number is FF_H (255_{10}). In signed numbers, the range is from -128 to +127 in decimal or 80_H to $7F_H$ in Hex (as discussed in Chapter 1). Whenever a result falls outside this range, the flag is set to indicate that the result is inaccurate. The overflow flag will be explained with examples later in the section. The instruction set has two branch instructions (BOV—Branch on Overflow and BNOV—Branch on No Overflow) based on this flag.

- N—Negative Flag. The N flag indicates the status of the most significant bit (MSB) D7 of the result of an arithmetic/logic operation. If bit D7 is 1, this flag is set; otherwise, it is reset. This flag is used for signed numbers and the instruction set has two branch instructions (BN—Branch on Negative and BNN—Branch on No Negative) based on this flag.

EXAMPLE 3.1

Explain the status of the data flags if the MPU adds the following two numbers: $9F_H$ and 58_H.

Solution:

The addition of two Hex numbers is as follows (To review the Hex arithmetic, see Appendix D):

```
        9 F
   +    5 2
   CY   1
   ---------------
        F 1        N = 1, OV = 0, Z = 0, C = 0
```

Flag Interpretation: The MPU sets or resets the flags based on the result $F1_H$. The result remains the same regardless of whether these are signed numbers or unsigned numbers; however, the interpretation varies.

If the numbers are assumed to be unsigned numbers, N (negative) and OV (overflow) flags are irrelevant. Therefore, the answer is $F1_H$ (Unsigned value = 241_{10}). There is no carry generated by the addition; therefore, the carry (C) flag is reset. The result is not zero; therefore, the zero (Z) flag is also reset.

If the numbers are assumed to be signed numbers, the result $F1_H$ is negative, and the N flag is set. There is no overflow; therefore, the overflow (OV) flag is reset. This can be justified with decimal calculations as follows:

1. $9 F_H$ is a negative number because the most significant bit (D7) is 1 and all negative numbers are represented in 2's complement.
   The 2's complement of $9 F_H = 6 1_H = (-9 7_{10})$.
2. $5 2_H$ is a positive number $= 82_{10}$. The addition of -97_{10} and 82_{10} results in -15_{10}.
3. The result F 1 is negative, and 2's complement of $F1 = 0F_H = -15_{10}$.
   The detailed explanation of the overflow flag is given in Chapter 5, Section 5.2.

3.1.4 File Select Registers (FSR)

There are three registers, FSR0, FSR1, and FSR2, that hold 12-bit addresses of data registers (data memory) and are used as pointers for indirect addressing. (The concept of pointers is discussed in Chapter 5. In brief, they are registers that hold a memory address, and the MPU looks up this address in these registers for a data copy operation.) To hold a 12-bit address FSR, registers require two 8-bit registers shown as FSRH and FSRL, and each FSR register is associated with an INDF register for indirect addressing. There are various ways these FSR registers can be used as pointers. For example, the addresses in FSR registers can be incremented before or after the data transfer. Similarly, these registers can be decremented before the data transfer. Furthermore, the address in the FSR can be offset by adding the contents of the W register. Table 3-1 lists the addresses of the associated registers with FSR2 ($FD9_H$–FDF_H), FSR1 ($FE1_H$–$FE7_H$), and FSR0 ($FE9_H$–FEF_H). Applications of these registers are illustrated in Chapter 5.

TABLE 3-1 Special Function Registers

File Address	File Name	Description
F80	PORTA	Read input pins and write to data latch
F81	PORTB	Read input pins and write to data latch
F82	PORTC	Read input pins and write to data latch
F83	PORTD	Read input pins and write to data latch
F84	PORTE	Read input pins and write to data latch
F89	LATAA	Read from and write to port data latch
F8A	LATAB	Read from and write to port data latch
F8B	LATAC	Read from and write to port data latch
F8C	LATAD	Read from and write to port data latch
F8D	LATAE	Read from and write to pins RE2, RE1, and RE0

TABLE 3-1   Special Function Registers (continued)

File Address	File Name	Description
F92	TRISA	Data direction control register for PORTA
F93	TRISB	Data direction control register for PORTB
F94	TRISC	Data direction control register for PORTC
F95	TRISD	Data direction control register for PORTD
F96	TRISE	Data direction control bits for pins RE2, RE1, and RE0
F9D	PIE1	Peripheral Interrupt Enable Register1
F9E	PIR1	Peripheral Interrupt Request Register1
F9F	IPR1	Peripheral Interrupt Priority Register1
FA0	PIE2	Peripheral Interrupt Enable Register2
FA1	PIR2	Peripheral Interrupt Request Register2
FA2	IPR2	Peripheral Interrupt Priority Register2
FA6	EECON1	EE (PROM) Configuration1 – control register for EEPROM memory accesses.
FA7	EECON2	EE (PROM) Configuration2 – used exclusively for memory write and erase sequences
FA8	EEDATA	EE (PROM) Read/Write Data Register
FA9	EEADR	EE (PROM) Address Register
FAB	RCSTA	Receive Status and Control Register
FAC	TXSTA	Transmit Status and Control Register
FAD	TXREG	USART1 Transmit Register
FAE	RCREG	USART1 Receive Register
FAF	SPBRG	USART1 Baud Rate Register
FB1	T3CON	Timer3 Configuration Register
FB2	TMR3L	Timer3 Low Byte Register
FB3	TMR3H	Timer3 High Byte Register
FBA	CCP2CON	Capture/Compare/PWM Register2 Configuration

TABLE 3-1   Special Function Registers (continued)

File Address	File Name	Description
FBB	CCPR2L	Capture/Compare/PWM Register2 Low Byte
FBC	CCPR2H	Capture/Compare/PWM Register2 High Byte
FBD	CCP1CON	Capture/Compare/PWM Register1 Configuration
FBE	CCPR1L	Capture/Compare/PWM Register1 Low Byte
FBF	CCPR1H	Capture/Compare/PWM Register1 High Byte
FC1	ADCON1	A/D Control Register1
FC2	ADCON0	A/D Control Register0
FC3	ADRESL	A/D Result Low Register
FC4	ADRESH	A/D Result High Register
FC5	SSPCON2	Serial Synchronous Port Control Register2
FC6	SSPCON1	Serial Synchronous Port Control Register1
FC7	SSPSTAT	Serial Synchronous Port Status Register
FC8	SSPADD	Serial Synchronous Port Address Register
FC9	SSPBUF	Serial Synchronous Port Buffer Register
FCA	T2CON	Timer2 Control Register
FCB	PR2	Timer2 Period Register
FCC	TMR2	Timer2 Register
FCD	T1CON	Timer1 Control Register
FCE	TMR1L	Timer1 Low Byte
FCF	TMR1H	Timer1 High Byte
FD0	RCON	Reset Control Register
FD1	WDTCON	Watchdog Timer Control Register
FD2	LVDCON	Low Voltage Detect Control Register
FD3	OSCCON	Oscillator Control Register
FD5	T0CON	Timer0 Control Register
FD6	TMR0L	Timer0 Low Byte
FD7	TMR0H	Timer0 High Byte
FD8	STATUS	Status or Flag register

TABLE 3-1 Special Function Registers (continued)

File Address	File Name	Description
FD9	FSR2L	File Select Register2 Low Byte
FDA	FSR2H	File Select Register2 High Byte
FDB	PLUSW2	Adds FSR2 and W and uses as a pointer for data memory
FDC	PREINC2	Increments FSR2 and uses the address as a memory pointer
FDD	POSTDEC2	Uses FSR2 as a memory pointer and decremented
FDE	POSTINC2	Uses FSR2 as a memory pointer and incremented
FDF	INDF2	Index Register2
FE0	BSR	Bank Select Register
FE1	FSR1L	File Select Register1 Low Byte
FE2	FSR1H	File Select Register1 High Byte
FE3	PLUSW1	Adds FSR1 and W and uses as a pointer for data memory
FE4	PREINC1	Increments FSR1 and uses the address as a memory pointer
FE5	POSTDEC1	Uses FSR2 as a memory pointer and decremented
FE6	POSTINC1	Uses FSR2 as a memory pointer and incremented
FE7	INDF1	Uses FSR1 to address data memory
FE8	WREG	Working Register
FE9	FSR0L	File Select Register0 Low Byte
FEA	FSR0H	File Select Register0 High Byte
FEB	PLUSW0	Adds FSR20 and W and uses as a pointer for data memory
FEC	PREINC0	Increments FSR0 and uses it as memory pointer
FED	POSTDEC0	Uses address in FSR0 as memory pointer and decrements FSR0
FEE	POSTINC0	Uses address in FSR0 as memory pointer and increments FSR0
FEF	INDF0	Uses address in FSR0 as memory pointer to Program Memory
FF0	INTCON3	Interrupt Configuration Register3
FF1	INTCON2	Interrupt Configuration Register2
FF2	INTCON	Interrupt Configuration Register
FF3	PRODL	Product Register Low Byte
FF4	PRODH	Product Register High Byte
FF5	TABLAT	Program Memory Table Latch
FF6	TBLPTRL	Program Memory Table Pointer Low Byte <7:0>

TABLE 3-1   Special Function Registers (continued)

File Address	File Name	Description
FF7	TBLPTRH	Program Memory Table Pointer High Byte <15:8>
FF8	TBLPTRU	Program Memory Table Pointer Upper Byte <20:16>
FF9	PCL	Program Counter Low Byte <7:0>
FFA	PCLATH	Program Counter Latch High Byte <15:8>
FFB	PCLATU	Program Counter Latch Upper Byte <20:16>
FFC	STKPTR	Stack Pointer
FFD	TOSL	Top of Stack Low Byte <7:0>
FFE	TOSH	Top of Stack High Byte <15:8>
FFF	TOSU	Top of Stack Upper Byte <20:16>

3.1.5 Program Counter (PC)

The Program Counter (PC) is a 21-bit register that functions as a counter and provides capability of addressing 2 MB (2^{21}) of memory (refer to addresses FF9$_H$, FFA$_H$, and FFB$_H$ in Table 3-1). The PC consists of three 8-bit registers—PCL, PCH, and PCU. The lower byte of a 21-bit address <7-0> is stored in PCL, the higher byte <15-8> in PCH, and the remaining bits <20-16> in PCU. The MPU uses this counter to sequence the execution of the instructions. For example, when the system is reset, the program counter is cleared, and the program execution begins at the memory location 000000$_H$. When the processor fetches (reads) the instruction stored in location 000000$_H$, the program counter is incremented to the address of the next instruction.

3.1.6 Table Pointer

These are 21-bit registers (see addresses FF6$_H$, FF7$_H$, and FF8$_H$ in Table 3-1) that are used as memory pointers to copy bytes between program memory and data memory.

3.1.7 Stack and Stack Pointer (SP)

The stack is a group of 31 word-sized registers that are used for temporary storage of memory addresses during the execution of a program. These registers are neither a part of program memory nor data memory. To identify these 31 registers, we need a 5-bit address, and the stack pointer register (address FFC$_H$ in Table 3-1) uses 5 bits to indicate where the addresses are stored. There are special instructions that are used to store and retrieve information from the stack and the storing of binary information in the stack requires a detailed discussion, which is included in Chapter 7.

3.1.8 Special Function Registers (SFRs)

In Figure 3-1, ten blocks are shown as special function registers representing registers associated with the I/O ports, support devices, and processes of data transfer described in Chapter 2.

This section provides an overview of how SFRs are associated with various devices in the MCU. We will broadly examine overall functions of these registers. The details of these registers are discussed in other chapters of this book.

- **I/O Ports:** There are five I/O ports, A to E, in the PIC18F452/4520 microcontroller, and these I/O ports have been assigned the addresses from F80$_H$ to F84$_H$. The number of I/O ports ranges from 3 to 10 depending on the version of PIC18 family. These I/O ports are bidirectional and can be set up either as input ports or output ports. The TRIS registers (F92$_H$–F96$_H$) are used to setup I/O ports. The user needs to write appropriate bits in TRIS registers to set up these ports.
- **Interrupts (F9D$_H$–FA2$_H$):** The PIC18F452/4520 microcontroller has three pins on PORTB that can be set up as interrupt requests from external sources. The interrupt registers listed in Table 3-1 are used to enable or disable the interrupts and set up priorities.
- **EEPROM (FA6$_H$–FA9$_H$):** These registers are used to read from, write to, and erase EEPROM registers.
- **Serial I/O (FAB$_H$–FAF$_H$):** These registers are used to set up serial communications.
- **Timers:** This microcontroller has four timers. The associated registers with timer functions are listed in Table 3-1.
- **CCP Registers (FBA$_H$ - FBF$_H$):** Capture/Compare/PWM modules are set up by CCP registers to perform various functions.
- **A/D Converter (FC1$_H$–FC4$_H$):** The 18F452 microcontroller includes eight, and the 18F4520 includes thirteen inputs for analog signals. In 18F452, four registers are associated with A/D conversion. The first two registers (FC1$_H$ and FC2$_H$) are used to set up various parameters for the conversion and the remaining two registers (FC3$_H$–FC4$_H$) are used to save the converted reading. The 18F4520 uses an additional register; ADCON2CFCO$_H$.
- **Synchronous Serial I/O (FC5–FC9):** In Chapter 2, Figure 2–7 shows that the PIC18F microcontroller includes a module labeled as MSSP (master synchronous serial port), and it is used in serial communication with other peripherals and microcontrollers. The registers with addresses FC5–FC9 are used in setting up various parameters of the serial communication.
- **Miscellaneous Function Registers:** The PIC18F452/4520 microcontroller includes various features such as watchdog timer, low-voltage detection, and various oscillator options. These functions must be set up using the configuration registers listed in Table 3-1

3.2 INTRODUCTION TO PIC18 INSTRUCTION SET

The PIC18 microcontroller includes 77 instructions; 73 are one word long (16-bit) and the remaining four instructions are two words long (32-bit). The newer four-digit versions XXXX such as PIC18F4520 include an additional eight instructions called the extended set. The instructions of the extended set are listed in Appendix A, but will not be discussed in the text.

The instruction set is divided into seven groups according to the functions: move (data copy) and load, arithmetic, logic, program redirection (branch/call), bit manipulation, table read-write, and machine control. The manufacturer, however, has classified these instructions according to the type of operations such as byte, bit, or literal. We will discuss the manufacturer's classification later in this section.

The instruction set of a processor determines the capability of its operations, the power of its data manipulation, and the ease of programming. However, to study the sheer number of instructions can be an intimidating and overwhelming task. Instead, our intent here is to expose you to the overall capability of the PIC18 microcontroller. In subsequent chapters, these instructions will be discussed again with examples and application programs will be illustrated.

To write an assembly language program, the user should understand the syntax (grammar or format of an instruction) and its function. Each instruction is divided into two parts: opcode (operation such copy or add) and operand (data to be operated on). The method of specifying an operand is called an **addressing mode**. The PIC18 instruction set is divided in the following groups based on their functions.

1. Move (Data Copy) and Load
2. Arithmetic
3. Logic
4. Program Redirection (Branch/Jump)
5. Bit Manipulation
6. Table Read/Write
7. Machine Control

3.2.1 Move (Data Copy) and Load Operations

The data copy operations are frequently used in processing data. Table 3-2 shows the data copy instructions. In the PIC microcontrollers these instructions use the mnemonics MOV (Move) that create an erroneous impression that the data byte is transferred from one register to another register; in reality, the byte is copied. In explaining these instructions, we refer to these operations as data copy.

TABLE 3-2 Move or Copy Instructions

Mnemonics	Operand	Description	Word	Cycle	Status bits affected
MOVLW	8-bit	Load 8-bit number into W register	1	1	
MOVWF	F, a	Copy W register into F (Data) register	1	1	
MOVFF	Fs, Fd	Copy source data register (Fs) into destination (Fd) data register	2	2	
MOVF	F, d, a	Copy F register into itself or W register	1	1	N Z
CLRF	F, a	Clear F (File/Data) register	1	1	Z
SETF	F, a	Set all bits to 1 in F (Data) register	1	1	
SWAPF	F, d, a	Exchange low-order four bits with high order four bits in File (Data) register	1	1	
LFSR	F, 12-bit	Load 12-bit address into File Select register	2	2	
PUSH		Copy address of next instruction on stack	1	1	
POP		Discard address on top of stack	1	1	

In PIC18 controller, there are three primary ways data are copied: (1) loading a number, called literal, in W register (also in FSR and BSR), (2) copying between W register and another data register, and (3) copying from one data register into another data register.

1. Loading a number (literal) directly in the W register. In this type of an instruction, we know the destination where the number (literal) is being loaded. The operand in this instruction is a number that follows the opcode; therefore, the **addressing mode** is called either **Literal** or **Immediate Addressing mode**. The format of the instruction is as follows:

	Opcode	Operand	Binary Code	Hex Code
Syntax:	MOVLW	8-bit	0000 1110 XXXX XXXX	0E XX
Example:	Load the 8-bit number (byte, literal) $F2_H$ in the working register			
Instruction:	MOVLW	0xF2[2]	0000 1110 1111 0010	0 E F 2 $_H$

Interpretation: Let us read this instruction from left to right by expanding the abbreviations. Move/Copy literal (number) into the W register, followed by the 8-bit number $F2_H$ to be copied (the term literal means a constant or a number). In assembly language, a Hex number is written in the format of 0xByte as shown in the previous section. The details of other formats are discussed in Chapter 4.

This is a one-word instruction. The opcode MOVLW represents an operation to be performed and the operand (8-bit) is data to be operated on. The instruction copies $F2_H$ into the W register, and its binary code is $0E_H$, which is followed by the operand code $F2_H$.

2. Copying between the W register to another data register. In this type of instruction, we need to know the location of the data register—specifically, the bank of the register. The address of the operand register is part of the instruction; therefore, this is called **Direct** or **Register Direct addressing mode**. PIC18 has the capability of having 4096 data registers that are divided into 16 banks from 0 to F, each having 256 registers, as discussed in Chapter 2. However, not all banks are being used by the present devices. For example, PIC18F242 has only 768 data registers with three banks. Therefore, to specify the bank of the register, the format of the instruction is as follows. The format includes two letters "F" and "a"; "F" represents the File or data register from data memory and "a" specifies the bank of the register. If a = 0, the data register is located in the Access bank and if a = 1, the address of the bank is specified by the contents of the BSR register.

---

[2]Hexadecimal numbers are shown as 0 x number or number followed by the letter H. In assembly language instructions and programs, we will follow the notation 0 x number, and in the text we will show hex numbers as number followed by subscript H.

Syntax: MOVWF³ F,a   ;Move W into F and the bank is specified 'a'.

Example: Copy contents of WREG into data register in Access bank with the address 25_H

Instruction: MOVWF 0x25, 0

Interpretation: Move/Copy W register into File/Data register 25_H, which is in the access bank because a = 0. If a = 1, the bank is selected based on the value in the BSR (Bank Select Register). See Chapter 2 for an additional explanation of banks and data registers.

3. Copying from one data register to another data register. This is a two-word instruction that copies the contents of one data register called source into another data register called destination. This instruction can copy from any of the 4096 data registers (as a source) to any destination register; it includes 12-bit addresses of the registers within the instruction. The format of the instruction is as follows.

Syntax: MOVFF fs, fd   1100 ffff ffff ffffs   Two Words

                                       1111 ffff ffff ffffd

Example: Copy contents of data register 0140_H into data register 0170_H

Instruction: MOVFF 0x140, 0x170   Hex code: C 1 4 0 F 1 7 0

Interpretation: This instruction copies contents of register 140_H into register 170_H. The first word of the binary code shows that the Hex 'C' followed by the 12-bit address represents the source register and the second word shows that the Hex digit 'F' followed by the 12-bit address represents the destination.

The copy instructions in PIC18 microcontroller are listed in Table 3-2. The MOV instructions are used frequently, and a thorough understanding of their format will help you to understand other instructions. The CLR and SET instructions shown in Table 3-2 are self-explanatory. The remaining instructions are included here to show the overall capability of the PIC18F controller. The LFSR instruction refers to File Select Registers and PUSH/POP instructions relate to the stack, and will be discussed again in later chapters in the context of their applications.

Points to Remember

1. Each instruction has two parts: opcode and operand. The operand of an instruction is selected by the user with a few exceptions. In these few exceptions, the operand is embedded in the instruction itself.

2. A one-word (16-bit) instruction requires two bytes of memory for storage, and a two-word (32-bit) instruction requires four (bytes) memory locations.

³The term file register refers to data register in data memory.

3. When Move instructions copy data from one register called source to another register called destination, the contents of the source register are not modified; only the contents of the destination register are modified.

4. In general, these instructions do not affect flags except two instructions: CLRF (Clear) and MOVF F (See Table 3-2).

3.2.2 Arithmetic Operations

The PIC18 instruction set includes various types of arithmetic operations such as addition, subtraction, increment (add one), and decrement (subtract one), 1's and 2's complement, and multiply. These instructions are listed in Table 3-3.

TABLE 3-3   Arithmetic Instructions

Mnemonics	Operand	Description	Word	Cycle	Status bits affected
ADDLW	8-bit	Add 8-bit number to W register	1	1	N OV Z DC C
ADDWF	F, d, a	Add WREG to File (Data) register and save result in W or F	1	1	N OV Z DC C
ADDWFC	F, d, a	Add WREG, File (Data) register, and Carry and save the result in W or F	1	1	N OV Z DC C
SUBLW	8-bit	Subtract WREG from 8-bit number	1	1	N OV Z DC C
SUBWF	F, d, a	Subtract WREG from File (Data) register and save result in W or F	1	1	N OV Z DC C
SUBWFB	F, d, a	Subtract WREG and Borrow from File (Data) register and save result in W or F	1	1	N OV Z DC C
SUBFWB	F, d, a	Subtract File (Data) register, Borrow, from WREG and save result in W or F	1	1	N OV Z DC C
INCF	F, d, a	Increment File (Data) register and save result in W or F	1	1	N OV Z DC C
DECF	F, d, a	Decrement File (Data) register and save result in W or F	1	1	N OV Z DC C
COMF	F, d, a	Complement File (Data) register and save result in W or F1	1	1	N Z
NEGF	F, a	Take 2's complement File (Data) register	1	1	N OV Z DC C
MULLW	8-bit	Multiply WREG and 8-bit and save result in PRODH:PRODL	1	1	

TABLE 3-3   Arithmetic Instructions (continued)

Mnemonics	Operand	Description	Word	Cycle	Status bits affected
MULWF	F, a	Multiply WREG and File (Data) register and save result in PRODH:PRODL	1	1	
DAW		Decimal Adjust WREG for BCD operations	1	1	C

Addition and Subtraction

There are three types of addition (as well as subtraction) instructions: (1) addition of an 8-bit number (literal) to the W register, (2) addition of a file/data register to the W register, and (3) addition of the W and the File register and a carry generated by the previous operation. When a data register is added to the W register, the sum can be saved either in the W register or in the data register. To specify the destination of the sum, the instruction format uses the letter "d" as shown in examples 2 and 3. If d = 1 or F, the sum is saved in the data register and if d = 0 or W, the sum is saved in the W register. If the user does not specify any value for 'd,' the sum is saved in the W register as default. Some of the examples are as follows:

1. Addition of an 8-bit number to the W register

Syntax: ADDLW   8-bit   ;Add 8-bit literal to W

Example: Add the byte $3F_H$ to the contents of WREG.

Instruction: ADDLW   0x3F

Interpretation: This instruction adds $3F_H$ to the contents of the W register and saves the result in W register. It affects all the flags, and the flags are set or reset based on the result of the addition.

2. Addition of W register to a File (data) register

Syntax: ADDWF F, d, a   ;Add W to F and save result in W or F

Example: Add the data register 05_H to WREG and save the result in register 05_H.

Instruction: ADDWF   0x05,1,0   Alternative Format: ADDWF 0x05, F, 0

Interpretation: This instruction adds the contents of the W register to the File/Data register 05_H from the Access bank and saves the result in the F register because the destination bit d = 1. If d = 0, the result is saved in the W register. The alternative format is better for clarity.

The bank of the register 05_H is specified by the bit "a." If a = 0, the bank is the Access bank, and if a = 1, the bank is specified by the value in BSR. The instruction affects all the flags, and the flags are set or reset based on the result of the addition.

3.   Addition of the W register to a File (data) register and Carry

Syntax: ADDWFC   F, d, a   ;Add W, F, and C (carry) and save result in W
                                                                            or F

Example: Add data register 02_H to W and the carry generated by the previous addition.

Instruction: ADDWF   0x02,0,0   Alternative Format: ADDWF   0x02, W,0

> Interpretation: This instruction adds the contents of the W register to the File/Data register 02_H from Access bank and saves the result in the W register. This instruction is used for multi-byte addition. For example in addition of 16-bit numbers, if low-order bytes generate a carry, this instruction will add carry to the addition of high-order bytes. Because the destination bit d = 0, the result is saved in W; otherwise, it would be saved in 02_H register. The instruction affects all the flags, and they are set or reset based on the result of the addition.

4.   Increment File (data) register by one

Syntax: INCF   F,d,a   ;Increment F by one and save either in W or F

Example: Add one to the contents of data register $2F_H$ and save the result
                            in WREG.

Instruction: INCF   0x2F, 0, 0   ;Increment data register $2F_H$ by one and save
                                                                            ;result in W

                            INCF   0x2F, W, 0   ;Alternative format

> Interpretation: This instruction adds 1 to the contents of register $2F_H$ from Access bank and saves the result in the W register. The instruction affects all the flags.

5.   Decrement File (data) register by one

Syntax: DECF   F,d,a   ;Decrement F by one and save either in W or F

Example: Subtract one from the contents of data register 10_H and save the result in register 10_H.

Instruction: DECF   0x10, 1, 0   ;Decrement data register 10_H, by one and save
                                                                            ;result in data register

                            DECF   0x10, F, 0   ;Alternative format

> Interpretation: This instruction subtracts 1 from the contents of register 10_H, which is in the Access bank and saves the result in the same data register 10_H. The instruction affects all the flags.

Table 3-3 also includes instructions such as taking 1's and 2's complements of register contents, multiply, and decimal adjust. The multiply instruction multiplies two unsigned 8-bit numbers. These numbers can be literal and the contents of W or the contents of W and F, and the product is saved in two registers: PRODH and PROL. The Decimal Adjust W (DAW) instruction is used in Binary Coded Decimal (BCD) operations. These instructions will be discussed again in later chapters in the context of their applications.

Points to Remember

1. The arithmetic instructions can perform operations on contents of the W register and 8-bit numbers (literal) and save the result in the W register.
2. The arithmetic instructions can perform operations on the contents of the W register and a File (data) register and save the result in W or F (data) register. The ability to save the result in any of the source registers (W or F) is a very powerful feature enabling the data registers to function almost as the working register.
3. In general, arithmetic instructions affect all flags with a few exceptions. Refer to Table 3-3 for specific details.

3.2.3 Logic Operations

The instruction set includes three types of Boolean logic operations: AND, IOR (Inclusive OR), and XOR (Exclusive OR). Any 8-bit number or the contents of a File (data) register can be ANDed, ORed, and Exclusive ORed with the contents of W register. The result can be stored either in the W register or the data register. The list of logic instructions is shown in Table 3-4.

TABLE 3-4   Logic Instructions

Mnemonics	Operand	Description	Word	Cycle	Status bits affected
ANDLW	8-bit	Logically AND 8-bit number with WREG	1	1	N Z
ADDWF	F, d, a	Logically AND WREG with File (Data) register and save result in W or F	1	1	N Z
IORLW	8-bit	Logically OR 8-bit number with WREG	1	1	N Z
IORWF	F, d, a	Logically OR WREG with File (Data) register and save result in W or F	1	1	N Z
XORLW	8-bit	Exclusive OR 8-bit number with WREG	1	1	N Z
XORWF	F, d, a	Exclusive OR WREG with File (Data) register and save result in W or F	1	1	N Z

The following two examples illustrate these instructions.

1. Logically ANDing of an 8-bit number with the W register

Syntax: ANDLW   8-bit   ;And 8 bit literal to W

Instruction: ANDLW   0x9F   ;AND 8 bit number 9FH with the W register

Interpretation: This instruction logically ANDs each bit of 8-bit number $9F_H$ with the corresponding bits of WREG and saves the result in the W register. The result of ANDing operation affects two flags: N and Z.

2. Logically ORing W register with a File register

Syntax: IORWF   F, d, a   ;OR W with F and save result in W or F

Instruction: IORWF   0x12,1,0   ;OR W with File register 12_H and save result in 12_H register

              IORWF   0x12, F,0   ; Alternative format

Interpretation: This instruction ORs each bit of the W register with the corresponding bits of File (data) register 12_H from the Access bank and saves the result in the data register because the destination bit d = 1. If d = 0, the result is saved in the W register. The bank of the register 12_H is specified by the bit 'a.' If a = 0, the bank is the Access bank, and if a = 1, the bank is specified by the value in BSR. The result affects two flags: N and Z

Points to Remember

1. The logic instructions can perform operations on the contents of the W register and 8-bit numbers (literal) and save the result in the W register.
2. The logic instructions can perform operations on the contents of the W register and a File (data) register and save the result in the W or F register.
3. The logic instructions affect only two flags: N and Z (see Table 3-4).

3.2.4 Branch Operations

The processor is a sequential machine meaning it executes the instructions in a sequence stored in program memory until it is directed to change its sequence by a branch instruction. The branch instructions can be classified into two groups: unconditional and conditional. The conditional branch instructions are based on four data flags: N, OV, Z, and C, and there are two instructions associated with each flag. These branch instructions are further classified as absolute and relative branch. In absolute branch instructions, the jump location is specified with a complete memory address, and in relative branch instructions, the jump location is a specified number of locations forward or backward from the address in the program counter (PC). In addition, the instruction set includes Call and Return instructions related to subroutines and interrupts. The explanation of Call and Return instructions require understanding of the concepts related to subroutine and stack. These instructions are explained in Chapter 7.

Table 3-5 lists the entire set of instructions that redirect the program execution. Some of these instructions are explained in the following examples.

TABLE 3-5   Program Redirection (Branch and Call Instructions)

Mnemonics	Operand	Description	Word	Cycle	Status bits affected
BC	n	Branch if Carry = 1 within ± 64 Words	1	1/2	
BNC	n	Branch if Carry = 0 within ± 64 Words	1	1/2	
BZ	n	Branch if Z = 1 within ± 64 Words	1	1/2	
BNZ	n	Branch if Z = 0 within ± 64 Words	1	1/2	
BN	n	Branch if N = 1 within ± 64 Words	1	1/2	
BNN	n	Branch if N = 0 within ± 64 Words	1	1/2	
BOV	n	Branch if OV = 1 within ± 64 Words	1	1/2	
BNOV	n	Branch if OV = 0 within ± 64 Words	1	1/2	
BRA	nn	Branch unconditionally within ± 512 Words	1	2	
GOTO	Address	Go to 20-bit address unconditionally	2	2	
Call and Return Instructions					
RCALL	nn	Call subroutine within ± 512 words	1	2	
CALL	20-bit, s	Call subroutine. Save W, STATUS, & BSR in their shadow registers if s = 1	2	2	
RETURN,	s	Return from subroutine. Restore W, STATUS, & BSR from shadow registers if s = 1	1	2	
RETLW	8-bit	Return 8-bit number to W	1	2	
RETFIE	s	Return from interrupt and if s =1, W, STATUS, & BSR values are restored	1	2	

1. Redirect (Branch) the program execution to a specified location if Carry flag is set.

Syntax: BC   n   ;Branch on Carry to location PC + 2 + 2n

Instruction: BC   5   ;Branch on Carry to location PC + 2 + 10 (memory
                                ;addresses)

Interpretation: This instruction checks for C (carry flag), and if it is set, the program jumps forward if n is a positive number or jumps backward if n is a negative number. This instruction redirects the program execution forward to memory location by 12 bytes in reference to the memory address in the program counter. The operand n is an

8-bit signed number and given in words (rather than bytes); therefore, it is multiplied by 2 to calculate the jump location. Similarly, to calculate the jump location, 2 is added to the address in the PC because when the processor fetches the instruction BC, the PC is incremented by 2 to fetch the next instruction. These calculations are illustrated in Chapter 5.

2. Redirect (Branch) the program execution to a specified location if there is No Carry.

Syntax: BNC   n   ;Branch unconditionally to location PC + 2 + 2 n

Instruction: BNC   0xFA   ;Branch unconditionally to location PC + 2 + 2 x (-6)

Interpretation: This instruction redirects the program backward by 10 bytes from the address of the BNC instruction. In this instruction the jump is backward because the value of n is negative, given in 2's complement, and the 2's complement of the byte FA_H is -6. These calculations are illustrated again in Chapter 5.

3. Redirect (Branch) the program execution unconditionally to a specified location.

Syntax: BRA   n   ;Branch to location PC + 2 + 2 x n

Instruction: BRA   0x0100   ;Branch to location PC + 2 + (2 x100H)

Interpretation: This is a relative branch (jump) instruction, and the operand n represents a 12-bit signed number, expressed in words (not in bytes). This instruction redirects the program unconditionally (without checking any flag) within −1024 to + 1023 byte-range that is equivalent to ± 512 words.

Points to Remember

1. The instruction set includes eight conditional relative branch instructions based on four flags. The offset byte n is a signed number limited to the range of ± 64 words <-128 to + 127> bytes.
2. The range of the unconditional relative branch instruction BRA is ± 512 words <-1024 to + 1023>.
3. If the operand n is positive, the jump is forward, and if n is negative, the jump is backward; all negative values of n are given in 2's complement.

3.2.5 Bit Manipulation Operations

These instructions can be classified into two groups. One deals with setting, resetting, and complementing (toggling) an individual bit in a File (data) register. The other deals with groups of 8 bits together, as in the "rotate" instruction, whereby each bit is shifted either right or left (see Table 3-6). Some of the examples are as follows.

TABLE 3-6 Bit Manipulation (Bit Set, Reset, Toggle, and Rotate)

Mnemonics	Operand	Description	Word	Cycle	Status bits affected
BCF	F, b, a	Clear bit b of register F where b = 0 to 7	1	1	
BSF	F, b, a	Set bit b of register F where b = 0 to 7	1	1	
BTG	F, b, a	Toggle bit b of register F where b = 0 to 7	1	1	
RLCF	F, d, a	Rotate File (data) register Left through Carry and save result in F or W	1	1	N Z C
RLNCF	F, d, a	Rotate File (data) register Left excluding Carry and save result in F or W	1	1	N Z
RRCF	F, d, a	Rotate File (data) register Right through Carry and save result in F or W	1	1	N Z C
RRNCF	F, d, a	Rotate File (data) register Right excluding Carry and save result in F or W	1	1	N Z

    1.   Clear (Reset) Bit in a File register.

Syntax: BCF   F,b,a   ;Reset bit b in File (data) register F

Instruction: BCF   0x2, 7, 0   ;Reset bit 7 in data register 02H

Interpretation: This instruction clears bit 7 of the register 02_H from the Access bank and saves the result in data register. It does not affect any flags.

    2.   Rotate each Bit of a File register to the right through Carry.

Syntax: RRCF   F,d,a   ;Rotate each bit of a File register to the right and save
                                         ;the result either in W or F

Instruction: RRCF   0x5,1,0   ;Rotate each bit of File (data) register 05_H to the right
                                         ;through carry and save the result in register 05_H

Interpretation: This instruction shifts each bit of register 05_H from Access bank to the right position, bit B0 is shifted into the carry and the carry is shifted into bit B7, and is saved in register 05_H because the destination bit d = 1. The result affects three flags: N, Z, and C.

Points to Remember

    1.   Any bit in any File (data) register can be set, reset, or complemented.
    2.   There are two types of rotate instructions: 8-bit and 9-bit rotation. In 9-bit rotation, carry is considered the 9th bit. Bits of any File (data) register can be rotated to the right or the left and saved in the W or F register.
    3.   The 8-bit rotate instructions affect N and Z flags, the 9-bit rotate instructions affect N, Z, and C flags, and bit set/reset instructions do not affect any flags (see Table 3-6).

3.2.6 Test and Skip Operations

These instructions combine bit manipulation, arithmetic operations, and branch operations. This group includes three types of instructions: The first tests a bit in a File register for 0 or 1; the second compares F register with the W register and checks whether F = W, F > W, or F < W; the third increments or decrements register F and checks whether the result = 0 (see Table 3-7). If a specified condition is met, the processor skips the next instruction. These instructions are coupled with the unconditional Branch (BRA) instruction to implement branch operations equivalent to Branch on Zero and Branch on No Zero. Some of the examples are as follows.

TABLE 3-7   Skip Instructions (Check Data Condition and Skip Next Instruction)

Mnemonics	Operand	Description	Word	Cycle	Status bits affected
BTFSC	F, b, a	Test bit b of register F where b = 0 to 7; skip the next instruction if bit = 0	1	1/2	
BTFSS	F, b, a	Test bit b of register F where b = 0 to 7; skip the next instruction if bit = 1	1	1/2	
CPFSEQ	F, a	Compare F with W register, skip if F = W	1	1/2	
CPFSGT	F, a	Compare F with W register, skip if F > W	1	1/2	
CPFSLT	F, a	Compare F with W register, skip if F < W	1	1/2	
TSTFSZ	F, a	Test F; skip if F = 0	1	1/2	
DECFSZ	F, d, a	Decrement F and save either in W or F; skip if result = 0	1	1/2	
DECFSNZ	F, d, a	Decrement F and save either in W or F; skip if result ≠ 0	1	1/2	
INCFSZ	F, d, a	Increment F and save either in W or F; skip if result = 0	1	1/2	
INCFSNZ	F, d, a	Increment F and save either in W or F; skip if result ≠ 0	1	1/2	

1. Test bit in a register, and if it is zero, skip the next instruction.

Syntax: BTFSC   F,b,a   ;Test bit b in File register F

Instruction: BTFSC   0x2, 7, 0   ;Test bit B7 in File (data) register 02H, and if
                                             ;B7 = 0, skip the next instruction

Interpretation: This instruction checks bit B7 of the register 02_H from the Access bank, and if bit B7 = 0, the processor skips the next instruction This instruction does not affect any flags.

2. Compare contents of F register with the contents of W register. If F > W, skip the next instruction.

Syntax:	CPFSGT	F,a	;Compare F with W, and if F > W, skip the ;next instruction
Instruction:	CPFSGT	0x5,0	;Compare register 05H with W, and if F > W, ;skip the next instruction

Interpretation: This instruction compares the contents of register 05_H from the Access bank with the contents of W, and if the byte in 05_H register is greater than the contents of W register, the processor skips the next instruction. This instruction does not affect any flags.

Points to Remember

1. Any bit in a File (data) register can be tested for 0.
2. A File register can be compared with W register for equality, for greater than and less than, and a File register can also be tested for zero.
3. A File register can be incremented or decremented, and tested for zero.
4. If a condition is met after testing, the processor skips the next instruction. These instructions are coupled with the unconditional branch instruction BRA to implement branch operations. These instructions do not affect any flags (see Table 3-7).

3.2.7 Table Read/Write Operations

The Table Read/Write Operations group includes eight instructions that copy (read/write) 8-bit data bytes from program memory into data memory (File registers) and vice versa (Table 3-8). These instructions use Table Pointer registers that can hold 21-bit addresses of the program memory. These instructions use indirect addressing and will be explained with illustrations in Chapter 5.

TABLE 3-8   Table Read and Write

Mnemonics	Operand	Description	Word	Cycle	Status bits affected
TBLRD*		Read Table from Program Memory pointed to TBLPTR into TABLAT	1	2	
TBLRD* +		Read Table from Program Memory pointed to TBLPTR into TABLAT and increment TBLPTR	1	2	
TBLRD* -		Read Table from Program Memory pointed to TBLPTR into TABLAT and decrement TBLPTR	1	2	
TBLRD +*		Increment TBLPTR and read Table from Program Memory pointed to TBLPTR into TABLAT	1	2	

TABLE 3-8   Table Read and Write (continued)

Mnemonics	Operand	Description	Word	Cycle	Status bits affected
TBLWT*		Write Table into Program Memory pointed to TBLPTR from TABLAT	1	2	
TBLWT*+		Write Table into Program Memory pointed to TBLPTR from TABLAT and increment TBLPTR	1	2	
TBLWT*-		Write Table into Program Memory pointed to TBLPTR from TABLAT and decrement TBLPTR	1	2	
TBLWT +*		Increment TBLPTR and write Table into Program Memory pointed to TBLPTR from TABLAT	1	2	

3.2.8 Machine Control Instructions

These instructions are concerned with the operations of the processor. Table 3-9 lists these instructions, and includes instructions such as Reset (establish or reestablish initial conditions of the MPU that includes clearing registers and flags), Sleep (turn down power and wait), and CLR-WDT (clear watchdog timer).

TABLE 3-9   Control Instructions

Mnemonics	Operand	Description	Word	Cycle	Status bits affected
CLRWDT		Clear Watchdog Timer	1	1	
RESET		Reset all registers and flags	1	1	N OV Z DC C
SLEEP		Go into standby mode	1	1	
NOP		No operation	1	1	

3.3 INSTRUCTION FORMAT

The manufacturer has divided the instruction set into four operation categories: byte-oriented, bit-oriented, literal, and control (branch), and the format of the instruction depends on its operation. As mentioned before, seventy-three instructions are one word in length. The opcode and the operand must fit into this one-word length, and the chip designer determines these formats based on underlying electronic circuits. In this processor, the operand can be an 8-bit constant or a register with addresses ranging from 8 bits to 20 bits. We will discuss the instruction format briefly to provide some background in understanding instructions and the importance for the destination of the result of an operation. However, the understanding of how an instruction is designed is not critical to writing programs.

3.3.1 Byte-Oriented Operations

Figure 3-3 shows the format of a byte-oriented operation ADDWF, which adds the W register to File register. It has a 6-bit opcode (B15-B10): 1 bit (B9) to specify whether the result should be stored in WREG or file register, 1 bit (B8) to specify the bank of data memory register, and 8 bits B7-B0 to specify the address of the data register being used.

B15 B10	B9	B8	B7 B0
6-bit Opcode	d	A	F = 8-bit File Register Address
0 0 1 0 0 1	1	0	0 0 0 0 0 0 0 1

FIGURE 3-3   Instruction Format of Byte-Oriented Instruction - ADDWF F, d, a

EXAMPLE 3.2

Example:	ADDWF 0x1, 1, 0 Code: 2 6 0 1 $_H$ (see Figure 3-3)
Interpretation:	Add contents of W and data register 01 $_H$ from Access bank and save the result in register 01$_H$.

If d = 0, the result is saved in WREG and if d = 1, the result is saved in file (data) register
If a = 0, file register is in Access bank, and if a = 1, the value in BSR selects the data memory bank.

3.3.2 Bit-Oriented Operations

Figure 3-4 shows the format of a bit-oriented operation BCF that clears the specified bit in a given data memory register. It has 4-bit opcode (B15-B12), 3 bits (B11-B9) to specify the position of the bit to be cleared, 1 bit (B8) to specify the bank of data memory register, and 8 bits (B7-B0) to identify the address of the data memory register being used.

B15 B12	B11 B9	B8	B7 B0
4-bit Opcode	b b b	a	F = 8-bit File Register Address
1 0 0 1	1 1 1	0	0 0 0 1 0 1 0 1

FIGURE 3-4   Instruction Format of Bit-Oriented Instruction - BCF F, b, a

EXAMPLE 3.3

Example: BCF   0x15, 7, 0   Code: 9 E 1 5 H (see Figure 3-4)

Interpretation: Clear bit 7 from register 15H in Access bank.
                  If a = 0, file register is in Access bank, and if a = 1, the value in BSR register
                  selects the data memory bank.

3.3.3 Literal Operations (Loading an 8-Bit Number)

Figure 3-5 shows the format of an instruction that loads a byte in the W register. The term literal is synonymous to a number or a constant. The instruction MOVLW loads the 8-bit operand that follows the W register. It has an 8-bit opcode (B15-B8) and 8 bits (B7-B0) to specify an 8-bit literal.

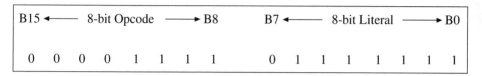

FIGURE 3-5   Instruction Format of Literal Instruction - MOVLW

EXAMPLE 3.4

Example: MOVLW   0x7F   Code: 0 E 7 F$_H$ (see Figure 3-5)

Interpretation: Load the 8-bit number 7F$_H$ in W register.

3.3.4 Branch Operation

Figure 3-6 shows the format of the branch operation: Branch on Carry. It is a one-word instruction, and it redirects the program if the carry flag is set. It has an 8-bit opcode (B15-B8) and an 8-bit (B7-B0) signed number that specifies the jump location relative to the program counter.

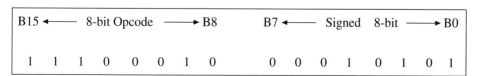

FIGURE 3-6   Instruction Format of Branch Instruction - BC n

> ## EXAMPLE 3.5
>
> Example: BC   0x15   Code: E 2 1 5$_H$ (see Figure 3-6)
>
> Interpretation: Redirect (jump) to location by 15 words forward relative to the program counter.

3.4 ILLUSTRATION: ADDITION

In the previous sections, we examined the programming model of the PIC18F microcontroller and provided an overview of its instruction set. Now we will take a simple example that illustrates a few instructions and their execution.

3.4.1 Problem Statement

Write instructions to load two bytes (37_H and 92_H) in data registers REG0 and REG1. Add the bytes and store the sum in REG2.

3.4.2 Problem Analysis

This problem is similar to the program we wrote in Chapter 2, Section 2.6 except that in this problem the result is stored in a data register rather than displayed at PORTC. This problem can be divided into three steps: (1) loading the bytes, (2) adding the bytes, and (3) storing the sum.

1. To load the bytes in data registers, REG0 and REG1, we need to examine the list of MOV instructions in Table 3-2. There is no instruction in the list that can load a byte directly into a data register. However, a byte can be copied from the W register into any data register. Therefore, we have to load the bytes one at a time in the W register and then copy them into the data registers.
2. To add bytes, we can examine the list of arithmetic instructions in Table 3-3. Here also we find that we cannot add the data registers directly, but we can add the W register and a data register. Therefore for the addition of two bytes, we need to have one byte in the W register and add the second byte from the data register.
3. To store the sum, the result from the W register can be copied into register REG2.

3.4.3 Assembly Language Program

```
1.  MOVLW    0x37       ; Load the first byte in W register
2.  MOVWF    REG0, 0    ; Copy the first byte from the W register into REG0
3.  MOVLW    0x92       ; Load the second byte in W register
4.  MOVWF    REG1, 0    ; Copy the second byte from the W register into REG1
5.  ADDWF    REG0,0     ; Add REG0 to REG1 and save the sum in W
6.  MOVWF    REG2, 0    ; Save the sum in REG2
7.  SLEEP               ; Go into power down mode
```

Each of these instructions is 16 bits (1 word) wide and requires two memory locations. Now let us store these instructions in program memory starting from memory location 000020_H.

We are starting our program at locations 000020_H for two reasons: (1) To demonstrate that a program can be written in any available memory location, and (2) To reserve the initial memory locations for interrupt vectors (explained in Chapter 10). From now on, all our programs will be written starting at memory location 000020_H. If a program begins at the memory location 00000_H, it will be redirected to the location 000020_H by an unconditional jump instruction. The program memory has 21-bit addresses, but we will ignore the first two zeros as before and use only the remaining four Hex numbers. We can find the binary code by looking up the instruction set or by assembling the program as described in Chapter 4. Our instructions will be stored in memory as follows:

Memory Address	Memory Contents Binary Code	Hex	Mnemonics	Comments
0020	0 0 1 1 0 1 1 1	3 7	MOVLW 0x37	Load first byte in W
0021	0 0 0 0 1 1 1 0	0 E		
0022	0 0 0 0 0 0 0 0	0 0	MOVWF REG0,0	Save first byte in REG0
0023	0 1 1 0 1 1 1 0	6 E		REG0 address[1]
0024	1 0 0 1 0 0 1 0	9 2	MOVLW 0x92	Load second byte in W
0025	0 0 0 0 1 1 1 0	0 E		
0026	0 0 0 0 0 0 0 1	0 1	MOVWF REG1,0	Save second byte in REG1
0027	0 1 1 0 1 1 1 0	6 E		REG1 address[2]
0028	0 0 0 0 0 0 0 0	0 0	ADDWF REG0,0 0	Add bytes and save sum in
0029	0 0 1 0 0 1 0 0	2 4		W register
002A	0 0 0 0 0 0 1 0	0 2	MOVWF REG2,0	Save sum in REG2
002B	0 1 1 0 1 1 1 0	6 E		REG2 address
003C	0 0 0 0 0 0 1 1	0 3	SLEEP	Power Down
003D	0 0 0 0 0 0 0 0	0 0		

3.4.4 Program Execution and Instruction Pipelining

When the processor is reset, its Program Counter is cleared and holds the 21-bit address 000000, and the program execution begins. However, our program begins at location 000020_H. Therefore, a jump instruction should be written at the location 000000 to redirect the program execution to the location 000020_H. In a simulator (Chapter 4) or a trainer, any starting location can be specified.

The execution of an instruction takes place in two stages: (1) fetch a 16-bit instruction from program memory, and (2) execute the task specified in that instruction (Figure 3-7a). However, in the second stage, when the execution is taking place internally, the next instruction is fetched as shown in Figure 3-7b. This is called instruction pipelining, and is done to increase the speed of execution. To illustrate the pipelining and execution, let us take an illustration from the program we wrote in the previous section.

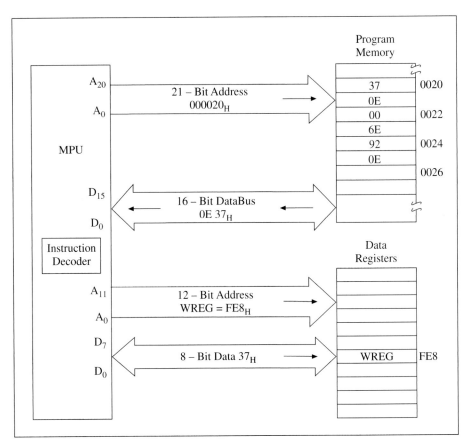

FIGURE 3-7a   Execution of MOVLW 0x37 Instruction

Instruction Cycles:

	Instruction Cycle 0	Instruction Cycle 1	Instruction Cycle 2	Instruction Cycle 3
	$Q_1 Q_2 Q_3 Q_4$	$Q_1 Q_2 Q_3 Q_4$	$Q_1 Q_2 Q_3 Q_4$	$Q_1 Q_2 Q_3 Q_4$
Fetch	$0E\ 37_H$	$6E\ 00_H$	$0E\ 92_H$	
Execute		$37_H \longrightarrow W$	$W \longrightarrow Reg0$	$92_H \longrightarrow W$
	Fetch 1	Fetch 2	Fetch 3	
		Execute 1	Execute 2	Execute 3

FIGURE 3-7b   Pipeline Fetch and Execution

Let us examine the instruction stored at location 000020_H. The MPU places the address 0020_H (the initial two zeroes are ignored for the discussion) on the address bus of the program memory and fetches the 16-bit instruction $0E\ 37_H$ via the data bus and places that instruction in the instruction decoder. In the next instruction cycle, when the 8-bit number 37_H is being loaded in W register, the MPU begins fetching the next instruction $6E\ 00_H$ stored in locations 0022 and 0023_H (Figure 3-7b). Even if each instruction takes two instruction cycles, it appears that the execution takes place in one cycle. For example, to execute ten instructions, the MPU requires eleven instruction cycles because there is no execution in the first cycle. Now the question becomes: What happens after fetching a Branch instruction? When a Branch instruction is fetched, the MPU clears the pipeline and begins a new sequence from a new branch location.

SUMMARY

This chapter described a PIC18F microcontroller programming model and provided an overview of the instruction set and its capabilities. The microprocessor (MPU) is a binary device capable of interpreting only binary language. The important concepts are summarized as follows.

1. The PIC18F programming model includes the following registers: working register (WREG), STA-TUS register, bank select register (BSR), file select register (FSR), program counter (PC), stack pointer (SP), and data (file) registers.

2. The W register is a working register, generally known as an "accumulator" in other microprocessors and microcontrollers, and is used for storing 8 bits and performing arithmetic and logic operations.

3. The MPU includes the STATUS register that stores five data flags (C—carry/borrow, Z—zero, OV—overflow, N—sign, and DC—digit carry) to indicate the nature of the result after the execution of an instruction. The first four flags are used by branch instructions to make decisions to redirect the program, and the digit carry flag is used internally by the MPU for BCD operations.

4. The MPU has a 12-bit wide address bus for data memory (registers); thus, it is capable of addressing 4096 data registers. The data memory (4096 registers) is divided into 16 banks from 0 to F_H; each bank has 256 registers. The beginning data registers from 00 to $7F_H$ in Bank 0 and the end registers from 80_H to FF_H in Bank F are identified as the access bank and can be accessed without the bank select register. The registers with the addresses from 80_H to FF_H in the Access bank are called special function registers (SFRs) and are used to configure support devices (such as timers and A/D converter) and define addresses of ALU registers.

5. The PIC18F instruction set has seventy-seven instructions; all but four instructions are one-word in length. The MPU executes many of these instructions in one instruction cycle. The branch, call, table read/write, and a few copy instructions require two cycles.

6. The instruction set can be divided into seven functional categories: data copy, arithmetic, logic, branch, bit manipulation, table read/write, and machine control. Similarly, the set can be categorized according to type of operation: byte-oriented, bit-oriented, literal, and control.

7. The instructions dealing with data memory (registers) specify the bank in the instruction using the letter 'a.' When a = 0, the bank specified is the Access bank, and when a = 1, the bank is specified by the bits (nibble) in the BSR register.

8. One of the most powerful features of the PIC18F microcontroller instruction set is the ability of its MPU to save the result either in the W register or the operand register after an arithmetic or logic operations. To save the result in the WREG or data register is specified by the letter 'd' in the instruction. If d = 0 (or W), the result is saved in the WREG and if d = 1 (or F), the result is saved in the data register.

QUESTIONS AND ASSIGNMENTS

3.1   In PIC18F programming model, what is the difference between the W register and data registers?

3.2   Specify the size of the program counter and its function.

3.3   List the registers and their addresses that are included in the access bank.

3.4   If the BSR register holds the byte 01, identify the data register in which the following instruction copies the contents of the W register: MOVWF   0x7F, 0

3.5 If the BSR register holds the byte 04, explain the result of the following instruction:
MOVWF 0x7F, 1

3.6 Explain the result of the following instruction and the status of the flags affected:
MOVF 0x10, 0, 1 if the data register 10_H holds the byte $9F_H$.

3.7 Explain the result of the following instruction and the flags affected: MOWF 0x80, 0 if W contains 00. Identify the device that is associated with the address 80_H.

3.8 Explain the result after the execution of the following instructions if the BSR holds 01_H.

MOVLW   0xF7

MOVWF   0x32, 1

3.9 Specify the expected result in W register after the execution of the following instructions, and specify the flags that are set after the addition.

MOVLW   0x89

ADDLW   0x77

3.10 Identify the contents of the W register and the status of the flags by filling the blanks as these instructions are being executed.

	W	N	OV	Z	C
MOVLW 0xFA					
ADDLW 0x38					

3.11 In Q. 3.10, if the numbers are unsigned, explain the result after the addition.

3.12 In Q. 3.10, if the numbers are signed, explain the result.

3.13 The following set of instructions is expected to load two bytes ($A7_H$ and 92_H) in data registers 01_H and 02_H, add the bytes and save the sum in register 03_H. Read the following instructions and calculate the sum of these two bytes.

1. MOVLW   0xA7
2. MOVWF   0x01, 0
3. MOVLW   0x92
4. MOVWF   0x02, 0
5. ADDWF   0x01, 1, 0
6. MOVWF   0x03, 0

3.14 In Q. 3.13, the ADD instruction sets the overflow and carry flag. Explain why the overflow flag is set and interpret the result if the numbers are signed numbers.

3.15 In Q.3.13, what is the total sum if the numbers are unsigned?

3.16 In Q. 3.13, explain why the W and 03_H register have the byte 92_H at the end of the program. Does the overflow discard the final answer and store the previous byte from W into register 03_H?

3.17   In Q. 3.13, identify the location where the sum is saved.

3.18   Explain the concept and the advantages of pipelining instruction.

3.19   If each one-word instruction requires two instruction cycles—fetch and execute—explain the statement that each one-word instruction (with a few exceptions) is executed in one cycle.

3.20   Explain why one-word branch instructions require two cycles for execution.

SIMULATION EXERCISES

Using PIC18 IDE

3.1 SE   Enter the following Hex code in Program Memory starting at the memory address 000020_H. Execute the program using Single Step as described in Appendix F.1.2. Observe the sum in the WREG and REG2. Identify the flags that are affected after the addition.

MemoryAddresses	Hex	Mnemonics	Comments
0020	0E 3 7	MOVLW 0x37	Load first byte in W
0022	6 E 00	MOVWF REG0,0	Save first byte in REG0
0024	0 E 92	MOVLW 0x92	Load second byte in W
0026	6 E 01	MOVWF REG1,0	Save second byte in REG1
0028	2 4 0 0	ADDWF REG0,0 0	Add bytes and save sum in W
002A	6 E 0 2	MOVWF REG2,0	Save sum in REG2
002C	0 0 0 3	SLEEP	Power Down

3.2 SE   In 3.1SE, replace the byte 92_H by 52_H. Calculate the sum and specify the flags that will be set/reset after the addition. Execute the program and verify the result. Explain why the OV flag is set.

CHAPTER **4**

PROGRAMMING AND PROBLEM SOLVING

OVERVIEW

This chapter introduces assembly language programming as a problem-solving technique. It discusses how to: (1) analyze a problem, (2) represent that problem in a flowchart, and (3) convert the flowchart into assembly language. An assembly language program can be assembled using the hand-coding technique or an assembler. Nowadays, using an assembler is almost a universal practice in assembling a program. The chapter illustrates the steps in writing, assembling, executing, and debugging a program using an Integrated Development Environment (IDE), called MPLAB® IDE, available from Microchip[1].

---

[1]MPLAB® IDE software can be downloaded from Microchip's website: http://www.microchip.com

OBJECTIVES

- Given a problem statement, divide the statement into small segments that can be performed by the processor.
- Draw a flowchart to specify the tasks to be performed and their sequence.
- Convert a flowchart into programming statements.
- List the steps in writing and executing an assembly language program.
- List assembler directives and their functions.
- Describe the functions of various programs such as an editor, assembler, linker, and debugger used in developing assembly language software.
- List the steps in writing, assembling, executing, and debugging a program in an Integrated Development Environment (IDE) called MPLAB® IDE available from Microchip.
- Write and execute a simple assembly language program using MPLAB IDE.
- Demonstrate the following debugging techniques in troubleshooting assembly language programs: single-step, breakpoint, and tracing code.

4.1 APPROACH TO PROBLEM SOLVING WITH PROGRAMMING

Programming can be viewed as a problem-solving technique using the capability and strength of a processor. By examining the instruction set, one can surmise that the processor is fairly limited in its instruction set, but its strength is in its speed and accuracy, and its ability to repeat steps. Therefore, programming involves analyzing and breaking down a given problem into small tasks, and specifying the sequence in which the tasks are to be performed. This is represented in a block diagram called a flowchart. This flowchart is then converted into instructions the processor can understand.

4.1.1 Problem Analysis and Modular Approach

To write a program or a set of instructions for a given problem, we need a well thought-out plan, just as an architectural drawing is needed to construct a custom-built house. The first step in writing a program is the problem analysis, which begins with a careful examination of the problem statement. Several clues can be found in the problem statement that can be used to divide the problem into various tasks, and, if necessary, these tasks can be further subdivided into smaller units. These units can be written independently as **subroutines** (discussed in Chapter 7) and then combined. This approach of writing programs is called the **modular design approach**. This is similar to building a house by subcontracting various independent segments of the house to different contractors and bringing them together at the location of the house.

4.1.2 Flowcharting

A flowchart is a graphical representation of processes (tasks) to be performed and the sequence to be followed in solving a computational problem. For example, the addition of two numbers is a simple sequential process. However, to add ten numbers, the microprocessor needs to repeat the process ten times because it can add only two numbers at a time. Here the programmer must take advantage of the processor's capability of repeating steps by writing addition instructions once and setting up a loop with an appropriate branch (jump) instruction (Chapter 5 explains the loop technique). When the addition is complete, the processor must terminate the addition process and continue with the next process, such as displaying the result at an output port. Therefore, after each addition process (in a loop), the processor should make a decision whether the process should be repeated or terminated. The symbols shown in Figure 4-1 are used commonly to draw a flowchart. Their description is given in the following section.

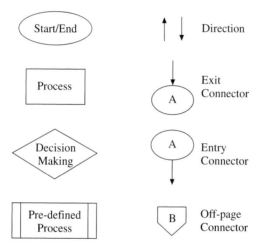

FIGURE 4-1   Flowchart Symbols

- Oval: indicates the beginning or end of a program.
- Arrow: indicates the direction of the program execution.
- Rectangle: represents a process or a task to be performed.
- Diamond: shows the decision point at which the processor must make a decision whether to change the sequence of the execution based on the result of the previous process.
- Double-Sided Rectangle: is a predefined process, written independently, and can be used or accessed by any program.
- Circle with Arrow: shows a continuation or an exit to a different column. A continuation to a different page is as shown with a letter "B."

Now we will illustrate an analysis and a flowchart with a simple problem as shown in Example 4.1.

EXAMPLE 4.1

Write instructions to load two bytes (37_H and 92_H) in data registers REG0 and REG1. Add the bytes and store the sum in REG2.

Solution:

Problem Analysis: This problem is the same as the problem in Chapter 3, Section 3.4 and is repeated here to show flowcharting. This is a simple problem, and we do not really need a processor to perform this addition. However, the analysis of simple problems can provide an understanding of the details needed to write a simple set of instructions. This problem was divided into the following steps:

1. Load the two bytes in data registers REG0 and REG1.
2. Add the bytes.
3. Save the sum in data register REG2.

Flowcharting:

The above three steps are processes and are represented in three blocks of the flowchart shown in Figure 4-2. In this problem, no decision is required. This is a straight line (linear) flowchart and it uses only two symbols: oval to show the beginning and the end and rectangle to show the copying and addition processes. The translation of this flowchart into assembly language is also straightforward, and has been performed in Chapter 3. The next section illustrates the use of the decision-making symbol.

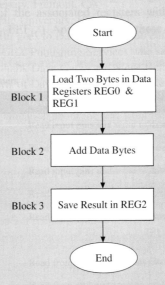

FIGURE 4-2   Linear Flowchart for Example 4.1

4.2 ILLUSTRATIVE PROGRAM: ADDITION WITH CARRY CHECK

The following program adds two bytes, and if the result generates a carry, it clears a specified register. It begins with the problem analysis, followed by the flowcharting and converting the flowchart into PIC18F instructions. The major emphasis of this simple program is on redirecting the program execution for a given data condition such as carry.

4.2.1 Problem Statement

Write instructions to load two bytes, Byte1 ($F2_H$) and Byte2 (32_H), in data registers REG0 and REG1 respectively and add the bytes. If the sum generates a carry, clear the data register REG2; otherwise, save the sum in REG2. The addresses of REG0, REG1, and REG2 are 00_H, 01_H, and 02_H, respectively, as in Chapter 3, Section 3.4.

4.2.2 Problem Analysis

The first part of the problem is the same as in Example 4.1. In the problem statement, the necessary steps are specified. The first step is to load two bytes in REG0 and REG1 and add these two bytes. After the addition, we need to make a decision based on the result. If the result generates a carry, we need to clear the data register REG2; otherwise, the sum should be stored in the same data register. Here we need to check the carry and make a decision whether to save the result or clear REG2. These steps are shown in the flowchart of Figure 4-3. If there is a carry, the program follows the straight path and goes to Block 4. If there is no carry, the program bypasses Block 4 and goes to Block 5; the sequence of the program execution changes. This change in sequence is called redirecting the program for a given condition of the carry and is also known as branching or jumping. The thorough understanding of this step is critical. If you examine the flowchart, it shows that the program jumps forward if the answer to the question is "No" and it indicates that the appropriate instruction to redirect the program is Branch on No Carry (BNC).

4.2.3 Writing Instructions

Before we write this program, let us briefly review the instructions we need to write this simple program (Refer to instructions listed in Chapter 3). The flowchart suggests that we have three types of operations—Copy, Addition, and Program Redirection (Branch)—for a given condition. We have already discussed the first two operations in Section 3.4. At this point, you are not expected to know all the instructions but simply be somewhat familiar with them after reading Chapter 3. These instructions will be discussed in detail in subsequent chapters. By examining the instructions in Chapter 3, we find the following instructions that may perform the operations we want.

1.	MOVLW	8-bit	;Load 8-bit literal (number) in W register
2.	MOVWF	F, a	;Copy W into Data (File) register
3.	ADDLW	8-bit	;Add 8-bit number to W register
4.	ADDWF	F, d, a	;Add W to Data (File) register and save the result in W or F
5.	BC	Address	;Branch on Carry to memory address

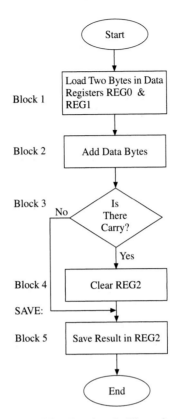

FIGURE 4-3   Flowchart for Illustrative Program: Addition with
Carry Check

6.	BNC	Address	;Branch on No Carry to memory address
7.	CLRF	F, a	;Clear Data (File) register

We do not need all the above instructions, but they will enable us to approach the problem in various ways using the same flowchart.

4.2.4 Converting Flowchart into Mnemonics

In the flowchart, the first block represents loading two numbers and the second shows the addition. Now we need to check for carry. If there is no carry, we will redirect the program forward to save the result.

Block 1.	MOVLW	0xF2	;Load F2H into W register
	MOVWF	0x00,0	;Save F2H in REG0
	MOVLW	0x32	;Load 32H into W register
	MOVWF	0x01	;Save 32H in REG1
Block 2:	ADDWF	0x00	;Add byte in REG0 to REG1

Block 3:		BNC	SAVE	;If there is no carry, branch to location SAVE
Block 4:		MOVLW	0x00	;Load 00 in W
Block 5:	SAVE:	MOVWF	0x02	;Save Result or clear REG2

Points to Remember in Translating the Flowchart into Mnemonics

1. The first block requires four instructions; the remaining blocks require one instruction each. The number of instructions that are necessary for a block depends on the process included in the block. The flowchart should show the logic of the program and not each instruction.
2. Block 3 is a decision-making block. It is a forward branch (jump) instruction, and we do not know the address of the memory location where the processor should jump. However, we can write the appropriate branch instruction with a label representing the memory location that can be calculated during the assembly process.

4.2.5 Assigning Memory Addresses and Code Look-up

Now we should look up the codes for each mnemonic found in PIC18F data manual or the instruction list on the inside cover of this text. We have already looked up the codes in Chapter 3 and selected the first memory address (000020_H). Once the starting address is selected, all machine codes must be entered in a sequence, and we will refer the memory addresses by eliminating the first two zeroes such as 0020_H as shown in the following table. We will repeat the coding of the first four instructions to make a note that the instructions are always stored beginning with an even-numbered address. The least significant byte of an instruction is always stored in even-numbered address (low-order address), and the most significant byte is stored in odd-numbered address as illustrated in the following table, in which the Hex code is in reverse order when stored in memory.

Memory Address	Hex Code	Memory Storage	
		Even Address	Odd Address
0020	0E F2	1 1 1 1 0 0 1 0	0 0 0 0 1 1 1 0
0022	6E 00	0 0 0 0 0 0 0 0	0 1 1 0 1 1 1 0
0024	0E 32	0 0 1 1 0 0 1 0	0 0 0 0 1 1 1 0
0026	6E 01	0 0 0 0 0 0 0 1	0 1 1 0 1 1 1 0
0028	24 00	0 0 0 0 0 0 0 0	0 0 1 0 0 1 0 0

We will assemble the entire program in Section 4.3 using the assembler in MPLAB.

4.2.6 Executing Assembly Language Programs

In the previous example, we wrote a simple assembly language program, looked up the Hex code for a few instructions, and assigned memory locations to each code (on paper). Now those binary codes (instructions) must be entered in assigned memory locations starting from address 0020_H in a target system. In college laboratories, training boards are used as target systems. The binary code of the assembled program must be transferred into the target system, and the program can be executed in that system. In the next section, we will illustrate how to write, assemble, debug, and execute a program using MPLAB simulator. Before we conclude this section, however, let us review the steps involved in writing and executing assembly language programs.

Review of Important Steps

In writing and executing assembly language programs, the following steps must be followed.

1. **Analyze the problem.** Divide the problem statements into small segments. List the initial conditions or preparatory steps that must be performed before the processor can process the data.
2. **Draw a flowchart.** Design a plan that specifies the tasks to be performed and their sequence. Draw a block diagram using the flowcharting symbols.
3. **Convert the flowchart into mnemonics.** Write instructions (mnemonics) that will enable the processor to perform the tasks. This is also called editing the program.
4. **Look up Hex code and assign memory addresses.** Find the Hex code for each mnemonic, check the number of bytes required from the manufacturer's instruction set, and assign the memory addresses based on available R/W memory in the training (target) system.
5. **Enter the Hex code into memory** using the keypad of a college laboratory training board.
6. **Execute the program** using the commands that can be entered through the keypad.
7. **Debug the program** if necessary.

The steps are listed here to show similarities between a program assembled manually and using computer resources. In the next section, we will use the resources of MPLAB and a PC to perform steps 3 through 7.

4.3 INTEGRATED DEVELOPMENT ENVIRONMENT (IDE)

In the last section, we listed seven steps necessary to successfully write and execute an assembly language program. The first step, analysis, requires thought, and the flowcharting, the second step, can be sketched manually or drawn using a computer program. The remaining five (or major segments of them) can be accomplished using a computer and related software. We will discuss the process of developing assembly language software using the Integrated Development Environment (IDE) provided by Microchip, the manufacturer of PIC controllers. The IDE is a collection of integrated programs, also called tools, that enables the user to write, assemble, debug, and execute programs for a specific embedded system. Microchip provides software called MPLAB IDE. Let us revisit the steps covered in the previous section because those concepts are central to developing software using IDE.

1. **Editing.** This is the same as writing instructions (mnemonics) in the text format. This is accomplished by using a text editor that displays the text on the screen as you type and the text is stored in memory under a file name. The program or the file is called the Source Code or Source Program. Common text editors used in PC systems are Notepad and WordPad. The user can also use full-screen editors such as Microsoft Word or WordPerfect, but the file must be saved as a text file. MPLAB also includes its own editor that offers many features to make the writing the assembly language programs easier.

2. **Assembling.** The process of converting mnemonics into Hex and binary codes and assigning memory addresses is known as assembling. This function is similar to looking up codes and assigning memory addresses in manual assembly. The programs used for assembling are known as assemblers or cross-assemblers. The assembler that translates PIC mnemonics (but operates under a different processor such as Intel Pentium in a PC) is called a cross-assembler. However, we will use the general term assembler for both of these programs.

   The assembler translates source code into binary code called the **Object Code** and generates a file called the **Object File**. In addition to translating mnemonics and assigning memory addresses, the assembler performs many other functions such as error checking, listing, and identifying types of errors. It also generates a list file and a Hex file. The list file is used for documentation that includes the source file, memory addresses, Hex codes, and comments. The Hex file is used to download a program from a PC to a trainer or a target system that is being designed. These concepts and files are illustrated using MPLAB assembler in the next section.

3. **Linking.** The Linker is a program that takes the object file generated by the assembler and other necessary files that are provided by the MPLAB and generates a file in binary code with the extension .COD that is executable by the microcontroller. These files are also shown with .COM and .EXE extensions for other microcontrollers.

4. **Downloading.** The process of transferring (copying) binary code from a PC into Read/Write (R/W) memory of a target system (such as a microcontroller training unit) is called downloading. The assembler creates a special file called a Hex file that is used for downloading a program.

5. **Executing.** To execute a program means to enable the processor (MPU) in the target system to locate the binary code (instructions), read the code, and perform the tasks specified in the instructions.

6. **Simulation.** A simulation is the use of a program, called simulator, that enables the user to execute a program on a PC without having a target system. The simulator in the MPLAB is known as MPLAB SIM. The simulator emulates the execution of the instructions with the operational speed of the PC; it is not a real-time execution as in hardware. You can observe the changes in registers as instructions are being executed and use the debugging techniques described in the following section.

7. **Debugging.** When the program does not function or does not achieve the expected results, it must be checked for errors. The program that enables the user to check for errors is called Debugger. It offers many techniques such as Single-Step, Break Point, Examine Register and Memory. These are discussed in the next section of MPLAB. The program should also be verified on a hardware system. Microchip provides in-circuit debugger called MPLAB® ICD 2 which is discussed in Appendix B.

4.3.1 Writing a Program Using an Assembler

In the previous section, we wrote a simple addition program using the mnemonics of PIC18F microcontroller and looking up the Hex code in the instruction list. It was then hand assembled starting at location 0020_H. Now we will accomplish the same steps by using an assembler. Therefore, we need to write our program in the format the assembler can understand. The program assembled by an assembler will appear similar to the program we wrote in the previous section, but with additional instructions called directives or pseudocodes, which are not translated into binary code of the MPU. An example of a directive is an instruction that tells the assembler the starting location of assembling the program. The assembly language program includes:

1. The user program in mnemonics.
2. Assembler directives that tell the assembler such items as the starting location and label definitions (see next section for details).
3. Comments that explain the logic of the program. The assembler simply reproduces the comments from the source program for documentation. The comments are optional as far as the assembly process is concerned, but essential to good programming practices.

4.3.2 Assembly Language Format

A typical statement of an assembly language source code has four fields: label, opcode (operation code), operand, and comment as follows:

Label	Opcode	Operand	Comment
START:	MOVLW	0xF2	;Load F2H in W register
↑	↑	↑	↑
Space or Colon	Space	Space	Semicolon

The requirements and explanation of these four fields are as follows.

Label Field

This field begins in column 1 and ends in a space or colon. A label can be up to thirty two characters long and must begin with an alphabetic character, or underscore, and may contain any alphanumeric characters. The labels in the label field represent memory addresses of the source code lines, and these memory addresses are assigned to the labels by the assembler in the assembly process. The label field is generally left blank for an instruction (source code line), unless we need to specify a jump location. The primary advantage of the label is to identify any memory location by name without knowing its address in advance.

The term *label* is also used for operands. For example, in our source line above we could have used a label such as BYTE1 for the data byte $F2_H$ and defined BYTE1 equal to 32_H by using the Equate (EQU) directive (explained later).

Operation (Opcode) Field

This field includes PIC18F mnemonics such as MOVLW and MOVWF or assembler directives and is separated by a space or a colon from the label field. If mnemonics are written starting from column one, the assembler assumes that they are labels and generates an error message. Generally, a tab key is used to separate the label and the operation fields to make the program easy to read.

Operand Field

If an instruction includes an operand, this field follows the operation field and must be separated from the operation field by at least a space. Multiple operands are separated by commas. The operand can be a byte, a word (16 bits), or a label representing a number or an address. A number can be written in binary, Hex, octal, or decimal format; the prefixes necessary for these formats are shown in Table 4-1. The assembler in MPLAB assumes a number without a prefix is a Hex number.

TABLE 4-1   List of Frequently Used Directives in Assembling a Program

Directives	Examples	Description
ORG: Originate	ORG 000020H	Assemble the program starting at location 000020_H
END: End	END	End of assembling
EQU: Equate	COUNT EQU 0x20	The label COUNT is equal to 20_H
SET: Define	REG10 SET 0x10	Define register REG10 as a variable
#INCLUDE	#include <P18F452.inc>	Header file for PIC 18F452
DB: Data Byte	DB 0x32, 0xF2, 0x45	Store next bytes in consecutive memory locations
DW: Data Word	DW 0x167F, 0x12A2	Store next two words, each requiring two memory locations—low-order byte first followed by high-order byte

Comment Field

This field is optional. In a source line, a comment begins with a semicolon (;). It is separated by a space from the operand field or it can begin anywhere (including column one) in the source line. Comments are used to communicate program logic to the user or other programmers and are viewed as a critical segment of the program documentation. The assembler does not interpret any characters after the semicolon but reproduces them as written.

4.3.3 Assembler Directives

The Assembler Directives, called pseudocodes or pseudo-ops, are instructions to the assembler. These instructions define constants and labels, reserve memory space for data, and tell the assembler where to assemble a program in memory. These directives are neither translated into machine code of the processor nor assigned any memory space in the object file. The assembler provided in MPLAB by Microchip has various types of directives. We limit our discussion to the directives that are commonly used and needed to develop our programs.

- ORG Origin—Begin assembly at this memory location. This directive specifies the starting address of the assembly, and the user can have more than one ORG statement in a program.
- END End—This directive must be included at the end of the program (including data storage). This directive is different from Halt or Sleep instructions to the processor, it informs the assembler that it should look no further for mnemonics.
- EQU Equate—Defines labels and constants.
- SET Set—Defines an assembler variable.
- #INCLUDE Include resources from available library—This directive adds source code or a file from the MPLAB library required to assemble a program for a particular device.
- RADIX Number format—The default radix for numbers in MPLAB is Hex. (See Table 4-2 for a complete list.)
- DB Define Byte—Store data (numbers) in bytes (8 bits).
- DW Define Word—Store data in words (16 bits).
- CBLOCK—Define a block of constant.
- ENDC—End a block of constant.
- RES—Reserve memory.

Table 4-1 includes a selected list of directives including the above directives with examples, and Table 4-2 lists the format of numbers with different bases that are used in assembly language programs.

TABLE 4-2   Format of Radixes (Numbers with Different Bases) and ASCII Characters Used in Assembly Language Programs

Type	Format	Examples Description
Hexadecimal	• 0x4F • H'4F' • 4F • 4FH and 0F7H[2]	Any of the four formats are recognized by MPLAB Assembler. Any number without any postscript or prescript is recognized as Hex number
Decimal	D'100'	This is 100 decimal that will be converted into 64_H by the assembler
Binary	B'10011000'	Binary number equivalent to 98_H
ASCII	'Susan' or A 'Susan'	ASCII characters

4.4 ILLUSTRATIVE PROGRAM: ADDITION WITH CARRY CHECK

This is the same program we wrote in Section 4.2 except that it illustrates the process of writing the program using the MPLAB editor.

---

[2] Any Hex number that begins with A through F must be preceded by zero in this format.

4.4.1 Problem Statement

Write a source program using MPLAB Editor illustrated in Section 4.2: Load the two bytes, Byte1 and Byte2 (defined as $F2_H$ and 32_H), in data registers REG0 and REG1 and add the bytes. If the sum generates a carry, clear the data register REG2; otherwise, save the sum in REG2. Assemble the program starting at location 0020_H and use the directive Equate (EQU) to define BYTE1, BYTE2, and data registers.

4.4.2 Creating the Source Code File Using MPLAB Editor

The first step is to install MPLAB IDE on your PC either from a CD or by downloading from Microchip.com. Though in actuality you would need to create a Project to assemble, debug, and execute a program using MPLAB IDE, for now we will focus only on: (1) creating a source code file that can be used later by the assembler, (2) using the necessary directives to write the program, and (3) coming up with a template that can be used to write other programs. The concept of creating a Project is discussed in the next section.

To write the program using the MPLAB Editor, open up MPLAB IDE by double clicking on the icon of MPLAB IDE on your desktop or begin with:

Start →[3] Programs → Microchip MPLAB IDE → → MPLAB IDE.

You will see two screens, Untitled workspace and Output, with the toolbars on the top similar to that of other application programs under the Windows operating system. The title bar shows MPLAB IDE v 7.41 and the menu bar shows categories such as File, Edit, View, Project, Debugger, Programmer, Tools, Configure, Windows, and Help (see Figure 4-4). This menu bar includes categories we have not yet discussed, such as Project and Debugger. The functions of these categories will be illustrated in later sections when we assemble and debug a program. For now, we will focus on writing a program using MPLAB Editor, which is similar to writing text using Notepad in Windows.

Source Code

To write the Source Code for our Illustrative Program: Addition with Check Carry, select File New and type the program Source Code as shown in Figure 4-4. The screens, Untitled workspace and Output, can be closed or remain behind the Edit screen. To save the program, create a folder called MyProj\Ch04\ and save the program in a file. Name the file **Addition with Carry Ckeck.asm** on C:\ drive. The extension of the file should be **.asm** indicating that this is an assembly language program file.

Now let us examine the Source Code in Figure 4-4. In the first section, we identified the PIC controller we are using by the directive #include. It provides a header file that defines various special function registers (SFRs) needed by the assembler. We can also use the directive such as Title to identify the name of the program for documentation as shown in Figure 4-4.

In the second section, we defined the values of constants BYTE1 and BYTE2, and the addresses of data registers identified as REG0, REG1, and REG2, using the directive EQU; these statements must begin in Column one of the monitor screen. The digits followed by the word REG represents the Hex address of the data register. This format is used for easy readability; you

---

[3] The symbol → represents a single click using a mouse. The symbol → → represents a double click using a mouse.

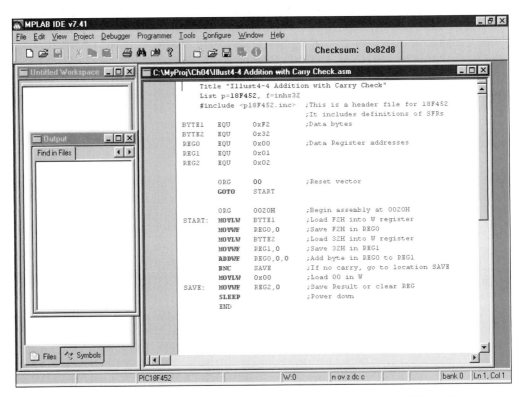

FIGURE 4-4   MPLAB IDE Screens Including Source Code for Illustrative Program: Addition with Carry Check

may use any label, but you are strongly advised to use labels that describe the function of a register or the process being used.

The third section begins with the ORG directive that specifies the starting address as 0000, but our program begins at memory location 0020_H. The first instruction at the location 0000 is GOTO START, which takes the program to location START with the memory address 0020_H (the first two zeros in the address 0020_H are not shown). When the PIC controller is turned on or reset, it begins at memory location 0000. Figure 4-4 shows the labels such as START and SAVE beginning in Column one. The label START is used primarily for documentation; it suggests the beginning of the program, and it is not referred to by any instruction in this program. On the other hand, the label SAVE identifies the memory address where the program should be redirected when there is No Carry. The last statement in the program is END, which tells the assembler the end of the source code.

Now the question is: do we really need these labels? Why do we not use actual addresses and constants in place of labels? For small programs such as those shown here, using actual addresses will not make any difference. But imagine a program that has used the value of REG1 in ten different instructions. If someone else in the project team is using that register, and you need to change the address of the register, you will have to go back in the program and find all the instructions that have used REG1 and rewrite them. If you use the directive EQU to define REG1, you need to make one change only at the beginning of your program.

4.5 ASSEMBLING, DEBUGGING, AND EXECUTING A PROGRAM USING MPLAB IDE

MPLAB IDE enables the user to write, assemble (or compile), debug, and execute an assembly language program in one environment, the IDE (Integrated Development Environment), and it can also be used to compile programs written in high-level languages such as C. In the process of assembling the program, MPLAB IDE generates many files, and all these files are stored under one heading, the Project Title. In using MPLAB IDE, the user will come across many new terms related to Project, which are becoming common in IDE. In Section 4.3, we listed seven steps to develop an assembly language program. The IDE provides the necessary programs to accomplish those steps. In Section 4.4, we already used built-in editor to write the Illustrative Program under the file name: Addition with Carry Check.asm and saved it on the C drive with the following path: C:\MyProj\Ch04\. The folder names and the file names used here are by choice for identification of the problem in the text.

To assemble, debug, and execute the Illustrative Program Addition with Carry Check.asm the following steps are necessary:

1. Create a new project: Create a folder and work space (provided by MPLAB) in which all files will be stored.
2. Add files: Add source code file/s to the project folder.
3. Build the project: Assemble the program and generate the necessary files such as List file for documentation, Hex file for downloading, and object code file for execution.
4. Run and Debug: Execute the program using the simulator MPLAB SIM and debug the program if necessary.
5. Download the program: Copy (transfer) the executable file in the Hex format to a target system such as a project you are building or a trainer such as PICDEM™2 PlusKit description is included on the CD that is available from Microchip, see (Appendix B), and run the program.

These steps are described in the following section with our illustrative program.

4.5.1 Creating a New Project

To create a new project, open MPLAB IDE by double clicking on its icon and follow the steps given below:

1. Select Project → Project Wizard. The Welcome screen is displayed. → Next.
2. Select Device. Use pull-down menu to select PIC18F452 (or 4520), or whatever device you are using (Figure 4-5a). → Next.

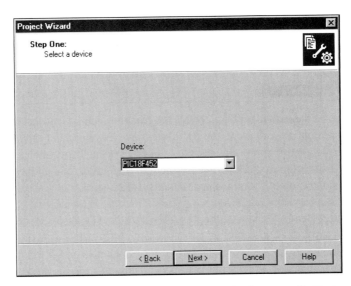

FIGURE 4-5 (a)   Screen Shot – Selecting a PIC Microcontroller Device

3. Select a Language Toolsuite. Use the pull-down menu to select Microchip MPASM Toolsuite. The complete path is shown at Location of Select Tool on the screen (Figure 4-5b). If it is incorrect or empty, use Browse to locate mpasmwin.exe file. → Next.

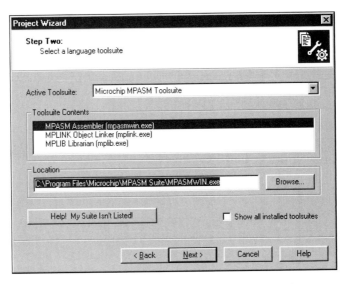

FIGURE 4-5 (b)   Screen Shot – Selecting a Language Toolsuite

4. Name Your Project. For our Illustrative Program 4-4, type Illust 4-4 Addition with Carry Check. For Project Directory, select Browse and find MyProj\Ch04. → Next (The screen display is shown in Figure 4-5c).

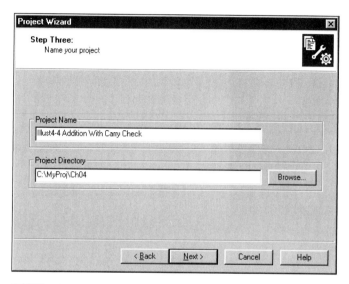

FIGURE 4-5 (c)   Screen Shot – Selecting a Project Name

5. Add any existing files. Select file Addition with Carry Check, which is stored under MyProj\Ch04 and → Add. (Figure 4-5d) → Next.

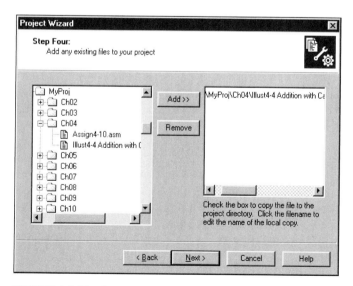

FIGURE 4-5 (d)   Screen Shot – Adding Source File

6. Project Wizard Summary. Figure 4-5e. → Finish. Figure 4-5f shows Project Window screen.

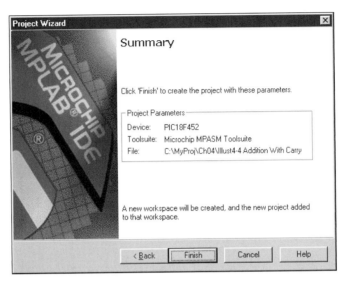

FIGURE 4-5 (e)   Screen Shot – Project Summary

FIGURE 4-5 (f)   Screen Shot – Project Window

After this initial set up, to assemble a new program, you can skip steps 1–3 and use the following steps.

1. Select Project → New. The screen shows Project Name and Project Directory. Write a new title for Project Name and use Browse to find the path of your Source code.
2. Select Project → Add Files to Project or right-click on Source Files (Figure 4-5f). Add the file of your new Source code. This is an important step.

Now you are ready to build the Project, which is also known as assembling the program.

4.5.2 Building the Project

Figure 4-5f shows the project window that lists the names of various folders. The Source folder also shows our asm file: Addition with carry Check.asm. Now to assemble the program and observe various files:

1. Select Project and → Build All. If there are no errors, you will get an output message "Build Succeeded." If there are errors, you will get an output message "Build Failed," and it will list all the errors. If you double-click on one of the errors, the screen will show the Source Code with an arrow pointing to an error. You must correct those errors and rebuild the project.

2. To see all the files generated after assembling the program (building the project), go to the tool bar of MPLAB IDE v7.40 and Select File → Open. It will take you back to Addition with Carry Check.asm folder (Figure 4-6). In the pull-down menu of Files of types, select ALL Files (*.*), and you will see files with extension such as: asm (original source file), lst (list), Hex, err (error), cod (object), mcp (project), and mcw (work space). Here we will focus on the list file to understand the translation of Source Code into Object Code. The assembled list file of our program is shown in the following section.

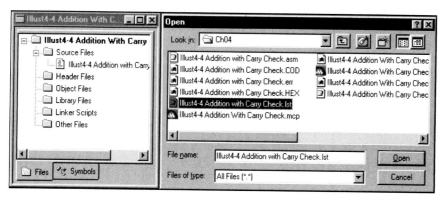

FIGURE 4-6   List of Files Generated by the MPLAB Assembler

4.5.3 Understanding the List File:

```
MPASM  5.04  ILLUST4-4 ADDITION WITH CARRY CH   8-5-2006  13:34:37    PAGE  1


LOC   OBJECT CODE     LINE SOURCE         TEXT
      VALUE

                      00001              Title "Illust4-4 Addition with Carry Check"
                      00002              List p=18F452, f=inhx32
                      00003              #include <p18F452.inc>  ;This is a header
                                                                 ;file for 18F452
                      00001              LIST
                      00002              ;P18F452.INC  Standard Header File,
                                         ;Version 1.4  Microchip Technology, Inc.
```

```
                               00983             LIST
                               00004

000000F2                       00005  BYTE1    EQU      0xF2       ;Data bytes
00000032                       00006  BYTE2    EQU      0x32
00000000                       00007  REG0     EQU      0x00       ;Register addresses
00000001                       00008  REG1     EQU      0x01
00000002                       00009  REG2     EQU      0x02
                               00010
000000                         00011           ORG      00         ;Reset vector
000000 EF10 F000               00012           GOTO     START
                               00013
000020                         00014           ORG      0020H      ;Begin assembly at 0020H
000020 0EF2                    00015  START:   MOVLW    BYTE1      ;Load F2H into W register
000022 6E00                    00016           MOVWF    REG0,0     ;Save F2H in REG0
000024 0E32                    00017           MOVLW    BYTE2      ;Load 32H into W
register26 6E01                00018           MOVWF    REG1,0     ;Save 32H in REG1
000028 2400                    00019           ADDWF    REG0,0,0   ;Add byte in REG0 to REG1
00002A E301                    00020           BNC      SAVE       ;If no carry, go to SAVE
00002C 0E00                    00021           MOVLW    0x00       ;Load 00 in W
00002E 6E02                    00022  SAVE:    MOVWF    REG2,0     ;Save Result or clear REG
000030 0003                    00023           SLEEP               ;Power down
                               00024           END
```

The list file is generated primarily for documentation and includes contents of source file with comments, Hex code, and memory addresses where the binary code is stored. The list file provides all the necessary information we need to understand the translation of mnemonics into processor code and their storage locations. The list file shown above has seven columns: memory addresses, Hex code (shown as object code), sequential line numbers (starting Line as 00001), labels, opcodes, operands, and comments. In the above list file, the column labels have been changed from the original file generated by MP LAB IDE for readability. We will explain the contents of the list file in reference to the line numbers.

Lines 1–9: These lines are used to specify the PIC microcontroller being used in the Project and define labels with EQU (Equate) that are used in the program. The "include" directive provides the files that define various parameters of PIC18F452 microcontroller (such as registers and SFRs) and lines 5 to 9 define labels of data bytes and data registers. However, no memory space is assigned to these directives.

Line 11: This line shows the directive ORG 00 (Begin assembly at location 0000) followed by the branch instruction GOTO that directs the program execution to our starting location 0020_H. The initial memory locations are reserved for interrupt vectors (explained in Chapter 10). Assembly can begin at location 0020_H, but starting assembly at 0000 emphasizes the fact that when PIC microcontroller is reset, the program counter is cleared and the execution begins at memory location 0000. The programmer must redirect the program to the starting location such as 0020_H.

Line 15: This the first line where the instruction MOVLW BYTE1 is translated into Hex codes 0E F2 and assigned memory locations $0020\text{-}21_H$. This device has a 21-bit program counter; therefore, the memory addresses are shown with a six-digit Hex code. However, this device (18F452) has a 32 K program memory extending the address up to $7FFF_H$; therefore, in our discussion we will ignore the two most significant 0s of the six-digit address. In this instruction, opcode is $0E_H$ and BYTE1 is $F2_H$ according to the EQU shown in the beginning part of the program.

Lines 16-19: These lines represent the translation of mnemonics up to the instruction ADDWF.

Line 20: This is a Branch instruction (BNC) with relative addressing (see Chapter 5 for further explanation of relative addressing and calculation of branch addresses). The opcode is E3, and the operand is 01, which represents a jump forward by one word (two bytes) if the previous addition does not generate a carry. These Hex codes are stored in memory locations $002E_H$ and $002F_H$.

Symbol Table and Messages: The list file includes a symbol table with labels and their values and messages of memory usage. However, the symbol table lists all SFR registers and their values, and it is too long to include here.

4.5.4 Executing a Program Using Simulator

The next step after building a program is to execute (run) the program using the available simulator in MPLAB and examine the output. If we do not get the expected output, we need to debug the program. The Debugger in IDE provides a simulator tool called MPLAB SIM that can simulate the execution of a program on a computer and assist the user in debugging it.

When a program is successfully built (assembled), it is free of errors due to syntax of the assembler; however, it may have logical errors such as wrong jump location. These errors can be searched and corrected by using MPLAB SIM as a debugging tool. This simulator provides many techniques such as Run, Animate, Single-Step, Breakpoint, and Watch (Examine Registers). We will discuss some of the techniques using the example of our Illustrative Program: Addition With Carry Check. The steps in setting up the MPLAB simulator are as follows:

1. Select Debugger → Select Tool → MPLAB Sim.
2. Select Debugger → Settings. It allows you to set clock frequency of the simulator and breakpoints; however, these features are not necessary for this program. If you want to change the default frequency (generally 4 MHz), type the number → Apply and → OK. There are also other settings related to stack and trace, and we will discuss those as needed.
3. Select View → Watch. The View provides menu options such as Program Memory, File Registers, Special Function Registers, and Watch. To observe changes in register values, you need to select Watch. There are two types of registers you can select to watch: Special Function Registers (SFR) such as WREG and STATUS and Symbols (labels) defined in your program. The commands are Add SFR and Add Symbol with a down arrow for selection. Or you can type in the names of the registers you want to observe in the space provided. The screen shows the Address of a register selected, Symbol Name, and its Value. You can also delete any of the registers selected by using the Delete key on the keyboard. In our Illustrative program, we are using five registers: working register WREG, STATUS, and three data registers REG0, REG1, and REG2. The STATUS register should be selected to check the carry flag. To observe the changes in these registers, you need to add these registers in the Watch option. For this program, add or type WREG, STATUS, REG0, REG1, and REG2. At this point, you should

resize the Watch screen and Source Program (asm) screen as shown in Figure 4-7 so that you can observe both screens.

4. Select Debugger → Reset. Select Debugger → Run. At the bottom of the screen, you will see the message: Running; however, you will not see any changes in the registers until the program stops. When the program execution is complete, you should observe the following: WREG = 00, STATUS = 01, REG0 = F2$_H$, REG1 = 32$_H$, REG2 = 0. In this program, the addition of two bytes generates a carry, which is indicated by the contents (01) of the STATUS register. The carry is indicated by bit B0 in the STATUS register (see Figure 3-2 for bit positions of the flags). Therefore, the program clears REG2 and is working as expected. If you do not observe the expected output, you need to use the single-step technique, which is described in the following section.

5. Select Debugger → Reset → Processor Reset (F6). Set the cursor at the first memory location. → Select Debugger → Animate. You should see the changes in registers as the instructions are being executed.

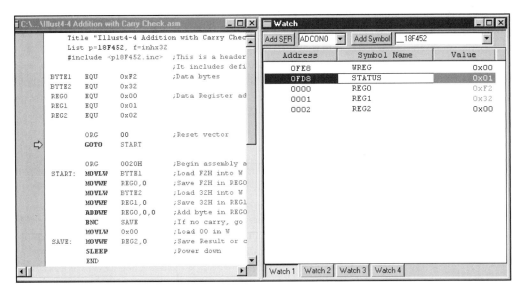

FIGURE 4-7   Screen Shot – View Registers and Source Program in MPLAB Simulator

4.5.5 Debugging a Program Using the Single-Step Technique

The single-step technique enables the user to execute one instruction at a time and observe the changes in the selected registers. This simulator includes three options: Step Into, Step Over, and Step Out. As the names suggest, the Step Into executes the instruction where the cursor is set, the Step Over skips the instruction, and the Step Out exits the program. The flowchart in Figure 4-3 shows the sequence of execution.

Before we begin the single step, we need to decide which registers to observe. In our Illustrative Program 4-4, we should observe all the registers we are using (WREG, REG0, REG1, and REG2) as well as the STATUS register to note the changes in the flags. In the previous section, we have already added these registers in the Watch list. The steps in using the single-step technique are as follows:

1. Select View → Watch. If you do not have WREG, STATUS, REG0, REG1, and REG2 in the Watch list from the previous section, add these registers to the Watch list.

2. Click on value column of the Watch screen and clear each of these registers by inserting 00 if they are not already cleared.

3. Select View → Program Memory. The Program Memory screen shows line numbers, addresses, Hex code, and mnemonics of the Illustrative Program. We could also use the Source Program to single step each instruction and observe the changes.

    We have assembled the program starting at location 0020_H; therefore, the initial memory locations (after the GOTO instruction) show the Hex code $FFFF_H$. Please note the Hex code of the instruction GOTO; it is EF 10. The hexadecimal number 10 is in words that represent the memory location 20, which is given in bytes. The first instruction is shown on line seventeen. It shows the memory addresses 0020_H (0021_H is assumed) with Hex code 0E F2 and the mnemonics MOVLW 0xF2. Resize all three screens: Source Code (.asm), Program Memory, and Watch as shown in Figure 4-8.

4. Select Debugger → Reset (or press F6). A green arrow will appear on the first instruction indicating that the program counter has the address 0000_H. The single-step can begin at this location. If you want to begin single stepping at any other location (such as 0020_H), select the instruction in either the Program Memory or Source Code screen and right click on it. Click on Set PC at the Cursor. The advantage of starting at location 0000 is that the program counter (PC) does not need to be set.

5. Select Debugger. → Step Into or Click on the icon or press the key F7. The simulator executes the first instruction GOTO and the program goes to the memory location 0020_H. If you press F7 again, the first instruction of our program MOVLW 0xF2 is executed. If you examine the Watch screen, you will see $F2_H$ in WREG in red and no change in the STATUS register; the instruction MOVLW does not affect any flags.

6. Continue to click on Step Into icon or press key F7. Observe the changes in the registers. When the instruction ADDWF is executed, WREG shows the result 24_H, and the STATUS register shows 01_H indicating that the carry flag is set; bit B0 in the STATUS register represents the carry flag (Figure 4-9).

7. In the next step, the BNC instruction—Branch on No Carry—is executed. Because the carry is set, the program does not branch and executes the next instruction, which clears the W register. In the next instruction, the simulator clears REG2. This is a small program, and it executes as expected. Other techniques, breakpoint, and code tracing are discussed in the following sections.

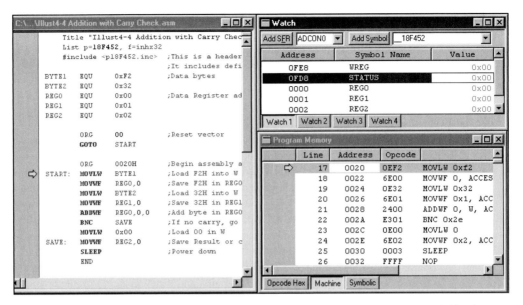

FIGURE 4-8   Screen Shot – View Registers, Source Program, and Program Memory in MPLAB Simulator

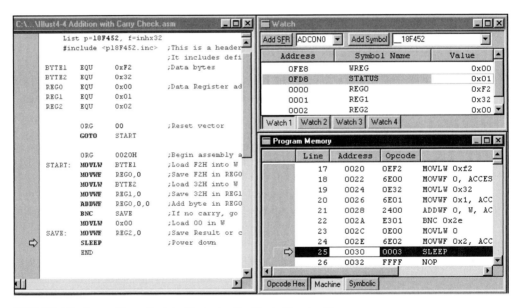

FIGURE 4-9   Screen Shot – Register Contents after Single-Stepping Program

Points to Remember in Single-Step

1. You should know what result to expect after each instruction.
2. Select appropriate registers to watch.
3. Open up the Program Memory.
4. Reset the simulator and set the cursor at the starting location.

4.5.6 Debugging a Program Using Breakpoints

The breakpoint technique enables the user to execute a group of instructions rather than a single instruction as in the single-step technique; thus, the user can debug the program in segments. You can set up a breakpoint or multiple breakpoints anywhere in the program and run the program. The simulator executes the program up to the breakpoint and halts. At that point, you can check all the registers in the Watch window, and if the contents of the registers are what you expect, this segment of the program can be assumed to be error free. If you have multiple breakpoints, you can run the program from the first breakpoint to the next breakpoint by pressing the key F9 (or clicking on the icon Run or selecting the Run menu option in Debugger), and check the contents of registers again. This is a powerful technique to check programs with multiple subroutines (see Chapter 7 for details). We will demonstrate this technique using the Illustrative Program Addition with Carry Check.asm, even if it is too small to use the breakpoint technique effectively. We will select the instruction BNC SAVE to set up a breakpoint because it is a decision-making point and can check the carry flag. The following steps are necessary to set up and run the program using the breakpoint technique:

1. Select the instruction BNC SAVE in the source code (Figure 4-10) and right-click on the instruction. A pop-up menu opens with an option to set up a breakpoint. Click on Set Breakpoint, and a symbol "B" appears on the code line. Another option is to select the instruction and double-click on the selected instruction.
2. Reset the program using the function key F6. Select the first instruction (START label) in the program, right-click on it, and select PC at Cursor. The green arrow appears at START or you may begin at location 0000. Clear all the registers in Watch window.
3. Select Debugger and click on Run or press the function key F9. The green arrow moves to the breakpoint, and new values appear in registers of the Watch window as in Figure 4-9 (WREG = 24_H and STATUS = 01_H indicating that the carry flag is set).
4. Press F9 to run the program again. The green arrow moves to the end of the program, indicating the completion of the execution.

4.5.7 Debugging a Program by Tracing Code

When a program is executed, MPLAB SIM has a facility called Simulator Trace to track the execution of each instruction and display the data as shown in Figure 4-11. The user can now examine the entire execution, called Trace, for errors. The information displayed is similar to that the user obtains through the single-step technique. Figure 4-11 shows the execution of all the instructions of the Illustrative Program 4-4, including memory addresses, Hex code, and instructions. In addition, it shows SA—Source Address, SD—Source Destination, DA—Data Address, and DD—Data Destination.

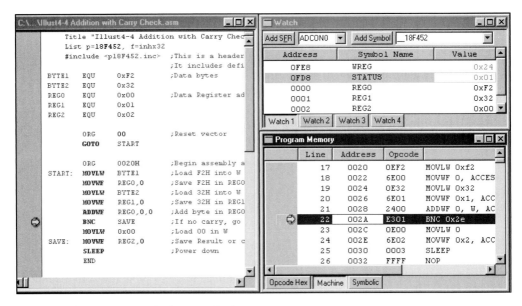

FIGURE 4-10   Screen Shot –Setting a Breakpoint

Line	Addr	Op	Label	Instruction	SA	SD	DA	DD
-10	0000	EF10		GOTO 0x20	----	--	----	--
-9	0002	F000		NOP	----	--	----	--
-8	0020	0EF2	START	MOVLW 0xf2	W	--	W	F2
-7	0022	6E00		MOVWF 0, ACCES	----	--	0000	F2
-6	0024	0E32		MOVLW 0x32	W	--	W	32
-5	0026	6E01		MOVWF 0x1, ACC	----	--	0001	32
-4	0028	2400		ADDWF 0, W, AC	0000	F2	W	24
-3	002A	E301		BNC 0x2e	----	--	----	--
-2	002C	0E00		MOVLW 0	W	--	W	00
-1	002E	6E02	SAVE	MOVWF 0x2, ACC	----	--	0002	00
0	0030	0003		SLEEP	----	--	----	--

FIGURE 4-11   Screen Shot – Code Tracing

For example, the trace of the add instruction ADDWF is interpreted as follows:

Source Address: Register 0000, Source Data: $F2_H$, Data Address: WREG, and Destination Data: 24_H. It shows that the byte $F2_H$ from REG0 is added to the contents of WREG, and the sum 24_H is stored in the W register. It does not provide any information about the flags. The Trace is capable of displaying data for 8192 (8 K) cycles. The Trace can be useful if combined with the breakpoint technique to find errors. To set up this Simulator Trace option, the following steps are necessary.

1. Select Debugger → Settings → Trace/Pins → Check both options: Trace Enable and Break on Trace Buffer Full.

2. Reset (press F6). Select the starting location and Run the program (F9).
3. Select View → Simulator Trace. The screen appears as in Figure 4-11.

SUMMARY

This chapter examined how programming is used to solve computing problems. The processor has a limited number of instructions, but it is fast and accurate. A problem needs to be divided into small units, and each small unit can be represented in a symbolic block diagram called a flowchart. Each block of this flowchart is converted into assembly language mnemonics (statements), and these statements are translated into binary code that can be executed by the processor. This process can be done manually or using the computer resources called "tools." This chapter focused primarily on how to use available tools to write, assemble, debug, and execute an assembly language program. Today, manufacturers of microcontrollers provide a set of tools or a group of programs called the integrated development environment (IDE) to assist the user in writing programs for microcontrollers. The IDE called MPLAB IDE provided by Microchip includes the following:

- Editor: A program (similar to Notepad in PC) that enables the user to write, edit, and save the user's program under a file name with the extension asm. This program is known as Source code. In addition to program mnemonics, the user program includes directives (pseudocodes) that provide the necessary information to the assembler.

- Assembler: This program converts the mnemonics of the Source code into binary code, assigns memory addresses, and checks for syntax errors. The assembler generates several files, such as an object file, Hex file, error file, and list file.

- Linker: This program combines the object code and the information necessary from the header file of the microcontroller and generates an executable file.

- Simulator: This program simulates the execution of the user's program on a computer. This is an alternative to executing the program on the user's target system. The user can run (execute) the program on the simulator and check the result. If the result is not what is expected, the user can use the Debugger (see the next paragraph) to troubleshoot the program.

- Debugger: This is a part of the simulator program that enables the user to find errors in the program by observing register contents as the execution of the instructions is being simulated. It includes debugging techniques such as single-step, register watch, and breakpoint.

The chapter concludes with a simple illustrative program that demonstrates how to write, assemble, debug, and execute a program using MPLAB IDE.

QUESTIONS AND ASSIGNMENTS

4.1   What is a flowchart?

4.2   What is an assembler?

4.3   What is the function of a header file in an assembly language program?

4.4   Explain the term directive (pseudocode) in an assembler and describe its function.

4.5   What does a label (excluding labels used for equates) represent in a program?

4.6   What is the function of the ORG statement in an assembly language program and can the program have more than one ORG statement?

4.7   List the parts of an assembly language statement and indicate the parts that are optional.

4.8   List the number of files generated by the assembler in the MPLAB after assembling a program.

4.9   What is the purpose of the List File of an assembled program and what does it include?

4.10   Specify the memory location where the program will be assembled if the address in ORG statement is as follows: ORG 0040

4.11   START: MOVLW   0x67

                           ADDLW   0x33

                           SLEEP

     a.   Specify the result you expect in the W register.

     b.   Specify the flags that are set after the addition.

     c.   Specify the byte you would observe in the STATUS register.

     Assemble the program using the editor and the assembler of MPLAB starting at location 0020_H. Build the project and execute the instructions using the single step and verify your answers

4.12   START: MOVLW   0x42

                           SUBLW   0x33

                           SLEEP

     a.   Specify the result you expect in the W register.

     b.   Specify the flags that are set after the addition.

     c.   Specify the byte you would observe in the Status register.

     Assemble the program using the editor and the assembler of MPLAB starting at location 0020_H. Build the project and execute the instructions using the single step and verify your answers

4.13   The following set of instructions is expected to load two bytes ($A7_H$ and 92_H) in data registers 01_H and 02_H, add the bytes, and save the sum in register 03_H. Calculate the sum of these two bytes and identify the flags that are set.

         1.   MOVLW   0xA7

         2.   MOVWF   0x01, 0

         3.   MOVLW   0x92

         4.   MOVWF   0x02, 0

         5.   ADDWF   0x01, 1, 0

         6.   MOVWF   0x03, 0

4.14   In Q. 4.13, the add instruction sets overflow and carry flag. Explain why the overflow flag is set and interpret the result if the numbers are signed numbers.

4.15   In Q. 4.13, explain why registers W and 03_H have the byte 92_H at the end of the program.

4.16   In Q.4.13, identify the location where the sum is saved.

SIMULATION EXERCISES

Using MPLAB IDE

4.1 SE   The following program adds two bytes, BYTE1 and BYTE2. If the sum generates an overflow, it clears REG10; otherwise save the sum in REG10. Answer the following questions.

BYTE1	EQU	0x34	;Data bytes
BYTE2	EQU	0x56	
REG10	EQU	0x10	;Register to save the sum
	ORG	0040	;Begin assembly at 0040H
START:	MOVLW	BYTE1	;Load first data byte
	ADDLW	BYTE2	;Add second data byte
	BNOV	SAVE	;Is there an overflow?
	MOVLW	0	;If yes, clear W and REG10
SAVE:	MOVWF	REG10,0	;If no, save sum
	SLEEP		End

   a.   Specify the sum of the data bytes.

   b.   Specify the contents of the STATUS register.

   c.   Specify the contents of REG10.

   d.   Assemble and single step the program and verify your answers.

4.2 SE   In 4.1SE, change BYTE1 = 35_H and BYTE2 = $2A_H$, and answer the following questions.

   a.   Specify the sum for new data bytes.

   b.   Specify the contents of the STATUS register.

   c.   Specify the contents of REG10

   d.   Assemble and single step the program and verify your answers.

4.3 SE   In 4.1SE, change BYTE1 = $F2_H$ and BYTE2 = 38_H, and answer the following questions.

   a.   Specify the sum for new data bytes.

   b.   Specify the contents of the STATUS register.

   c.   Specify the contents of REG10

   d.   Assemble and single step the program and verify your answers.

4.4 SE   In 4.1SE, change BYTE1 = $F7_H$ and BYTE2 = 88_H, and answer the following questions.

   a.   Specify the sum for new data bytes.

   b.   Specify the contents of the STATUS register.

      c.   Specify the contents of REG10

      d.   Assemble and single step the program and verify your answers.

4.5 SE The following instructions add two 16-bit numbers (3267_H and $42F2_H$) stored in four data registers. REG3-REG0. Registers REG1-REG0 hold the number 3267_H (low-order number 67_H in REG0 and high-order number 32_H in REG1), and REG3-REG2 hold the number $42F2_H$. Calculate the sum.

REG0	EQU	0x00	;Low-order byte 67_H of first 16-bit number
REG1	EQU	0x01	;High-order byte 32_H
REG2	EQU	0x02	;Low-order byte $F2_H$ of second 16-bit number
REG3	EQU	0x03	;High-order byte 42_H
REG10	EQU	0x10	;Low-order byte of sum
REG11	EQU	0x11	;High-order byte of sum
	ORG	0x0040	;Begin assembly at 0040H
START:	MOVF	REG0, 0,0	;Get 67_H in W register
	ADDWF	REG2,0,0	;Add $F2_H$
	MOVWF	REG10,0	;Save in REG10
	MOVF	REG1,0,0	;Get 32_H
	ADDWFC	REG3,0,0	;Add 42_H with carry from first addition
	MOVWF	REG11,0	;Save in REG11
	SLEEP		
	END		

Assemble the program. Add all registers in the watch window and store the respective values in REG0-REG3. Using the View menu, open File Registers and enter the 16-bit data in registers REG0-REG3. Single step the program and watch the result in REG11-REG10.

Using PIC18 IDE

4.6 SE Open PIC18 IDE. Select Tools and ➡ Program Memory Editor. Enter the Hex code of Illustrative Program: Addition with Carry Check shown in Section 4.2.5 in memory starting from the address 000020H. Single step the program and verify the result.

4.7 SE Assemble the program # 4.1SE and verify the answers

4.8 SE Assemble and load the program #4.5SE. Select the Rate: Single Step and start simulation. Enter the data in registers 00-03 by clicking on respective spaces in front the register addresses. Single step the program and check the result in registers REG11-REG10,

CHAPTER **5**

INTRODUCTION TO DATA COPY (MOVE), ARITHMETIC, AND BRANCH INSTRUCTIONS

OVERVIEW

In Chapter 3, we examined the PIC18 programming model and its capability in terms of its instruction set. We classified the instruction set into seven groups: data copy (move), arithmetic, logic, branch, bit manipulation, table read/write, and machine control. In this chapter, we will discuss instructions in detail from four groups: data copy, table read/write, arithmetic, and branch. We selected these groups based on the applications and illustrative examples discussed in this chapter.

As mentioned in Chapter 4, the capability of the PIC18 instruction set is limited, but the processor's strength is its ability to execute simple tasks repeatedly, with speed and accuracy. The processor can be programmed to repeat tasks under certain data conditions indicated by flags, and by using the programming techniques such as looping, indexing, and counting. We will discuss these techniques with various illustrations in this chapter.

OBJECTIVES

- List various ways data are copied in the PIC18 microcontroller, and explain how the source and destination registers/memory locations are affected.

- Explain the concepts of memory pointers and indexing.

- Explain how data are copied between the program memory and data (memory) registers.

- Explain the functions of arithmetic instructions (add, subtract, increment, decrement, 1's and 2's complement, and compare) and how flags are affected.

- Write a set of instructions to illustrate the use of data copy and arithmetic instructions.

- Explain the functions of branch (jump) instructions and the difference between absolute and relative branch instructions.

- Explain how flags are used in making programming decisions and setting up loops.

- Calculate the time needed for MPU to execute instructions in a loop, and write instructions to generate time delays.

- Write programs to generate wave forms, copy data, perform arithmetic operations on a given set of data, and search for a specific character in a data set.

5.1 DATA COPY (MOVE) AND SET/CLEAR OPERATIONS

The data copy instructions, known as MOV[1] instructions, are used frequently in programming. The various components inside the PIC18 microcontroller, such as program memory, data registers, W register in Arithmetic Logic Unit (ALU), and I/O ports that store and/or use data are shown in Figure 5-1. The data copy operations of the PIC18 controller can be classified as follows:

1. loading 8-bit data directly in WREG (W register)
2. copying data between WREG and a data (file) register including I/O ports
3. copying data from one data (file) register to another data (file) register
4. clearing or setting all bits in a data (file) register
5. exchanging low-order four bits (nibble) with high-order four bits in a data (file) register

---

[1]The term MOV, often used in mnemonics, creates an erroneous perception that a data byte is removed from one register called source and placed in another register called destination. In reality, it is a copy operation from one register to another register; the source register is not affected by this operation.

6. copying data between program memory and data (file) registers
7. copying data between registers and memory stack locations.

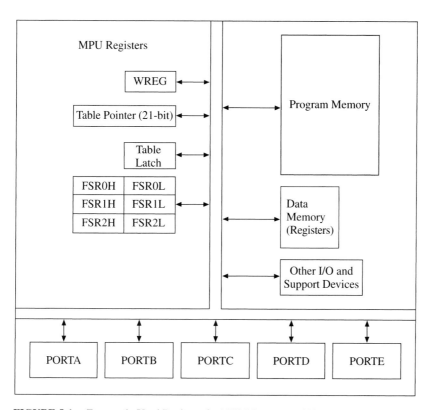

FIGURE 5-1   Frequently Used Registers in ALU, Memory, and I/O Ports For Data Copy

We have already discussed the first two copy operations in previous chapters, and will postpone the discussion of data copy operations related to the stack (Item #7) until we discuss the stack concepts in Chapter 7. We will focus here on the remaining four data copy operations from Item #3 to Item #6 in the above list. We will also identify these operations in terms of the addressing modes: direct addressing and indirect addressing.

5.1.1 Addressing Modes: Direct and Indirect Addressing

In Chapter 3, we defined the addressing mode as the way of specifying an operand, and based on manufacturer's literature, we identified the addressing modes such as literal, immediate, and register direct. These labels are unimportant. What is important is to understand the difference between the direct addressing and the indirect addressing. In the direct addressing mode, the operand is a part of the instruction, and in the indirect addressing mode, the MPU has to look up the address of an operand in one of the registers of the MCU. An analogy may clarify the difference between the two. If we want to telephone a friend, and if we know the number, we dial it directly; this is similar to direct addressing. If we do not know the telephone number, we need to look up the number in the directory and then dial it; this is similar to indirect addressing. In MCU operations, the advantage of indirect addressing is that the register that is being used to look up the address can be modified (incremented or decremented) and used again to look up the next address and a loop can be set up to perform many operations. The concept of indirect addressing is further clarified in Sections 5.1.3 and 5.1.4, and illustrated by the program in Section 5.6.

5.1.2 MOV (Data Copy) and SET/CLR Instructions

The instructions in this data copy group, listed in Table 3-1, are again listed here with examples for easy reference.

Instruction	Examples	Description
MOVLW 8-bit[2]	MOVLW 0xA2	;Load 8-bit number A2H in W register
MOVLB 4-bit[2]	MOVLB 0x0F	;Load 4-bit in the Bank Select Register
LFSR F, 12-bit[2]	LFSR 1, 0x3F2	;Load 12-bit address in FSR1 which is made up of ;two 8-bit registers FSR1H (high) and FSR1L (low) – ;the most significant 4 bits of FSR1H are always zero
MOVWF F, a	MOVWF REG1, 0	;Copy W into REG1 from the Access bank (a = 0)
MOVF F, d, a	MOVF REG1, 0, 0	;Copy File REG1 from Access bank into W because ;d = 0
	MOVF REG1,W,A	;Same as above but in a different format
	MOVF REG1,1,0	;Copy REG1 into itself to check for flags
MOVLB 8-bit	MOVLB 0x0F	;Copy 0FH in Bank Select Register (BSR)
MOVFF Fs, Fd	MOVFF REG1, REG2	;Copy REG1 (Fs = Source Register) into ;REG2 (Fd = Destination Register)
CLRF F, a	CLRF REG1, 0	;Clear REG1 in Access bank
SETF F, a	SETF REG1, 0	;Set all bits to 1 in REG1 located in Access bank
SWAP F, d, a	SWAP REG1, 1	;Exchange low-order and high-order nibbles and save ;the result in REG1 because d = 1

---

[2]The manufacturer lists all literals of various sizes with one common letter k; we will show the number of bits used to load a register.

Points to Remember

1. The MOV instructions copy data from one register, called the source register, into another register, called the destination register, without affecting the contents of the source register.
2. The MOV instructions do not affect flags—with a few exceptions such as CLRF and MOVF, F, F/W.
3. Most of the data copy instructions are one word in length and executed in one instruction cycle, except MOVFF Fs, Fd and LSFR; these are two-word instructions.
4. The letter "d" in the instruction format represents the destination where the result should be saved, and it can be written in various formats. If d = 0 or omitted, the result is saved in the WREG. In place of the letter "d", the letter "W" or the data register can also be used. The letter "a" represents the Access bank. If a = 0 or omitted, the data register is assumed to be from the Access bank. If d = 1, the contents of the BSR register specifies the bank.

The following example illustrates the instruction MOVFF—copying data from one data register to another data register.

EXAMPLE 5.1

Write instructions to store the number $7F_H$ in data register 01_H and the number 82_H in data register 02_H and copy both numbers in register 10_H and 11_H, respectively. Define the data registers with the labels as REG1, REG2, REG10, and REG11. All data registers are in the Access bank.

Solution:

```
REG1      EQU          0x01            ;Define data registers 01H, 02H, 10H, and 11H
REG2      EQU          0x02
REG10     EQU          0x10
REG11     EQU          0x11

          MOVLW        0x7F            ;Load first number 7FH in W register
          MOVWF        REG1,0          ;Copy 7FH from W into Register 01H
          MOVLW        0x82            ;Load second number 82H in W register
          MOVWF        REG2,0          ;Copy 82H from W into Register 02H
          MOVFF        REG1, REG10     ;Copy first number in Register 10H
          MOVFF        REG2, REG11     ;Copy the second number in Register 11H
          END
```

Description:

The purpose of this example is to review MOV instructions and demonstrate the use of MOVFF instruction. The instruction set does not include any instruction that loads a byte directly in a data register. Therefore, the number $7F_H$ is loaded first into WREG and copied into register 01_H, and the same process is repeated for the second number. The instruction MOVFF copies these numbers from REG1 and REG2 directly into REG10 and REG11 respectively rather than going through W register. The MOVFF is a very efficient instruction. The example illustrates copying a few bytes. To copy a large block of bytes, we need to combine this instruction with indexing and memory pointers as discussed in the following section.

5.1.3 Using File Select Registers (FSRs) as Pointers

The concepts of memory pointers and indexing are critical in copying or operating on a block of data. The memory pointer is a register that holds an address of a data register, and when a byte in the data register is operated on for operations (such as copying, arithmetic, or logic), the MPU looks up the address in the memory pointer and performs the specified operation. This process of pointing or indexing a data register using another register is called **indirect addressing.**

As described in Chapter 3, the PIC18 microcontroller has three File Select Registers (FSR0, FSR1, and FSR2; see Table 3-1). Each FSR is made up of two 8-bit registers—FSRH (high) and FSRL (low)—and associated with an index register (INDF) as shown in Table 5-1. The FSRs are used as pointers to data registers. The user can load an address in one of the FSRs, and the MPU uses the address in the FSRs to access and operate on a byte in that data register. In the previous section, 5.1.1, we listed the instruction LFSR that loads a 12-bit address in two 8-bit registers—FSRH and FSRL. The four most significant digits in FSRH are always zero. The primary advantage of the indirect addressing is that the address in the FSR register can be incremented or decremented or offset by a signed number in W register. Table 5-1 shows that four Special Function Registers—POSTINC, POSTDEC, PREINC, and PLUSW—are used to modify the addresses in the FSRs. These four registers can be used as operands with other instructions, as shown in Example 5.2. The Illustrative Program: Copying a Block of Data from Program Memory to Data Registers (Section 5.6) shows an application of the indirect addressing to copy data from one group of data registers to another group of data registers.

TABLE 5-1 Indirect Addressing of Data Registers

Operation	FSR0	FSR1	FSR2	Example	Explanation
Access address from FSR, perform operation, but do not change FSR	INDF0	INDF1	INDF2	MOVFF INDF0,INDF1	Copy byte from register shown by FSR0 to register shown by FSR1 and do not modify FSRs
Access address from FSR and increment FSR after an operation	POSTINC0	POSTINC1	POSTINC2	ADDWF POSTINC,1	Add W and register shown by FSR1, save sum in register, and increment FSR1
Access address from FSR and decrement FSR after an operation	POSTDEC0	POSTDEC1	POSTDEC2	ADDWF POSTDEC2,0	Add W and register shown by FSR2, save sum in W, and decrement FSR2
Increment address in FSR and then access address to perform an operation	PREINC0	PREINC1	PREINC2	SUBWF PREINC0,0	Increment FSR0, subtract W from register shown by FSR0, save result in W

TABLE 5-1   Indirect Addressing of Data Registers (continued)

Operation	FSR0	FSR1	FSR2	Example	Explanation
Add WREG to FSR and access address to perform an operation	PLUSW0	PLUSW1	PLUSW2	MOVFF PLUSW1,PORTC	Add W to FSR1, copy byte from that address to PORTC, and keep initial value in FSR1

EXAMPLE 5.2

Explain the result of the following instructions assuming that the data registers with the addresses 120_H and 121_H hold the bytes $F2_H$ and $A3_H$ as shown in Figure 5.2.

```
LFSR     FSR1, 0x0120        ;Load 12-bit address 120H in FSR1
LFSR     FSR2, 0x0150        ;Load 12-bit address 150H in FSR2
MOVFF    POSTINC1, POSTINC2  ;Copy the byte in data register 120H into data
                             ; register 150H and increment FSR1 to 121H and
                             ; FSR2 to 151H
```

Solution:

The LFSR instructions load the 12-bit addresses 120_H and 150_H in registers FSR1 and FSR2 respectively. When the MPU reads the instruction MOVFF, it looks up the addresses in FSR1 and FSR2 and copies the byte $F2_H$ from the data register 120_H into the register 150_H and increments the addresses in FSR1 and FSR2 to 121_H and 151_H respectively as shown in Figure 5-2. After the execution of the instruction MOVFF, the MPU is ready to copy the next byte $A3_H$ from 121_H into 151_H register.

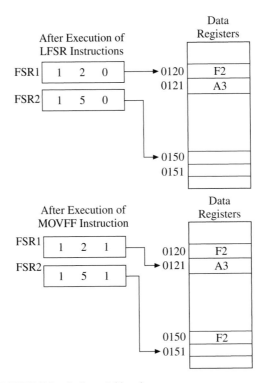

FIGURE 5-2   Indirect Addressing

5.1.4 Using Table Pointers to Copy Data Between Program Memory and Data Registers

When an assembler assembles a source program, it stores data bytes in the program memory. For example, to add five data bytes, the user includes those bytes in the source program with the DB directive (see Chapter 4, Section 4.3.3 for the explanation of DB). Bytes must be copied from program memory into data registers in order to perform any arithmetic or logic operations on them. However, the program memory and data registers are incompatible in their sizes: program memory is 16-bit wide and the data registers are 8-bit wide. In addition, the address bus of the program memory is 21-bit wide.

The solution to this problem is shown in Figure 5-3a. The PIC18 has two Special Function Registers: TBLPTR (Table Pointer) and TABLAT (Table Latch). The TBLPTR consists of three registers—TBLPTRU (Upper), TBLPTRH (High), and TBLPTRL (Low)—that hold a 21-bit address, and the MPU uses the TBLPTR as a memory pointer for program memory. The TAB-LAT is an 8-bit intermediate register, and the instruction TBLRD* (listed in Table 5-2) copies a byte from the program memory into TABLAT that can be copied into a data register. Table 5-2 shows additional three TBLRD* instructions that can post-increment, decrement, and pre-increment the Table Pointer. Thus the TBLRD* instructions can be used for indexing or pointing to a block of memory.

Instruction TBLRD* Copies Data From Program Memory to Table Latch

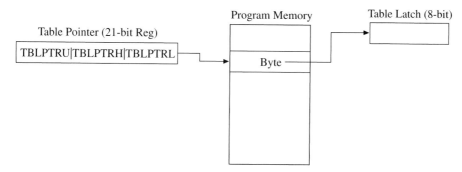

FIGURE 5-3 (a)   Copying Data From Program Memory to Table Latch Using TBLRD* Instruction

Instruction TBLWT* Copies Data From Table Latch to Program Memory

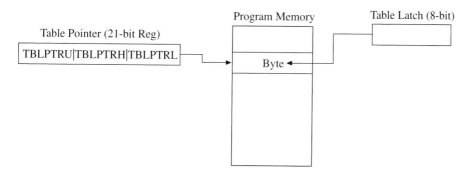

FIGURE 5-3 (b)   Copying Data From Program Table Latch to Program Memory Using TBLWT* Instruction

TABLE 5-2   Copying Between Program Memory and Table Latch

Instruction	Description	Example: Initial Values as follows: Program Memory 01251_H holds byte 76_H and 01252_H holds $F2_H$ TBLPTR = 01251_H follows: Program TABLAT = $3F_H$
TBLRD*	Copy from program memory location pointed by the address in TBLPTR (Table Pointer) to TABLAT (Table Latch)	$01251_H = 76_H$ TABLAT = 76_H TBLPTR = 01251_H
TBLRD*+	Copy from program memory location pointed by the address in TBLPTR to Table Latch (Table Latch) and increment TBLPTR	$01251_H = 76_H$ TABLAT = 76_H TBLPTR = 01252_H
TBLRD*-	Copy from program memory location pointed by the address in TBLPTR (Table Pointer) to TABLAT (Table Latch) and decrement TBLPTR	$01251_H = 76_H$ TABLAT = 76_H TBLPTR = 01250_H

TABLE 5-2   Copying Between Program Memory and Table Latch (continued)

Instruction	Description	Example: Initial Values as follows: Program Memory 01251_H holds byte 76_H and 01252_H holds $F2_H$ TBLPTR = 01251_H follows: Program TABLAT = $3F_H$
TBLRD+*	Increment TBLPTR (Table Pointer) and then copy from program memory location pointed by the address in TBLPTR to (Table Latch)	$01252_H = F2_H$ TBLPTR = 01252_H TABLAT = $F2_H$
TBLWT*	Copy from TABLAT (Table Latch) to program memory location pointed by the address in TBLPTR (Table Pointer)	TABLAT = $3F_H$ $01251 = 3F_H$ TBLPTR = 01251_H
TBLWT*+	Copy from TABLAT (Table Latch) to program memory location pointed by the address in TBLPTR (Table Pointer) and increment TBLPTR	TABLAT = 3F $01251 = 3F_H$ TBLPTR = 01252_H
TBLWT*-	Copy from TABLAT (Table Latch) to program memory location pointed by the address in TBLPTR (Table Pointer) and decrement TBLPTR	TABLAT = $3F_H$\$ $01251 = 3F_H$ TBLPTR = 01250_H
TBLWT+*	Increment TBLPTR (Table Pointer) and then copy from TABLAT (Table Latch) to program memory location pointed by the address in TBLPTR.	TBLPTR = 01252_H TABLAT = $3F_H$ $01252 = 3F_H$

Figure 5-3b shows how to copy a data byte from a data register into program memory by using the instruction TBLWT* (Table Write). This instruction can also post-increment, decrement, and pre-increment the Table Pointer. Thus, the TBLWT instruction is also used in copying a block of data from data registers into program memory. Example 5.3 illustrates how to copy a data byte from program memory into a data register, and Illustrative Program (Section 5.6) shows how to copy a block of data from program memory to data registers.

EXAMPLE 5.3

The program memory location BUFFER with the address 000040_H holds the byte $F6_H$. Write instructions that can be assembled using MPLAB editor to copy the byte $F6_H$ in the data register REG10.

Solution:

To copy the byte from the program memory into the data register, we need to load the memory address 000040_H in the TBLPTR using the three Special Function Registers—TBLPTRU, TBLPTRH, and TBLPTRL (upper, high, and low)—and use the TBLPTR as a memory pointer for the instruction TBLRD*. The address 000040_H is labeled as BUFFER, and the MPLAB assembler recognizes the labels such as UPPER, HIGH, and LOW BUFFER to the identify entire address as shown in the following instructions.

continued

Instructions:

```
        Title "Example 5-3"
        List p=18F452, f =inhx32
        #include <p18F452.inc>    ;This is a header file

REG10   EQU     0x10              ;Define data register address

        MOVLW   UPPER BUFFER      ;Load the upper bits of BUFFER in
        MOVWF   TBLPTRU           ;TBLPTRU
        MOVLW   HIGH BUFFER       ;Load high-byte in TBLPTRH
        MOVWF   TBLPTRH
        MOVLW   LOW BUFFER        ;Load low-byte in TBLPTRL
        MOVWF   TBLPTRL
        TBLRD*                    ;Copy byte F6H in Table Latch
        MOVF    TABLAT,W          ;Copy F6H from Table Latch into W
        MOVWF   REG100            ;Copy byte F6H from W into REG10

        ORG     0X40
BUFFER  DB      0xF60             ;Store the byte F6H in 000040H
        END
```

5.2 ARITHMETIC OPERATIONS

The PIC18F MPU is capable of performing many arithmetic operations such as add, subtract, multiply, compare, complement (1s and 2s), and increment/decrement (a complete list is shown in Chapter 3; see Table 3-3 and Appendix A). This is an 8-bit processor and performs most arithmetic operations on 8-bit data. However, the MPU also includes some instructions that enable the user to perform arithmetic operations on numbers larger than 8 bits wide. The arithmetic operations can be performed on literal (numbers) and contents of WREG and data (File) Registers. Most arithmetic operations affect all the flags. The frequently used instructions are discussed in the following sections with examples, and the remaining instructions such as Multiply are discussed in Chapter 6.

5.2.1 Addition and Subtraction

Instruction	Example	Description
ADDLW 8-bit	ADDLW 0x A2	;Add A2H to WREG and save the result in W
ADDWF F, d, a	ADDWF REG1, 0, 0	;Add W and REG1 and save the sum in W
	ADDWF REG1,W,A	;Both instructions are the same but in a ; different format
	ADDWF REG1,1,0	;Add W and REG1 and save the result ;in REG1
	ADDWF REG1,REG1,A;	This is the same as above but in a different format
ADDWFC F, d, a	ADDWFC REG2, 0,0	;Add W, REG2, and Carry from previous ; operation and save the result in W; d =0
SUBLW 8-bit	SUBLW 0 x F2	;Subtract W from 8-bit number F2H and save ;the result in W
SUBWF F, d, a	SUBWF REG1,0,0	;Subtract W from REG1 and save the result ;in W; d = 0
	SUBWF REG1,1.0	;Subtract W from REG1 and save the result in ;REG1; d = 1
SUBFWB F, d, a	SUBFWB REG1,0,0	;Subtract REG1 and Borrow from W and ;save in W
	SUBWFB REG1, 1,0	;Subtract W and borrow from REG1 and ;save in REG1

Points to Remember

1. Add and subtract instructions affect all the flags, and the status of the flags is determined by the result of the operation.

2. The result of an operation can be stored either in WREG or data (File) register based on the parameter d. This is one of the powerful features of this processor that enables the programmer to use any data register as almost as an accumulator. In most other processors, the result is generally stored in the accumulator (similar to WREG).

3. The instruction SUBLW is confusing and contrary to the convention used—left to right—in other instructions. The mnemonic SUBLW appears to convey the meaning that a literal is subtracted from WREG. However, the instruction subtracts WREG from the literal. This is in contrast to other subtract instructions. For example, the instruction SUBWF subtracts WREG from data (File) register.

4. The instruction ADDWFC is used for addition of numbers larger than 8 bits. For example, in the addition of 16-bit numbers, if the sum of low-order 8 bits generates a carry, the carry is added to the sum of high-order 8 bits (see Example 5.6). This instruction enables the user to add large numbers of any size. However, this instruction should not be used to account for carries generated by the addition of multiple 8-bit numbers.

5. The instructions SUBWFB and SUBFWB are used for the subtraction of the numbers larger than 8 bits; the borrow generated by low-order bytes is subtracted from the next subtraction of high-order bytes.

EXAMPLE 5.4

Write instructions to store two 8-bit Hex numbers 48_H and $3F_H$ in data registers 20_H and 21_H, respectively. Add the numbers and save the sum in register 30_H. The addition of these numbers does not generate a carry. Explain the status of all the flags after the addition and interpret the result.

Solution

Given two Hex numbers are constants (literal) and must be loaded in WREG one at a time, use MOVLW instructions discussed in the previous section, and store them in registers known as File Registers 20_H and 21_H respectively. Once the numbers are stored, they can be added and the result can be saved in Register 30_H.

continued

```
REG20       EQU       0x20        ;Register 20H is labeled as REG20
REG21       EQU       0x21        ;Register 21H is labeled as REG21
REG30       EQU       0x30        ;Register 30H is labeled as REG30

            MOVLW     0x48        ;Load the first Hex number 48H in WREG
            MOVWF     REG20       ;Store 48H in File Register 20H in data memory
            MOVLW     0x4F        ;Load the second number 4FH in WREG
            MOVWF     REG21       ;Store 4FH in File Register 21H in data memory
            ADDWF     REG20, 0, 0 ;Add 4FH which is still in WREG to 48H
                                  ; and save the sum in WREG
            MOVWF     REG30       ;Save the sum in File Register 30H
            END
```

Results:

The ADDWF instruction adds the two numbers as follows:

```
WREG     0  1  0  1  1  1  1     (4FH)
+
REG20    0  1  0  0  1  0  0     (48H)
 Carry   1        1
```
_____
```
WREG     1  0  0  1  0  1  1  1 (97H)
```

Flag Status: N = 1, OV = 1, Z = 0, DC = 1, C = 0
Explanation: The result 97 $_H$ determines whether a flag is set or reset.

- C = 0.   There is no carry generated from Bit7; therefore the carry flag is reset.
- Z = 0.   The result is not zero, therefore, the zero flag is reset.
- DC = 1.   There is a carry from Bit3 to Bit4; therefore the digital carry is set.
- N = 1.   Bit7 of the result 97 $_H$ is 1 indicating it is a negative result; therefore the sign flag is set.
- OV = 1.   For signed numbers, the result generates an overflow; therefore, this flag is set. This is the most confusing flag to interpret and is discussed in the next section.

The processor is unable to recognize whether numbers entered by the user are signed numbers or unsigned numbers. It considers both possibilities and determines the flag status. Assuming the numbers are unsigned, the result is unsigned 97_H, and it is correct. If the numbers are unsigned, the user should ignore the sign and overflow flags.

If we assume the numbers are signed numbers, only seven bits of a register are used for the magnitude, and the eighth bit, Bit7, is reserved for a sign. This means the largest positive number the register can accommodate is $7F_H$. The numbers 48_H and $4F_H$ are positive numbers.

After the addition, the result is 97_H and the overflow flag is set. Therefore, this is a wrong result for signed numbers. The setting of the overflow flag can be reasoned in various ways as follows:

1. the result 97_H is larger than $7F_H$, and it cannot be accommodated within seven bits
2. the carry from Bit6 changes the sign, and there is no carry from Bit7 to the Carry bit. to change it back to original sign.
3. the addition of two positive numbers cannot generate the negative result.

EXAMPLE 5.5

Write instructions to load the number $7F_H$ in WREG (W register), and subtract it from the number 28_H. Display the result at PORTC and also save it in Register 30_H. Explain the status of all the flags after the subtraction.

Solution

The number to be subtracted ($7F_H$) must be copied into WREG, and it can be subtracted from 28_H by using the instruction SUBLW. In addition, we need to initialize PORTC to display the result.

```
REG30      EQU         0x30          ;Register 30H is labeled as REG30

           MOVLW       0 x 00        ;Initialize all pins of PORTC as outputs
           MOVWF       TRISC
           MOVLW       0x7F          ;Load the subtrahend 7FH in WREG
           SUBLW       0x28          ;Subtract 7FH from 28H and save the result in WREG
           MOVWF       PORTC         ;Turn on LEDs
           MOVWF       REG30         ;Save the result in data register 30H
           END
```

Results:

The SUBWF instruction subtracts the two numbers by using the 2's complement procedure. It takes the 2's complement of the **subtrahend** (the number to be subtracted) and adds to the **minuend** (the number to be subtracted from).

WREG	0 1 1 1 1 1 1 1	(7FH - Subtrahend)
2's Com.	1 0 0 0 0 0 0 1	(81H - 2's Complement of 7FH)
of 7FH +		
	0 0 1 0 1 0 0 0	(28H - Minuend)
WREG	1 0 1 0 1 0 0 1	(A9H)

continued

Flag Status: N = 1, OV = 0, Z = 0, DC = 0, C = 0
Explanation: The result A9H determines whether a flag is set or reset.

- C = 0.   The MPU performs the subtraction using 2's complement process. In the above example, the addition of 28H and 81H does not generate a carry.
- Z = 0.   The result is not zero, therefore, the zero flag is reset.
- DC = 0.   There is a no carry from Bit3 to Bit4; therefore the digital carry is reset.
- N = 1.   Bit7 of the result $A9_H$ is 1 indicating it is a negative result; therefore the sign flag is set.
- OV = 0.   There is a no carry from Bit6 to Bit7, thus resulting in no change to the sign of the result.

If we assume the numbers are unsigned, a larger unsigned number is subtracted from a smaller number; therefore, the result is in 2's complement. If we assume that the numbers are signed, the result is also a negative number because a larger signed number is subtracted from a smaller signed number.

EXAMPLE 5.6

Two 16-bit numbers $29F2_H$ and 3587_H are stored in data registers with addresses from 0010_H to 0013_H as shown: low-order byte followed by the high-order byte.

REG10—0010   F2
REG11—0011   29
REG12—0012   87
REG13—0013   35

Write instructions to add these 16-bit numbers and save the sum in registers 0020_H and 0021_H.

Solution

PIC18 is an 8-bit microcontroller that has an 8-bit working register. Therefore, to add 16-bit numbers, we need to add low-order byte (8 bits) of each number first and then add the high-order byte. If the sum of the low-order bytes generates a carry, the instruction ADDWFC will add the carry to the sum of the high-order bytes. With this instruction we can add a number with any size. The equates defining the registers are not shown in the following instructions. In addition, to run this program in the MPLAB IDE, the above numbers must be loaded one byte at a time in the respective registers.

continued

MOVF	REG10,W	;Copy low-order byte F2H in WREG
ADDWF	REG12,0,0	;Add low-order byte 87H and save sum in WREG
MOVWF	REG20	;Save sum of low-order bytes in REG20
MOVF	REG11,W	;Copy high-order byte 29H in WREG
ADDWFC	REG13,0,0	;Add high-order byte 35H and save sum in WREG
MOVWF	REG21,0	;Save sum of high-order bytes in REG21

Result:

The sum is as shown:

$$\begin{array}{r} 2\ 9\ F\ 2 \\ +\ 3\ 5\ 8\ 7 \\ \text{Carry}\quad 1 \\ \hline 5\ F\ 7\ 9 \end{array}$$

The sum of the low-order bytes is 79_H with a carry; the sum 79_H is saved in REG20. The next addition is performed by the instruction ADDWFC which adds 29_H, 35_H, and the carry generated by the previous sum. The sum, $5F_H$ is saved in REG21.

5.2.2 Increment, Decrement, and Complement Operations

Instruction	Example	Description
INCF F, d, a	INCF REG1, 1, 0	;Increment contents of REG1 by one ; and place the result in REG1 because d = 1
DECF F, d, a	DECF REG1, 0, 0	;Decrement contents of REG1 by one ; and place the result in W register because d = 0
COMF F, d, a	COMF REG1, 0, 0	;Complement each bit in REG1 and save in W
NEGF F, a	NEGF REG1, 0	;Take 2's complement of byte in REG1

Points to Remember

1. These instructions except COMF affect all the flags; the COMF affects only N & Z flags.
2. The result of an operation can be stored either in the WREG or the data (File) Register based on the parameter d except for NEGF instruction.
3. Increment and decrement instructions are commonly used in counter (counting) applications.

EXAMPLE 5.7

Write instructions to store the number $7F_H$ in data register 20_H and find the 2's complement of the byte $7F_H$ by two different processes: (1) without using the instruction NEGF and save the result in register 20_H, and (2) use the instruction NEGF and save the result in the same register 21_H.

Solution:

The objective of Example 5.5 is to demonstrate that one can find 2's complement (negative of a number) without having an instruction in an MPU. This is particularly useful if you are using previous series of PIC controllers. We can follow the same procedure that is taught in Digital Logic class: take 1's complement and add 1. We can also demonstrate that both procedures achieve the same result.

```
REG20       EQU      0x20        ;Define data register addresses
REG11       EQU      0x21

            ORG      0x20        ;Begin assembly at program memory 20H

            MOVLW    0 x7F       ;Load number 7FH in W register
            MOVWF    REG20       ;Copy 7FH from W into Register 20H
            COMF     REG20,1,0   ;Take 1's complement of 7FH and save in REG20
            INCF     REG20,1     ;Increment to obtain 2's complement
            MOVWF    REG21       ;Save 7FH in REG21
            NEGF     REG21,0     ;Take 2's complement
            END
```

Result:

The COMF instruction complements all the bits of the number 0111 1111 ($7F_H$) resulting into 1000 000 (80_H). The increment instruction adds 1 making the byte 1000 0001 (81_H). This is 2's complement of the number $7F_H$, and it is stored in REG20. These instructions also demonstrate that any general purpose data register can be used as an accumulator.

REG20:	0 1 1 1 1 1 1 1	(7FH)
1's Complement	1 0 0 0 0 0 0 0	(80H)
Add 1	+ 0 0 0 0 0 0 0 1	(01H)
REG20	1 0 0 0 0 0 0 1	(81H)

The last instruction NEGF takes the 2's complement of the number $7F_H$ and stores the result (81_H) in REG21.

5.3 REDIRECTION OF PROGRAM EXECUTION (BRANCH AND SKIP OPERATIONS)

The processor is a sequential machine. Once it begins processing (executing an instruction), it goes from one memory location to the next memory location and executes the binary commands (instructions) stored in memory. It continues the sequential execution unless it is told to change the sequence of execution or to stop (suspend). In PIC MCU, there are three groups of instructions that change the sequence of execution: Call, Branch, and Skip instructions. The Call instructions redirect the MPU to perform subtasks called subroutines, and after the completion of the subtasks, the MPU returns to the previous sequence of execution. These instructions and subroutines are discussed in Chapter 7. We focus here first on branch instructions and discuss skip instructions in Section 5.3.3.

The processor performs one operation at a time such as copying or adding only one number at a time. The strength of the processor comes from its ability to repeat operations with high speed. The processor can be set up to repeat the operations by using branch instructions, which is called "setting up a loop."

The branch (also known as jump) instructions can be classified into two categories: conditional and unconditional, as shown in Figure 5-4. A "conditional" branch instruction changes the sequence of program execution based on the status of a flag (set or reset) by an operation prior to the branch instruction. In contrast, an "unconditional" branch instruction (such as GoTo in PIC processor) always changes the sequence of execution independent of the flag status to the memory location specified in the operand. These branch instructions are further divided into two categories: absolute addressing and relative addressing, based on its method of specifying the address where the processor should go next. As an analogy, if someone in your neighborhood is looking for your friend's house, you can give two types of directions such as: your friend lives at 19 Victoria Street (the absolute address) or her house is ten houses down from your house (the relative address).

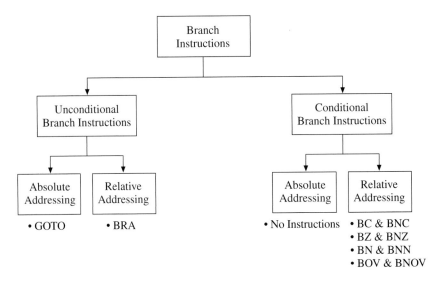

FIGURE 5-4   Branch Instructions in PIC18 Family

5.3.1 Unconditional Branch Instructions

The PIC18F processor has two unconditional branch instructions—GoTo with absolute addressing and BRA (Branch) with relative addressing as shown in Figure 5-4.

Instruction	Description
BRA n	This is a one-word unconditional relative branch instruction where n is a signed number that can be used to redirect the program within the range of -1024 to +1023 memory locations where the memory is organized in bytes. However, in this instruction, "n" is specified in words instead of bytes. A negative number (n) represents a backward jump and a positive number represents a forward jump. In a source program, a jump location is written with a label, and the assembler calculates the relative address. The calculations of relative addresses are shown in the section on Conditional Branch instructions.
GOTO k	This is a two-word unconditional absolute branch instruction. This is used to redirect the program anywhere in 2 MB memory locations.
	Example. GOTO Label

5.3.2 Conditional Branch Instructions

The PIC18F processor includes eight conditional branch instructions with the relative addressing. These branch instructions are based on four data flags—N, OV, Z, and C—with two instructions per flag. There are no branch instructions based on the Digital Carry (DC) flag; it is used internally by the processor for Binary Coded Decimal (BCD) operations. The jumps in relative addressing are limited to -128 to +127 (total of 256) memory locations in relation to the

Program Counter (PC). On the other hand, the instruction such as GoTo, with absolute addressing, can redirect the program anywhere in the memory map.

A branch instruction includes an 8-bit signed operand that specifies the memory address where the processor should go next for execution. This address can be a forward memory location indicated by the positive value (00 to $7F_H$) of the operand or a backward memory location indicated by the negative value (80_H to FF_H) of the operand. In assembly language programs, these addresses are specified by using labels, and the assembler calculates these addresses in relation to the address in the program counter plus two (PC +2). The 2 is added to the program counter because when the processor is executing one-word (two-byte) branch instruction, the address in the PC is already incremented by 2 pointing to the address of the next instruction. The format of some these branch instructions is shown in the following examples. Examples 5.8 and 5.9 illustrate the calculations of relative addressing.

Instruction	Example	Description
BC n	BC 0x05	;If C = 1, jump forward to location = PC + 2 + (05H x 2)
BNC n	BNC 0xF8	;If C = 0, jump backward to location = PC + 2 + (2's Com. F8H x 2)
BZ n	BZ 0x82	;If Z = 1, jump backward to location = PC + 2 + (2's Com 82H x 2)
BNZ n	BNZ 0x72	;If Z = 0, jump forward to location = PC + 2 + (72H x 2)

Points to Remember

1. The branch instructions use relative addressing.
2. These instructions do not affect any flags, but use the flag status set by the previous instruction to make decisions (to either change the sequence of execution or continue).
3. The MPU branches to the location if the condition is met; otherwise, the execution continues to the next instruction in the sequence.
4. The branch memory location is calculated by adding the address of the branch instruction plus two and the operand multiplied by two (see Examples 5.8 and 5.9).
5. If the operand byte is negative, it indicates a backward jump, and if the operand byte is positive, it indicates a forward jump.

EXAMPLE 5.8

A set of instructions that adds two bytes is assembled starting at location 0020_H. Prior to the addition, data registers 10_H and 11_H hold the bytes BYTE1 (97_H) and BYTE2 ($A2_H$) respectively. After the addition, if the sum generates a carry, the MPU is directed to branch forward to program memory location DISPLAY. Examine the Hex code shown and (1) calculate the sum, (2) determine whether it is a forward or backward jump and calculate the memory location DISPLAY, and (3) identify the memory location where the program is redirected if there is a no carry.

continued

Memory Address	Hex Code	Mnemonics	Comments
0020	50 10	MOVF 0x10, 0, 0	;Copy BYTE1 from REG10 into W
0022	26 11	ADDWF 0x11, 0,0	;Add BYTE2 and save the sum in W
0024	E2 05	BC DISPLAY	;Branch to memory location DISPLAY ; if the sum generates a carry
0026		Next Instruction	

Solution

1. The addition of the two bytes generates a carry as shown.
   $97_H + A2_H = 139_H$: W = 39_H and C = 1

2. The Hex code for the instruction BC (Branch on Carry) is E2 05. The Hex code E2 is the opcode for the instruction, and 05 is a positive number. Therefore, the processor will jump forward with respect to the address in program counter (PC). When the MPU begins to execute the code BC, the PC holds the address of the next instruction 0026_H. Therefore, the branch location is calculated in reference to address 0026_H. The code 05_H represents the number of words; therefore, the number of bytes (05 x 2) = $0A_H$. The program is then redirected to the location 0030_H as follows:
   Branch Location Display = PC + 2 + 2n = $0024_H + 2 + 2 \times 5_H = 0026_H + 0A_H$
   $$= 0030_H$$

3. If the sum does not generate a carry, the program will go to the next location 0026_H and continue its execution from that location on.

This type of a problem provides understanding of how the processor calculates relative addresses—forward and backward branch locations—by using signed numbers and 2's complement method. In practice, the assembler will calculate these locations and include them in the list file.

EXAMPLE 5.9

Calculate the value of the operand XX of the following branch instruction that redirects the program backward from location 0028_H to location 0020_H.

0028	D7 XX	BRA 0x20	;Branch to location 0020H
002A	Next Instruction		

continued

Solution:

This operand represents the number of words the program should jump backward from the address of the next instruction $002A_H$. We will calculate this operand as follows:

Subtract the difference between the two locations to get the number of bytes and then divide the number of bytes by two to get the number of words.

$002A_H - 0020_H = 0A_H / 2 = 05_H$

This is a backward jump; therefore, 05 should be represented as a negative number in 2's complement = FB_H.

5.3.3 Arithmetic Instructions with Skip

The PIC18 instruction set includes arithmetic/test instructions (listed in Table 3-7, Chapter 3) that perform arithmetic operations and check for a certain data condition. If the condition is met, the processor skips the following instruction and continues the execution. These instructions are generally followed by the unconditional branch instruction (BRA) to redirect the program beyond the 64-word jump provided by relative branch instructions such as BZ or BNZ. Some of these instructions are discussed in the following paragraphs.

Instructions	Description
CPFSEQ	Compare File Register (F) with W—Skip if F=W. This instruction checks if the File Register (data register) is equal to the WREG by subtracting W from F. If the result is zero, the processor skips the execution of the next instruction; neither contents of F or W are modified, the subtraction is performed internally for comparison.
DECFSZ	Decrement File Register—Skip if File Register = 0. This instruction decrements the contents of the File Register (data register), and if F = 0, the processor skips the execution of the next instruction.

EXAMPLE 5.10

Explain the operation of the following instructions stored in program memory starting at location 0020_H.

COUNTER	EQU	0x01		;Set up Register 01 as a counter
0020	0E 05	MOVLW	0x05	;Load 05H in WREG
0022	6E 01	MOVWF	COUNTER	;Load 05 in Counter Register 01
0024	2E 01	DECFSZ	COUNTER	;Decrement Counter. When Counter
				;≠ 0, go to location 0028H
0026	D7 FE	BRA	0x0024	;Go back to location 0024H
0028	XX XX	Next Instruction		

continued

Solution:

The first two instructions set up data register 01 as a counter by loading 05_H into the register. The DECFSZ instruction decrements the counter by one, and the BRA instruction takes the processor back to the DECFSZ instruction and repeats the loop until COUNTER (Reg. 01) becomes zero. When Reg. 01 becomes zero, the processor skips the BRA instruction and goes to the location of the Next Instruction 0028_H.

5.4 GENERATING TIME DELAYS

Time delays are frequently used in embedded systems. There are two methods of generating time delays: using hardware (timers) and software (instructions). The time delays using timers are discussed in Chapter 11; here we discuss software delays. A commonly used technique to generate software time delays is to load a register with a number (count) and set up a loop to decrement that number until the register becomes zero; thus, the processor is kept busy executing the loop. The time the processor takes to execute that entire loop is a time delay, which is determined by the instructions in the loop and the frequency of the processor. These concepts are illustrated in Example 5.11.

Applications: Time delays are required in many applications such as clock, traffic light, waveform generation, and timing functions of many appliances. Software time delays, based on instructions, tend to be less accurate than those of hardware timers. Modern microcontroller chips include multiple timers, and are commonly used in delay applications.

EXAMPLE 5.11

Calculate the time required for the 18F452 PIC controller to execute the LOOP1 in the following program if the operating frequency of the controller is 40 MHz (Figure 5.5).

```
COUNT1    EQU     D'250'
REG10     EQU     0x10


          MOVLW   COUNT1    ;Load decimal count in W
          MOVWF   REG10     ;Set up REG10 as a counter
```

continued

```
LOOP1:      DECF        REG10, 1      ;Decrement REG10 - 1W / 1C / 4 CLK³
            BNZ         LOOP1         ;Go back to LOOP1 if REG10 ≠ 0
            END                       ;1W / 2C / 8 CLK⁴ when it branches; otherwise
                                      ;1C / 4CLK
```

Solution

The time required to execute this loop is also called loop delay. The term loop delay refers to the total time the MPU requires—from entering the loop to exiting the loop. The first instruction loads the decimal number 250 (COUNT1) in the WREG, and the second instruction sets up REG10 as a counter. The LOOP1 consists of two instructions: DECF and BNZ. The processor executes the instruction DECF in one cycle with four clock periods, and the instruction BNZ in two cycles with eight clock periods. However, the processor requires only one cycle to execute BNZ when it does not branch—the last execution of the loop.

1. Total instruction clock cycles (I_{CLK} in the LOOP1 = 12 CLK

2. Processor Frequency is 40 MHz. Therefore, the time period T is:

$$T = 1/F = 1/40 \times 10^6 = 25 \text{ nS}$$

3. Time to execute the loop once = 12 CLK x 25 nS = 300 nS

4. The count for the loop N_{10}=250. The loop is executed 250 times; therefore, total time delay T_L = 250 x 300 nS = 75 μS

5. To be exact, we should subtract 100 nS because the last cycle in the loop takes only eight clock cycles rather than 12 clock cycles.

6. Ignoring the difference in the execution of the last cycle, we can calculate the execution time or time delay T_L as:

$$T_L = \text{Number of Instruction Clock Cycles in a loop x Clock Period x Count}$$
$$= I_{CLK} \times T_c \times N_{10} = 12 \times 25 \text{ nS} \times 250 = 75 \text{ μS}$$

5.4.1 Generating Time Delays Using Nested Loop (Loop-Within-a-Loop) Technique

In the previous example, we used one register and the decimal count of 250, and that generated 75 μS delay. With one register, the largest number of time the loop can be repeated is 256_{10}. If we use two registers, one after another, we can get double the delay. However, if we use the nested loop (loop-within-the-loop) technique with two registers as shown in Figure 5.6, we can multiply the delay in the inner loop by the count in the outer loop, and add the delay due to the instructions outside the inner loop as shown in the next example.

---

[3]The abbreviation 1W / 1C / 4 CLK refers to one word instruction with one cycle with four clock periods

[4]Some instructions, such as BNZ, have two cycles requiring 8 clock periods when the condition is true and the MPU redirects the program. In the last execution, when MPU goes to the next instruction in the program, it takes only one cycle (4 clock periods).

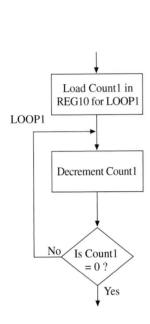

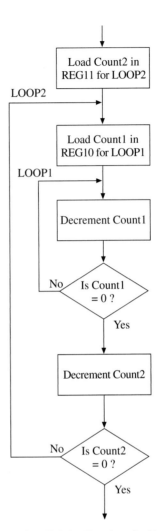

FIGURE 5-5   Flowchart for Delay Loop

FIGURE 5-6   Flowchart for Delay Using Nested Loop Technique

EXAMPLE 5.12

Calculate the time required for the PIC18F controller to execute both loops: LOOP1 and LOOP2 in the following program if the operating frequency of the controller is 40 MHz. Ignore the time difference in execution of the instruction BNZ in the last cycle.

```
        COUNT1  EQU       D'250'
        REG10   EQU       0x10
        REG11   EQU       0x11


        CLRF    REG11       ;Set up REG11 for 256 execution
LOOP2:  MOVLW   COUNT1      ;Load decimal count in W
        MOVWF   REG10       ;Set up REG10 as a counter
LOOP1:  DECF    REG10, 1    ;Decrement REG10 - 1W/1C/4CLK
        BNZ     LOOP1       ;Go back to LOOP1 if REG10 ≠0
        DECF    REG11,1     ;Decrement REG11
        BNZ     LOOP2       ;Go back to load 250 in REG10 and start LOOP1 again
        END
```

Solution

The instructions in LOOP1 are the same as in Example 5.11, and it provides 75 µS time delay; let us call it T_{L1}. If we replace LOOP1 with its time delay $T_{L1} = 75\,µS$, as shown below, we will have one loop again, and we can calculate the total delay the same way we calculated before.

```
        CLRF    REG11             ;Set up REG11 for 256 execution
LOOP2:  MOVLW   D'250' - 4 Clk    ;Load decimal count in W
        MOVWF   REG10 – 4 Clk     ;Set up REG10 as counter
           ↓
        75 µS
           ↓
        DECF    REG11,1 – 4 Clk   ;Decrement REG11
        BNZ     LOOP2 - 8 /4 Clk  ;Go back to load 250 in REG10 and start
                                   LOOP1 again
```

1. The time required to execute LOOP2 once: $T_{L1} + T_{O2}$ (Delay in LOOP2 instructions)
   = 75 µS + (20 Cycles x 25 nS) = 75.5 µS

2. REG11 is used as a counter for LOOP2, and it begins with 00. The instruction DECF REG11 decrements the register to FF_H. Thus, the loop is repeated 256 times.
   Therefore, the total delay in LOOP2 including LOOP1
   = 75.5 µS x 256 = 19328 µS = 19.328 mS

5.5 ILLUSTRATIVE PROGRAM: GENERATING WAVEFORMS

In previous examples, we have seen that we can generate time delays by keeping the processor in a loop for a given period of time. Now we can generate square or rectangular waves by turning on and off a bit of an output port at a given delay. The following illustrative program generates a square wave of 10 kHz frequency.

5.5.1 Problem Statement

Write a program to generate a square wave of 10 kHz frequency by turning on and off Bit0 (RC0) of PORTC if the clock frequency is 40 MHz.

5.5.2 Problem Analysis

1. To generate a square wave, we are turning Bit0 of PORTC on and off; therefore, PORTC must be initialized as an output port as in Example 5.5.
2. To generate 10 kHz frequency, we need to calculate the time period; half of that time period will be on time and the other half will be off time. The time period of 10 kHz frequency is: $T = 1/f = 1/ (10 \times 10^3) = 100 \, \mu S$. Therefore, on-time should be $= 50 \, \mu S$ and off-time $= 50 \, \mu S$. To generate $50 \, \mu S$ we need one loop as shown in Example 5.11.
3. Time Delay Calculations:

    If we use the loop in Example 5.11 that has 12 cycles, we can calculate the count as follows:

    Time Delay T_L = Number of Instruction Cycles x Clock Period x Count

    $$50 \, \mu S = 12 \times 25 \, nS \times Count$$
    $$\text{Therefore, Count} = \frac{50 \times 10^{-6}}{12 \times 25 \times 10^{-9}} \cong 166$$

    The flowchart of this analysis is shown in Figure 5-7.

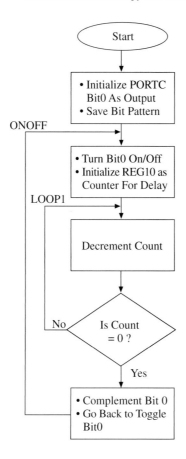

FIGURE 5-7 Flowchart for Illustrative Program (Section 5-5)
– Generating Square Wave

5.5.3 PROGRAM

```
Title "IP5-5 Generating 10 kHz Square Wave"
List p=18F452, f =inhx32
#include <p18F452.inc>        ;This is a header file and must be
                             ;included in the Source Program to
                             ;define SFRs and other parameters
REG1     EQU 0x01            ;Address of Data Register1
REG10    EQU 0x10            ;Address of Data Register10
COUNT    EQU D'166'
         ORG    0x20
START:   MOVLW  B'11111110'  ;Byte to set up Bit0 as an output
         MOVWF  TRISC        ;Initialize Bit0 as an output pin
         MOVWF  REG1         ;Save Bit pattern in REG1

ONOFF:   MOVFF  REG1,PORTC   ;(8 Clk) Turn on/off Bit0
         MOVLW  COUNT1       ;(4 Clk) Load decimal count in W
         MOVWF  REG10        ;(4 Clk)Set up REG10 as a counter
LOOP1:   DECF   REG10,1      ;(4 Clk) Decrement REG10
         BNZ    LOOP1        ;(8/4 Clk)Go back to LOOP1 if
                             ;REG 10 ≠ 0
         COMF   REG1,1       ;(4 Clk)Complement Bit Pattern
         BRA    ONOFF        ;(8 Clk)Go back to change LED
         END
```

5.5.4 Program Description

The above program can be assembled using MPLAB IDE editor. The header file provides the definitions of SFR registers such as PORTC and TRISC. The first three lines define the addresses of the data registers, and the directive ORG instructs the assembler to assemble the program at memory location 000020_H. The bit pattern 1111110 sets up only the Bit0 pin as an output; other bits are irrelevant in this problem. We need the same bit pattern to turn on/off Bit0, and data register REG1 is used to save that bit pattern. The instruction MOVLW loads decimal number 166 in the W register, and the instruction MOVWF sets up data register REG10 as a counter for the delay loop. The delay loop is the same as in Example 5.11 except for the count. After the delay loop, the instruction COMF complements (inverts) the bit pattern that sets Bit0 to 1. The BRA instruction takes the processor back to turn on Bit0.

5.5.5 Delay Recalculation

If you examine the program, you will notice that there are five instructions outside the delay loop LOOP1, and the execution time for those instructions must be accounted for in our delay calculation; otherwise, the On/Off time will be larger than 50 µS

$$\text{On/Off Time} = 50\,\mu S \;=\; (\text{Loop Instruction Cycles I}_{CLK} \times \text{CLK} \times \text{Count N}_{10})$$
$$+\; (1 \times \text{Cycles Outside Loop} \times \text{CLK})$$

$$= (12 \times 25\,\text{nS} \times \text{Count}) + (1 \times 28 \times 25\,\text{nS})$$

$$= (300\,\text{nS} \times \text{Count}) + 700\,\text{nSx}$$

$$\text{Count} = \frac{50\,\mu S - 0.7\,\mu S}{300\,\text{nS}} = \frac{49.3 \times 10^{-6}}{300 \times 10^{-9}} \approx 164$$

Therefore, the Count should be 164 rather than 166, which we calculated in our initial calculation. These calculations can be refined further by accounting for the lesser number of cycles needed for the last execution of the BNZ instruction in LOOP1.

5.5.6 Program Execution and Troubleshooting (Debugging)

The MPLAB IDE (Integrated Development Environment) includes all the resources such as Editor, Assembler, and Debugger to write, assemble, and debug a program as described in Chapter 4. It also includes a simulator program called MPLAB SIM as a debugging tool that can simulate the execution of a program on a computer. This simulator provides many techniques such as Run, Animate, Single-Step, Breakpoint, and Watch (Examine Registers). These techniques were described briefly in Chapter 4. Now we can use some of these techniques to execute and troubleshoot the Illustrative Program—Generating Waveforms.

To execute and debug a program, we need to know what we expect as the final output of a program and what the outputs should be at different stages of its flowchart. In our Illustrative Program-Generating Waveforms, we are looking for a square wave of 10 kHz frequency. We can observe the square wave only if we build the hardware and measure the waveform by connecting pin RC0 of PORTC to an oscilloscope.

On the simulator, we can see only the changes in register contents. In this program, we are turning RC0 on and off at a given interval; thus, we should be able to see 00 and 01 at some interval, which will be different from 50 μS because the computer executes the instructions in the simulator at a much slower rate than that of real hardware. To observe the output of PORTC, we need to set up MPLAB simulator as described in Chapter 4. Some of the steps of setting up MPLAB simulator are repeated here for easy reference.

How to Set Up MPLAB Simulator and Use Simulation

Before we set up MPLAB Simulator, we are assuming that MPLAB IDE is open, the Square Wave generation program has been written and assembled, and the Project is built successfully. Now we will set up MPLAB Simulator to execute and debug the program. To set up MPLAB Simulator:

Step 1: Click on Debugger → Select Tool → MPLAB Sim

Step 2: Click on Debugger → Settings. You can change the frequency of the simulator to match the frequency of the controller, but you will still not be able to measure the output.

Step 3: Click on View → Watch. A window is opened with two options: Add SFR and Add Symbol (as shown in Figure 5-8). Type WREG in the space shown next to Add SFR or use down arrow to find WREG and select it. Click on Add SFR, and WREG will appear under the symbol column with its address and value in the respective columns. Repeat the same process for adding TRISC and PORTC. Similarly, add REG1 and REG10 from the symbol option. Figure 5-8 shows a screen shot that includes the source program and five registers in Watch window.

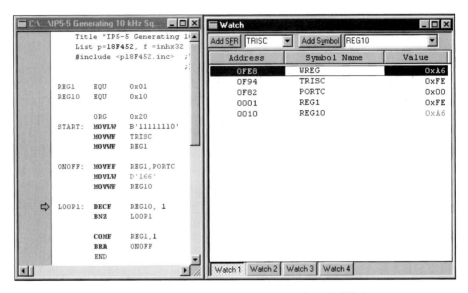

FIGURE 5-8   Screen Shot Displaying Source Program and Registers in Window

Step 4: Select the first instruction with the label START in the source program and place the cursor on that instruction. Right click on the label START and select the option Set PC on Cursor and the arrow appears in the margin indicating that the program execution begins at that instruction.

Step 5: Select Debugger → Animate and watch the contents of PORTC and REG10, which is set up as a counter. If PORTC displays 00 and 01 alternatively and REG10 begins at $A6_H$ (166_{10}) and goes to 0 and begins again at $A6_H$, your program is working as expected. If you do not observe these expected outputs, you need to use the single-step technique described next.

Step 6: The single-step technique allows you to execute one instruction at a time and observe the changes in the selected registers. The Source Program shown in Section 5.5.3 shows various segments of the program, and each segment can be examined using the single-step technique. To execute one instruction at a time, select Debugger and click on Step Into (or F7 on the keyboard) or select the icon for Step Into and execute instructions up to the label LOOP1. The following values should be displayed in registers to indicate appropriate initial values: WREG = A6 $_H$ (count =166), PORTC = 0, RGE1 = FE $_H$, and REG10 = A6 $_H$ (see Figure 5.8). As you continue to click on Step Into, the value in REG10 should be decremented. Now you can Step Out of the loop and continue the execution of the remaining instructions. If the Step Out function does not work, you can change the count by double clicking on the count in the value column and entering a smaller number. When the execution goes back to the label ONOFF, and when the instruction MOVFF is executed, the value in PORTC should change to 01. If you do not observe these results, you should make appropriate changes in the program and reassemble the program (Click on Project and Quick Build Project). This process should be continued until the expected output is observed.

Potential Sources of Errors

Several sources of logical errors exist in these types of programs as listed below.

1. Selecting an incorrect destination location between a data register and the W register for a MOV instruction. For example, the instruction DECF REG10, 1 stored the result after decrementing the count in REG10. If we change the instruction to:
   DECF REG10, 1 → DECF REG10, 0
   The count will be stored in the W register, and the count in the counter will always be between A6$_H$ and A5$_H$ (166$_{10}$ and 165$_{10}$). This is will set up an infinite loop, and the output will remain the same.
2. Selecting an incorrect data condition: If we write the instruction BZ (Branch on zero) instead of BNZ, the loop will be executed once, and the delay will be insignificant.
3. Selecting an incorrect flag: If we use the instructions BC (Branch on Carry) or BNC (Branch on No Carry), the execution of the delay loop will depend on the status of the carry flag, and we will get an erroneous result.
4. Specifying an incorrect branch location: If we write the label LOOP1 at the instruction MOVWF REG10, the counter register REG10 will be reinitialized at the count 166$_{10}$ in the execution of each cycle, thus setting up an infinite loop.

5.6 ILLUSTRATIVE PROGRAM: COPYING A BLOCK OF DATA FROM PROGRAM MEMORY TO DATA REGISTERS

When a source program is assembled, the data associated with the program is stored in program memory using the DB directive. For example, if we are adding ten data bytes, the bytes will be stored in the program memory. In PIC18 processors, to perform any operation such as addition, the bytes must be copied into data registers. In Example 5.3, we used TABLPTR as a (memory) pointer and TABLRD* instruction to copy one byte from program memory into the data register, and in Example 5.2, we used FSR (File Select Register) as a pointer for data registers and MOVFF instruction to copy the data byte from one data register to another data register. The following illustrative program uses the concepts from these two examples to copy a block of data from program memory into data registers (memory).

Applications

Copying blocks of data is a task that is frequently performed in computer operations. For example, we need to copy data between a storage device (disk) and system memory whenever we open or save a file. When we turn on a microcontroller-based embedded system, the system program generally checks the functioning of memory by loading test data into each register.

5.6.1 Problem Statement

Write a program to copy the following five data bytes stored in program memory with the starting location 000050_H called SOURCE to the data registers labeled as BUFFER with the beginning address 0010_H. When the copying process is complete, indicate the completion by turning on all LEDs at PORTC.

Data Bytes: $F6_H$, 67_H, $7F_H$, $A9_H$, and 72_H

5.6.2 Problem Analysis

As illustrated in Example 5.3, we need to set up one pointer (TBLPTR) to identify the beginning address of the data bytes, SOURCE, in the program memory and another pointer (FSR) to identify the beginning address of the data register BUFFER. A byte can be copied from program memory to a data register by using these pointers. This copying process needs to be repeated five times to copy five bytes; therefore, we need to set up a counter and a loop to count five repetitions, and indicate the end of the looping by checking the zero flag. To turn on LEDs, PORTC should be initialized as an output port. The flowchart of these steps is shown in Figure 5-9, and the source program is shown in the next section.

5.6.3 PROGRAM

```
        Title "IP5-6 Data Copy from Program Memory to Data Registers"
        List p=18F452, f =inhx32
        #include <p18F452.inc>;The header file

BUFFER      EQU   0x10      ;Define the beginning data register address
COUNTER     EQU   0x010     ;Set up register 01 as a counter

            ORG   0x000     ;Begin assembly at memory location 0000
            GOTO START0      ; Jump to location 0020H

            ORG   0x200     ;Begin Program assembly at 0020H
```

continued

```
START:      MOVLW       0x00            ;Byte to initialize port as an output port
            MOVWF       TRISC           ;Initialize PORTC as an output port
            MOVLW       0x05            ;Count for five bytes
            MOVWF       COUNTER         ;Set up counter
            LFSR        FSR0, BUFFER    ;Set up FSR0 as pointer for data registers
            MOVLW       UPPER SOURCE    ;Set up TBLPTR pointing to Source
                                        ;address at 000050H
            MOVWF       TBLPTRU
            MOVLW       HIGH ADDRESS
            MOVWF       TBLPTRH
            MOVLW       LOW ADDRESS
            MOVWF       TBLPTRL
NEXT:       TBLRD*+                     ;Copy byte in Table Latch and increment
                                        ;source pointer
            MOVF        TABLAT,0        ;Copy byte from Table Latch in W
            MOVWF       POSTINC0        ;Copy byte from W into data register
                                        ;BUFFER and increment FSR0
            DECF        COUNTER,1,0     ;Decrement count and store it in Counter
            BNZ         NEXT            ;Is copying complete? If not go back and
                                        ;copy next byte
            MOVLW       0xFF            ;Load completion indicator byte
            MOVWF       PORTC           ;Turn on all LEDs at PORTC
            ORG         0x50            ;Store data bytes starting from 0050H
SOURCE:     DB          0xF6,0x67,0x7F,0xA9,0x72
            END
```

5.6.4 Program Description

The statements in the comment column explain the logic behind each instruction, and the flow-chart in Figure 5-9 shows the sequence of the execution. You should examine the flowchart and instructions associated with each block of the flowchart. In addition, you should read the comments carefully to understand the logic and the sequence of the program. Block 1 is the initialization block; the instructions in this block set up pointers for program memory, data registers, and a counter and initialize PORTC as an output port. Block 2 is a process block whereby a byte pointed by the TBLPTR is copied from program memory into TABLAT and from TABLAT into the data register identified by the FSR0. Both pointers are also incremented for the next operation.

The next instruction decrements the counter, and the instruction in Block 4 checks whether the counter = 0. If it is not zero, the MPU changes the sequence of execution and goes back to the location NEXT which is a part of the loop (Figure 5-9), and then the loop is repeated until the counter is zero. This is a critical decision-making block. The flowchart shows the sequence of the execution is changed if the counter ≠ 0, and the decision is based on the Zero flag. The understanding of the condition that changes the direction of the execution and selecting the correct branch instruction is an important key in writing error-free assembly language programs (see the additional explanation below in Section: Potential Errors). Once the flag is set, the MPU goes to the next block of instructions that display FF_H at PORTC, indicating the completion of the copying process.

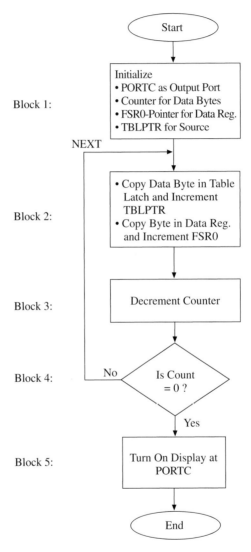

FIGURE 5-9   Flowchart for Illustrative Program: Copying Data From Program Memory to Data Registers (sections 5 and 6)

5.6.5 Program Execution and Troubleshooting (Debugging)

This program includes many important concepts such as pointers, indexing, and setting up a loop, and you should have a thorough understanding of these concepts and how they are to be analyzed through debugging techniques. The results of the execution of each instruction can be examined critically through single-stepping. You should single-step the entire program and watch the following registers as each instruction is being executed: TBLPTR, FSRL, COUNTER, TABLAT, and WREG. The steps are as follows:

- Set up the MPLAB Simulator as described in Section 5.5.6, Illustrative Program: Generating Waveforms. Single-step the instructions until the instruction TBLRD*+.
- The following values should be displayed in the registers:
  TBLPTR = 000050_H, FSRL = 10_H, TABLAT = 0, and COUNTER = 5
- Single-step the next four instructions: from TBLRD*+ to DECCF, the values in the above registers should be as follows:
  TBLPTR = 000051_H, FSRL = 11_H, TABLAT = $F6_H$, and COUNTER = 4
- The next critical instruction is the branch instruction: BNZ. When the BNZ instruction is executed, the program goes back to the location NEXT to execute TBLRD*+ again and places the next byte 67_H in TABLAT.
- In the execution of the next cycle, the MPU increments TBLPTR, FSRL, and decrements COUNTER. This loop is repeated five times, and when COUNTER = 0, the DECF (decrement file) sets the zero flag. Then the MPU skips the loop and goes to the next block of instructions to display FF_H.
- The final result—the contents of the data registers REG10–REG14—can be verified two ways: (1) displaying the contents of data (file) registers, and (2) using the simulator trace. To display the contents of data registers, select View and → File Registers. You can observe the contents of registers REG10–REG14. The second method—using the simulator trace—is described at the end of this section.

Potential Errors (bugs)

1. Selecting an incorrect flag: The flowchart in Figure 5-9 indicates that the decision to change the direction of the execution sequence is based on the status of the zero flag. If we use the instructions BC (Branch on Carry) or BNC (Branch on No Carry), the execution of the loop will depend on the status of the carry flag, and we will get an erroneous result.
2. Selecting an incorrect data condition: After deciding the zero flag in Step 1, the next question is: when should the MPU change the direction of the execution? Is it when the flag is zero or not zero? This is a critical question that is answered in the flow chart. The diamond symbol tells us that the direction of the execution is changed when the answer is "No". Therefore, the correct instruction is BNZ (Branch on No Zero). If we write the instruction BZ (Branch on zero) instead of BNZ, the instructions in the loop will be executed once, and only the first byte will be copied.
3. Selecting an incorrect location for a branch instruction: If we write the label NEXT at the instruction MOVWF TBLPTRL, one instruction above the present instruction, the MPU will copy byte $F6_H$ in the first execution, but in the next cycle, the MPU will copy $F6_H$ from W register into TBLPTR making the address in TBLPTR as $0000FF_H$. The default byte in program memory that is not being used is always FF_H. Therefore, the

MPU copies FF$_H$ from memory location 0000FF$_H$ into next data register and increments TBLPTR to 000100. At the end of the program, data registers 11$_H$ to 14$_H$ will have the byte FF$_H$.

If we write the label NEXT at the instruction MOVF TABLAT, the instruction TABLRD*+ will be executed once, and the pointer will not be incremented. Therefore, the MPU will copy the first byte F6$_H$ in all the data registers.

4. Selecting an incorrect instruction: If we use the instruction TBLRD* instead of TBLRD*+, the pointer will hold the address 000050$_H$ throughout the execution of the loop, and only the first byte F6$_H$ will be copied in all the registers.

5. Saving the result in an incorrect register. In this program, the instruction DECF COUNTER stores the result after decrementing the count in the data register COUNTER by setting the destination bit d = 1. If we were to write the instruction "DECF COUNTER, 0, 0", the MPU will store the count 4 in the W register instead of the COUNTER register. In the next cycle, the count will again decrement from 5 to 4, thus setting up an infinite loop.

6. Overlapping data storage with program instructions. In this program, we selected the starting location 000050$_H$ for data storage without knowing the length of the program. If we were to choose the location 000040$_H$, some of our last instructions would have been overwritten by the data. However, the assembler in MPLAB is designed to recognize such an overlap and produces an error message. Some other assemblers may not recognize such an overlap. The guaranteed error-free solution to this problem is not to define the location of the data by using the directive ORG 0x50, but to simply enter the data with the label SOURCE and the directive DB at the end of the program. The assembler will store the data bytes in locations immediately after assembling the program without any overlap.

Using Simulator Trace

After the execution of the program, we assumed that the bytes of the data block were copied in data registers 0010$_H$ to 0014$_H$, but we were not able to verify the contents of data registers using the Watch window. However, we can verify the contents of these data registers by examining the Simulator Trace as shown in Figure 5-10. To display the Simulator Trace left by the execution of this program, select View in MPLAB and click on Simulator Trace.

In Figure 5-10, the DA (Destination Address) and the DD (Destination Data) columns show the address of the data registers and their contents. You can verify the contents of all registers by examining the entire trace on the computer screen. In addition, you can observe the execution of each instruction and its results.

Trace

Addr	Op	Label	Instruction	SA	SD	DA	DD
0038	0009	NEXT	(Table Read)	----	--	----	--
0039	0000		(Table Read)	----	--	----	--
003A	50F5		MOVF 0xff5, W,	0FF5	67	W	67
003C	6EEE		MOVWF 0xfee, A	----	--	0011	67
003E	0601		DECF 0x1, F, A	0001	04	----	--
0040	E1FB		BNZ 0x38	----	--	----	--
0041	0000		NOP	----	--	----	--
0038	0009	NEXT	(Table Read)	----	--	----	--
0039	0000		(Table Read)	----	--	----	--
003A	50F5		MOVF 0xff5, W,	0FF5	7F	W	7F
003C	6EEE		MOVWF 0xfee, A	----	--	0012	7F
003E	0601		DECF 0x1, F, A	0001	03	----	--
0040	E1FB		BNZ 0x38	----	--	----	--
0041	0000		NOP	----	--	----	--
0038	0009	NEXT	(Table Read)	----	--	----	--
0039	0000		(Table Read)	----	--	----	--
003A	50F5		MOVF 0xff5, W,	0FF5	A9	W	A9
003C	6EEE		MOVWF 0xfee, A	----	--	0013	A9
003E	0601		DECF 0x1, F, A	0001	02	----	--
0040	E1FB		BNZ 0x38	----	--	----	--
0041	0000		NOP	----	--	----	--
0038	0009	NEXT	(Table Read)	----	--	----	--
0039	0000		(Table Read)	----	--	----	--
003A	50F5		MOVF 0xff5, W,	0FF5	72	W	72
003C	6EEE		MOVWF 0xfee, A	----	--	0014	72
003E	0601		DECF 0x1, F, A	0001	01	----	--

FIGURE 5-10 Simulator Trace: Execution of Illustrative Program Section 5.6 – Copying a Block of Data

5.7 ILLUSTRATIVE PROGRAM: ADDITION OF DATA BYTES

The Illustrative Program in Section 5.6 copies five data bytes from program memory into data registers. Now we can perform various arithmetic operations on these data bytes. For example, we can add all the bytes, find the average, or sort them. The following program illustrates the addition of these bytes.

Applications

A microprocessor (or a microcontroller) is a computing device; therefore, applications of arithmetic operations are obvious. The PIC18 instruction set has instructions such as add, subtract, 1's and 2's complement, increment, decrement, compare, and multiply. These are used in many mathematical operations in our daily lives such as: summing, averaging, multiplying and dividing numbers, unit conversions, searching for the largest or the smallest byte, and sorting numbers in any order.

5.7.1 Problem Statement

Write a program to add the following five data bytes that are stored in data registers from 0010_H to 0014_H. The sum is a 16-bit number that should be displayed at PORTB and PORTC. Data Bytes: $F6_H$, 67_H, $7F_H$, $A9_H$, and 72_H stored in data registers from 0010_H to 0014_H.

5.7.2 Problem Analysis

This is an addition problem similar to a data copy problem. The program is expected to add five bytes that are already stored in registers from 0010_H to 0014_H. Therefore, in the planning stage, we need to set up a pointer and a counter. The problem statement also suggests that an addition may generate a carry; therefore, we need a data register to keep track of carries. The flowchart in Figure 5-11 shows that the CYREG (Carry Register) is incremented if there is a carry after an addition. This flowchart includes two decision-making points: one when an addition generates a carry and the other when the counter becomes zero. When an addition does not generate a carry, the MPU skips the carry register and jumps forward. As the flowchart suggests, the appropriate instruction here is BNC. Similarly, the flowchart shows a decision-making point when the counter is decremented. If the counter is not zero, the MPU is redirected back to add the next byte, suggesting that the appropriate instruction is BNZ.

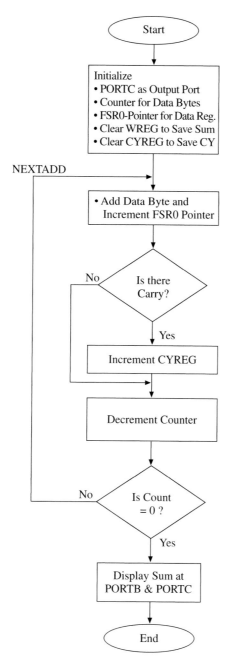

FIGURE 5-11 Flowchart for Illustrative Program: Addition of Data
Bytes (section 5)

5.7.3 PROGRAM

```
                Title "IP5-7Addition of Data Bytes"
                List p=18F452, f =inhx32
                #include <p18F452.inc>          ;The header file

BUFFER    EQU      0x10                 ;Define the beginning data register address
COUNTER   EQU      0x01                 ;Set up register 01 as a counter
CYREG     EQU      0x02                 ;Set up register 02 to save carries

          ORG      0x00
          GOTO     0x20

          ORG      0x20                 ;Begin program assembly at 0020_H
START:    MOVLW    0x00                 ;Byte to initialize port as an output port
          MOVWF    TRISB                ;Initialize Ports B &C as an
                                         output ports
          MOVWF    TRISC
          MOVLW    0x05                 ;Count for five bytes
          MOVWF    COUNTER              ;Set up counter
          CLRF     CYREG                ;Clear carry register
          LFSR     FSR0,BUFFER          ;Set up FSR0 as pointer for data registers
          MOVLW    0x00                 ;Clear W register to save sum
NEXTADD:  ADDWF    POSTINC0,W           ;Add byte and increment FSR0
          BNC      SKIP                 ;Check for carry - if no carry jump
                                         to SKIP
          INCF     CYREG                ;If there is carry, increment CYREG
SKIP:     DECF     COUNTER,1,0          ;Next count and save count in register
          BNZ      NEXTADD              ;If count ≠ 0, go back add next byte
          MOVWF    ORTC                 ;Display low-order sum at PORTC
          MOVFF    CYREG, PORTB         ;Display high-order byte at PORTB
          END
```

5.7.4 Program Description

The first six instructions initialize various registers, setting up PORTB as an output port, FSR0 as a pointer to the first data register 0010_H (labeled as BUFFER), the counter, and the carry register, and directives where to assemble the program. The addition begins at the location labeled NEXTADD. After the addition, the MPU checks for a carry, and it bypasses the instruction INCF CYREG if there is no carry. The next decision block checks to see if the counter is 0. If the counter is not zero, the MPU goes back to ADDNEXT to add the next byte. When the counter is zero, the low-order sum ($F7_H$) in the W register is displayed at PORTC and the high order number representing the carries (2_H) is displayed at PORTB. The I/O ports B and C together display the 16-bit number $02F7_H$.

5.7.5 Program Execution and Troubleshooting (Debugging)

Before you execute a program or begin troubleshooting, you should know what results to expect after the addition of the first two bytes and when an addition generates a carry. The final sum after adding these bytes is $2F7_H$ which indicates that the addition generates two carries. You can run this program as suggested in the following steps.

Step 1: To run the program using MPLAB SIM, set up the MPLAB Simulator as described in Section 5.5.6, and add PORTB and PORTC in the Watch Window. Select View and click on File Registers; the display will show File Registers as in Figure 5-12. Enter the data bytes in File Registers (data registers) from 0010_H to 0015_H as shown in Figure 5-12 and run the program (key F9). Halt (key F5) the program and examine the output at PORTB and PORTC. If the sum displayed is $02F7_H$, you have no errors in the program.

Step 2: Assuming that the sum displayed is different from $02F7_H$, you can troubleshoot the program by using the breakpoint technique or the single-step technique or the combination of both. To use the breakpoint technique, you need to select instructions that will provide critical information to assess whether the previous segment is working properly.

- Set three breakpoints as shown in Figure 5-12. To set a breakpoint, select the instruction you want to set the breakpoint, and double-click on the instruction and the letter "B" in red will appear in the margin.

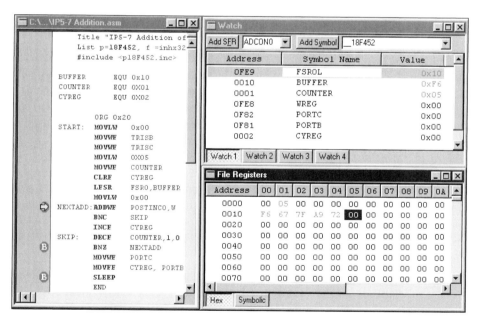

FIGURE 5-12 Screen Shot Displaying Breakpoints

- Add the following registers in the Watch Window: FSR0L, BUFFER, COUNTER, WREG, PORTC, PORTB, and CYREG.
- Select the first instruction at the label START and right click on the instruction to set PC at the cursor.

Step 3: Select Debugger and click on Run, or in the menu bar, click on the icon Run, or press key F9. The MPU executes the instructions until the first breakpoint and stops at the instruction ADDWF POSTINC0, W. The contents in the registers of the Watch Window should be:

FSROL = 10_H, BUFFER = $F6_H$, COUNTER = 05_H, WREG = 0, and CYREG = 0 (see Figure 5-12)

This observation should verify that: the data register pointer is pointing to the first data register 0010_H, the first data byte $F6_H$ is in BUFFER, and all other registers are initialized correctly.

Step 4: Click on Run (or F9) again. The MPU executes the program until the second breakpoint and the registers in the Watch Window should display the following:

FSROL = 11_H, COUNTER = 04_H, WREG = $F6_H$, and CYREG = 0

This observation indicates that the first byte $F6_H$ is added to the WREG (which was cleared), the FSR is incremented, and the counter is decremented, and there is no carry in the first addition.

Step 5: Click on Run (or F9) again. The MPU repeats the execution of the loop and stops again at the breakpoint, and the registers in the Watch Window should display:

FSROL = 12_H, COUNTER = 03_H, WREG = $5D_H$, and CYREG = 1

This observation verifies that the sum of the first two bytes is $15D_H$: the low-order byte is in the WREG and the carry flag is set. This is the correct sum. These readings confirm that the second decision point—BNC SKIP—is working correctly.

Step 6: To confirm the final result, you have two options: (1) run the loop until the zero flag is set, and when it falls out of the loop, run the program until the third breakpoint and observe the output at PORTB and PORTC; (2) alternatively, remove the second breakpoint by double-clicking on the breakpoint instruction and run the program until the third breakpoint and observe the outputs at PORTB and PORTC. The primary reason to set the third breakpoint is to stop the execution at the end; otherwise, the program keeps running, and you need to stop the program by using the Halt command (click either on icon or press F5). You can also observe the execution of each instruction by selecting Simulator Trace in View.

5.8 ILLUSTRATIVE PROGRAM: SEARCHING FOR SPECIFIC CHARACTERS IN A DATA SET

The following program searches for a specific character in a given data set. The end of the data set is indicated by a null character such as 00.

Applications

The technique of searching for characters and counting them has many practical applications. For example, to find the number of words in a file, we can search for space (ASCII 20_H) and carriage returns ($0D_H$) in the file. Similarly, with Compare and Skip instructions, we can arrange a data string in ascending or descending order.

5.8.1 Problem Statement

Write a program to search for a character byte 20_H in a string stored in data registers beginning at BUFFER0. The end of the data string is indicated by the character 00. When the character byte is found, display at PORTB and PORTC the 12-bit address of the data register where the character is found. If the character is not found in the data string, display all 1's at PORTC.

5.8.2 Problem Analysis

In the previous Illustrative Programs (Section 5.7) we knew the size of the data and we could set a counter to count the number of operations. In this problem, the size of the character string is variable. The MPU will know the end of the data search when it encounters 00 in the data set. Therefore, the program should check at the beginning whether the character is 00, and when the MPU finds 00, it is the end of the search. When a character is not zero, the program should check whether it is 20_H and continue to check until it finds 20_H and then display its address. Otherwise, it should display all 1's at PORTC. Figure 5-13 shows the flowchart of the program. It has two decision points as well as two end points.

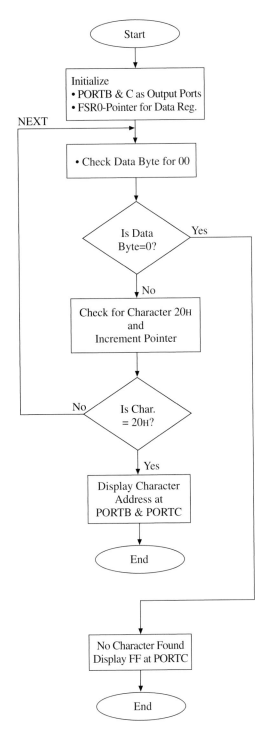

FIGURE 5-13   Flowchart for Illustrative Program: Searching for Specific Characters (Section 5.8)

5.8.3 PROGRAM

```
            Title "IP 5-8 - Searching for Character"
            List p=18F452, f =inhx32
            #include <p18F452.inc>                    ;The header file

    BUFFER0  EQU      0x10                    ;Define the data register addresses
    BUFFER1  EQU      0x11                    ;where data string is stored
    BUFFER2  EQU      0x12
    BUFFER3  EQU      0x13
    BUFFER4  EQU      0x14

             ORG 00
             GOTO     START

             ORG      0x20                    ;Begin program assembly at 0020H
    START:   MOVLW    0x00                    ;Byte to initialize port as an output
             MOVWF    TRISB                   ;Initialize PORTB as an output
             MOVWF    TRISC                   ;Initialize PORTC as an output
             LFSR     FSR0,BUFFER0            ;Initialize FSR0 as a pointer
    NEXT:    MOVLW    0x00                    ;Test byte for end of string
             CPFSGT   INDF0                   ;Is this end of string?
             BRA      STOP                    ;Go to display FFH
             MOVLW    0x20                    ;Test byte to check
             CPFSEQ   POSTINC0,W              ;Check if this 20H and increment pointer
             BRA      NEXT                    ;If this is not 20H, go back and check
                                              ;next byte
             DECF     FSR0L                   ;Decrement pointer, which is one address
                                              ;ahead
             MOVFF    FSR0L, PORTC            ;Display register address at PORTC
             MOVFF    FSR0H, PORTB            ;Display register address at PORTB
             SLEEP                            ;End
    STOP:    MOVLW    0xB'11111111'           ;No test byte in string, display all 1's
             MOVWF    PORTC
             SLEEP
             END                              ;End of assembly
```

5.8.4 Program Description

The data string has only five characters stored in registers from 10_H to 14_H (BUFFER0 to BUFFER4), and the last character is 00 stored in BUFFER4. The labels BUFFER1 to BUFFER4 are included in the source program to load the data string during the simulation (see Section 5.8.6); otherwise, these assignments are unnecessary for the execution of the program. The first four instructions initialize ports B and C as output ports and FSR as the pointer where the character string begins. The instruction CPFSGT (Compare File register with WREG, and if greater, skip the next instruction) checks if the character is greater than 00, and if it is greater than zero, the MPU skips the next BRA STOP (Branch to Stop) instruction. If it is 00, the MPU executes the BRA instruction and jumps to display 1's at PORTC. The next instruction CPFSEQ (Compare File register with WREG, and if equal, skip the next instruction) checks whether the character byte is equal to 20_H, and if it is 20_H, the MPU skips the next BRA NEXT (Branch to Next) instruction, decrements the pointer, and displays the data register address. If the character byte is not equal to 20_H, the MPU goes back to check the next character. This program does not check whether there are any additional 20_H characters in the string, and will not work properly if the addresses of data registers cross the boundary FF_H because the instruction DECF FSR0L decrements only the low-order address of the pointer.

5.8.5 Program Execution and Troubleshooting

This program has two decision-making points, and both are implemented with Skip and Branch instructions. If the condition is true, the MPU skips the branch instruction and executes the following instructions. Therefore, the program should be tested when (1) character if $\neq 20_H$, (2) character = 20_H, and (3) character = 0, which allows us to verify the three segments of the program. The addresses of additional data registers BUFFER1 to BUFFER4, defined in the program, ranges from 0011_H to 0014_H. These are the addresses of data registers where the character string can be stored. The reasons to include the labels of these registers are as follows: (1) any data string can be loaded into these registers for simulation, (2) data registers can be examined in the Watch window, and (3) various segments of the program can be properly tested. An alternative to defining the registers BUFFER1 to BUFFER4 in the source program is to select the option File Registers in the View menu and display all the file registers. You can manually enter the data in registers 0011_H to 0014_H. To test the various segments of the program, perform the following steps.

Step 1: The program has two termination points indicated by the instruction SLEEP at two locations. Set two breakpoints at the instructions SLEEP and add the registers in the Watch window as shown in Figure 5-14a.

Step 2: Insert the following data string in the value column of the Watch window:

$7A_H$, $6F_H$, 20_H, 34_H, and 00 in BUFFER0 to BUFFER4 data registers. This data string can be inserted by double-clicking on the value column. In this string, 20_H is in register 0012_H (Figure 5-14a).

Step 3: Reset (F6) the MPU. Place the cursor on the START label and right-click to select PC at the Cursor option (Figure 5-14a).

Step 4: Run (F9) the program. When the program is executed, the MPU checks the first two bytes by going back to the location NEXT twice, and in third cycle, it finds 20_H and skips the BRA NEXT instruction and displays the register address 0012 at PORTB and PORTC as shown in Figure 5-14b. It then stops at the first breakpoint. If you do not observe this expected output, you should single-step the program to check whether you have a wrong combination of Compare and Skip and Branch instructions.

Step 5: To execute the remaining segment of the program, delete 20_H from the string by inserting some other byte, such as 97_H, as shown in Figure 5-14c.

Step 6: Run (F9) the program. When the MPU finds 00, it executes the instruction BRA STOP, skips the first breakpoint, and jumps to display all 1's at PORTC as shown in Figure 5-14c, and stops at the second breakpoint.

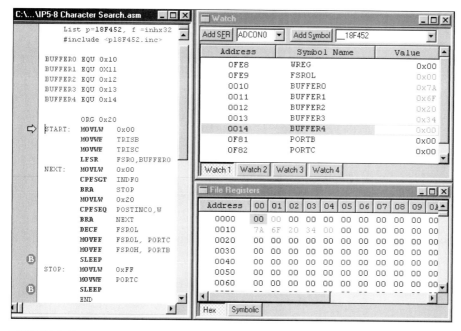

FIGURE 5-14 (a) Setting Up for Program Execution: Searching for Specific Characters in Data set (Section 5.8)

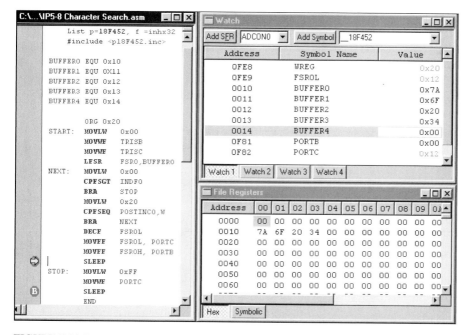

FIGURE 5-14 (b) Program Execution when Characters in Data Set (Section 5.8)

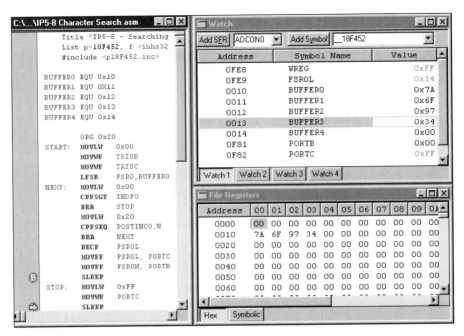

FIGURE 5-14 (c)   Program Execution when Character Is Not in Data Set (Section 5.8)

SUMMARY

In this chapter, we selected the following four types of computing operations and related instruction groups: data copy (MOVE), table read/write, arithmetic, and redirection of program execution (branch and skip). Most of these instructions are one word wide and executed in a one-instruction cycle with four clock periods. Some instructions, such as branch and skip, require two instruction cycles for execution. The important features of these instructions are as follows:

1. The MOVE instructions copy the contents of the source register into the destination register without modifying the contents of the source. In general, the MOVE instructions do not affect the flags (with few exceptions).

2. The table read/write instructions (TBLRD/WT) are designed to copy data between the program memory and data registers. However, the address bus and data bus of the program memory and data memory (registers) differ in size. The program memory has a 21-bit address bus and a 16-bit data bus, and the data memory (registers) has 12-bit address bus and 8-bit data bus. Therefore, TBLPTR (Table Pointer) is used as a pointer for program memory and FSR (File Select Register) is used as a pointer for data registers. To resolve the differences between the size of data (word versus byte), the instructions such as TBLRD/WT are used to copy bytes in the intermediary 8-bit register TABLAT (Table Latch). These instructions do not affect flags.

3. The arithmetic instructions (Add and Subtract) can operate on an 8-bit number (literal) as well as contents of a data register in conjunction with W register (WREG). However, the result of the operation can be saved either in the W register or data register. This is a very powerful feature in the PIC18

microcontroller family that enables data registers to function almost as accumulators. The arithmetic instructions affect all flags. The result of the operation determines the status (set/reset) of each flag.

4. The increment/decrement and 2's complement instructions (also classified as arithmetic instructions) operate on any data register and save the result in either the W register or data register. These instructions also affect all flags.

5. The program redirection instructions (branch) are classified in two groups, unconditional and conditional, and each group is further classified into two addressing modes: absolute branch and relative branch. The conditional branch instructions are based on four flags: sign (S), overflow (OV), zero (Z), and carry (C). There are two instructions associated with each flag, one when the flag is set and the other when the flag is reset.

   The absolute branch instructions include a complete address in the instruction where the MPU should branch, and the relative branch instructions include a signed 8-bit displacement indicating the number of locations by which the MPU should jump with respect to the program counter.

6. The skip instructions are also used in conjunction with unconditional branch instructions to redirect the program execution.

7. To write programs, this chapter focuses on three concepts: (1) use of pointers, (2) indexing, and (3) looping. A pointer is a register that holds an address, and the MPU uses that address to identify that memory location. Indexing is the process of pointing by incrementing or decrementing the addresses in the pointer, and looping is setting up a condition that enables the MPU to repeat a group of instructions.

   The PIC18 microcontroller instruction set has two pointers: TBLPTR (Table Pointer) and FSR (File Select Register). The TBLPTR, made up of three registers TBLPTRU, TBLPTRH, and TBLPTRL, is used as a pointer to hold the 21-bit address of program memory; and the FSR (File Select Register), made up of two registers FSRH and FSRL, is used as a pointer for data registers. The PIC18 includes three FSRs (FSR0, FSR1, and FSR2) that are used as pointers. All of these pointers are also associated with indexing registers that can be incremented and decremented. The conditional branch instructions are used to set up loops based on the status of the flags, and the unconditional branch instructions are used to set up continuous loops and with skip instructions.

8. The chapter includes four programs that demonstrate applications of instructions from the above four categories and the concepts of pointers, indexing, and looping.

QUESTIONS AND ASSIGNMENTS

Data Copy Operations

5.1 Write instructions to load $F8_H$ in data register REG0 (address 00), $3A_H$ in REG1 (address 01_H), and 62_H in REG2 (address 02_H).

5.2 Write instructions to load the byte 00 in WREG. Does the instruction set the Z flag? Copy the byte in REG1 and set the Z flag.

5.3 Write instructions to load 97_H in REG10 (address 10_H) and set the N flag.

5.4 Write instructions to copy bytes in Q.5.1 from REG0, REG1, and REG2 to REG10, REG11, and REG12 (with addresses 10_H, 11_H, and 12_H) respectively.

5.5   Write instructions to copy bytes in Q.5.1 from REG0 into REG2, REG1 into REG3 (03 $_H$), and REG2 into REG4 (04 $_H$).

5.6   Write instructions to set up I/O lines of PORTB: RB0–RB3 as input lines and RB4–RB7 as output lines.

5.7   Write instructions to set up PORTB as an input port and PORTC as an output port.

5.8   Write instructions to set up PORTC as an output port and display 55 $_H$ at PORTC.

Arithmetic Operations

5.9   Write instructions to add two unsigned numbers $3F_H$ and 47_H and save the result in data register REG3 (address 03 $_H$). Identify the status of Z and C flags.

5.10   In Q.5.9, what is the status of N and OV flags. Explain the significance of these flags to the sum.

5.11   If we assume that the numbers in Q.5.9 are signed numbers, explain the result and the significance of OV flag.

5.12   Write instructions to add two numbers 8A $_H$ and 76 $_H$. Calculate the result (manually) and identify the status of N, OV, Z , DC, and C flags.

5.13   In Q.5.12, explain the result and the significance of Z and C flags if data bytes were signed numbers.

5.14   In Q.5.12, explain the result and the significance of Z and C flags if data bytes were unsigned numbers.

5.15   Write instructions to subtract the byte $7F_H$ from the byte 25_H. Show your calculations using 2's complement method and identify the status of the flags.

5.16   Write instructions to add two bytes: 78_H and $F2_H$ and identify the status of all the flags. Explain the result and the significance of OV and C flags if the bytes were unsigned.

5.17   In Q.5.16, explain the result and the significance of OV and C flags for signed numbers.

5.18   Write instructions to subtract 72 $_H$ from the result of Q.5.16 and identify the status of N, OV, Z, and C flags.

5.19   Add the following three 16-bit numbers stored in data registers from 0050_H to 0055_H: $12F5_H$, $273A_H$, $5A23_H$, and store the sum in data registers 0056_H and 0057_H.

5.20   Add the following two 24-bit numbers stored in data registers 0020_H to 0025_H, and store the sum in data registers 0030_H, 31_H, and 32_H: $158FA2_H$, 249264_H.

5.21   Identify the contents of WREG and REG5 and the status of the flags as the following instructions are executed by PIC MCU.

		REG5	W	N	OV	Z	C
SETF	REG5, 0						
INCF	REG5, 1						
COMF	REG5, 0						
DECF	REG5, 1						

Branch Operations

5.22   How many times is LOOP1 executed in the following instructions?

	MOVLW	0x32	;Load count 50 10 in W
	MOVWF	REG2, 0	;Copy count in REG2
LOOP1:	DECF	REG2, 1 - 4 Cyl	;Decrement REG2 and save in REG2
	BNZ	LOOP1 - 8 / 4 Cyl	;If REG2 ≠ 0, go back to decrement

5.23   How many times is LOOP1 executed in Q.5.22 if the instruction BNZ (Branch on No Zero) is changed to BZ (Branch on Zero)?

5.24   In Q.5.22, how many times is LOOP1 executed if the instruction BNZ is changed to BNC (Branch on No Carry)?

5.25   In Q.5.22, how many times is LOOP1 executed if the instruction DECF REG2, 1 is changed to DECF REG2, 0?

5.26   The Hex code for the instruction BZ (Branch on Zero) is stored in memory locations 0020_H and 0021_H as: E0 12. Calculate the address where the program is redirected when Z flag is set.

5.27   The Hex code for the instruction BNC (Branch on No Carry) is stored in memory locations 0052_H and 0053_H as: E3 F6. Calculate the address where the program is redirected when C flag is reset.

5.28   The code for the BRA instruction is written in locations 0030_H and 0031_H as D0 FA. Is this a forward or backward jump and by how many words? Hint: BRA instruction can redirect the MPU from −1024 to +1023 (2 K) memory locations.

5.29   Calculate the time delay in LOOP1 (Q.5.22) if the MCU clock frequency is 10 MHz by ignoring the difference in the execution of the last cycle of the BNZ instruction.

5.30   In Q.5.29, how much delay will be reduced if you account for the difference in the execution of the last cycle of the BNZ instruction.

5.31   In Q.5.22, calculate the Count (in place of 32_H) to obtain the delay of 150 μS.

5.32   Identify the error/s in the following delay loop.

	MOVWF	REG10	;Set up REG10 as a counter
LOOP1:	DECF	REG10,0	;Decrement REG10
	BNZ	LOOP1	;Go back to LOOP1 if REG 10 ≠ 0

5.33   Identify the error(s) in the following delay loop.

	MOVLW	0x64	;Count = 100 decimal
LOOP1:	MOVWF	REG10	;Set up REG10 as a counter
	DECF	REG10,1	;Decrement REG10
	BNZ	LOOP1	;Go back to LOOP1 if REG 10 ≠ 0

5.34   Identify the error(s) in the following delay loop.

	MOVLW	32H	;Count = 32 decimal
	MOVWF	REG10	;Set up REG10 as a counter
LOOP1:	DECF	REG10,1	;Decrement REG10
	BZ	LOOP1	;Go back to LOOP1 if REG 10 ≠ 0

5.35   The following program is similar to the Illustrative Program—Generating Wave forms (Section 5.5). However, it generates the square wave without using the instruction MOVFF; it uses WREG to save the bit pattern. Find and correct the errors using the MPLAB SIM and without using the instruction MOVFF (appropriate header instructions should be added before using MPLAB—see the Simulation Exercise 5.1SE).

REG1	EQU	0x01	;Address of Data Register1
REG10	EQU	0X10	;Address of Data Register10
START:	MOVLW	B'11111110'	;Byte to set up Bit0 as an ;output
	MOVWF	TRISB	;Initialize Bit0 as an output pin
	MOVWF	REG1	;Save Bit pattern in REG1
ONOFF:	MOVWF	PORTB	;Turn on/off Bit0
	MOVLW	D'166'	;Load decimal count in W
LOOP1	MOVWF	REG10	;Set up REG10 as a counter
	DECF	REG10,1	;Decrement REG10
	BNZ	LOOP1	;Go back to LOOP1 if ;REG 10 ≠ 0
	COMF	REG1,0	;Complement Bit Pattern & ;save in W
	BRA	ONOFF	;Go back to change LED
	END		

5.36   Calculate the total delay in LOOP2 including LOOP1 if the clock frequency = 10 MHz. Ignore the difference in the execution of the last cycle of BNZ instruction.

	MOVLW	D'60'	
	MOVWF	REG11	;Set up REG11 for 256 execution

```
          MOVLW     D'60'

LOOP2:    MOVLW     0x64        ;Load 64H WREG

          MOVWF     REG10       ;Set up REG10 as a counter

LOOP1:    DECF      REG10, 1    ;Decrement REG10 - 1W/1C/4CLK

          NOP                   ;Increase delay by 4 CLK

          BNZ       LOOP1       ;Go back to LOOP1 if REG 0 if
                                ;REG10 ≠ 0

          DECF      REG11,1     ;Decrement REG11

          NOP                   ;Increase delay

          NOP

          BNZ       LOOP2       ;Go back to load REG10 and start
                                ;LOOP1 again

          END
```

5.37   Recalculate the delay in Q.5.36 by considering that the execution of the last cycle of the BNZ instruction takes only 4 cycles instead of 8 cycles.

5.38   Write a program to generate a rectangular wave of 200 if REG10 µS on-time and 350 µS off-time if the clock frequency is 10 MHz. Use Bit7 of PORTC to generate the waveform.

5.39   Write a program to copy the following seven data bytes from program memory to data registers starting from REG26 (0x26) in the reverse order.

Data Bytes (H): 72, F2, 82, 68, 49, 7F, 9C

5.40   Write a program to copy only the positive numbers from the data string given in Q.5.39 in data registers starting from RGE30 (0x30).

5.41   Write a program to add the positive numbers that are copied to, starting from REG30 in Q.5.40 and display the result at PORTB and PORTC.

5.42   Write a program to copy the following data string, that is terminated by the ASCII character 'carriage return (0DH)', from program memory to data registers starting from REG10 (0x10).

Data Bytes (H): 78, F7, 5A, 68, 4F, 7F, 35, 9C, 5A, A7, 0D

5.43   In Q.5.42, 10 data bytes are copied from REG10 to REG19. Shift the entire data string to register starting from REG15 (0x15).

5.44   After copying ten data bytes as in Q.5.42, write a program to add the positive numbers and display the sum at PORTB and PORTC.

5.45   Write a program to add the positive numbers of the data string of 10 bytes in Q.5.42, count the numbers that are added, and display the sum at PORTB and PORTC and the count at PORTA.

5.46   Write a program to search for ASCII characters "Z" in the data string of 10 bytes copied in Q.5.44, count the number of "Z" characters, and display the count at PORTB.

SIMULATION EXERCISES

5.1 SE   Assemble and run the program listed in Q.5.35 using MPLAB SIM. If the program does not run as expected, debug it using the break point and single-step techniques.

5.2 SE   Assemble and execute the program Q.5.39 and verify your answer.

5.3 SE   Assemble and execute the program Q.5.40 and verify your answer.

5.4 SE   Assemble and execute the programs Q.5.43 and Q.5.44 and verify your results.

Using PIC18 IDE

5.5 SE   Build the system using PIC18F 452 microcontroller, Program Memory Editor, and the 8 x LED board. Connect the LED board to PORTC. Write and assemble instructions to set up PORTC as an output port and display 55H. This assignment is the same as 5.8. Execute the program to turn on the alternated LEDs of the LED board.

5.6 SE   In Assignment Q.5.5SE, remove the 8 x LED board and connect the Oscilloscope. Set up Bit0 of PORTB as an output. Assemble and debug the program in Q.5.35 and execute the program and observe the square wave on the scope.

CHAPTER 6

INTRODUCTION TO LOGIC, BIT MANIPULATION, AND MULTIPLY-DIVIDE OPERATIONS

OVERVIEW

In Chapter 3 we classified the instruction set of PIC18 microcontroller into seven groups, and in Chapter 5 we introduced commonly used instructions from four groups: data copy, table read/write, arithmetic, and branch along with their applications. This chapter focuses on the remaining groups, instructions related to logic and bit operations and their applications. One of the applications includes how to find the highest reading in a given data set. In addition, we have also discussed multiply and divide operations and illustrated an example of finding the average of a given data set.

OBJECTIVES

- List the format of three types of logic instructions: AND, IOR (inclusive OR), and XOR (exclusive OR) and explain the results and flag status after the execution of these instructions.
- Explain how logic instructions are used to set, reset (mask), and toggle bits in a given data register.
- List the format of bit set, reset, and toggle instructions and explain their operations.
- List the format of rotate instructions and explain the difference between 8-bit and 9-bit rotations.
- Explain how rotate instructions are used to perform multiply and divide operations and demonstrate these operations for unsigned numbers larger than 8 bits.
- List the format of multiply instructions and explain their operations.
- Explain and demonstrate the division of unsigned numbers using algorithms.
- Write programs to demonstrate how to find: (1) the highest and the lowest reading in a data set and (2) the average of a set of data bytes.

6.1 LOGIC OPERATIONS

The PIC18 MPU performs all the logic functions that are generally performed using logic gates: AND, OR, XOR (Exclusive OR), and NOT (Complement). However, the MPU performs these functions at the register level; a logic instruction performs the logic function equivalent of eight gates. In addition, these instructions are used at the bit level to set, reset, and toggle a bit. The instructions related to these logic functions (except NOT function), and their applications are described in the following sections. The complement function is discussed in Chapter 5 in the context of the "2's complement" arithmetic operation.

6.1.1 Logic Instructions: AND, Inclusive OR (IOR), and Exclusive OR (XOR)

As mentioned in Chapter 3, the PIC18 instruction set includes three types of logic instructions: AND, IOR (Inclusive OR), and XOR (Exclusive OR). Each instruction includes two types of operands: (1) any 8-bit logically operated with the WREG saving the result in the WREG, and (2) any data (file) register operated with the WREG saving the result either in the WREG or data register.

The logic instructions function similar to two input gates, but these instructions operate at the register level. Figure 6–1 shows the AND operations of two registers: WREG and the data register REG10. The WREG holds the byte 77_H and REG10 holds the byte 81_H. Each bit of the

WREG is ANDed with the corresponding bit of REG10, and the result 01_H is saved in WREG. The operation is equivalent to two-input eight AND gates. The three types of the logic instructions are listed in the following section with examples.

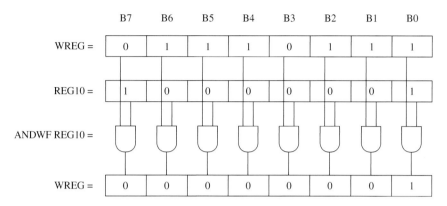

FIGURE 6-1 AND Operation of Two Registers

Instructions

Format	Example	Description
ANDLW 8-bit	ANDLW 0x9A	;Logically AND 9AH with W and save the ; result in W
IORLW 8-bit	IORLW 0x3F	;Logically OR 3FH with W and save the ; result in W
XORLW 8-bit	XORLW 0x78	;Logically exclusive OR 78H with W and ; save the result in W
ANDWF F, d, a	ANDWF REG1, 0, 0	;Logically AND REG1 and W and save the ; result in W because d = 0, and REG1 is ; from the Access bank because a = 0
IORWF F, d, a	IORWF REG1, 1, 0	;Logically OR REG1 and W and save the ; result in REG1 because d = 1, and REG1 is ; from the Access bank because a = 0
	IORWF REG1, W, 0	;This is the same instruction as above except ; in a different format
XORWF F, d, a	XORWF REG2, 1, 1	;Exclusive OR REG2 and W and save the ; result in REG2 because d = 1, and REG2 is ; specified by the contents of BSR

Points to Remember

1. Each bit <7-0> of the W register is logically operated with the corresponding bit of the operand. The operation is the equivalent of 8 logic gates.
2. When the operand is a data register, the result can be saved either in the W register or the data register by selecting the value of "d" parameter.
3. These instructions affect only N and Z flags.

Applications

1. The AND instructions are used to perform AND logic when multiple conditions need to be valid. For example, a car does not start unless a key is in the ignition and the gear shift is in the park position. Both of those conditions must be met before one can proceed. At the bit level, these AND instructions are used to reset bits in a given register (see Example 6.2) because when ANDing a bit with 0, the result is always 0 and when ANDing with 1, the source bit is unchanged.
2. The OR instructions are used to perform OR logic functions. For example, a two-way switch can turn on a light from either switch. At the bit level, these instructions are used to set a bit (see Example 6.3) because when ORing a bit with 1, the result is 1 and when ORing with 0, the source bit is unchanged.
3. The XOR instructions are used to perform exclusive OR (either OR but not both) functions. For example, an air conditioner and a heater should never be on at the same time in a given house. The XOR instructions prevent conflicts of this nature. At the bit level, these XOR instructions are used to toggle bits. When XORing a bit with 1, the source bit is complemented (inverted) and when XORing with 0, the bit is unchanged.
4. The PIC18 instruction set includes special bit set-reset-toggle instructions, but logic instructions will be more efficient to set-reset-toggle multiple bits than the special bit instructions.

EXAMPLE 6.1

Write instructions to AND 38_H with the W register that holds the byte 87_H and identify the status of the flags.

Solution:

Instruction	ANDLW	0x38

$$\begin{array}{lll} \text{WREG} & = & 1\,0\,0\,0\,0\,1\,1\,1 = 87_H \\ & \text{AND} & \\ \text{8-bit} & = & 0\,0\,1\,1\,1\,0\,0\,0 = 38_H \\ \text{Literal} & & \overline{} \\ \text{WREG} & = & 0\,0\,0\,0\,0\,0\,0\,0 = 0\,0 \qquad \text{Flag Status: N = 0, Z = 1} \end{array}$$

EXAMPLE 6.2

The W register holds a packed BCD number 68_H. Write instructions to mask high-order four bits <7-4>, preserve the logic levels of bits <3-0>, and save the result in data register REG1 with the address 01_H.

Solution:

The concept of masking a bit is similar to resetting a bit, and a bit can be reset by ANDing the bit with 0. In this problem, we need to reset bits <7-4> without affecting bits <3-0> for any byte (not necessarily just 68_H). The logic level of a bit can be preserved by ANDing the bit with 1. Therefore, the instructions are as follows:

```
ANDLW    B'00001111'    ;Masking bits to reset <7-4> bits and preserve bits <3-0>
MOVWF    REG1           ;Save result in REG1
```

Result:

After ANDing the byte 68_H with $0F_H$, the result is 08_H, which is saved in REG1.

$$
\begin{array}{ll}
\text{WREG} = & 0\ 1\ 1\ 0\ 1\ 0\ 0\ 0 = 68_H \\
\quad\quad\text{AND} & \\
\text{Masking} = & 0\ 0\ 0\ 0\ 1\ 1\ 1\ 1 = 0F_H \\
\text{Byte} & \text{--------------------------} \\
\text{WREG} = & 0\ 0\ 0\ 0\ 1\ 0\ 0\ 0 = 08_H
\end{array}
$$

EXAMPLE 6.3

The W register holds the byte 82_H. Write instructions to set bits <1-0>, and toggle the bits <7-6>. This simulates an example in which the output port is connected to various devices, and we want to turn on the devices connected to bits <1-0> and change the operations of the devices connected to bits <7-6>.

Solution:

A bit can be set by using the OR function and can be toggled by using the XOR function, keeping the remaining bits at logic 0. Therefore, the instructions are as follows:

```
IORLW    B'00000011'    ;Binary number to set bits <1-0> bits
XORLW    B'11000000'    ;Binary number to toggle bits <7-6>
```

continued

Result:

$$
\begin{array}{ll}
\text{WREG} = & 1\,0\,0\,0\,0\,0\,1\,0 = 82_H \\
& \text{IOR} \\
\text{Setting} = & 0\,0\,0\,0\,0\,0\,1\,1 = 03_H \\
\text{Byte} & \rule{3cm}{0.4pt} \\
\text{WREG} = & 1\,0\,0\,0\,0\,0\,1\,1 = 83_H \\
& \text{XOR} \\
\text{Toggling} = & 1\,1\,0\,0\,0\,0\,0\,0 = C0_H \\
\text{Byte} & \rule{3cm}{0.4pt} \\
\text{WREG} = & 0\,1\,0\,0\,0\,0\,1\,1 = 43_H
\end{array}
$$

Description: The WREG holds the byte 82_H of which Bit1 is already set. By inclusive ORing the byte with 03_H, both bits <1-0> are set (83_H). The next instruction XORLW $C0_H$ toggles the bits <7-6>. Bit7 and Bit6 were 1 0, and after the instruction, they are 0 1.

6.2 BIT OPERATIONS

Microcontrollers tend to focus on bit level I/O operations rather than large computational applications. In PIC18 microcontroller, these operations can be divided into three types of instructions: (1) set, reset, and toggle a bit, (2) test the logic level of a bit, and (3) rotate bits in a register. These instructions and their applications are described in the following sections.

6.2.1 Bit Set, Clear, and Toggle Instructions

The instruction set includes three instructions that can set, reset, or toggle any bit in a data register. The instructions are as follows:

Instructions

Format	Examples	Description
BCF F, b, a	BCF REG1, 7, 0	;Clear Bit7 in REG1 from the Access bank (a = 0)
BSF F, b, a	BSF REG2, 4, 1	;Set Bit4 in REG2 which is identified by ; the value in BSR (a = 1)
BTG F, b, a	BTG REG5, 0, 0	;Toggle Bit0 in REG5 from Access bank (a = 0)

Points to Remember

1. No flags are affected by these bit instructions.
2. Any bit in a data register can be set, reset, or toggled by these instructions. To operate on multiple bits, instructions need to be repeated. As mentioned in the previous section, to operate on multiple bits, logic instructions are more efficient than the bit instructions.

Applications

These bit instructions are commonly used to:

- Turn on and off peripherals connected to I/O ports.
- Enable or disable processes such as interrupts and serial communication by manipulating bits in Special function registers (SFRs).

EXAMPLE 6.4

The W register holds the bit pattern that turns on and off machines connected to PORTC. Write instructions to turn off Bit1 and toggle Bit0 and send output to PORTC assuming that PORTC is already initialized as an output port.

Solution:

The bit instructions in PIC18 microcontroller, discussed above, can manipulate bits in the data register including the W register. The instructions are as follows:

```
BCF     WREG, 1, 0      ;Clear Bit1 in WREG
BTG     WREG, 0, 0      ;Toggle Bit0 in WREG
MOVWF   PORTC           ;Output to PORTC to turn off Bit1 and toggle Bit0
```

6.2.2 Bit Test and Skip Instructions

The instruction set includes two instructions that test a bit in a data register: one checks whether the bit is clear and the other checks whether the bit is set. If the desired condition is met, the MPU skips the following instruction. The instructions with examples are as follows:

Instructions

Format	Example	Description
BTFSC F, b, a	BTFSC REG1, 7, 0	;Test Bit7 in REG1 and if the bit is zero, skip the ;next instruction, REG1 is in Access bank.
BTFSS F, b, a	BTFSS REG1, 5, 1	;Test Bit5 in REG1 and if the bit is one, skip the ;next instruction, REG1 is specified by BSR

Applications — These instructions test a bit and are generally followed by branch instructions to redirect the program. This is illustrated in Example 6.5. These instructions are used extensively in checking:

- a bit at an I/O port
- a bit in a SFR register
- a flag status to make a decision

EXAMPLE 6.5

A switch is connected to Bit1 and a LED is connected to Bit0 of PORTC (as shown in Figure 6–2). Write instructions to continuously monitor the switch when it is open (logic 1). When the switch is closed (logic 0), turn on the LED. Assume that individual bits are properly initialized for I/O.

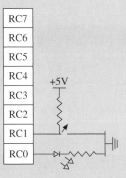

FIGURE 6-2   Checking a Switch

Solution:

To monitor the switch, the MPU needs to check the logic level of Bit1. If it is open (logic 1), the MPU should continue to check it. When the switch is closed, the MPU should turn on the LED connected at Bit0.

Instructions:

CHECK:	BTFSC	PORTC, 1, 0	;Test switch logic. If it is 0, skip the ;next instruction
	BRA	CHECK	;If switch logic = 1, continue ;checking
	BSF	PORTC, 0, 0	;Turn on LED by outputting logic 1

6.2.3 Bit Rotation

The instruction set includes rotate instructions whereby each bit in a data register is shifted to the adjacent position. Bits can be shifted either to the left or to the right. Each of these rotations are further classified in two groups (see Figure 6–3): 8-bit rotation and 9-bit rotation through Carry. In 9-bit rotation, a register is viewed as a 9-bit register with the carry flag as the ninth bit. The instructions are as follows:

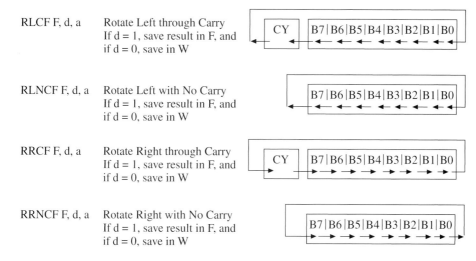

RLCF F, d, a   Rotate Left through Carry
If d = 1, save result in F, and
if d = 0, save in W

RLNCF F, d, a   Rotate Left with No Carry
If d = 1, save result in F, and
if d = 0, save in W

RRCF F, d, a   Rotate Right through Carry
If d = 1, save result in F, and
if d = 0, save in W

RRNCF F, d, a   Rotate Right with No Carry
If d = 1, save result in F, and
if d = 0, save in W

FIGURE 6-3   Rotate Instructions

Points to Remember

1. The 9-bit rotations affect three flags: N, Z, and C. The 8-bit rotations affect only two flags: N and Z. The flags are determined by the value of the byte after rotation.
2. The result of a register rotation can be saved either in the register or W register.

Applications

1. The rotation concept is used in multiplying and dividing in powers of 2. For example, when a byte is rotated left once, it is equivalent to multiplying the byte by two and when rotated twice, it is multiplied by four. On the other hand, when the byte is rotated right, it is equivalent to dividing by two. Nine-bit rotation is commonly used for multiplying or dividing the numbers larger than 8 bits, as shown in Example 6.8.
2. Bit rotation is used in serial communication wherein one bit is sent out on a single line.

EXAMPLE 6.6

The data register labeled as REG1 holds the byte 43_H with the carry flag set. Illustrate the contents of the register and the flag status after the execution of the following two rotate left instructions: RLCF and RLCNF shown in Figure 6-4.

continued

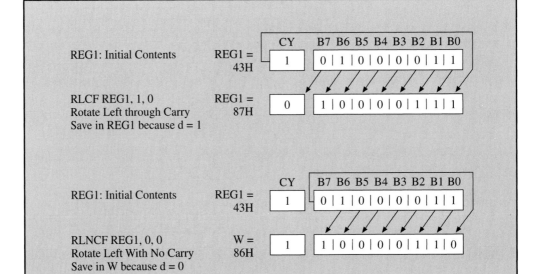

FIGURE 6-4   Rotate Left and Register Contents

Description: After the execution of the RLCF (Rotate Left through Carry), each bit shifts to the adjacent left position, Bit7 (= 0) goes into CY (carry), and CY (= 1) goes into Bit0. The result is 87_H, placed in the data register REG1 because the instruction specifies d = 1, and Bit7 resets CY = 0.

Flag Status: N = 1, Z = 0, and C = 0

After the execution of the RLNCF (Rotate Left with No Carry), each bit shifts to the adjacent left position; however, Bit7 (= 0) goes into Bit0 and CY is not affected. The result is 86_H, placed in W register because the instruction specifies d = 0, and CY remains 1.

Flag Status: N = 1, Z = 0, and C is not affected.

EXAMPLE 6.7

The data register REG1 with the address (01_H) holds the byte 43_H with the carry flag reset. Illustrate the contents of the registers after the execution of the following two rotate right instructions: RRCF and RRCNF (Figure 6–5).

continued

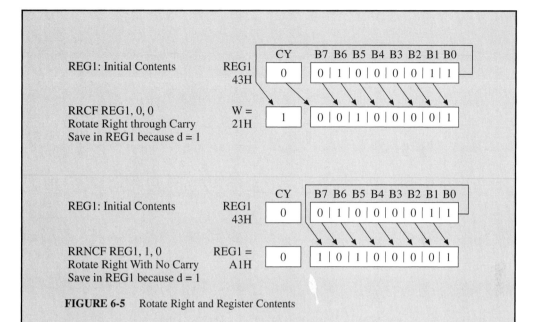

REG1: Initial Contents   REG1 43H

RRCF REG1, 0, 0
Rotate Right through Carry
Save in REG1 because d = 1   W = 21H

REG1: Initial Contents   REG1 43H

RRNCF REG1, 1, 0
Rotate Right With No Carry
Save in REG1 because d = 1   REG1 = A1H

FIGURE 6-5   Rotate Right and Register Contents

Description: The instruction RRCF (Rotate Right Through Carry) shifts each bit to the adjacent right position, places Bit0 (= 1) into CY (carry), and CY (= 0) into Bit7. The MPU places 21_H in W register because d = 0 and sets CY = 1 because Bit0 = 1.

The instruction RRNCF (Rotate Right with No Carry) shifts each bit to the adjacent right position; however, Bit0 (= 1) goes into Bit7 and CY is not affected. The result is $A1_H$, placed in REG1 because the instruction specifies d = 1, and CY remains 0.

EXAMPLE 6.8

Write instructions to load a packed BCD byte 37_H in REG1 (address 01_H). To display the byte at two seven-segments LEDs, unpack the byte and save in two data registers BUFFER1 and BUFFER2 (addresses 10_H and 11_H respectively).

Solution:

A seven-segment LED can display only one digit; therefore, the byte 37_H must be unpacked as 03_H and 07_H. Once the byte is unpacked, the seven-segment code can be found for individual digits (see Chapter 9, Section 9.3.2), and two digits are displayed at two different LEDs. The byte can be unpacked by using the masking technique (ANDing) and rotate instructions as follows:

continued

```
          Title "Ex6-8 - Unpacking a Byte"
          List p=18F452, f =inhx32
          #include <p18F452.inc>  ;The header file
BUFFER1      EQU   0x10            ;Define data register addresses
BUFFER2      EQU   0X11
REG1         EQU   0x01


             ORG   0x00
             GOTO  START


             ORG   0x20            ;Begin assembly at 0020H
START:       MOVLW 0x37            ;Load the packed byte
             MOVWF REG1            ;Save packed byte in REG1
             ANDLW 0x0F            ;Mask high-order nibble 3 of byte 37H
             MOVWF BUFFER1         ;Save 07H in BUFFER1
             MOVF  REG1,W,0        :Get the byte again
             ANDLW 0xF0            ;Mask low-order nibble 7 of byte 37H
             RRNCF WREG,0,0        ;Rotate high-order nibble four times
             RRNCF WREG
             RRNCF WREG
             RRNCF WREG
             MOVWF BUFFER2         ;Save high-order nibble as 03H
             END
```

Description: The first two instructions load the packed byte 37_H and save it in REG1. The AND instruction masks the high-order nibble and saves as 07_H in register BUFFER1. The next AND instruction masks the low-order nibble and modifies the byte as 30_H. The next four rotate instructions shift digit 3 into low-order nibble position and save it as 03_H in register BUFFER2.

EXAMPLE 6.9

Write instructions to multiply 16-bit number 0181_H, stored in data registers REG2 (high-01) and REG1 (low-81), by four.

Solution:

To multiply the number by four, we need to meet the following conditions:

- rotate left two times each register by setting a counter or repeating all the instructions twice

continued

- Push 0 into Bit0 of REG1 for every rotation; therefore, C flag should be cleared before every rotation
- Begin rotation with low-order byte (REG1)

Instructions are as follows:

```
Title "Ex6-9 - Multiplication Using Rotate Instructions"
List p=18F452
#include <p18F452.inc>   ;This is a header file

REG1    EQU   0x01            ;Define registers
REG2    EQU   0x02
C       EQU   0               ;Bit0 is C (carry flag)in STATUS


        ORG   00
        GOTO  MAIN


        ORG   0x020           ;Begin assembly at 0020H
MAIN:   MOVLW 0x81            ;Load 0181H in REG1 and REG2
        MOVWF REG1,0
        MOVLW 0x01
        MOVWF REG2,0
        MOVLW 0x02            ;Set up W as a counter for 2
REPEAT: BCF   STATUS,C,0  ;Clear carry flag
        RLCF  REG1,1,0      ;Multiply REG1 by 2
        RLCF  REG2,1,0      ;Multiply REG2 by 2
        DECF  WREG,0,0      ;Reduce counter by one
        BNZ   REPEAT         ;If counter ≠ 0, go back and repeat
        END
```

Description: The first four instructions load the 16-bit number 0181_H in REG1 and REG2 (Figure 6–6a), followed by setting WREG as a counter to repeat the process twice. The segment starting from the label REPEAT begins the multiplication process. The instruction BCF (Bit Clear File) clears the carry flag. Figure 6–6a shows that the carry flag is cleared and registers are ready for rotation. The first rotate instruction (RLCF) pushes Bit7 in C and sets the carry flag and shows 02_H in REG1 (Figure 6–6b). The next RLCF instruction rotates bits in REG2 and pushes carry into Bit0. Figure 6–6c shows that the result is 0302_H. The decimal conversion (Figure 6–6c) shows that the initial number is multiplied by 2.

continued

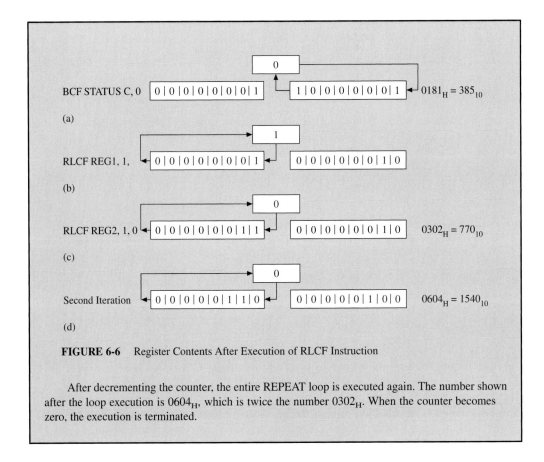

FIGURE 6-6   Register Contents After Execution of RLCF Instruction

After decrementing the counter, the entire REPEAT loop is executed again. The number shown after the loop execution is 0604_H, which is twice the number 0302_H. When the counter becomes zero, the execution is terminated.

6.3 MULTIPLY AND DIVIDE OPERATIONS

The PIC18 instruction set includes two instructions dealing with multiplication of two 8-bit unsigned numbers. However, it does not have an instruction that performs division; therefore, other techniques such as repeated subtraction or rotate right must be used to perform division.

6.3.1 Multiply Instructions

In PIC18 instruction set, the multiply instructions multiply two 8-bit unsigned numbers in two ways: (1) multiply 8-bit number (literal) with WREG, and (2) multiply WREG and a data register. The MPU places the 16-bit product in the register pair PRODH and PRODL, with the high-order byte in PRODH and the low-order byte in PRODL. The result does not affect any flags.TFSC

Instructions

Format	Example	Description
MULLW 8-bit	MULLW 0x08	;Multiply W and 08H and place the product ; in PRODH:PRODL
MULLWF F, a	MULLWF REG1,0	;Multiply W and REG1 and place the product ; in PRODH:PRODL

EXAMPLE 6.10

Multiply the following two unsigned bytes 81_H and 04_H and save the result in data registers REG10 and REG11 with the addresses 10_H and 11_H respectively.

Solution:

These two bytes can be multiplied by loading one byte in W register and using the other one as a 8-bit number as shown in the following:

```
REG10      EQU    0x10
REG11      EQU    0x11


           MOVLW 0x81                ;Load first byte in W
           MULLW 0x04                ;Multiply 81H by 04H
           MOVFF PRODL, REG10        ;Save low-order byte in REG10
           MOVFF PRODH, REG11        ;Save high-order byte in REG11
```

6.3.2 Division

The PIC18 instruction set does not include an instruction that divides two numbers. Therefore, we need to devise algorithms by using the available instructions to divide two numbers. The following two algorithms are commonly used: (1) use of rotate right instruction, and (2) repeated subtraction. Other algorithms such as emulation of manual (paper-and-pencil) method are also used (Chapter 8).

1. Rotate Right: When a byte is rotated right once, it is equivalent to dividing the number by two. Therefore, to divide the number by powers of two (such as 2, 4, and 8), we need to rotate the number multiple times based on the powers of 2. For example, to divide a number by 4, we need to rotate the number twice and by 8 three times. However, to divide by a number that is not a power of two requires a different algorithm, such as repeated subtraction. Example 6.11 illustrates the division of a 16-bit number by 8.

2. Repeated Subtraction: Conceptually, this is a simple process. For example, to divide 17_{10} by 5_{10}, we need to subtract 5 three times. Therefore, 3 will be the quotient and 2 will be the remainder. The algorithm includes a counter that counts the number of subtractions until a borrow is generated. The count is adjusted for the last increment and if necessary, the remainder is found by adding the divisor for the last subtraction.

EXAMPLE 6.11

Write instructions to divide the unsigned 16-bit number 0794_H by 8. The number is stored in data registers REG11 (high-order 07_H) and REG10 (low-order 94_H) with the register addresses 10_H and 11_H respectively.

Solution:

The 16-bit number should be rotated right three times starting from high-order register (REG11) to divide the number by eight. The carry flag should be cleared before the rotation because in the first rotation of REG11, the carry flag is shifted into Bit7 and that should be logic 0. In the first rotation of REG10, Bit0 of REG11, which is shifted into carry flag previously, becomes Bit7 of REG10 and Bit0 of REG10 shifts into carry flag. Instructions are as follows:

```
        Title "Ex6-10 Division by Rotation"
        List p=18F452
        #include <p18F452.inc>  ;This is a header file

REG10     EQU   0x10            ;Define registers
REG11     EQU   0x11
C         EQU   0               ;Bit0 is C (carry flag)in STATUS


          ORG   00
          GOTO MAIN

          ORG   0x020           ;Begin assembly at 0020H
MAIN:     MOVLW 0x94            ;Load 0794H in REG1 and REG2
          MOVWF REG10,0
          MOVLW 0x07
          MOVWF REG11,0
          MOVLW 0x03            ;Set up W as a counter for 3
REPEAT:   BCF   STATUS,C,0      ;Clear carry flag
          RRCF  REG11,1,0       ;Divide REG11 by 2
          RRCF  REG10,1,0       ;Divide REG10 by 2
          DECF  WREG,0,0        ;Reduce counter by one
          BNZ   REPEAT          ;If counter ≠ 0, go back and repeat
          END
```

Description: The first four instructions load the 16-bit number 0794_H in REG10 and REG11, followed by setting the WREG as a counter to repeat the process three times. The segment, starting from the label REPEAT, begins the division process. The instruction BCF (Bit Clear File) clears the carry flag. The first rotate instruction (RRCF REG11) pushes Bit0 in C and C (logic 0) in Bit7 of REG11. The second rotate instruction shifts C (which is Bit0 of REG11) into Bit7 of REG10 and Bit0 into C, which is cleared before the next cycle of rotation begins. In this process of dividing by shifting right, we lose the remainder but the quotient is calculated correctly.

6.4 ILLUSTRATIVE PROGRAM: FINDING THE HIGHEST TEMPERATURE IN A DATA STRING

The following program searches for the largest positive number in a given data string stored in data registers. Finding a largest number, smallest number, or organizing data in the ascending or descending order are tasks of common occurrence in microcontroller applications.

6.4.1 Problem Statement

A set of temperature readings are stored in data registers starting from register 10_H located in the Access bank. The set includes both positive and negative readings (numbers), and it is terminated in byte 00. Write instructions to find the highest positive temperature reading in the data set. This problem assumes that the data set will not include 00 as a legitimate temperature reading.

6.4.2 Problem Analysis

The data string is terminated in 00 indicating that the number of bytes vary from one string to another; therefore, we cannot set up a counter. We need instead to check every byte, starting from the first, to see if it is zero. Once the byte is checked for a non-zero value, it should be checked to see if it is positive by testing Bit7 of the data byte. If Bit7 is zero, the number is smaller than 80_H; therefore, it is a positive reading. Otherwise, it is a negative reading. Once we find a positive reading, it is first compared with zero and the larger reading is saved. In the subsequent cycle, the new reading is compared with the saved reading, and if new reading is larger than the previous reading, it is saved; otherwise, the program goes back to get the next reading. Figure 6–7 shows the flow chart with three decision-making points as discussed above. This comparison continues until we come across the zero byte.

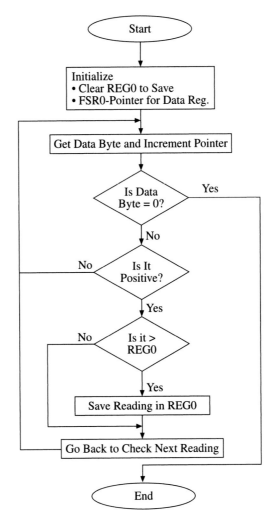

FIGURE 6-7   Flowchart for IP6-4: Finding Highest Temperature in Data String

6.4.3 PROGRAM

```
        Title "IP 6-4: Finding Highest Temperature"
        List p=18F452, f =inhx32
        #include <p18F452.inc>        ;The header file

REG0        EQU    0x00               ;Define data register addresses
BUFFER      EQU    0x10               ;Data stored starting from register 10H

            ORG    0x20               ;Begin assembly at 0020H
START:      CLRF   REG0               ;REG0 is used to store maximum
                                      ;temperature reading
            LFSR   FSR0,BUFFER        ;Set up FSR0 as pointer for data
                                      ;registers
NEXT:       MOVF   POSTINC0,W         ;Get data byte in WREG
            BZ     FINISH             ;Is byte = 0? If yes, this is end
                                      ;of data string
            BTFSC WREG,7              ;Is byte positive? If yes, skip
                                      ;next
            BRA    NEXT               ;Byte is negative - go back
            CPFSLT REG0,0             ;Is byte larger than previous one?
                                      ;If yes, save
            BRA    NEXT               ;If byte is smaller - get next one
            MOVWF REG0                ;Save larger byte
            BRA    NEXT               ;Go back and check next byte
FINISH:     NOP
            END
```

6.4.4 Program Description

The first instruction clears REG0 to begin comparison with the smallest number, and the LFSR instruction sets up the File Select Register0 (FSR0) as a pointer for the register BUFFER where the first temperature reading is stored. The MOVF instruction gets the data byte in WREG, and if it is zero, the program jumps to FINISH on the zero flag. The next instruction BTFSC checks Bit7, and if it is cleared, the program goes to compare two bytes: one in WREG and the other one previously stored in REG0. If the byte in WREG is larger than the byte in REG0, the program replaces the contents of REG0 with those of WREG. If the byte in WREG is smaller than the byte in REG0, the program goes back to check the next data byte without replacing the contents of REG0. This process is repeated until the string terminal byte 00 is found.

6.4.5 Program Execution and Troubleshooting (Debugging)

To execute and troubleshoot the program, you need to check three conditions: (1) in comparison of two numbers, the higher number is saved, (2) rejection of any negative numbers (numbers 80_H and higher, and 3) termination of the program when the data byte is 00. To perform simulation using the MPLAB IDE, enter the suggested data string and follow the steps.

Step 1: Assemble the program using MPLAB IDE, and add the following registers in the Watch window: WREG, REG0, BUFFER, and FSR0.

Step 2: Go to View and click on File Registers. Insert the following data string of five bytes in the data registers starting from BUFFER (Address 10_H to 14_H): 32_H, 47_H, 87_H, 35_H, and 00. These data bytes are selected to test various conditions. However, you may enter any other data set.

Step 3: Single-step each instruction and watch the contents of various SFR registers. For the first data byte 32_H, the program should skip all the jump instructions, and the byte should be saved in REG0. Similarly, for the second round, the byte 47_H should replace 32_H.

Step 4: For the third byte 87_H, when the program checks Bit7 in WREG, the C flag is set, and the program executes the next BRA instruction and jumps to the label NEXT to check the next byte.

Step 5: In the next round, the program checks the next byte 35_H. This byte is smaller than the previously saved byte 47_H; therefore, the program skips the next step and goes back to NEXT to check the next byte.

Step 6: The last byte is 00. Therefore, the program should jump to the end after the first check.

6.5 ILLUSTRATIVE PROGRAM: FINDING AN AVERAGE TEMPERATURE OF DATA READINGS

The concept in finding the average of data readings is simple: add all the bytes and divide the sum by the number of the readings. However, the PIC18 instruction set does not include a divide instruction; therefore, the division process should be implemented by using algorithms based on available instructions. The following program demonstrates two methods of dividing a number: (1) using shift right instructions, and (2) repeated subtraction.

6.5.1 Problem Statement

Sixteen 8-bit temperature readings are stored in data registers. Find the average temperature.

6.5.2 Problem Analysis

To find the average of the 16 readings, we need to add them and divide the sum by 16. The total sum of the readings is likely to be more than 8 bits; therefore, we need two data registers to save the 16-bit sum (as in Section 5.7.3, Chapter 5). To divide the sum by 16, we can employ two methods: (1) shift right the sum four times because the divisor is powers of two, and (2) subtract the number repeatedly until a borrow is generated, and the number of subtractions becomes the quotient. Both methods are demonstrated in the following program using the same sum.

6.5.3 PROGRAM

```
      Title "PIC18F452 IP6-5: "Finding Average Temperature "
      List p=18F452, f =inhx32
      #include <p18F452.inc>        ;The header file


LO_SUM      EQU  0x01               ;Define addresses of data registers
HI_SUM      EQU  0x02
AVERAGE     EQU  0x03
LO_BYTE     EQU  0x04
HI_BYTE     EQU  0x05
QUOTIENT    EQU  0x06
COUNTER     EQU  0x07
BUFFER      EQU  0x10               ;Data stored starting from register 10H
CFLAG       EQU  0
            ORG  0x20               ;Begin assembly at 0020H
START:      MOVLW D'16'             ;Set up counter for number of readings
            MOVWF COUNTER
            LFSR  FSR0,BUFFER       ;Set up FSR0 as pointer for data  registers
            CLRF  WREG,0
            CLRF  LO_SUM            ;Registers to save sum
            CLRF  HI_SUM
            ;::::::::::::::::::::::::::::::::::::::::::::::::::::::::::::::::::::::
            ;The next segment adds 8-bit numbers resulting onto 16-bit sum in   :
            ;registers HI_SUM and LO_SUM. The sum is also saved HI_BYTE and     :
            ;LO_BYTE registers to use it again to demonstrate the second        :
            ;algorithm of repeated subtraction.                                 :
            ;::::::::::::::::::::::::::::::::::::::::::::::::::::::::::::::::::::::
SUM:        ADDWF POSTINC0,0,0      ;Add data byte and save sum in WREG
            BTFSC STATUS,C          ;Check C flag. If C = 0, skip  next
                                    ;instruction
            INCF  HI_SUM            ;If C = 1, increment HI_SUM
            DECF  COUNTER,1,0       ;One addition done - decrement count
            BNZ   SUM               ;If count ≠ 0, go back to add next byte
            MOVWF LO_SUM            ;Save low-order byte in LO_SUM
            MOVFF LO_SUM, LO_BYTE   ;Save 16-bit sum in HI_ and LO_ byte
            MOVFF HI_SUM, HI_BYTE
            ;:::::::::::::::::::::::::::::::::::::::::::::::::::::::::::::::::::::
            ;The following instructions divide the sum by using the            :
            ;instruction Rotate Right through Carry four times. The            :
            ;average of 8-bit numbers is always 8-bit number it is saved       :
            ;in register AVERAGE                                               :
            ;:::::::::::::::::::::::::::::::::::::::::::::::::::::::::::::::::::::
AVG1:       MOVLW 0x04              ;Shifting right four times - set up
                                    ;counter for 4
```

continued

```
              MOVWF COUNTER
  CLEARCY:    BCF    STATUS, CFLAG    ;Clear C flag
              RRCF   HI_SUM           ;Divide byte in HI_SUM by 2
              RRCF   LO_SUM           ;Divide byte in LO_SUM by 2
              DECF   COUNTER,1,0      ;Decrement rotation count
              BNZ    CLEARCY          ;If counter ≠ 0, go back to next divide
              MOVFF  LO_SUM AVERAGE   ;Save result in register AVERAGE
              ;:::::::::::::::::::::::::::::::::::::::::::::::::::::::::::::::::::
              ;The following instructions demonstrate the division by repeated   :
              ;subtraction. finds the average The final result is saved in       :
              ;QUOTIENT register. It should be the same as the result in AVERAGE  :
              ;:::::::::::::::::::::::::::::::::::::::::::::::::::::::::::::::::::
  AVG2:       MOVFF  LO_BYTE, LO_SUM  ;Get the saved sum again
              MOVFF  HI_BYTE, HI_SUM
              CLRF   QUOTIENT         ;Clear QUOTIENT register
  DIVIDE:     MOVLW  D'16'            ;Place 16₁₀ in WREG as a divisor
              INCF   QUOTIENT         ;Begin counting number of subtractions
              SUBWF  LO_SUM           ;Subtract 16₁₀ from LO_SUM
              BTFSC  STATUS, 0        ;If C flag =0, skip next instruction
              GOTO   DIVIDE           ;If C flag =1, go back to next  subtraction
              DECF   HI_SUM           ;Subtraction is beyond LO-SUM,  decrement
  HI_SUM
              BTFSC  STATUS, 0        ;If C flag = 0, skip next instruction
              GOTO   DIVIDE           ;If C flag =1, go back to next subtraction
              DECF   QUOTIENT         ;Adjust the last subtraction
              END
```

6.5.4 Program Description

The initialization block (START segment) sets up a counter for 16 to add 16 bytes, initializes the FSR0 register as a pointer for BUFFER where data storage begins, and clears WREG and registers LO_SUM and HI_SUM where the sum is saved.

The SUM segment adds the bytes similar to that in Illustrative Program: Addition of Data Bytes (Section 5.7) except that this program uses the bit test instruction (BTFSC) to check Bit0 (which is Carry) in the STATUS register, and if there is no carry, the program skips the next instruction and does not increase the HI_SUM register.

To find the average of the total sum, the next segment AVG1 divides the sum by 16 using the instruction Rotate Right through Carry four times and saves the result in register AVERAGE. This method is explained in Example 6.11 and is applicable when the divisor is a power of two.

The AVG2 segment finds the average by the method of repeated subtraction. The beginning instructions get the sum saved in registers HI_BYTE and LO_BYTE and clears the register QUOTIENT to count how many times subtraction has been performed. The divisor 16_{10} is loaded in WREG and the divisor is subtracted repeatedly from LO_SUM until a borrow is generated. Then the program decrements the high-order byte in HI_SUM, checks for borrow from high-order byte, and returns to the label DIVIDE again to continue the subtraction from

LO_SUM. The borrow flag in the decrement of the HI_SUM indicates the end of the subtraction. The number in the QUOTIENT register must be adjusted by one because it is already incremented by one in the beginning.

6.5.5 Program Execution and Troubleshooting (Debugging)

To execute the program using the MPLAB IDE, we need 16 bytes of data entered into data registers starting from BUFFER. However, we know that the segment SUM, the addition of 16 data bytes, functions correctly based on the result of the Illustrative Program: Addition of Data Bytes (Chapter 5, Section 5.7). The result is a 16-bit sum saved first in registers HI_SUM and LO_SUM and copied into registers HI_BYTE and LO_BYTE to be used again for the second time. We can assume an arbitrary number as a 16-bit sum for the simulation. For example, if we assume the sum of 16 data bytes is 484H ($= 1156_{10}$), the average can be found by dividing the sum by 16, which is equal to 72.25_{10} in decimal (Quotient is 72_{10} and the remainder is 4_{10}). In our processes of finding an average, we should expect the result to be 72_{10}, which is equivalent to 48_H; the above algorithms cannot account for a fraction or a remainder. Now follow the steps as suggested here.

Step 1: Assemble the program using MPLAB IDE, and add the following registers in the Watch window: WREG, STATUS, LO_SUM, HI_SUM, AVERAGE, and QUOTIENT or you can open File Registers and, knowing the addresses of these registers, you can watch the changes.

Step 2: Insert the sum 0484_H as the assumed result of the addition of 16 data bytes—04 in HI_SUM and 84 in LO_SUM in the value column of the Watch window and set the cursor at the instruction MOVFF LO_SUM, LO_BYTE in the segment SUM. The execution of the program will begin at the location where the cursor is set.

Step 3: Set the breakpoint at the instruction MOVFF LO_SUM, AVERAGE, the last instruction in the segment AVG1.

Step 4: Run the program from the cursor, and it will stop at the breakpoint. The result in registers LO_SUM and AVERAGE should be 48_H.

Step 5: Set the second breakpoint at the instruction DECF HI_SUM and run the program from the previous break point. At this point, the program has gone through one complete cycle of dividing the low-order sum 84_H (132_{10}) eight times (132/16 = 8.25) plus one additional cycle that goes beyond the number and generates a borrow. You should observe: QUOTIENT = 9, LO_SUM = $F4_H$.

Step 6: Now it is time to decrement the high-order byte in HI-SUM from 4 to 3. Single-step the next instruction DECF HI_SUM and run the program. The program stops again at the breakpoint displaying QUOTIENT = 19 and LO_SUM = $F4_H$.

Step 7: Single-step again the instruction DECF HI_SUM and run the program, and repeat the process until the HI_SUM = 00. And in the last iteration, QUOTIENT = 49_H. In this step, the C flag will be cleared and the program will skip the instruction GOTO DIVIDE, and the next instruction adjusts the QUOTIENT = 48 by decrementing.

Step 8: Now you can verify that the averages by using the Rotate Right Instruction and repeated subtraction method are the same.

SUMMARY

In this chapter, we have discussed the instructions related to the following operations: logic, bit, and multiply-divide. The important features of these instructions are as follows:

1. We discussed three type of logic instructions: AND, IOR (inclusive OR), and XOR (exclusive OR). These instructions perform logic functions at the register level, thus making them equivalent to the logic operations of eight two-input gates. For each logic operation, the PIC18 instruction set includes two instructions: logic operation of (1) 8-bit (literal) with WREG, and (2) WREG with any data (file) register. In the second case, the result can be stored either in the WREG or in data register. Logic instructions affect only Z and N flags.

    Applications of logic instructions include such operations as: masking, bit set, reset, toggle, and logic operations. In a given register, a bit or multiple bits can be reset, set, or toggled. AND instructions are used for resetting a bit, IOR instructions for setting, and XOR for toggling.

2. The instruction set includes three instructions—BCF, BSF, and BTG—that operate at the bit level in a given register. The BCF clears, BSF sets, and BTG toggles a bit. These instructions do not affect any flags.

3. The instruction set includes two instructions that test a bit for logic 0 (clear) or 1 (set) condition. If the condition is valid, the instruction skips the next instruction. These instructions are used to redirect the program.

4. The rotate instructions can shift bits to the right or left position. These instructions are further classified in to two groups: 8-bit rotation and 9-bit rotation. In the 9-bit rotation, the carry flag is used as the ninth bit. Applications of these instructions include serial data transfer and arithmetic operations such as multiplication and division. One rotate left operation multiplies the byte in a register by two and the rotate right operation divides the byte by two.

5. The instruction set includes two multiply instructions of 8-bit unsigned numbers: multiplication of (1) 8-bit literal and WREG, and (2) WREG and a data (file) register. The product is saved in two registers—PRODH (Product High) and PRODL (Product Low). The set does not include a divide instruction. A division should be performed by using rotate instructions or repeated subtraction.

QUESTIONS AND ASSIGNMENTS

Logic Operations

6.1   Write instructions to logically AND bytes 97_H and 68_H and identify the status of the flags.

6.2   Write instructions to load $F8_H$ and 31_H in data registers REG1 and REG2 (addresses 01 and 02 respectively), AND the bytes, and save the result in REG2. Identify the status of the flags.

6.3   Write instructions to load a byte in WREG and mask bits Bit7 and Bit0.

6.4   Write instructions to reset Bit7 of the byte in REG1 (address 01_H).

6.5   The lowercase ASCII letter z ($7A_H$) is stored in REG10 (address 10_H). Write instructions to convert the letter as uppercase and save it in REG10.

6.6   Write instructions to logically OR the bytes 37_H and $F8_H$ and identify the result and the status of the flags.

6.7   Write instructions to set bits Bit0 and Bit1 of PORTC, assuming PORTC is already initialized as an output port.

6.8   Write instructions to exclusive OR the bytes 31_H and $C8_H$ and identify the result and the status of the flags.

6.9   Write instructions to toggle bits Bit0 of PORTB assuming the PORTB is already initialized as an output port.

6.10   Write instructions to read PORTA and toggle bits Bit0 and Bit1

Bit Operations

6.11   Write instructions to load 37 in REG5 (address 05_H) and set Bit7.

6.12   Write instructions to toggle bits Bit3 and Bit4 of PORTC.

6.13   Write instructions to load $F2_H$ in REG0 (address 00) and reset Bit7.

6.14   Register REG10 holds the ASCII character uppercase A. Write instructions to convert 'A' into lowercase 'a'.

6.15   Write instructions to reset the OV (overflow) flag in the STATUS register.

6.16   Write instructions to modify the bits of PORTA as follows: set Bit0, reset Bit1, and toggle Bit7.

6.17   Register REG20 holds the byte F8 and the C flag is reset. Identify the contents of REG20 and the flag status after the execution of the instruction RLCF (Rotate Left through Carry) once.

6.18   In Q. 6.17, identify the contents REG20 and the flag status if the instruction used is RLNCF (Rotate Left with No Carry) instead of RLCF.

6.19   The RLCF instruction in 6.17 should multiply the contents of REG20 by 2. Explain why the result does not reflect the expected product.

6.20   Write instructions to load the byte 93_H in the WREG and execute the instruction RRCF (Rotate Right through Carry) twice. Identify the result and the status of the flag.

6.21   In Q. 6.20, the result after the execution of the instruction RRCF twice should be equal to the unsigned byte 93_H divided by 4 (ignoring the remainder). Explain why the result is inaccurate.

6.22   Modify the instructions in 6.20 to divide the byte 93_H by 4.

6.23   Register REG5 (with the address 05_H) holds the unsigned byte 85_H. Write instructions to multiply the byte by 10_{10} and save the result in registers REG5 and REG6.

6.24   The 16-bit number $F247_H$ is stored in registers REG10 and REG11 (with the addresses 10_H and 11_H) with the low-order number in REG10 and high-order number in REG11. Multiply the number by 8 and store the product in registers REG10, REG11, and REG12.

6.25   The 16-bit number $14F2_H$ is stored in registers REG0 and REG1 with the addresses 00_H and 01_H, respectively; the high-order byte 14_H is stored in REG1 and the low-order byte $F2_H$ in REG0. Divide the number by 16_{10}.

6.26   Registers REG0 and REG1 hold the unsigned bytes 78_H and $2A_H$, respectively. Write instructions to multiply these bytes and save the product in registers REG10 and REG11.

SIMULATION EXERCISES

Using MPLAB IDE or PIC18 IDE

6.1 SE   Modify the Illustrative Program in Section 6.4 if the number of signed data bytes to be searched is equal to 10_{10}, stored in data registers starting from BUFFER.

Data (H): 56, 85, 67, 78, 6F, 7F, F2, 98, 72, 65

6.2 SE   Modify the program in 6.1SE to count the number of positive readings.

6.3 SE   Modify the program in 6.1SE to count the number of positive readings that are larger than 68_{10}.

6.4 SE   Modify the Illustrative Program in Section 6.4 to find the minimum temperature for the following data set in signed numbers. Assemble and execute the program. (Hint: the minimum temperature is the largest negative reading).

Data (H): 72, 85, F2, 37, A2, 92, 68, 62, 00

6.5 SE   The following unsigned temperature readings are stored in data registers starting from REG10 (address 10_H). Find the average temperature.

Data (H): 40, 45, 4A, 32, 52, 38, 3A, 44

CHAPTER 7

STACK AND SUBROUTINES

OVERVIEW

In Chapters 5 and 6, we introduced the assembly language instructions of PIC18 MCU and used the flowcharting technique to analyze and write simple programs. However, at the time, we did not have any tools to handle large programs. This chapter introduces the technique of dividing a large program into small segments or modules, writing these modules as independent units called subroutines, and finally bringing the subroutines together to build a large program.

A subroutine is a set of instructions that performs a simple task similar to programs we discussed in previous chapters. These subroutines are written as independent units and stored anywhere in memory. Now we need instructions that can tell the MPU to go out to these different subroutines in a given sequence and come back after completing the task of a given subroutine. These instructions are known as CALL and RETURN instructions. A CALL instruction is similar to a branch instruction, which was discussed in Chapter 5, but it remembers to go back (return address) where the CALL is initiated. To remember these return addresses, we need a temporary storage called the STACK.

This chapter introduces the CALL and RETURN instructions in PIC18 instruction set and discusses the concept of the stack and how it is used to implement the subroutines. In addition, the concept of the macro is introduced. The macro in an assembly language is a group of instructions that can be identified by a single name or label. It is a shortcut to combine and automate a group of tasks or instructions. An analogy is a group of e-mail addresses identified as a list. To prepare a list of addresses, we label the list with a name and type all the e-mail addresses in the group. When we want to send a message to all the members of the group, we simply type the name of the list, and the e-mail program expands the list to include all the names in the list. Finally, an illustrative program is written to integrate the concepts of subroutine, macro, and the stack.

OBJECTIVES

- Define the terms stack and stack pointer, and explain how they are used.
- Explain the functions of the instructions PUSH and POP.
- Define the terms subroutine and macro, and list advantages and disadvantages of each.
- Explain the instructions CALL and RETURN and how the stack pointer and the stack are used in the execution of these instructions.
- Explain the difference between two formats of the CALL instruction CALL and CALL FAST.
- Illustrate how a subroutine and a macro are used in writing assembly language programs.
- Illustrate how the subroutine concept is used in modularizing and troubleshooting assembly language programs.

7.1 STACK

A **Stack** is a temporary storage space used during the execution of a program and can be a part of R/W memory or a specially designed group of registers or both. It is called stack because of how the information is stored in these registers. When we write a program, it is stored in memory in the increasing order of addresses, and the MPU executes those instructions in the same increasing sequence by incrementing the program counter. For example, if we begin a program at location 2000_H, the next location is 2001_H. However, for the stack, in most MCUs, the information is stored in the reverse order. For example, if the stack is defined at 2075_H, the next location of the stack is 2074_H. This storage technique gives an appearance of stacking items. However, this raises a new question: how does the MPU identify the stack locations? The MPU uses a register as a pointer; called the stack pointer (SP), it is similar to the program counter (PC). The program counter holds the address of the memory location of the next code to be executed, and it is modified as the instructions are being executed. Similarly, the stack pointer holds the address of the top of the stack location. As the information is stored or retrieved from the stack, the address in the stack pointer is modified (incremented or decremented) to always show the top of the stack.

The next question is about how the information is stored and retrieved from the stack. The stack space is shared by the MPU and the user. The MPU stores the contents of the program counter (the memory address) automatically when it executes a CALL instruction for a subroutine and retrieves that address when it executes a RETURN instruction in the subroutine—this will also be explained later. The MPU also stores the address of the program counter on the stack when it is interrupted and retrieves that address when it returns from the service routine (discussed in Chapter 10). The user stores the contents of registers on the stack by using special instructions called PUSH and POP instructions (explained later). When the MPU executes these instructions, it modifies the stack pointer accordingly to point to the top of the stack.

7.1.1 PIC18 Microcontroller Stack

The stack in the PIC18 microcontroller consists of 31 registers, called the hardware stack; each register is 21 bits wide to hold memory addresses. The stack space (31 registers) is neither a part of program memory nor data registers. To identify these 31 registers, we need only 5 bits ($2^5 = 32$), and the PIC18 uses one of the special function register, called STKPTR (Stack Pointer), to identify the address of the stack. Five bits (SP4–SP0) of the STKPTR register are used to identify the available stack location (see Figure 7–1). If the user attempts to use more than 31 registers to store addresses on the stack, it is called an overflow condition and indicated by setting Bit7 in STKPTR. If the user attempts to retrieve more addresses than are stored, this is called stack underflow and is indicated by setting Bit6 in STKPTR (Figure 7–1).

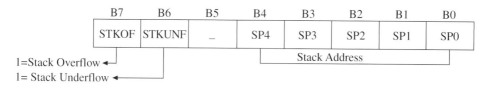

FIGURE 7-1   STKPTR (Stack Pointer) Register

7.1.2 Instructions to Store and Retrieve Information from the Stack

The PIC18 instruction set includes only two PUSH and POP instructions that are used to store the contents of program counter on the stack or discard information from the stack. These instructions are used primarily to set up a software stack, a process we will discuss later. In other microcontrollers such as the Motorola 68HC series or the Intel 8051 series, where a part of R/W memory is used as stack, the instruction sets include many more instructions related to the stack.

Instructions

PUSH   ; Increment the memory address in the stack pointer (by one) and store the
                ; contents of the program counter (PC + 2) on the top of the stack.

POP   ;Discard the address of the top of the stack and decrement the
                ; stack pointer (by one).

The contents of the stack (21-bit address), pointed to by the stack pointer, are copied into three special function registers: TOSU (Top-of-Stack Upper), TOSH (High), and TOSL (Low). The contents of the stack pointer and TOS registers can be accessed (read) and modified (written into).

EXAMPLE 7.1

The following program consists of two PUSH and POP instructions assembled beginning at location 0020_H. Identify the contents of PCL (Program Counter Low) and TOS and STKPTR registers as these instructions are executed.

Solution:

	PCL	TOSU	TOSH	TOSL	STKPTR
	Hex	Hex	Hex	Hex	Binary
					00000
0020 PUSH	22	00	00	22	00001
0022 PUSH	24	00	00	24	00010
0024 POP	26	00	00	22	00001
0026 POP	28	00	00	22	00000

Description:

The first instruction PUSH is stored at location 0020_H. When that instruction is executed, PCL is incremented to 000022_H (PCU and PCH are not shown) pointing to the next instruction to be executed, and the 5-bit binary address in the stack pointer is incremented by one; 000022_H (PC +2) is also stored in TOS (Top of Stack) registers. When the next PUSH instruction is executed, the PCL goes to 24_H, STKPTR to 00010_B, and TOSL to 24_H. However, in the execution of the next POP instruction, the PCL is incremented by two, but TOSL shows the previous contents of the stack, and the STKPTR is reduced by one because when the MPU executes the POP instruction, it discards the contents of the top of the stack and decrements the stack pointer by one. Therefore, TOSL shows the address that is stored in stack location 00001. In the execution of the second POP instruction, the stack pointer goes back to 00000.

7.2 SUBROUTINE

A **Subroutine** is a group of instructions that performs a specified task. It is written independent of a main program and can be called multiple times to perform that task whenever needed by the main program or by another subroutine. The subroutine structure provides modularity and efficiency.

A large software program can be divided into small independent tasks or modules that are written independently as subroutines and brought together to build a large program. Subroutines become building blocks of a large program, similar to building a computer with various components such as the memory modules, video card, hard drive, power supply, and I/O boards, built independently by different manufacturers and brought together in a chassis to build a computer. In terms of efficiency, subroutines save memory space. For example, if we need a 100-ms delay five times in the main program, we can write a 100-ms subroutine once and call it five times.

When the main program calls a subroutine that is stored in memory at a separate location, the MPU transfers the program execution to the memory address of the subroutine. The MPU continues to execute the instructions in the subroutine, and when it finds a Return instruction, it returns to the main program. Therefore, the CALL instruction is equivalent to a jump or branch instruction, but capable of providing the return address after the completion of the subroutine. Similarly, the subroutine needs a RETURN instruction, equivalent to a branch or jump instruction, that finds the address at the top of the stack and returns the program execution to that address. These CALL and RETURN instructions use the stack to remember and find the return address.

When the MPU executes the CALL instruction, it stores the contents of the program counter, which is the address of the next instruction to the CALL, on the top of stack, and jumps to the subroutine location. When the MPU executes the RETURN instruction at the end of the subroutine, it finds the address on the top of stack and jumps to that address. This process of CALL and RETURN is explained in the next section with the instructions of PIC18 microcontroller.

7.2.1 PIC18 CALL and RETURN Instructions

The PIC18 instruction set includes two CALL instructions: one is a two-word instruction, capable of calling a subroutine anywhere in 2 MB memory range, and the other is a one-word relative CALL instruction that can call a subroutine within ±1K byte memory range. The set includes two RETURN instructions: one is used with the above CALL instructions and the other returns an 8-bit number (literal) to the calling program that is used primarily in older medium- to small-sized versions of PIC family of microcontrollers.

Instructions

CALL Label, s (0/1)   ;Call subroutine located at Label   2-W /2 Cyl
CALL Label, FAST

> ;If s = 0, increment the stack pointer and store the contents of the
> ; program counter (PC +4) on the top of the stack (TOS) and
> ; branch to the subroutine address located at Label.

> ;If s = 1, increment the stack pointer and store the contents of the
> ; program counter (PC + 4) on the top of the stack (TOS) and the
> ; contents of W, STATUS, and BSR registers in their respective
> ; shadow registers and branch to the location Label. When s = 1,
> ; the equivalent format of the instruction is CALL Label, FAST
> ; and the instruction can be written in the either format.

RCALL n   ;Relative call to subroutine within   1W/2Cyl
                 ;± 512 words (or 1K byte)

> ;Increment the stack pointer and store the contents of the program
> ; counter (PC + 2) on the top of the stack (TOS) and branch to the
> ; location Label within ± 512 words (or ±1K byte).

RETURN, s (0/1) RETURN FAST	;Return from subroutine	1W/1Cyl

;If s = 0, get the address from the stack (TOS) and place it in
;PC and decrement the stack pointer

;If s = 1, get the address from the stack (TOS) and place it in PC,
; retrieve the contents of W, STATUS, and BSR registers from
; their shadow registers and decrement the stack pointer. Instead
; of writing 1 (in place of s) , the instruction can be written in the
; Alternative format: RETURN FAST

RETLW 8-bit   ;Return 8-bit (literal) to W register

;Return 8-bit literal in WREG, get the address from the stack
; (TOS) and place it in PC and decrement the stack pointer

EXAMPLE 7.2

Rewrite Illustrative Program: Generating Waveforms (Section 5.5) to include 50 µs delay as a subroutine.

Solution:

Now we will divide the program in two sections: the main program and a 50 µs delay subroutine. The main program initializes the necessary registers and turns on (and off) Bit0 of PORTC and calls 50 µs delay subroutine. The delay subroutine is written separately beginning at the location 000040_H. The subroutine includes all the instructions of the previous delay segment; however, it is terminated in the instruction RETURN.

Program

```
REG1        EQU    0x01            ;Address of Data Register1
REG10       EQU    0x10            ;Address of Data Register10


            ORG    0x20
START:      MOVLW  B'11111110'     ;Number to set up BIT0 as an output
            MOVWF  TRISC           ;Initialize BIT0 as an output pin
            MOVWF  REG1            ;Save Bit pattern in REG1
ONOFF:      MOVFF  REG1,PORTC      ;Turn on/off BIT0
            CALL   DELAY50MC       ;Call 50 µs delay
            COMF   REG1,1          ;Complement Bit Pattern
            BRA    ONOFF           ;Go back to change LED
```

continued

```
        ;::::::::::::::::::::::::::::::::::::::::::::::::::::::::::::::
        ;Function: This subroutine provides 50 µs delay          :
        ;Input: None                                             :
        ;Output: None                                            :
        ;Registers modified: WREG and data register REG10
        ;::::::::::::::::::::::::::::::::::::::::::::::::::::::::::::::

               ORG   0x40            ;Begin subroutine at 00040H
DELAY50MC:     MOVLW D'166'          ;Load decimal count in W
               MOVWF REG10           ;Set up REG10 as a counter
LOOP1:         DECF  REG10,1         ;Decrement REG10
               BNZ   LOOP1           ;Go back to LOOP1 if REG 10 ≠ 0
               RETURN
               END
```

```
  Main Program                                  Subroutine

  0020 0EFE      START: MOVLW    B'11111110'  DELAY50MC:
  0022 6E94             MOVWF    TRISC         0040   0EA6   MOVLW   D'166'
  0024 6E01             MOVWF    REG1          0042   6E10   MOVWF   REG10
  0026 C001 FF82 ONOFF: MOVFF    REG1,PORTC    0044   0610   DECF    REG10,1
  002A EC20 F000        CALL     DELAY50MC     0046   E1FE   BNZ     LOOP1
  002E 1E01             COMF     REG1,1        0048   0012   RETURN
  0030 D7FA             BRA      ONOFF
```

FIGURE 7-2   Program Listing with Memory Addresses

Description:

Figure 7–2 shows the program of Example 7.2 with memory addresses and the Hex codes; it is divided into two segments: the main program and the subroutine. In the main program, the two-word CALL instruction is located at the memory address $002A_H$ and the next instruction COMF REG1 is located at the address $002E_H$.

The main program turns on Bit0 of PORTC and calls the delay. The MPU first increments the address in the stack pointer from 0000 to 0001 and stores the address of the next instruction ($00002E_H$ of COMF—the program listing does not show the first two zeros) on the stack and branches to the location 000040_H. If one were to single-step the program and examine the registers TOS (Top of the Stack), PCLAT (Program Counter Latch), and STKPTR (Stack Pointer) after the execution of the CALL instruction, the following would be observed: the address of the next instruction COMF -$002E_H$ on the stack, the program counter holding the address of the subroutine DELAY50MC (000040_H), and the 5-bit stack pointer incremented to 01.

TOS:	00002E	PCLAT:	000040	STKPTR:	01
	21-bit Register		21-bit Register		5-bit Register

The program execution continues from location 0040_H to 0048_H. When the MPU executes the RETURN instruction, the MPU goes to the top of the stack, gets the address $0002E_H$, places the address in the program counter, decrements the stack pointer to 00, and continues the execution from COMF REG1 in the main program.

7.2.2 Subroutine Documentation and Parameter Passing

In large software projects, the problem is divided into small segments, which are written as independent subroutines by various team members. Some subroutines require information from the calling program. For example, a subroutine that performs the addition of the number of bytes in data registers may require information concerning the number of bytes to be added as well as the starting address where the bytes are stored. This subroutine is expected to return the total sum, stored in specific registers, to the calling program. This exchange of information between the calling program and the subroutine is called parameter passing. Therefore, it is essential that each subroutine should provide the necessary information to the user of the subroutine. The subroutine documentation should include the following elements:

1. Function of a subroutine
2. Input parameters
3. Output parameters
4. Registers modified
5. List of subroutines called

The delay subroutine discussed in Example 7.2 illustrates one example of documentation.

Function of a subroutine

The function of a subroutine should describe briefly what a subroutine does. From this description, the user should be able to recognize the appropriateness of its use. In Example 7.2, the function is described in one sentence that it provides 50 μs delay.

Input

Input is the information that should be provided by a calling program. In Example 7.2, the calling program does not need to provide any information to the subroutine. In Example 7.3, discussed in the next section, the total delay can be obtained as multiples of 50 μs by setting the value of REG11 in the calling program. The information (the byte) in REG11 is the input parameter to the subroutine.

Output

Output is the information or the result provided by a subroutine and sent back to a calling program. For example, the subroutine that adds the number of bytes is expected to provide the sum to be sent back to the calling program; this will be described as an output parameter.

Registers Modified

Registers Modified is a list of registers that are changed by a subroutine. If a calling program is using the same registers as that of the subroutine, the calling program can save the contents of these registers before calling the subroutine and retrieve the contents after the end of the subroutine. Another option is that the subroutine can save the contents of the registers it plans to use on the stack at the beginning of the subroutine and retrieve the contents just before the end of the subroutine. In this option, no calling program registers will be modified. In PIC18 instruction set, the instruction CALL FAST (or s = 1) saves the contents of W, STATUS, and BSR registers in the shadow registers, and the instruction RETURN FAST (or s = 1) retrieves the contents of these registers before returning to the calling program.

List of Subroutines Called

If a subroutine is calling other subroutines, they should be listed. This information enables the user to check the other subroutines called and provide the necessary parameters.

EXAMPLE 7.3

Rewrite Illustrative Program: Generating Waveforms (Section 5.5) to generate a square wave of 2 kHz – 250 μs on-time–using the subroutine technique. Modify Example 7.2 to provide 250 μs delay – as a multiple of 50 μs delay by passing an appropriate parameter from the main program to the subroutine.

Solution:

The subroutine in Example 7.2 provides 50 μs of delay (clock frequency = 40 MHz). To obtain 250 μs delay, we need to repeat the loop 5 times, and this can be accomplished by having one more loop with a register outside the first loop (see Example 5.12). To provide the delay of multiples of 50 μs, the parameter for this outside loop can be provided by the calling program.

Program

```
REG1       EQU    0x01     ;Define addresses of Data registers
REG10      EQU    0x10     ;Register for 50µs delay count
REG11      EQU    0x11     ;Register to specify multiple of delay

           ORG    0x20
START:     MOVLW  B'11111110' ;Byte to set up Bit0 as an output
           MOVWF  TRISC    ;Initialize Bit0 as an output pin
           MOVWF  REG1     ;Save Bit pattern in REG1
           MOVLW  05       ;Count to get multiple of 50 µs delay
           MOVWF  REG11    ;Load count to get 250µs delay

ONOFF:     MOVFF  REG1,PORTC ;Turn on/off Bit0
           CALL   DELAY
           COMF   REG1,1   ;Complement Bit Pattern
           BRA    ONOFF    ;Go back to change LED

DELAY:     ;::::::::::::::::::::::::::::::::::::::::::::::::::::::::::::::::
           ;Function: This subroutine provides multiple of 50 µs delay  :
           ;Input:    Count of delay multiples in REG11                 :
           ;Output:   None                                             :
           ;Registers modified: WREG and data registers REG10 and REG11:
           ;::::::::::::::::::::::::::::::::::::::::::::::::::::::::::::::::

DELAY50MC: MOVLW  D'166'   ;Load decimal count in W
           MOVWF  REG10    ;Set up REG10 as a counter
```

continued

```
   LOOP1: DECF   REG10,1      ;Decrement REG10
          BNZ    LOOP1        ;Go back to LOOP1 if REG10 ≠ 0
          DECF   REG11,1      ;Decrement multiple
          BNZ    DELAY50MC
          RETURN
          END
```

Description:

The main program loads 05_{10} in REG11 and calls the DELAY subroutine. The count 05_{10} is an input parameter passed on to the subroutine. The label DELAY serves the same function as the label DELAY50MC; it is used here to differentiate this subroutine from the previous $50\,\mu s$ delay subroutine. The REG11 sets up a second loop outside the first loop. Every time the MPU falls out of LOOP1, it decrements REG11 and repeats the $50\,\mu s$ delay subroutine. Thus, the $50\,\mu s$ delay subroutine is repeated five times and provides delay slightly more than $250\mu s$, a discrepancy that occurs because the delay in the outside loop is not considered in the calculation. The delay subroutine provides approximately $250\,\mu s$ delay but the program does not generate a square wave or turn on/off LED as expected. (See Simulation Exercise 7.2SE.)

7.3 MACROS AND SOFTWARE STACK

A macro is a group of assembly language instructions that can be labeled with a name, and the name can be written in a program to represent the instructions. It is a shortcut provided by the assembler. The MPU has no instructions in its instruction set to call a macro. When the program is assembled, the assembler substitutes the macro label with the instructions it represents. For example, we can write a set of instructions once that provides $50\,\mu s$ delay, label it in the macro format (described later), and use that label as many times as needed. We can pass parameters to a macro and change the delay. Now the question is: what is the difference between a subroutine and a macro and what are the advantages and disadvantages of one over the other?

7.3.1 Subroutine versus Macro

As we have seen in the previous section, a subroutine is a group of instructions that performs a task, is written separately once, and can be called by using CALL instructions. When the subroutine is called, its return address is stored on the stack. If a subroutine is called three times, the MPU goes to the subroutine location three times and returns to the calling program. On the other hand, when a macro label is written three times in a program, the assembler will substitute all the instructions in the assembled program three times. The advantages and disadvantages are as follows:

1. A subroutine is based on the MPU instructions such as CALL and RETURN and requires the stack. A macro is based on the assembler; it is a shortcut in writing assembly code provided by the assembler and its pseudocodes. A macro is a command to the assembler and not to the processor. Not all assemblers will provide the macro pseudocodes.

2. The memory space required by a subroutine does not depend on how often it is called. The memory space needed by a macro, on the other hand, depends on the number of times it is used in the program.

3. In execution, a macro is more efficient than a subroutine because the macro does not have the overhead of executing Call and Return instructions. In addition, it is simpler to pass on parameters in a macro (as explained in the next example).

7.3.2 Macro Format and Directives

The macro format includes three parts:

1. Header: This includes a name (label) of the macro in the label field, MACRO as pseudo-op (or code), and a list of arguments in the operand field.

   Example: DELAY   MACRO   VAR1, VAR2, VAR3

2. Body of Instructions: A set of assembler instructions and pseudocodes that specifies the task of the macro (see Example 7.4).

3. Termination: ENDM (End of macro) is the pseudocode used to indicate the end of macro.

EXAMPLE 7.4

Rewrite the square wave generation program in Example 7.3 with a macro in place of the subroutine to provide 250 µs delay.

Solution:

Figure 7–3 shows the screen shot of the macro labeled as DELAY. The source file of the square wave generation program is on the left-hand side, and the contents of the program memory are on the right-hand side.

   The macro shown in Figure 7–3 includes all the three elements: (1) pseudocode MACRO with a variable COUNT, (2) the body of the macro, including the assembly language statements to generate delay, and (3) pseudocode ENDM to terminate the macro.

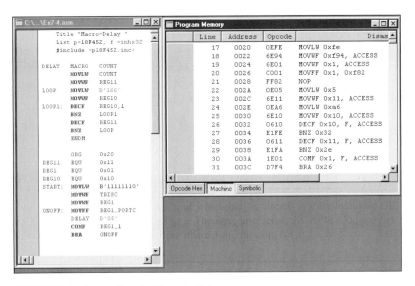

FIGURE 7-3 Screen Shot for Example 7-4

The program begins at the label START after defining all the registers with equates. The CALL instruction of Example 7.3 in Section 7.2.2 is replaced by the macro label DELAY with the parameter 05. The macro is a part of the assembler and not assigned any memory locations until it is assembled. Examine the contents of the Program Memory shown in Figure 7–3 and note the following points. The program assembly begins at location 0020_H, according to the ORG statement in the program.

1. The macro statement DELAY D'05' is replaced by the code of the macro assembly statements, and the value of the parameter COUNT is given as 05. Therefore, the delay loop of 50 µs is repeated five times to provide a 250 µs delay.
2. The assembly of the macro begins at location $002A_H$ and terminates at the location 0038_H. This is followed by the Hex code of the instruction COMF REG1.
3. The assembled program does not include any call and return instructions and the memory locations are assigned to the macro statements based on where they are written in the program. If this macro statement were to be used by some other program with the parameter 10, the macro would provide 500 µs delay and be assembled where the program is located.

7.3.3 Software Stack

Most microprocessors and microcontrollers have a register, called the stack pointer. It is the same size as the program counter, and these MCUs (and MPUs) use a part of R/W memory as the stack. The stack pointer holds the address of the top of the stack, and as information is saved (push) on the stack or retrieved (pop) from the stack, the address in the stack pointer is modified to always show the top of the stack. In these MCUs, the size of the stack can be almost unlimited, determined only by the size of R/W memory used for the stack. However, in the PIC18 MCU, a group of 31 registers, separate from program memory and data registers, are used as the stack. The group of these registers is called the hardware stack, which is used to store return addresses of the subroutines and interrupt service routines (ISRs), discussed in Chapter 10. When a subroutine is called, one stack location is needed to store the return address, but that location becomes available after the return instruction. If subroutines call other subroutines, multiple locations of the stack are required. In most applications of PIC MCUs, we are less likely to run out of the stack space. However, there are applications in which local variables are saved on the stack before calling a subroutine, and additional stack space may be necessary. In these situations, a software stack can be devised by using data registers and file select registers (FSRs). Any bank of the data registers can be used for a software stack. Generally, the highest memory locations available in the MCU are used for the software stack; therefore, we will use here the memory locations 500_H to $5FF_H$ for our illustration of a software stack. In Example 7.3, the delay subroutine uses REG10 and REG11. It is possible that these registers are also being used by other subroutines or the main program. To avoid any corruption or modification to the information located in other parts of the program, we can save the contents of these registers on the software stack before using them in the delay subroutine. In turn, we can retrieve the contents just before completion of the subroutine as illustrated in Example 7.5

EXAMPLE 7.5

Define the beginning of the software stack for PIC18F452 microcontroller. In Example 7.3, save the contents of REG10 and REG11 on the software stack and retrieve the contents before the completion of the subroutine.

Solution:

Instructions

	LFSR	FSR1, 0x500	;Initialize software stack pointer at data register 500H
	MOVFF	REG10, POSTINC1	;Save contents of REG10 in 500H and increment FSR1
	MOVFF	REG11,POSTINC1	Save contents of REG11 in 501H and increment FSR1
DELAY:			
DELAY50MC:	MOVLW	D'166'	;Delay subroutine
	↓		
	↓		
LOOP1:	DECF	REG10,1	
	BNZ	LOOP1	
	MOVFF	POSTDEC1, REG11	;Decrement FSR1
	MOVFF	INDF1, REG11	;Retrieve contents of REG11

continued

```
        MOVFF     POSTDEC1, REG10    ;Decrement FSR1
        MOVFF     INDF1, REG10       ;Retrieve contents of REG10
        RETURN
```

Description:

The first instruction sets up the FSR1 as a pointer to data register 500_H, thus initializing the software stack pointer. The second instruction MOVFF saves contents of REG10 in data register 500_H and increments the pointer to 501_H. The next instruction saves the contents of REG11 into data register 501_H and increments the pointer to 502_H. In the delay loop, REG10 and REG11 are used to generate a delay subroutine, and the next group of instructions retrieves the previous contents of REG10 and REG11. Here two points should be noted: (1) the information is retrieved in "last-in, first-out" order (LIFO); in this case, that would mean that information is retrieved in REG11 first followed by REG10. (2) The instruction set does not include a PRE-DEC procedure; therefore, we decremented the pointer by using the instruction POSTDEC1 and copied the contents of registers by using INDF1.

7.4 ILLUSTRATIVE PROGRAM: COPYING AND ADDING DATA BYTES

In previous sections, we discussed three concepts: stack, subroutine, and macro. These concepts helped us to break large programs into small modules and, in turn, to write these modules. The illustrative program below demonstrates how to integrate these three concepts when writing a program.

The illustrative program copies a number of bytes from the program memory to data registers, adds the data bytes, and displays the sum at the I/O ports. It consists of two subroutines: one for copying data bytes and the other for adding those data bytes. The subroutines are written as generally as possible so that they can be used in other programs.

7.4.1 Problem Statement

Write a program to copy a given number of unsigned 8-bit numbers from program memory into data registers starting from a specified data register, add the numbers, and display the 16-bit sum at PORTB and PORTC. The program should be subdivided into general-purpose subroutines.

7.4.2 Problem Analysis

This problem can be very easily divided into either two or three subroutines: data copy and addition, and the third one can be displayed. In analyzing this problem, the user must make a decision about the tradeoff between a general type subroutine that can be used by anyone for a similar problem, or a specific subroutine that is applicable only to this particular problem. The more general a subroutine, the more parameters must be passed on to the subroutine from a main program.

For this problem, we will divide the program into three segments: MAIN, and two subroutines: DATACOPY and ADDITION. The MAIN segment will initialize the I/O ports, call the subroutines, and display the result at the I/O ports. This segment also illustrates a macro that defines the number of bytes.

The DATACOPY subroutine is similar to the Illustrative Program in Section 5.6: Copying a Block of Data from Program Memory to Data Registers, except that the program is written here as a subroutine. The MAIN program must pass on the number of bytes to be copied and where to be copied to the subroutine.

The ADDITION subroutine adds the bytes as illustrated in Section 5.7: Addition of Data Bytes. Again, for this subroutine, the MAIN program must pass on the number of bytes to be added and where they are stored.

PROGRAM 7.4.3

```
;::::::::::::::::::::::::::::::::::::::::::::::::::::::::::::::::::::::::::::::
;This program copies unsigned data bytes from program memory into data     :
;registers and adds the bytes. The 16-bit sum is displayed at PORTB and    :
;PORTC. The program includes two subroutines and one macro. The main       :
;program provides the parameters: the number of bytes and their memory     :
;locations to the subroutines.                                             :
;::::::::::::::::::::::::::::::::::::::::::::::::::::::::::::::::::::::::::::::

        Title "IP7-4 Coping and Adding Data Bytes"
        List p=18F452, f=inhx32
        #include <p18F452.inc>      ;The header file

BUFFER  EQU     0x10                ;Define the beginning data register address
COUNTER EQU     0x01                ;Set up register 01 as a counter


        ORG     00

        GOTO    MAIN
BYTES   MACRO   COUNT               ;Macro to provide count for bytes
        MOVLW   COUNT
        MOVWF   COUNTER             ;Load counter with a number
        ENDM


        ORG     0x20                ;Begin assembly at 0020H
MAIN:   MOVLW   0x00                ;Byte to initialize port as an output port
        MOVWF   TRISB               ;Initialize Ports B &C as output ports
        MOVWF   TRISC
        LFSR    FSR0, BUFFER        ;Set up FSR0 as pointer for data registers
        BYTES   05                  ;Macro - Count = 05,
        CALL    DATACOPY            ;Copy bytes from program memory to data
                                    ;registers
```

                                                    continued

```
          LFSR    FSR0, BUFFER    ;Set up FSR0 as pointer for data registers
          BYTES   05              ;Macro - Count = 05, number of  bytes to
                                  ;be added
          CALL    ADDITION        ;Add bytes
          MOVWF   PORTC           ;Display low-order byte of sum at PORTC
          MOVFF   CYREG, PORTB    ;Display high-order byte of sum at PORTB
          SLEEP


DATACOPY:;::::::::::::::::::::::::::::::::::::::::::::::::::::::::::::::::::::::
          ;Function: This subroutine copies data bytes from the program    :
          ;          memory SOURCE to data registers BUFFER.               :
          ;Input:    The pointer to BUFFER and the number of bytes to be   :
          ;          copied                                                :
          ;::::::::::::::::::::::::::::::::::::::::::::::::::::::::::::::::::::::

          MOVLW   UPPER SOURCE    ;Set up TBLPTR pointing to Source
          MOVWF   TBLPTRU
          MOVLW   HIGH SOURCE
          MOVWF   TBLPTRH
          MOVLW   LOW SOURCE
          MOVWF   TBLPTRL
    NEXT:TBLRD*+                  ;Copy byte in Table Latch and increment
                                  ;source pointer
          MOVF    TABLAT,0        ;Copy byte from Table Latch in W
          MOVWF   POSTINC0        ;Copy byte from W into BUFFER and
                                  ;increment FSR0
          DECF    COUNTER,1,0     ;Decrement count and store it in Counter
          BNZ     NEXT            ;Is copying complete? If not go back and
                                  ;copy next byte
          RETURN
SOURCE:   DB   0xF6,0x67,0x7F,0xA9,0x72


ADDITION:;::::::::::::::::::::::::::::::::::::::::::::::::::::::::::::::::::::::
          ;Function: This subroutine adds data bytes stored in data        :
          ;          registers  BUFFER.                                    :
          ;Input:    The pointer to BUFFER and the number of bytes to be   :
          ;          copied                                                :
          ;Output:   The 16-bit sum in CYREG and WREG                      :
          ;::::::::::::::::::::::::::::::::::::::::::::::::::::::::::::::::::::::
CYREG     EQU     0x02            ;Define register to save carries
          CLRF    CYREG           ;Clear carry register
          MOVLW   0x00            ;Clear W register to save sum
NEXTADD:  ADDWF   POSTINC0,W      ;Add byte and increment FSR0
          BNC     SKIP            ;Check for carry-if no carry jump to SKIP
          INCF    CYREG           ;If there is carry, increment CYREG
SKIP:     DECF    COUNTER,1,0     ;Next count and save count in register
          BNZ     NEXTADD         ;If COUNT ≠ 0, go back to add next byte
          RETURN
          END
```

7.4.4 Program Description

The equats (EQUs) define the registers BUFFER, which is a starting location to which the data bytes are to be copied, and the COUNTER that specifies the number of bytes to be copied. The first part of this program is a macro labeled BYTES that defines the number of bytes to be copied and added. The macro is written here to demonstrate an application of a macro; otherwise, we could have written two instructions to initialize the counter.

The MAIN program initializes PORTB and PORTC as output ports and the address of the first data register where the copying of the bytes begins. It specifies the number of bytes in the macro by substituting 05 in place of the parameter COUNT and calls the subroutine DATACOPY. This subroutine takes the parameters for COUNTER and BUFFER and copies five data bytes starting from the data register 10_H. It then returns to the MAIN program. The same parameters are again passed on to the next subroutine, ADDITION, which adds the number of bytes and passes the sum to the MAIN, high-order byte in CYREG and low-order bytes in W register. The MAIN program displays the sum at PORTB and PORTC.

7.4.5 Program Execution and Troubleshooting

The program execution and troubleshooting in modularized programs is very systematic and logical. You should know the output of each module and test each module by setting up the breakpoint at the end of each module or the memory location in the main program where parameters can be observed.

To test this program, you should assemble the program using MPLAB IDE, set up two breakpoints in the MAIN program as suggested, and follow the steps listed below:

1. In MPLAB IDE, go to View and add the following registers in the Watch window: WREG, FSRL0, COUNTER, BUFFER, PORTB, and PORTC.
2. In View, click on File Registers, and the screen with register addresses will be displayed.
3. Set up two breakpoints in the source program: one breakpoint immediately after the instruction CALL DATACOPY and the second one after the instruction CALL ADDITION
4. Set up the cursor at START and single-step (F7) the program until the first call instruction. You should observe the following: COUNTER = 05_H. FSRL0 holds the address of the first register 10_H where the first byte will be copied.
5. Now run the program (press F9). The MPU will execute the first subroutine: DATACOPY. Then, if there is no error, all five data bytes should appear in the second row registers, with addresses in 10_H to 14_H and the COUNTER = 0. If these registers do not show the bytes from the source, you have an error in the DATACOPY subroutine. You should single-step the subroutine and correct the errors.
6. Now you can run (F9) the second subroutine and single-step the two instructions that display the output, and you should observe the sum $02F7_H$ at PORTB and PORTC and CYREG = 02. If the sum is different from the expected result, you should single-step the ADDITION subroutine.

7.4.6 Examining Hardware Stack

To understand how subroutines are executed, you should observe the hardware stack and the contents of the program memory. In the MPLAB IDE, you should go to View and click on Hardware Stack and Program Memory and two screens will appear; one will show the contents of the stack with 31 registers and the other will display the memory addresses and their contents. You should note the following points before you begin the execution of the program.

1. The program memory shows the addresses from memory location 0000, but our program is assembled at location 0020_H (the first two 00s of the 21-bit address are not shown) excluding the codes of the GOTO instruction, so you will begin to see the Hex code stored starting from location 0020_H. The first CALL is located at $002E_H$, and the address of the next instruction is 0032_H. Similarly, the second CALL is located at 0036_H, and the address of the next instruction to the second CALL is $003A_H$.

2. When you execute the first CALL instruction located at $002E_H$, the address of the next instruction 0032_H is stored on the Hardware Stack in register 1, indicating that register 1 is the top of the stack (TOS).

3. When the RETURN instruction is executed, the MPU finds the address 0032_H on the top of the stack, returns the execution back to the main program, and decrements the stack address to 0.

4. When the second CALL instruction is executed, the address $003A_H$ is placed on the stack in register 1, and this address is retrieved when the RETURN instruction is executed.

7.5 ILLUSTRATIVE PROGRAM: CALCULATING AVERAGE TEMPERATURE

In Section 7.4, we illustrated how to modularize a program by dividing it into small modules and writing those modules as subroutines. The main program supplies the necessary parameters to these subroutines and calls the subroutines. In this section, we continue the same concept but with a different example, extending the illustration to include a subroutine calling another subroutine.

This illustration copies a set of temperature readings from program memory to the data registers, adds the readings, finds the average of the readings, and displays the temperature at one of the I/O ports.

7.5.1 Problem Statement

A set of 8-bit temperature readings taken during a given autumn season are recorded and stored in program memory starting at location SOURCE. Copy the reading from program memory to data registers starting at location BUFFER, add the readings, find the average temperature, and display the temperature at PORTC.

7.5.2 Problem Analysis

A careful reading of the problem statement provides all the clues needed to divide the problem into four segments: (1) copy the temperature readings from program memory to data registers, (2) add the readings to find the sum, (3) find the average by dividing the sum by the number of the readings, and (4) display the average temperature at PORTC.

The first three segments can be written as subroutines. Displaying the temperature, which consists of one or two statements, can be included in the main program. The first segment, copying data from program memory to data registers, was discussed in Section 7.4.

The second segment, adding the readings, was also illustrated in Section 7.4. Rather than using the same method, however, we will use the method illustrated in Chapter 6, Section 6.5. We will convert the problem in Section 6.5 into two subroutines: SUM and AVG. The SUM subroutine adds all the readings and calls the subroutine AVG that finds the average of the readings.

PROGRAM 7.5.3

```
;::::::::::::::::::::::::::::::::::::::::::::::::::::::::::::::::::::::::::
;This program copies 8 temperature readings stored in program  memory at  :
;location SOURCE to data registers starting at BUFFER.                    :
;It adds the readings and finds the average temperature reading and       :
;displays at PORTC                                                        :
;::::::::::::::::::::::::::::::::::::::::::::::::::::::::::::::::::::::::::

        Title "IP7-5 Calculating Average Temperature"
        List p=18F452, f =inhx32
        #include <p18F452.inc>  ;The header file

BUFFER   EQU    0x10            ;Define the beginning data register address
COUNTER  EQU    0x01            ;Set up register 01 as a counter
         ORG    0x20            ;Begin assembly at 0020H
BYTES    MACRO  COUNT           ;Macro to provide count for bytes
         MOVLW  COUNT
         MOVWF  COUNTER
         ENDM

MAIN:    MOVLW 0x00             ;Byte to initialize port as an output port
         MOVWF TRISC
         LFSR  FSR0, BUFFER     ;Set up FSR0 as pointer for data registers
         BYTES 08               ;Macro that specifies the number of bytes
         CALL  DATACOPY         ;Copy data from program memory to data
                                ;registers
         LFSR  FSR0, BUFFER
         BYTES 08
         CALL  SUM              ;Find average temperature
         MOVFF TEMP,PORTC       ;Display temperature reading at PORTC
         SLEEP
```

continued

```
SOURCE:    DB     0x41,0x45,0x40,0x42
           DB     0x40,0x41,0x4F,0x50

DATACOPY:;::::::::::::::::::::::::::::::::::::::::::::::::::::::::::::::::::::::::::
           ;Function: DATACOPY copies the number of bytes from program      :
           ;          memory SOURCE to data registers starting from BUFFER  :
           ;Input:    The address of BUFFER in FSR0 and number of bytes in  :
           ;          COUNTER                                               :
           ;::::::::::::::::::::::::::::::::::::::::::::::::::::::::::::::::::::::::::

           MOVLW UPPER SOURCE ;Set up TBLPTR pointing to Source address
           MOVWF TBLPTRU
           MOVLW HIGH SOURCE
           MOVWF TBLPTRH
           MOVLW LOW SOURCE
           MOVWF TBLPTRL
  NEXT:    TBLRD*+            ;Copy byte in Table Latch and increment
                              ;source pointer
           MOVF  TABLAT,0     ;Copy byte from Table Latch in W
           MOVWF POSTINC0     ;Copy byte from W into data register BUFFER
                              ;and increment FSR0
           DECF  COUNTER,1,0  ;Decrement count and store it in Counter
           BNZ   NEXT         ;Is copying complete?  If not go back and
                              ;copy next byte
           RETURN

  SUM:     ;::::::::::::::::::::::::::::::::::::::::::::::::::::::::::::::::::::::::::
           ;Function:  The subroutine SUM adds 8-bit numbers resulting into :
           ;           16-bit sum in registers HI_SUM and LO_SUM and calls  :
           ;           another subroutine AVG to find the average           :
           ;Input:     The address of data registers in FSR0, where numbers :
           ;           are stored  and number of bytes in COUNTER           :
           ;Output:    The average temperature reading in data register     :
           ;           TEMP                                                  :
           ;Calls:     Subroutine AVG to find the average reading           :
           ;::::::::::::::::::::::::::::::::::::::::::::::::::::::::::::::::::::::::::

LO_SUM EQU    0x03           ;Define addresses of data registers
HI_SUM EQU    0x04           ;to store sum
TEMP   EQU    0x05           ;Data register to store average temperature
CFLAG  EQU    0              ;Bit 0 in STATUS register
       CLRF   WREG,0         ;Clear WREG for addition
       CLRF   LO_SUM         ;Clear registers to save sum
       CLRF   HI_SUM

  ADD:     ADDWF POSTINC0,0,  ;Add data byte and save sum in WREG
           BTFSC STATUS,C     ;Check C flag. If C = 0, skip next instruction
           INCF  HI_SUM       ;If C = 1, increment HI_SUM
           DECF  COUNTER,1,0  ;One addition done - decrement count
           BNZ   ADD          ;If count ≠ 0, go back to add next byte
```

continued

```
        MOVWF LO_SUM
        MOVLW 0x03            ;
Shifting right three times - set up counter for 3
        MOVWF COUNTER
        CALL  AVG
        RETURN


AVG:    ;:::::::::::::::::::::::::::::::::::::::::::::::::::::::::::::::::::::::
        ;Function: Finds the average of 8-bit data bytes. The total      :
        ;          number of bytes should be in powers of two. It divides :
        ;          the sum by using the instruction Rotate Right through  :
        ;          Carry                                                  :
        ;Input:    The total sum in HI_SUM and LO_SUM registers and      :
        ;          powers of two in COUNTER                              :
        ;Output:   The average in register TEMP                          :
        ;:::::::::::::::::::::::::::::::::::::::::::::::::::::::::::::::::::::::
CLEARCY: BCF   STATUS, CFLAG ;Clear C flag
         RRCF  HI_SUM            ;Divide byte in HI_SUM by 2
         RRCF  LO_SUM            ;Divide byte in LO_SUM by 2
         DECF  COUNTER,1,0       ;Decrement rotation count
         BNZ   CLEARCY           ;If counter ≠ 0, go back to next divide
         MOVFF LO_SUM, TEMP  ;Save result for display
         RETURN
         END
```

7.5.4 Program Description

The MAIN program initializes the necessary registers as in Section 7.4, loads the counter with 8 to copy eight data bytes, sets the pointer in FSR0 to BUFFER location, and calls the subroutine DATACOPY. During the execution of the subroutine DATACOPY, the address in FSR0 is modified. Therefore, the MAIN program sets the pointer again to BUFFER and calls the subroutine SUM. The SUM subroutine adds all the readings and stores the 16-bit sum in two registers: HI_SUM and LO_SUM. To calculate the average reading, the sum should be divided by eight. That can be accomplished by shifting the sum to the right three times. Therefore, before calling the subroutine AVG (Average), the counter is loaded with the number 3. The AVG subroutine uses the sum stored in HI_SUM and LO_SUM registers, calculates the average by shifting the 16-bit number three times to the right, and returns the reading in register TEMP. The SUM subroutine returns the execution to the MAIN program, which displays the average temperature reading at PORTC.

7.5.5 Program Execution and Troubleshooting

This program has three subroutines, and the output and the proper functioning of each can be verified by setting the breakpoints at appropriate locations. Assemble the program using MPLAB IDE and follow the suggested steps.

1. Go to View and click on File Registers. The screen with the addresses of the file registers will be displayed. From View menu, click on Watch and add the following registers in the Watch window: WREG, COUNTER, FSR0L, LO_SUM, HI_SUM, and PORTC.
2. Set up the first breakpoint in the MAIN program at the instruction CALL DATACOPY and run the program from the label MAIN. It will stop at the CALL location, and you can examine the initial conditions as follows: COUNTER = 8, FSR0L = 10.
3. Now set up the next breakpoint at the location immediately after the call at the location of LFSR FSR0 and run the program. The MPU executes the DATACOPY subroutine and displays the eight data readings in file registers starting from 10_H to 17_H. This will confirm that the DATACOPY subroutine functions properly.
4. Now set up the next breakpoint in the SUM subroutine at the location of CALL AVG and run the program. This completes the execution of the SUM subroutine that adds all the eight bytes. The sum should be 0228_H; it should be displayed at registers HI_SUM and LO_SUM (this can be verified by manually adding the numbers). If you find an error, you should go back to the previous breakpoint and single-step the instructions in the SUM subroutine.
5. Now go back to the MAIN program, set up the breakpoint at the output instruction MOVFF TEMP, PORTC, and run the program. The register TEMP should show that the average temperature is 45_H degrees.

7.5.6 Extending the Program to a Real Life Application

This program adds the temperature readings, finds the average reading in Hex, and displays it at PORTC in binary. However, in real life, we want temperatures to be displayed in BCD at seven-segment LEDs. Therefore, to display the temperatures in decimal at the seven-segment LEDs, the following steps are necessary:

• Convert the average temperature reading from 45_H into 69_{BCD}.
• Unpack the decimal reading 69 into 06 and 09.
• Find the seven-segment code for the digit 6 and digit 9.
• Display these digits at two seven-segment LEDs.

All of the above steps can be written as subroutines; some of these subroutines are illustrated in Chapter 8. This program can be further extended to conversions of temperature readings from Fahrenheit to Celsius and vice versa.

SUMMARY

This chapter discussed the concepts of stack, subroutine, and macro. The subroutine concept provides modularity: a large program can be divided into small modules, and these modules can be written as subroutines and tested individually. This approach provides flexibility in writing and ease in debugging the program. The important concepts discussed in this chapter are summarized as follows:

1. A stack is a temporary storage space, a part of R/W memory or a specially designed group of registers called hardware stack, used during the execution of a program.

2. A stack pointer is a register, similar to the program counter in function, that holds the address of the top of the stack. This address is modified as the information is stored or retrieved from the stack.

3. The PIC18 microcontroller has 31 registers, 21-bit wide, that are used as the stack, which are separate from program memory and data registers. The stack pointer that holds the address of the top of the stack (TOS) is 5 bits wide.

4. The PIC18 instruction set includes two instructions—PUSH and POP—that are used to store and retrieve information from the stack. The PUSH instruction increments the stack pointer and stores the contents of the program counter (PC + 2) on the stack and the POP instruction discards the contents of the stack and decrements the stack pointer by one.

5. A subroutine is a group of instructions that performs a task. It is written separately, independent of the main program, and can be called to perform that task when needed by the main program or another subroutine.

6. The PIC18 instruction set includes two CALL instructions: one is an absolute CALL instruction that can call a subroutine located anywhere in memory, and the other is a relative CALL instruction RCALL that is limited to call a subroutine within ±512 words. The CALL is a two-word instruction, and the RCALL is a one-word instruction.

7. When a CALL or RCALL instruction is executed, the MPU increments the address in the stack pointer, stores the address of the next instruction (return address) on the top of the stack, and jumps to the call location.

8. The PIC18 instruction set includes two return instructions: RETURN and RETLW. When the RETURN instruction is executed, the MPU goes to the top of the stack, gets the return address, decrements the address in the stack pointer, and jumps to the return address to continue the execution of the program. The RETLW instruction performs the same way as the RETURN instruction, except that it returns an 8-bit number (literal) to WREG.

9. The absolute CALL instruction has two formats: (1) CALL Label and (2) CALL Label, FAST (or CALL Label, 1). In the first format, the MPU stores only the return address on the stack, and in the second format, the MPU stores the return address on the stack and the contents of WREG and BSR registers in their shadow registers.

10. If the calling program is required to provide parameters (values) to a subroutine, the parameters are called input parameters. If the subroutine returns a value to the calling program, it is called an output parameter. This process of information exchange between the calling program and the subroutine is called parameter passing.

11. A macro is a group of assembly language instructions that can be labeled with a name, and the name can be written in a program to represent the instructions. It is a shortcut provided by the assembler.

The MPU has no instruction in its instruction set to call a macro. When the program is assembled, the assembler substitutes the macro label with the instructions it represents.

12. A subroutine is written once and can be called multiple times. On the other hand, the macro, when it is being assembled, is duplicated as many times it is used. The macro is somewhat more efficient in execution of the program but may require a lot more memory space.

QUESTIONS AND ASSIGNMENTS

7.1   Define the stack and the stack pointer.

7.2   How many registers are available as the stack in PIC18 microcontroller and what is the size of the stack pointer?

7.3   Explain what stack overflow and underflow are.

7.4   What is stored on the stack when the PUSH instruction is executed by the PIC18 MPU?

7.5   If the PUSH instruction, stored at location 000028_H, is executed, what is stored on the stack? What is the address in the stack pointer if the address was 00100_B before the execution?

7.6   In Question 7.5, what is stored in TOS register?

7.7   If the POP instruction is executed immediately after the PUSH instruction in Question 7.5, specify the contents of the stack pointer and the TOS register after the execution.

7.8   When a subroutine is called, what is stored on the stack?

7.9   If the CALL instruction, located at 0032_H, is executed, what is stored on the stack?

7.10 If the RCALL instruction, located at 0032_H, is executed, specify the address that is stored on the stack.

7.11 Explain the difference between the execution of the instructions CALL DELAY, CALL DELAY, FAST, and CALL DELAY, 1.

7.12 In Example 7.2, if the instruction "COMF REG1,1" in the main program is replaced by the instruction "COMF REG1,0", specify the output of PORTC.

7.13 In Example 7.2, replace the instruction "CALL DELAY50MC" by the instruction "RCALL DELAY50MC", specify the address that is stored on the TOS (Refer to Figure 7–2).

7.14 Explain the terms input and output parameters to a subroutine.

7.15 Explain the concept of the software stack in PIC18 microcontroller.

7.16 Define the term macro.

7.17 What is the advantage of writing a macro over a subroutine?

7.18 In Section 7.4.3 modify the macro BYTES to include the instruction LFSR FSR0, BUFFER.

7.19 Modify the main program to include the macro from Question 7.18

7.20 Illustrative Program 7.4 adds five bytes using the counter technique. Rewrite the program where the number of data bytes to be added is variable, but the data string is terminated in the byte 00.

7.21 In Section 7.5, the Illustrative Program includes two subroutines SUM and AVG that are written such a way that the average can be found only for eight readings. Modify the program that can find the average of any number of powers of two readings.

7.22   Modify the Illustrative Program in Section 7.5 to find the average of 24 readings. (Hint: Rewrite the subroutine AVG to find the average by dividing the sum by repeated subtraction.)

SIMULATION EXERCISES

Using PIC18 Simulator IDE.

7.1 SE   Assemble and load the program of Example 7.2 using PIC18 Simulator IDE. Connect the 8 x LED board to PORTC (→ on Tools and → 8 x LED Board) . Observe the LED turn on/off under various simulation rates.

7.2 SE   Assemble the program of Example7.3 at the memory location 00 and load the program in the simulator's memory. Open the LED board and connect Bit0 of PORTC to one of the LEDs. Execute the program with the simulation rate without refresh (Option –Ultimate). Does the program turn on and off the LED? If the LED is turned on once and remains on, troubleshoot the program by setting a breakpoint at the Return instruction of the delay subroutine.

7.3 SE   Modify the program of Example 7.3 to turn on/off LEDs alternately connected to Bit0 and Bit1 at the interval of 500 µs. The program should simulate the railway crossing lights that flash alternately.

Using MPLAB IDE

7.4 SE   Given a data string of signed numbers with a specific number of bytes, write a subroutine that adds only the positive numbers and returns the sum in data registers REG10 and REG11 (the 16-bit sum can be in the unsigned format). The main program should pass on the following parameters. The number of bytes in a string and the beginning address of the data string. Simulate the output of the subroutine using the following data. You can enter the data manually in File Registers.

Data (H): 0x32, 0x89, 0x78, 0x19, 0x37, 0xF5, 0x98, 0x69, 0x47, 0x57.

7.5 SE   Modify the Illustrative Program: Copying and Adding Data Bytes (Section 7.4) as follows: The number of given data bytes in a string is variable; however, the string is terminated in the null character 00. Copy the data bytes to data registers with the starting address BUFFER. Count the number of bytes copied and pass that parameter to the subroutine ADDITION. Assemble and simulate the execution of the program using MPLAB IDE.

7.6 SE   Set up the breakpoints to verify the output of the subroutines DATACOPY and ADDITION. Display the data bytes copied in File Registers (data registers) in View window. Display the sum at PORTB and PORTC.

Data(H): 0xF2, 0xA2, 0x69, 0x87, 0x9F, 0xF7, 0x76, 0x19, 0x00

CHAPTER **8**

APPLICATION PROGRAMS AND SOFTWARE DESIGN

OVERVIEW

In Chapters 5 and 6, we introduced frequently used instructions of the PIC18 instruction set. In addition, we wrote simple programs such as copying data, adding numbers, and finding the largest (or the smallest) number in a data set. In Chapter 7, we introduced the subroutine technique that provides modularity and flexibility in designing large programs.

In this chapter, we focus on writing additional subroutines that are essential and commonly used in communication between human beings and binary machines such as microcontrollers. We, as human beings, communicate in alphanumeric symbols (letters and numbers) and microcontrollers function in a binary language; therefore, we need to translate between these two worlds. The American Standard Code for Information Interchange (ASCII) and Binary Coded Decimal (BCD) codes are commonly used for human/machine communication. This chapter illustrates code conversion subroutines such as binary to BCD and binary to ASCII and vice versa.

The PIC18 MCU is an 8-bit microcontroller that performs arithmetic operations dealing primarily with 8-bit data; however, there are occasions for which numbers larger than 8 bits (such as 16 bits or 32 bits) should be processed. Similarly, this controller does not have instructions that provide for the division of two numbers. This chapter includes two subroutines that deal with multiplication of 16-bit and division of 8-bit numbers. Finally, a simple software design project is illustrated that adds the number of 8-bit temperature readings, finds the average, and converts the reading into ASCII format that can be displayed at an ASCII peripheral such as a printer, video screen, or Liquid Crystal Display (LCD).

OBJECTIVES

- Write subroutines that perform code conversions between binary, BCD, and ASCII characters.
- Write a subroutine that multiplies unsigned numbers larger than 8 bits.
- Write a subroutine that divides 8-bit unsigned numbers.
- Illustrate a design of a simple software project that employs the modular approach using various subroutines.

8.1 BCD TO BINARY CONVERSION

In many microcontroller-based products, data are entered in decimal format. For example, to use a microwave oven, we enter time in decimal numbers, and the microwave's system program accepts data in BCD numbers. However, because the numbers A through F are invalid in the BCD number system, it is inefficient to process the data in BCD numbers. Therefore, it is necessary to convert the BCD data into binary numbers. This illustration converts a two-digit 13 BCD number into its binary equivalent.

8.1.1 Problem Statement

Given a packed BCD number in the WREG, write a subroutine to unpack the number, convert it into its equivalent binary number, and return the number in the WREG.

8.1.2 Problem Analysis

The BCD numbers include only ten digits from 0 to 9. However, the value of the digit is based on its position in a given number, called positional weighting. For example, in decimal number 97, the value of 9 is 90. The number 97 is equivalent to $(9 \times 10) + 7$. To find the binary value of 97_{BCD}, we can convert the following steps into assembly language statements.

Packed 97_{BCD} = 1001 0111

Step 1: Unpack the number   Unpacked 97_{BCD} = 0 9 (00001001) and 07 (00000111)

Step 2: Multiply high-order digit by 10 and add low-order digit = (00001001) x 10 + (0000111)

The subroutine given in the next section converts this procedure in PIC18 assembly language.

8.1.3 PROGRAM

```
;Function: BCDBIN subroutine converts a two-digit BCD number into its binary equivalent   :
; Input:    Two-digit BCD number in WREG                                                    :
; Output:   Binary equivalent of the BCD number in WREG                                     :

BUFFER1   EQU   0x10         ;Define data register addresses
BUFFER2   EQU   0X11
REG1      EQU   0x01

BCDBIN:   MOVWF REG1         ;Save packed BCD byte in REG1
          ANDLW 0x0F         ;Mask high-order digit
          MOVWF BUFFER1      ;Save unpacked low-order digit in BUFFER1
          MOVF  REG1,W,0     ;Get the byte again in WREG
          SWAPF WREG,0       ;Swap the digits
          ANDLW 0x0F         ;Mask original low-order digit
          MOVWF BUFFER2      ;Save unpacked high-order digit
          MULLW D'10'        ;Multiply high-order unpacked digit by 10
          MOVFF PRODL,WREG   ;Copy the product in WREG
          ADDWF BUFFER1,0    ;Add low-order digit
          RETURN
```

8.1.4 Program Description

When the subroutine receives a BCD number in WREG, it saves the number in REG1. It unpacks the low-order digit by masking the high-order digit, and saves the low-order digit in register BUFFER1. It gets the number again, swaps the digits and unpacks the high-order digit by masking, and saves unpacked high-order digit in register BUFFER2. Next, the subroutine multiplies the high-order digit by ten, adds the low-order digit, and returns the binary equivalent of the BCD number in WREG.

8.1.5 Program Execution and Troubleshooting

This is a subroutine, not a complete program. To verify or execute this subroutine in MPLAB IDE, you should add the following information: the header file for the PIC controller you are using, the END statement, and a specific BCD number in WREG. You can load a number in WREG by adding the MOVLW instruction or inserting a number manually in WREG in the Watch window. Assemble the subroutine and add the following registers in the Watch window: WREG, BUFFER1, BUFFER2, and PROD (or PRODL). When you single-step the instructions for a given number, you will see the number being unpacked and converted to Hex value. For example, if you begin with the BCD number 65, you should see 06 in BUFFER2, 05 in BUFFER1, and 41_H in WREG.

8.2 BINARY TO BCD CONVERSION

Even if the data processing inside the microcontrollers is done in binary, displays are generally in BCD numbers. Typical examples of BCD displays include readouts on the dashboard of a car, microwave ovens, measuring instruments, and electronic clocks. It is therefore necessary that the binary output of the MCU be converted into BCD digits. This is accomplished by dividing the binary number by powers of ten, as illustrated in the following problem. This problem converts an 8-bit binary number into three unpacked BCD numbers that can be displayed at any seven-segment LEDs.

8.2.1 Problem Statement

Given an 8-bit binary temperature reading in WREG, write a subroutine to convert the reading into equivalent BCD numbers and store the numbers as unpacked digits in buffer registers BUFF0, BUFF1, and BUFF2.

8.2.2 Problem Analysis

A binary reading is converted into BCD numbers by dividing the reading by powers of ten until a remainder is found. The largest 8-bit number is FF_H, equivalent to 255_{BCD}. Therefore, we need three registers to save BCD digits for an 8-bit number, and two divisors of powers of ten (100 and 10). Let us take an example from the binary reading of $7C_H$, equivalent to 124_{10} in decimal, and divide the number by the repeated subtraction method. We begin the subtraction with the divisor 100 (this is the largest power of ten we need for our example) and continue until the number becomes less than the divisor. Once the number becomes negative, we will add the divisor to get the value of the previous subtraction, and adjust the number of subtractions by adding one (this is equivalent to canceling the last subtraction). The number of subtractions represents the most significant digit (BCD2 among three digits) in the quotient. Next, we will continue the same process with the divisor 10 and find BCD1. The remainder after dividing by ten represents the least significant BCD0. This process is shown below.

EXAMPLE 8.1

Number = 124 Divisor = 100	Number of Subtractions	Number = 24 Divisor = 10	Number of Subtractions	Number = 04 Divisor = 1
124		24		
- 100 = + 2 4	1	- 10 = + 14	1	Result is same
- 100 = - 7 6	2	- 10 = + 04	2	as the number
+ 100 = + 24	[1]	- 10 = - 06	3	
		+ 10 = +04	[2]	
BCD2 = 1		BCD1 = 2		BCD0 = 4

The subroutines written in the next section duplicate the above steps except that the numbers will be in binary (Hex).

8.2.3 PROGRAM

```
BUFF0    EQU       0x10            ;Define buffer registers to save BCD digits
BUFF1    EQU       0x11
BUFF2    EQU       0x12
TEMPTR   EQU       0x00            ;Register to hold temperature reading

BINBCD:  CLRF      BUFF0           ;Clear registers to save BCD digits
         CLRF      BUFF1
         CLRF      BUFF2

         MOVWF     TEMPTR          ;Save temperature reading in register
         LFSR      FSR0,BUFF2      ;Pointer to BUFF2 to save BCD2
         MOVLW     D'100'          ;Divisor – largest powers of 10 needed
         RCALL     DIVIDE          ;Find BCD2
         LFSR      FSR0, BUFF1     ;Pointer to BUFF1 to save BCD1
         MOVLW     D'10'           ;Divisor 10
         RCALL     DIVIDE          ;Find BCD1
         LFSR      FSR0, BUFF0     ;Pointer to BUFF0 to save BCD0
         MOVFF     TEMPTR,BUFF0    ;Remainder = BCD0
         RETURN

DIVIDE:  INCF      INDF0           ;Begin counting number of subtractions
         SUBWF     TEMPTR          ;Subtract power of ten from binary reading
         BTFSC     STATUS, C       ;If C flag =0, skip next instruction
         GOTO      DIVIDE          ;If C flag =1, go back to next subtraction
         ADDWF     TEMPTR ,1       ;Adjust the temperature by adding divisor
         DECF      INDF0           ;Adjust the number of subtraction
         RETURN
```

8.2.4 Program Description

The program consists of two subroutines: BINBCD and DIVIDE. The BINBCD subroutine gets the binary temperature reading in WREG and saves it in register TEMPTR. The subroutine initializes the FSR0 pointer to identify buffer registers, supplies the powers of ten, and finally calls the DIVIDE subroutine. The buffer register is incremented (INCF INDF0) for each subtraction in the loop. The instruction BTFSC (Test Bit 0 in STATUS register) checks the carry flag and repeats the subtraction until the carry flag is cleared. The ADDWF instruction cancels the last subtraction and the DECF instruction adjusts the subtraction count. This process is repeated for each power of ten. The final remainder in 1's value is in TEMPTR register that is simply copied into BCD0.

8.2.5 Program Execution and Troubleshooting

To execute these subroutines in MPLAB IDE, you need to add the following information: the header file for the PIC controller your are using, the END statement, and a specific binary number in WREG. You can insert an 8-bit number manually in WREG in the Watch window. Assemble the subroutine and add the following registers in the Watch window: WREG, BUFF2, BUFF1, BUFF0, and FSR0. When you single-step the instructions for a given number, you will observe the subtraction of 100 from the number you inserted in WREG and for each subtraction the corresponding buffer register is incremented. You can also observe how the DIVIDE subroutine adjusts the last subtraction.

8.3 ASCII CODE TO BINARY CONVERSION

As mentioned earlier, the microcontrollers operate in binary language, but human beings use alphanumeric symbols (letters and numbers) to communicate. Therefore, to communicate with a microcontroller, we need to translate between alphanumeric symbols and binary language. The commonly used code for such translation is ASCII (American Standard Code for Information Exchange). It is a 7-bit code that represents 128 characters, and its range is from 00 to $7F_H$; the eighth bit is always zero. In serial communication, characters are transmitted over a medium such as a telephone line; the eighth bit is used to check transmission errors and is called a parity check (see Chapter 13).

The ASCII keyboard is a standard input device in computers. When someone presses the key with digit 1 on the ASCII keyboard, the MPU receives the ASCII code for 1 which is 31_H. This program converts these ASCII codes for Hex digits 0 to F_H into their binary equivalent to be processed for arithmetic operations.

8.3.1 Problem Statement

Write a subroutine to convert an ASCII code for a Hex digit from 0 to F_H, provided in WREG, into its binary equivalent. The subroutine should also check for any invalid codes below 30_H.

8.3.2 Problem Analysis

The ASCII code for digits 0 to 9 is from 30_H to 39_H, in sequential order. Converting these codes into their binary equivalent can be accomplished by subtracting 30_H from the code. The code for digits A though F ranges from 41_H to 46_H; there is gap of seven digits ($3A_H$ to 40_H) from the code of the last digit 9 (39_H). Therefore, if we subtract 7 again (after subtracting 30_H) from the code, we can find the binary equivalent digits A through F.

8.3.3 PROGRAM

```
ASCII      EQU      0x00          ;Define registers that are used
CHECK1     EQU      0x01
CHECK2     EQU      0x02


ASCIIBIN: ;::::::::::::::::::::::::::::::::::::::::::::::::::::::::::::::::::::::::::::::
          ;Function: This subroutine takes an ASCII Hex digit, masks its parity bit. If the digit is :
          ;          less than 30H it sends out an error code; otherwise, it converts the number  :
          ;          into its binary equivalent                                                   :
          ;Input:    ASCII digit in WREG                                                          :
          ;Output:   Binary equivalent of ASCII digit in WREG                                    :
          ;Modifies the contents in WREG                                                         :
          ;::::::::::::::::::::::::::::::::::::::::::::::::::::::::::::::::::::::::::::::


          ANDLW    B'01111111'   ;Mask parity bit B7
          MOVWF    ASCII         ;Save Hex digit
          MOVLW    0x2F          ;Check invalid ASCII Hex digits
          MOVWF    CHECK1
          MOVLW    D'10'         ;Check for ASCII larger than digit 9
          MOVWF    CHECK2
          MOVF     ASCII,W       ;Get original ASCII characters
          CPFSLT   CHECK1        ;Is value less than 30H
          BRA      ILLEGAL       ;If yes, go to display illegal code
          MOVLW    0x30
          SUBWF    ASCII,0,0     ;Find ASCII digits from 0 to 9
          CPFSLT   CHECK2        ;If larger than 9, find A through F
          RETURN
          MOVWF    ASCII         ;Subtract additional 7 to find A to F
          MOVLW    0x07
          SUBWF    ASCII,0,0
          RETURN
ILLEGAL:  MOVLW    0xFF          ;Hex code less than 30H
          RETURN
```

8.3.4 Program Description

The subroutine gets an ASCII code in WREG from the calling program. First, it masks Bit7 because this bit does not carry any information about ASCII codes; it is used for other purposes. The subroutine saves the code in register ASCII (defined as 00), and loads $2F_H$ and 10 in registers CHECK1 and CHECK2 respectively for comparison. The ASCII code for 0 begins at 30_H; therefore, by comparing the code with $2F_H$, the subroutine rejects all codes below 30_H. After subtracting 30 from the code, it checks whether the result is larger than 9 by comparing it with 10. If the result is larger than 10, it subtracts 7 again to find the code for A_H though F_H.

8.3.5 Program Execution and Troubleshooting

To execute the subroutine in MPLAB IDE, you need to add the following information: the header file for the PIC controller, the END statement, and a specific ASCII code in WREG. You can insert the ASCII code manually in WREG in the Watch window. Assemble the subroutine and add the following registers in the Watch window: WREG, ASCII, CHECK1, and CHECK2. You should single-step the instructions for various ASCII codes that can verify different return conditions as follows:

1. Insert the code 2F (or smaller value) in WREG and see whether the subroutine returns with the illegal code FF_H.
2. Insert any ASCII code between 30_H and 39_H and see whether the subroutine returns appropriate binary value for the code.
3. Check the subroutine for the codes between 41_H and 46_H; you should get A through F values.
4. Check the subroutine for values from $3A_H$ to 40_H and beyond 46_H You will get the wrong binary code because this subroutine does not check for these values.

8.4 BINARY TO ASCII CODE CONVERSION

As mentioned in the previous section, the MPU performs data processing in binary language. But to display the results at output devices such as printers or video screens, the binary readings must be converted into ASCII characters. This program illustrates such a conversion.

8.4.1 Problem Statement

Write subroutines to convert an 8-bit binary number given in WREG into its ASCII characters.

8.4.2 Problem Analysis

An 8-bit number must be first unpacked to get two digits. The process of finding ASCII codes for these digits is just the opposite of what we discussed in the previous section. If the digit is between 0 and 9, add 30_H, and if it is larger than 9 (from A to F), add an additional 07_H.

8.4.3 PROGRAM

```
BUFFER1   EQU   0x10       ;Define data register addresses
BUFFER2   EQU   0X11
REG1      EQU   0x01
ASCII     EQU   0x02
```

continued

```
UNPACK:    MOVWF    REG1            ;Save packed byte in REG1
           ANDLW    0x0F            ;Mask high-order digit
           MOVWF    BUFFER          ;Save low-order nibble in BUFFER1
           MOVF     REG1,WREG       ;Get the byte again
           SWAPF    WREG,0          ;Exchange digits
           ANDLW    0x0F            ;Mask low-order digit
           MOVWF    BUFFER2         ;Save high-order nibble
           RETURN


BINASCII:  MOVWF    ASCII           ;Save the binary number
           MOVLW    D'10'           ;Load 10 in W for comparison
           CPFSGT   ASCII           ;If number is larger than 10, add 07
           BRA      ADD30
           MOVLW    0x07
           ADDWF    ASCII,1,0       ;Add 07H and save result in ASCII register
ADD30:     MOVLW    0x30
           ADDWF    ASCII,0,0       ;Add 30H and return the conversion in WREG
           RETURN
```

8.4.4 Program Description

This program includes two subroutines: UNPACK and BINASCII. The UNPACK subroutine, after getting a byte in WREG, saves the byte. It masks each digit one at a time by using the AND instruction and saves the unpacked digits in BUFFER1 and BUFFER2.

Similarly, the BINASCII (Binary to ASCII) subroutine should also receive an unpacked digit in WREG. The subroutine checks whether the digit is larger than 9. If it is larger than 9, it adds 07_H first and 30_H later. If the digit is between 0 and 9, it skips the segment that adds 07_H and goes directly to add 30_H. The subroutine returns the ASCII character in WREG.

8.4.5 Program Execution and Troubleshooting

The above program is not a complete program that converts a Hex number into two ASCII characters; it merely includes two independent subroutines that can be tested separately. You need to write a main module to obtain two ASCII characters for a given byte.

The main module should load a byte in WREG and call the UNPACK subroutine that will return two unpacked digits in BUFFER1 and BUFFER2. The main module should take the unpacked digits one at a time, load them in WREG and call the BINASCII subroutine, and save these ASCII characters back in BUFFER1 and BUFFER2 or some other registers.

8.5 ILLUSTRATIVE PROGRAM: MULTIPLICATION OF 16-BIT NUMBERS

The PIC18 is an 8-bit microcontroller, and most of its operations deal with 8-bit numbers. There are occasions, however, when we need to perform arithmetic operations for numbers larger than 8 bits. In Chapter 5, we illustrated that we can perform arithmetic operations such as additions and subtractions of 16-bit (or larger) numbers.

This illustration deals with multiplication of 16-bit unsigned numbers. The PIC18 microcontroller includes multiply instructions that multiply 8-bit numbers and store the results in the 16-bit registers PRODH and PRODL. To multiply 16-bit numbers, we can first split them into groups of 8 bits; we can then use the multiply instruction on 8 bits at a time and add the numbers as we add them in manual paper-and-pencil procedure. The process is explained in Section 8.5.2.

8.5.1 Problem Statement

Write subroutines to multiply two 16-bit numbers stored in registers NUM1H, NUM1L, NUM2H, and NUM2L and store the product in four data registers PROD0 (LSB) through PROD3 (MSB).

8.5.2 Problem Analysis

We are dealing with number systems with positional weightings, meaning the position of the digit in a given number determines its value. When we multiply numbers in a decimal system, we multiply two digits at a time and place the product in appropriate positions.

Let us take an example of multiplying two-digit decimal numbers:

$N1_H N1_L N2_H N2_L = 32_{10} \times 48_{10} = 1536_{10}$. The steps are shown below:

$$10^3 \ 10^2 \ 10^1 \ 10^0$$

Step 1: Multiply 2 x 8 =			1	6	: Product of digits in 1's position $= N1_L \times N2_L$
Step 2: Multiply 3 x 8 = +		2	4		: Product of digits in 10's and 1's position $= N1_H \times N2_L$
Step 3: Multiply 2 x 4 = +		0	8		: Product of digits in 10's and 1's position $= N1_L \times N2_H$
Step 4: Multiply 3 x 4 = + 1		2			: Product of both digits in 10's position $= N1_H \times N2_H$
CY		1			

$$\underline{\qquad\qquad\qquad}$$

$$1 \quad 5 \quad 3 \quad 6$$

In the above multiplication, we multiplied two digits at a time with all four combinations. The addition is done after shifting the multiplication results to the left position based on the positions of the digits. We will follow the same procedure to write assembly language instructions to multiply two 16-bit Hex numbers.

8.5.3 PROGRAM

```
NUM1LO    EQU     0x01                    ;Define registers to store
NUM1HI    EQU     0x02                    ; 16-bit numbers
NUM2LO    EQU     0x03
NUM2HI    EQU     0x04


WORD1     SET     0x10                    ;Define registers for
WORD2     SET     0x20                    ;multiplicand, multiplier,
RESULT    SET     0x30                    ;and product

SECT0:    MOVLW   NUM1LO                  ;Store 16-bit numbers as
          MOVWF   WORD1                   ;multiplicand and multiplier
          MOVLW   NUM1HI
          MOVWF   WORD1+ 1
          MOVLW   NUM2LO
          MOVWF   WORD2
          MOVLW   NUM2HI
          MOVWF   WORD2+ 1

SECT1:    MOVF    WORD1,W         ;NUM1LO x NUM2LO
          MULWF   WORD2
          MOVFF   PRODL, RESULT   ;Save product in RESULT and
          MOVFF   PRODH, RESULT+1 ;RESULT +1


SECT2:    MOVF    WORD1+1,W       ;NUM1HI x NUM2HI
          MULWF   WORD2+1
          MOVFF   PRODL, RESULT+2 ;Save product in RESULT+2
          MOVFF   PRODH, RESULT+3 ;and RESULT +3


SECT3:    MOVF    WORD1+1,W       ;NUM1HI x NUM2LO
          MULWF   WORD2
          MOVF    PRODL,W
          ADDWF   RESULT+1,1      ;Add previous product and save
                                  ;in RESULT+1

          MOVF    PRODH,W
          ADDWFC  RESULT+2,1      ;Add previous product and save
                                  ;in RESULT+2
          BTFSC   STATUS, C       ;If no carry, skip next
                                  ;instruction
          INCF    RESULT+3
```

continued

```
SECT4:   MOVF    WORD1,W          ;NUM1LO x NUM2HI
         MULWF   WORD2+1
         MOVF    PRODL,W
         ADDWF   RESULT+1         ;Add previous product and save
                                  ;in RESULT+1
         MOVF    PRODH,W
         ADDWFC  RESULT+2         ;Add previous product and save
                                  ;in RESULT+2
         BTFSC   STATUS, C        ;If no carry, skip next
                                  ;instruction
         INCF    RESULT+3
         NOP
         END
```

8.5.4 Program Description

This is a multiplication of two 16-bit numbers NUM1 and NUM2 which are split as two 8-bit numbers LO and HI. The result of the multiplication of two 16-bit numbers requires four registers for storage; this program uses data registers 30_H to 33_H labeled as RESULT, +1, +2, and +3. The least significant byte (LSB) is stored in the data register 30_H and the most significant byte (MSB) in data register 33_H. The 16-bit numbers are copied in data registers labeled WORD1, WORD1+1, WORD2, and WORD2+1.

The program is divided into five sections: SECT0 to SECT4. Each section performs a task. In SECT0, the program copies the numbers to be multiplied in data registers WORD1, WORD1+1, WORD2, and WORD2+1.

In SECT1, the program multiplies 8-bit numbers NUM1LO and NUM2LO resulting in a 16-bit product in PRODH and PRODL registers. It saves PRODL in register RESULT and PRODH in RESULT+1. In SECT2, the program repeats the same process and saves the result in RESULT+2 and RESULT+3.

In SECT3, the program multiplies NUM1HI and NUM2LO. Now to perform the addition in proper columns as discussed in the decimal example in Section 8.5.2, we need to add the PRODL to the register RESULT+1. This addition may generate a carry. Therefore, the next addition of PRODH and RESULT+2 uses the instruction ADDWFC (Addition with Carry). The program checks whether this addition generates a carry, and if it does, it increments RESULT+3; otherwise, it skips the increment instruction. In SECT4, the same process of SECT3 is repeated for the numbers NUM1LO and NUM2HI. The final result of 32-bit product is saved in four registers RESULT: RESULT+3 – LSB to MSB respectively.

8.5.5 Program Execution and Troubleshooting

To understand the multiplication process, you need to check the program in sections using MPLAB IDE. As an example, load one 16-bit number (0102_H) in registers 10_H and 11_H (WORD1 and WORD+1) and the second (0205_H) in 20_H and 21_H (WORD2 and WORD2+1). The numbers are entered in the reversed order—the low number followed by the high number. The beginning addresses of data registers WORD1, WORD2, and RESULT are selected as 10_H, 20_H, and 30_H respectively so we can observe these registers more easily during the execution of the program. Perform the following steps to execute and troubleshoot the program.

1. Add the header file and assemble the program. Set five breakpoints at each section, the first breakpoint at SECT1 and the last breakpoint at the NOP instruction. The NOP instruction is added to stop the execution after the last instruction.

2. Go to View and add the following registers in the Watch window: WREG, PRODL, PRODH, and STATUS. Go back to View and click on File Registers. Set the cursor at the starting instruction and run the program. It will stop at SECT1 – the first breakpoint. You should observe NUM1 (0102_H) in data registers 10_H and 11_H in the reversed order. Similarly NUM2 (0205_H) will be displayed in data registers 20_H and 21_H. The numbers selected for this illustration are small enough for you to verify the results manually.

3. Click on Run (or press F9) again. The program will stop at the second breakpoint SECT2. You should observe the product of two low bytes (02_H x 05_H = 00 $0A_H$); 05 in register 30_H, and 00 in 31_H. The numbers are not large enough to have any digit in PRODH register.

4. Run the program again. It executes SECT2 and stops at SECT3. This section multiplies two high bytes (01_H x 02_H = 02_H) and stores 02_H in register 32_H and 00 in register 33_H.

5. When you run the program again, the program executes SECT3. This section multiplies NUM1HI and NUM2LO (01 x 05 = 00 05_H) and adds 05 to the register 31 (RESULT+1) and 00 to register 32 (RESULT+2).

6. When SECT4 is executed, it multiplies NUM1LO and NUM2HI (02_H x 02_H = 00 04_H) and adds 04_H to register 31 making the number 09_H and *forcing* 00 to register 32. The final result 00 02 09 0A appears as 0A 09 02 00 in registers 30_H to 33_H. You can also verify the result by multiplying these Hex numbers on the Hex calculator or the calculator available under Accessories on a PC.

8.6 ILLUSTRATIVE PROGRAM: DIVISION OF TWO 8-BIT UNSIGNED NUMBERS

In Chapter 6, we found the average of a set of readings by finding the sum of the readings and dividing it by the number of readings. To divide the sum, we illustrated two techniques: (1) using the rotate right instruction and (2) repeated subtraction. The rotate right technique can work if the number of readings is a power of two. The repeated subtraction is a conceptually easy technique and simple to write, but it can be inefficient in execution as it is determined by the number of loops (subtractions) to be executed. In this section, we will attempt to emulate the paper-and-pencil technique that we use to divide two numbers manually.

8.6.1 Problem Statement

Write a subroutine to divide two 8-bit unsigned numbers. The calling program provides the dividend (the number to be divided) in the register labeled as DIVIDND and the divisor in the register DIVISOR. The subroutine should return the quotient in register DIVIDND and the remainder in register REMAINDR.

8.6.2 Problem Analysis

In the paper-and-pencil procedure, we perform the following steps to divide two numbers:

1. Begin with the most significant digit of a dividend and divide it by a divisor. If the divisor is larger than the dividend digit, we place a zero in the quotient. If the divisor is smaller than the dividend digit, we find the quotient and the remainder.
2. We bring the next digit from the dividend and place it in the next column of the remainder. Divide the adjusted remainder by the divisor and find the quotient and the remainder. This quotient (digit) is placed in the column next to the previous quotient.
3. This process is continued until the last digit of the dividend is brought in the division process and the remainder is less than the divisor.

In the above division process, the following points should be noted: (1) when the divisor is smaller than the dividend, zero is placed in the quotient; (2) after the completion of one division, the next digit is not added; but is placed in the next column of the remainder; and (3) the quotient digits are not added, but placed in the adjacent left column. This is illustrated in the next example of dividing decimal 537_{10} by 5_{10}.

	Divisor	Dividend	Quotient
Step 1: Get the most significant digit 5 from the dividend 537. Divide 5 by 5, Quotient = 1 and Remainder = 0	5	$\begin{array}{r}537\\ \overline{5}\\ -5\\ \hline R=0\end{array}$	= 1
Step 2: Add the digit 3 to the remainder's next column. Divisor is > Remainder, Quotient = 0, Remainder = 03	5	$\begin{array}{r}\overline{03}\\ R=3\end{array}$	= 10
Step 3: Add the digit 7 to 03 = 0 3 7 and divide by 5. The division is performed by subtracting 5 seven times. We do it manually through the multiplication table. Here Quotient = 7 and Remainder = 2	5	$\begin{array}{r}\overline{37}\\ 35\\ \hline R=2\end{array}$	= 107

In this example, the quotient is 107 and the remainder is 2.

In binary division, the process is somewhat simpler. Because when the divisor is subtracted, the quotient is always either 1 or 0. The MSD (most significant digit seventh Bit) of the dividend can be brought to the remainder by using the instruction Rotate Left through Carry. The above procedure is implemented in the following program.

8.6.3 PROGRAM

```
DIVIDND   EQU     0x10        ;Define registers
DIVISOR   EQU     0x11
REMAINDR  EQU     0x12
COUNT     EQU     0x13


MAIN:     MOVLW   0xFF        ;Number to be divided
          MOVWF   DIVIDND
          MOVLW   0x0A        ;Divisor
          MOVWF   DIVISOR
          RCALL   DIVIDE      ;Call divide routine
HERE:     BRA     HERE


DIVIDE:   MOVLW   0x08        ;Set up counter for 8 rotations
          MOVWF   COUNT
          CLRF    REMAINDR    ;Clear remainder register
LOOP:     BCF     STATUS,C,0  ;Clear carry flag
          RLCF    DIVIDND     ;Take seventh bit in Carry flag
          RLCF    REMAINDR    ;Place carry flag in remainder
          MOVF    DIVISOR,W   ;Get divisor in WREG
          SUBWF   REMAINDR,0  ;Subtract divisor from remainder
                              ;- save in WREG
          BTFSS   STATUS,C,0  ;Test carry flag
          BRA     RESET0      ;If dividend < divisor, go to
                              ;reset carry flag SET0:
          BSF     DIVIDND,0,0 ;If dividend > divisor, set
                              ;carry flag
          MOVWF   REMAINDR,0  ;Save remainder for next
                              ;operation
          BRA     NEXT
RESET0:   BCF     DIVIDND,0,0 ;No quotient - reset carry flag
NEXT:     DECF    COUNT,1,0   ;One operation complete -
                              ;decrement count
          BNZ     LOOP        ;Are eight rotations complete?
          RETURN              ;If not go back
          END
```

8.6.4 Program Description

The subroutine DIVIDE begins by loading 8 in the counter because we have an 8-bit dividend. It will be rotated eight times. To explain this subroutine, it is necessary to focus on the operation of the register REMAINDR. Initially, the carry flag and the REMAINDR register are cleared. The first rotate instruction RLCF DIVIDND places the MSD of the dividend in the carry flag and the next instruction RLCF REMAINDR places the carry flag (seventh Bit of dividend) as the first bit of the register REMAINDR; the divisor is subtracted from this bit the first time. In our example, the divisor > remainder; therefore, the program branches to reset Bit0 in the register DIVIDND. The subroutine decrements the counter, and jumps back to repeat the operation.

In the next operation, the sixth bit (which is in the seventh position after the rotation) of the dividend is placed in the remainder as Bit0. If the divisor is smaller than the remainder, the subroutine sets Bit0 in DIVIDND register and copies it into the REMAINDR register. This process is repeated eight times until the counter becomes zero and the last digit of the dividend is added to the remainder. At the end of the subroutine, the original byte in DIVIDND register is replaced by the quotient.

8.6.5 Program Execution and Troubleshooting

The program has three branch instructions that check three different conditions. To understand the division process, you need to single-step the program and check whether the program works for all three conditions. Similarly, single-stepping the entire program will help you to understand how the rotate instructions make use of the carry flag to combine bits in appropriate columns, for the quotient as well as for the remainder. Perform the following steps to execute and troubleshoot the program.

1. Add the header file and assemble the program.
2. Go to View, and add the following registers in the Watch window: WREG, DIVIDND, DIVISOR, REMAINDR, COUNT, and STATUS. Go back to View and click on File Registers, and if File Registers are not clear, go to Debugger and select Clear Memory, and Clear File Registers.
3. Single-step the MAIN segment, and you should observe the contents of DIVIDND and DIVISOR.
4. Single-step the DIVIDE subroutine until the instruction SUBWF. You should observe how rotate instructions are used to get the appropriate bits in DIVIDND and REMAINDR, and check the carry flag in the STATUS register and determine whether the program will set or reset Bit0 in the DIVIDND register.
5. At label NEXT, one operation is completed and the counter is decremented to 7. The program goes back to label LOOP and repeats the process again. You should single-step the program until Bit0 is set in the DIVIDND register.

8.7 SOFTWARE DESIGN

To facilitate interaction between human beings and computing binary machines, we wrote subroutines in previous sections that performed various code conversions. In addition, we also wrote subroutines for arithmetic operations such as multiply and divide. These subroutines were independent modules. Now we can combine these modules to design a simple project.

The following project deals with finding an average of temperature readings and converting the reading to a proper format that could be displayed on an ASCII peripheral such as a printer, monitor, or LCD.

8.7.1 Project Statement

Design a software program to meet the following specifications:

Given a string of temperature readings in unsigned numbers that terminates in a null character 00, find the average temperature. Convert the average temperature reading from BCD digits into ASCII characters to be displayed by some other subroutine. A new data set of temperature readings is recorded every 100 ms, and it is indicated by setting Bit0 (carry flag) in STATUS register.

8.7.2 Project Analysis

By examining the project statements carefully, we can find many clues that enable us to divide the project into small segments. Each of these small segments can be written as a single independent module or subroutine. In our first attempt, we can divide the project as follows:

1. A string of readings that terminates in the null character 00 is stored in data registers that need to be defined (labeled). To find the average reading, we need to add these readings that may generate carries, and the sum will require more than one 8-bit register. In Chapter 7, Section 7.4.3, we have already written a subroutine that adds the number of bytes specified by the main program. In this project, the number of temperature readings is variable and determined by the last null character. We can write a separate subroutine to check for the null character and count the number of bytes or modify the ADDITION subroutine (Section 7.4.3) to include this new feature of checking for the null character.
2. To find the average temperature reading, the sum must be divided by the number of readings. Therefore, the number of bytes in the string counted in Step 1 must be passed on to this division process. The number of bytes is variable, and the sum is likely to be a 16-bit number. We can use the repeated subtraction method to divide a 16-bit sum by an 8-bit divisor. In Section 6.5, we used this method in the Illustrative Program: Finding an Average Temperature of Data Readings.
3. The average temperature in Step 2 is in binary (Hex). We need to convert the reading into BCD numbers. The subroutine in Section 8.2 converts an 8-bit Hex number in three BCD numbers.
4. To display these digits at ASCII peripherals, the digits must be converted into ASCII codes using the subroutine BINASC discussed in Section 8.4.
5. The last sentence in the project statement indicates that a new data string is available every 100 ms. The availability of a new data set is indicated by setting the carry flag in the STATUS register. This suggests that we need to repeat this process by setting a delay loop for 100 ms. However, it may create a problem of synchronizing the arrival of new data and the execution of this program. This can be avoided if this program checks for the carry flag at the beginning before it begins execution. This program stays in a continuous loop until the carry flag is set.

8.7.3 PROGRAM

```
    Title " IP8-7 Temperature Display"
        List p=18F452, f =inhx32
        #include <p18F452.inc>        ;The header file


BUFFER      EQU       0x10
COUNTER     EQU       0x01
TEMP        EQU       0x02


MAIN:       BTFSS     STATUS,C        ;Is there a set of new data?
            BRA       MAIN            ;If yes, skip this instruction
            LFSR      FSR0, BUFFER    ;Set up pointer for where data begins
            RCALL     CHKNULL         ;Find number of bytes in data string
            LFSR      FSR0, BUFFER    ;Set up pointer for data again
            MOVFF     COUNTER,TEMP    ;Save the count in temporary register
            RCALL     ADDITION        ;Add temperature readings
            MOVFF     TEMP,COUNTER    ;Get the count again to find average
            RCALL     AVG             ;Find average temperature
            MOVF      QUOTIENT,W      ;Copy the average in WREG
            RCALL     BINBCD          ;Convert temperature reading in BCD
            MOVF      BUFF0,W
            RCALL     BINASC          ;Convert BCD numbers in ASCII codes
            MOVWF     BUFF0
            MOVF      BUFF1,W
            RCALL     BINASC
            MOVWF     BUFF1
            MOVF      BUFF2,W
            RCALL     BINASC
            MOVWF     BUFF2
            BRA       MAIN            ;Go back to check for new data


CHKNULL:    ;::::::::::::::::::::::::::::::::::::::::::::::::::::::::::::::::::::::
            ;Function: This subroutine counts the number of temperature      :
            ;          readings in the data string until it finds the null   :
            ;          character.                                            :
            ;Input:    Address in FSR0 to point to data string              :
            ;Output:   Count in COUNTER register representing the number of   :
            ;          bytes in data string                                  :
            ;Registers Changed: FSR0                                         :
            ;::::::::::::::::::::::::::::::::::::::::::::::::::::::::::::::::::::::
```

continued

```
          CLRF      COUNTER                ;Clear counter
AGAIN:    INCF      COUNTER,1              ;Begin counting
TEST:     TSTFSZ    POSTINC0,BUFFER        ;Check for 00 and increment pointer
          BRA       AGAIN                  ;Continue checking
          DECF      COUNTER,1              ;Adjust counter for extra increment
          RETURN


ADDITION: ;::::::::::::::::::::::::::::::::::::::::::::::::::::::::::::::::::::::
          ;Function: Adds the data bytes in the data string          :
          ;Input:    Address in FSR0 to point to data string and Count of   :
          ;          data bytes in COUNTER register                  :
          ;Output:   16-bit sum-High byte in CYREG and low byte in W  :
          ;Registers Changed: COUNTER and FSR0                       :
          ;::::::::::::::::::::::::::::::::::::::::::::::::::::;::::::::::::::::::::::


CYREG     EQU       0x03
          CLRF      CYREG                  ;Clear carry register
          LFSR      FSR0,BUFFER            ;Set up FSR0 as pointer for data registers
          MOVLW     0x00                   ;Clear W register to save sum
NEXTADD:  ADDWF     POSTINC0,W             ;Add byte and increment FSR0
          BNC       SKIP                   ;Check for carry-if no carry jump to SKIP
          INCF      CYREG                  ;If there is carry, increment CYREG
SKIP:     DECF      COUNTER,1,0            ;Next count and save count in register
          BNZ       NEXTADD                ;If COUNT ≠ 0 , go back to add next byte
          RETURN


AVG:      ;::::::::::::::::::::::::::::::::::::::::::::::::::::::::::::::::::::::
          ;Function: Finds the average of 16-bit sum by dividing the sum by  :
          ;          the number of data bytes                        :
          ;Input:    16-bit sum - High-byte in CYREG and low-byte in W  :
          ;          Count of data bytes in COUNTER register          :
          ;Output:   Average reading in QUOTIENT register            :
          ;Registers Changed: WREG                                   :
          ;::::::::::::::::::::::::::::::::::::::::::::::::::::::::::::::::::::::


DIVIDNDLO EQU       0x04                   ;Stores low byte of the sum
DIVIDNDHI EQU       0x05                   ;Stores high byte of the sum
QUOTIENT  EQU       0x06                   ;Saves the average
DIVISOR   EQU       0x07                   ;Stores divisor


          MOVWF     DIVIDNDLO              ;Get 16-bit sum
          MOVFF     CYREG, DIVIDNDHI
          MOVFF     COUNTER, DIVISOR       ;Place count in DIVISOR register
          CLRF      QUOTIENT               ;Clear to save result
```

continued

```
SUBTRAC:    MOVF      DIVISOR,W           ;Place count in WREG
            INCF      QUOTIENT            ;Add one to average reading
            SUBWF     DIVIDNDLO,1         ;Divide by subtracting
            BTFSC     STATUS, 0           ;Is low byte negative?
            GOTO      SUBTRAC             ;If not, go back to subtract again
            DECF      DIVIDNDHI,1         ;Decrement high byte
            BTFSC     STATUS, 0           ;Is high byte negative?
            GOTO      SUBTRAC             ;If not, go back to subtract again
            DECF      QUOTIENT,1          ;Adjust for extra addition in beginning
            RETURN


BINBCD:     ;::::::::::::::::::::::::::::::::::::::::::::::::::::::::::::::::
            ;Function: Converts a Hex data byte into three BCD digits    :
            ;Input: Hex data byte in W                                   :
            ;Output: BCD digits in BUFF0, BUFF1, and BUFF2              :
            ;Registers Changed: WREG                                     :
            ;Calls BCD subroutine                                        :
            ;::::::::::::::::::::::::::::::::::::::::::::::::::::::::::::::::

BUFF0       EQU       0x20                ;Define registers to save BCD digits
BUFF1       EQU       0x21
BUFF2       EQU       0x22
TEMPR       EQU       0x23

            CLRF      BUFF0               ;Clear registers for BCD digits
            CLRF      BUFF1
            CLRF      BUFF2

            MOVWF     TEMPR               ;Get the saved sum again
            LFSR      FSR0,BUFF2          ;Set up pointer to BUFF2
            MOVLW     D'100               ;Place 100 in WREG as a divisor
            RCALL     BCD                 ;Find BCD2 digit
            LFSR      FSR0, BUFF1         ;Set up pointer to BUFF1
            MOVLW     D'10'               ;Place 10 in WREG as a divisor
            RCALL     BCD                 ;Find BCD1
            LFSR      FSR0, BUFF0         ;Set up pointer to BUFF0
            MOVFF     TEMPR,BUFF0         ;Remainder is BCD0
            RETURN

BCD:        ;::::::::::::::::::::::::::::::::::::::::::::::::::::::::::::::::
            ;Function: Converts Hex number into BCD one at a time        :
            ;Input:    Hex data byte in TEMPR register Powers of ten     :
            ;          as divisor in W and pointer FSR0 where BCD digit  :
            ;          should be stored                                  :
```

continued

```
                ;Output: BCD digit in register pointed by FSR0          :
                ;Registers Changed: TMPR                                 :
                ;:::::::::::::::::::::::::::::::::::::::::::::::::::::::::::

                INCF       INDF0              ;Begin counting number of subtractions
                SUBWF      TEMPR              ;Subtract powers of ten
                BTFSC      STATUS,C           ;If C flag =0, skip next instruction
                GOTO       BCD                ;If C flag =1, go back to next
                                              ;subtraction
                ADDWF      TEMPR,1            ;One too many subtractions - adjust
                DECF       INDF0              ;Adjust the last subtraction
                RETURN

BINASC:         ;:::::::::::::::::::::::::::::::::::::::::::::::::::::::::::
                ;Function: Converts BCD digit into ASCII character        :
                ;Input:    BCD digit in WREG                              :
                ;Output:   ASCII character in WREG                        :
                ;Registers Changed: WREG                                  :
                ;:::::::::::::::::::::::::::::::::::::::::::::::::::::::::::
ASCII           EQU        0x33

                MOVWF      ASCII              ;Get BCD digit
                MOVLW      D'10'
                CPFSGT     ASCII              ;Is digit smaller than 9
                BRA        ADD30              ;If yes, branch to add 30H
                MOVLW      0x07               ;If no, add 07H
                ADDWF      ASCII,1,0

ADD30:          MOVLW      0x30
                ADDWF      ASCII,0,0          ;Add 30H
                RETURN
                END
```

8.7.4 Program Description

This program follows the steps discussed in the project analysis. It consists of five different subroutines and begins by checking the carry flag in the STATUS register. Here we are assuming that some other program is responsible for recording temperature readings and setting the carry flag when a set of data is stored in data registers starting from BUFFER. The above program stays in a continuous loop until the carry flag is set.

In this program, the number of readings in a data string is variable; therefore, the first subroutine CHKNULL begins by checking each data byte starting at BUFFER by pointing the register FSR0 at BUFFER. This subroutine continues to count each data byte in COUNTER register until it finds the null character 00 and passes the counter value to the next subroutine

ADDITION. The MAIN program saves the value of the counter in the TEMP register because the subroutine ADDITION destroys the value of the count during the addition process, and it would not be available to find the average in the next subroutine. The ADDITION subroutine generates a 16-bit sum and passes that result to the next subroutine.

The next subroutine AVG uses the value of the count to perform the repeated subtraction and finds the average that is passed on to the next subroutine BINBCD in the register QUOTIENT. The BINBCD subroutine receives the average value of the temperature reading in WREG, and it divides the reading by powers of ten (100 and 10) to find three BCD digits from 0 to 9. These BCD digits are stored in registers BUFF0, BUFF1, and BUFF2 as unpacked digits. The MAIN program supplies these digits to the subroutine BINASC one at a time, the subroutine BINASC returns the ASCII codes for each of these digits in WREG, and they are saved back in the same registers.

8.7.5 Program Execution and Troubleshooting

This program is designed by calling five subroutines that were written and discussed in previous chapters. Even if these were working subroutines, they may not work properly when combined, because of the interactions among them. The common errors are as follows:

1. Labeling: Labels become duplicated and undefined. When you combine various subroutines, the duplication of labels occurs very frequently. In addition, some labels remain undefined. However, these errors are identified by the assembler.

2. Parameter Passing: Subroutines are generally dependent on parameters passed on to them for proper functioning. When you begin to combine various subroutines, the necessary parameters should be passed on to the subsequent subroutines in appropriate registers.

3. Modifying registers: Registers that are used by a subroutine are modified during the execution. Because information needed by subsequent subroutines could be modified or destroyed, it should either be saved by the calling program or, alternately, the subroutine should save the contents of the registers it is using and retrieve that information just before the end of the subroutine. A good example in our program is the ADDITION subroutine that decrements the count in the counter until it becomes zero. However, the count is needed for the division process in the next subroutine AVG.

4. Logical errors: Many logical errors occur because of the interaction between various subroutines. These errors can be detected by setting breakpoints at the beginning and verifying expected outputs of the subroutines.

To execute and troubleshoot this program, you should assemble the program, correct the assembly errors, and verify the proper functioning of the subroutines by setting breakpoints at the beginning of each one. You should have the (File) registers and selected special function registers in the Watch window. To begin with, for illustration, you should load ten temperature readings in data (File) registers starting from data register 10_H as shown in Figure 8-1. For these data bytes the sum is 0335_H and the average temperature reading is 52_H, which is equal to 82_{BCD} (shown initially as unpacked BCD 08_H and 02_H). Finally, these BCD digits should be shown as 38_H and 32_H as ASCII characters (Figure 8-1—Registers 21 and 20). To verify the proper functioning of the program, perform the following steps.

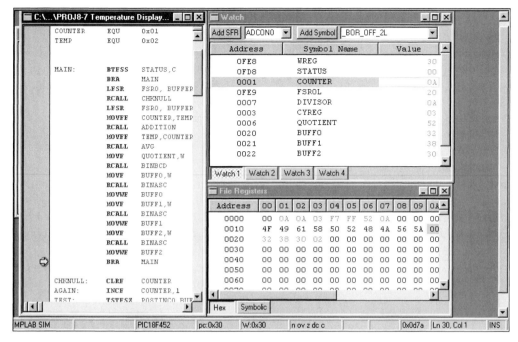

FIGURE 8-1

1. Add WREG, STATUS, and COUNTER registers in the Watch window, load 01 in the STATUS register to simulate the setting of the carry flag, and execute the first subroutine CHKNULL by setting the breakpoint at the instruction RCALL ADDITION. This subroutine checks and counts each data byte in the string until it finds the null character 00. At the end of this subroutine, you should have $0A_H$ in the COUNTER register that is also saved in the TEMP register for later use.

2. Add FSR0 and CYREG registers in the Watch window and execute the subroutine ADDITION. You should observe 03_H in CYREG as a high-byte and 35_H as a low-byte of the sum. You can also examine this sum in data registers at the appropriate addresses.

3. Add register QUOTIENT in the Watch window and execute the next subroutine AVG by setting the next breakpoint. This subroutine divides the sum 0335_H ($= 821_{10}$) by the number of bytes (10_{10}), and stores the result 52_H (82_{10}) in register QUOTIENT. You may also observe this result in data register 06_H.

4. Add registers BUFF0, BUFF1, and BUFF2 in the Watch window and execute the next subroutine BINBCD. This subroutine converts 52_H into 82_{BCD} and stores them as unpacked BCD digits 02 and 08 in BUFF0 and BUFF1.

5. Execute the next subroutine BINASC three times, as shown in the MAIN program. This subroutine converts 02 into 32_H, 08 into 38_H, and 00 into 30_H as ASCII characters, and saves them in BUFF0, BUFF1, and BUFF2.

8.7.6 Extending the Program to Applications for Embedded Systems

This program adds the data readings and finds the average reading for display on an ASCII device. In addition, it waits in a continuous loop for the carry flag to be set by some other program. However, for the MPU to wait idly is a waste of resources. In a real-life embedded system, such as the Time and Temperature Monitoring System (TTMS) described in Chapter 1, the program can be designed to be interrupt driven (Interrupts are discussed in Chapter 10). For example, the temperature readings are recorded by an A/D converter (Chapter 12) and stored in data registers. At the end of certain number of readings, the A/D converter can generate an interrupt, and the MPU can be directed to execute this program.

The conversion to ASCII characters is useful for such devices as an LCD or a printer that require ASCII codes. In many small systems, seven-segment LEDs (discussed in Chapter 9) are used. The data byte or code required to display a digit at the seven-segment LED is dependent on the type of LED and the connection of the data bus. There is no relationship between the codes required to display digits 0 to 9. Therefore, the technique used to display a seven-segment LED is called table look-up, whereby all codes are stored in memory in a sequence, and the code is extracted from memory using a memory pointer and the sequential number of the digit in the table (discussed in Chapter 9). To apply this program in an embedded system with seven-segment LEDs, we can replace the ASCII conversion with the table look-up technique.

SUMMARY

This chapter focused on writing various types of subroutines as independent modules that can be combined in designing a large program. The subroutines included code conversions such as binary to BCD, binary to ASCII code and vice versa, and arithmetic operations such as division of unsigned 8-bit numbers and multiplication of numbers larger than 8 bits. The chapter also illustrated the process of software design with an illustration of a simple program that uses many of these subroutines.

QUESTIONS AND ASSIGNMENTS

8.1   In Section 8.1, BCD to Binary conversion program, write a separate subroutine UNPACK that stores the unpacked digits in BUFFER1 and BUFFER2, and modify the BCDBIN subroutine to include the UNPACK subroutine.

8.2   Rewrite the UNPACK subroutine in Question 8.1 using rotate instructions in place of the SWAPF instruction.

8.3   Write a subroutine to convert a three-digit BCD number into its binary value.

8.4   Rewrite the program in Section 8.2.3 as follows: (1) Eliminate the subroutine DIVIDE; (2) BIN-BCD subroutine should include the functions of the subroutine DIVIDE so that when BINBCD is called, the calling program should get three BCD digits.

8.5   Write a subroutine to convert a 16-bit number, limited to a three-digit BCD number, into its equivalent BCD number.

8.6   Rewrite the subroutine ASCIIBIN (Section 8.3.3) to invalidate any number larger than 46_H.

8.7   Write a subroutine that checks an ASCII character in WREG and validates the characters between 0 to 9 and A to F.

8.8   Write a subroutine to divide two unsigned 16-bit numbers.

8.9   In Section 8.7 – Temperature Display – modify the ADDITION subroutine to add only positive readings.

8.10   Write a subroutine to convert an 8-bit temperature reading from 0 to 1000 degrees, given in Fahrenheit, into Celsius.

SIMULATION AND TROUBLESHOOTING EXERCISES

Using either MPLAB IDE or PIC18 Simulator IDE

8.1 SE   Assemble, troubleshoot, and execute the software design program described in Section 8.7 for the following summer temperature readings recorded in hexadecimal.

Readings (H): 52, 58, 5F, 42, 63, 5A, 41, 3F

8.2 SE   Modify the software design program described in Section 8.7 to calculate the average temperature reading for the following readings of a week given in BCD. The final result should be expressed in BCD.

Readings (BCD): 68, 64,72, 74, 60, 67, 71

CHAPTER 9

INPUT/OUTPUT (I/O) PORTS
AND INTERFACING

OVERVIEW

Microcontrollers inside the embedded systems communicate with the outside world through I/O ports connected to peripherals, such as keyboards and LEDs. Peripherals such as switches and keyboards are called input peripherals (devices) that provide binary data to the MPU, and peripherals such as LEDs, seven-segment LEDs, LCDs, and printers are called output peripherals (devices) that receive data from the MPU. I/O ports on a MCU are interfacing devices such as buffers and latches. Peripheral devices are connected to MCU through these I/O ports.

I/O ports are assigned binary addresses by decoding the address bus, and the MPU accesses these I/O ports through these addresses. The MPU transfers data bits using the data bus; it receives data bits from input peripherals and sends data bits to output peripherals This is called parallel I/O. When a single bit is transferred at a time over one signal line, it is called serial I/O. This chapter deals with parallel I/O and Chapter 13 deals with serial I/O.

This chapter discusses basic concepts in I/O interfacing and focuses on interfacing simple peripherals such as LEDs and switches to PIC18 I/O ports. Illustrations include: (1) designing a BCD counter with a seven-segment LED display, (2) interfacing a bank of push-button keys and a matrix keyboard, and (3) interfacing an LCD.

OBJECTIVES:

- Explain the basic concepts in I/O interfacing using a block diagram.

- Explain how binary addresses are assigned to I/O ports and how data bits are transferred using the data bus and control signals.

- Explain the term "multiplexed I/O ports" and list various functions performed by PIC18 I/O ports.

- Write instructions to initialize the PIC18 I/O ports as input ports and output ports.

- Interface LEDs and seven-segment LEDs to PIC18 I/O ports and write instructions to display data.

- Interface switches, push-button keys, and a matrix keyboard to PIC18 I/O ports and write instructions to read and process key inputs.

- Interface an LCD to PIC18 I/O ports, and write instructions to display a message.

- Illustrate the time-multiplex scanning technique used in interfacing multiple seven-segment LEDs.

9.1 BASIC CONCEPTS IN I/O INTERFACING AND PIC18 I/O PORTS

In Chapters 1 and 2, we discussed I/O devices as one of the essential components of the microprocessor-based systems. Input devices such as switches, a keyboard, a scanner, and a digital camera provide digital information to an MPU, and output devices such as LEDs, seven-segment LEDs, LCDs, and a printer receive digital information from an MPU. We labeled these devices as peripherals to avoid any confusion between the terms: I/O devices and I/O ports. In a microprocessor-based system, a peripheral is interfaced with the MPU through an I/O port, and this I/O port is assigned a binary number (called an address) using the address bus. The microprocessor transfers information to/from these peripheral devices using the data bus.

In Chapter 2, we discussed two modes of data transfer: parallel and serial, and three conditions for data transfer: unconditional, status-check (handshake), and interrupt. In this chapter, we focus on parallel mode; using eight data lines for data transfer. We will illustrate the interfacing of simple peripherals such as switches, keyboards, LEDs, and seven-segment LEDs to PIC I/O ports, and these peripherals are read or written into without checking their readiness (unconditional data transfer). The data transfer using the status-check condition is illustrated by an LCD interfacing and using the interrupt process is illustrated in Chapter 10.

9.1.1 I/O Ports: Interfacing and Addressing

In Chapter 2, we defined I/O ports as interfacing devices that are often confused with external peripherals, such as a keyboard or LEDs. I/O ports are registers that can function as buffers or latches. When an I/O port is set up as an input port, it functions as a buffer that allows the data to come inside the system from the outside world. When it is set up as an output port, it can function either as a buffer or a latch that takes data from the internal data bus and makes that data available to the outside peripherals. When the I/O port is set up as a latch, it latches (holds that data) for an outside peripheral such as LEDs. We can view I/O ports as gateways that can be opened and closed for data to get through.

Figure 9-1 shows two I/O ports interfaced with MPU: one is an output port (a latch such as 74LS373) and the other is an input port (a buffer such as 74LS244). The output port is connected to LEDs and the input port is connected to switches. The MPU address bus is decoded using a decoder, and specific addresses are assigned to both ports. The MPU communicates with peripherals by accessing these I/O ports through their addresses. The MPU uses Read/Write control signals to enable the ports, and the data are transferred using the data bus. In Figure 9-1, the same address lines and data lines are connected to multiple I/O ports. However, the MPU accesses (or enables) only one device at a time, and others are in the high impedance state. In microprocessor-based systems, the buses are like airplane runways in that only one plane can use the runway at a time.

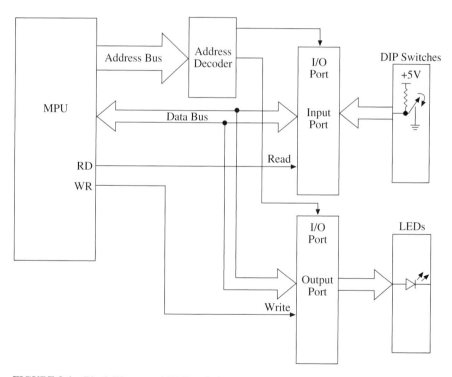

FIGURE 9-1   Block Diagram of I/O Interfacing

In a microcontroller, an I/O port is assigned an address using the same process, but it is done in the design stage of the microcontroller, and the user is given the address. For example, in PIC18F microcontroller series, PORTA and PORTB are assigned the addresses $F80_H$ and $F81_H$, respectively. The address bus, data bus, and control signals are internal and generally not available to the user. Therefore, in microcontroller-based systems, interfacing I/O peripherals amounts to designing the external hardware of a peripheral and writing instructions to set up these ports and transfer data as necessary.

9.1.2 Writing to and Reading from I/O Ports

When the MPU sends out or transfers data to an output port, it is called Writing to the port, and when the MPU receives data from an input port, it is called Reading the port.

To send data to an output port, we need to write an appropriate instruction with the address of the output port in the program memory. To execute the instruction, the MPU places the address of the output port on the address bus and selects the I/O port using the decoder, places the data to be sent out on the data bus, and asserts the Write control signal to enable the output port. The I/O port latches the data and displays at the LED as in Figure 9-1. To get or read the data from an input port, the MPU places an address of the input port, asserts the Read control signal to enable the input port, and receives the data via the data bus.

9.1.3 PIC I/O Ports

The various versions of the PIC18F family of microcontrollers have between two to ten I/O ports. They are identified alphabetically with designations such as PORTA, PORTB, and PORTC. Many of them have eight bidirectional pins and can be set up as input or output 8-bit ports. Most of the pins are multiplexed, meaning they are designed to be used for multiple purposes and can be set up as individual pins for I/O functions.

As an illustration look at PORTA (Figure 9-2). It is a bidirectional port with seven pins for digital signals—RA6–RA0 (Table 9-1). In addition, pins RA3–RA0 can be used for analog input signals. The RA4 pin is multiplexed with Timer0 clock input. Therefore, these I/O ports need to be programmed (or set up) to perform a specific task.

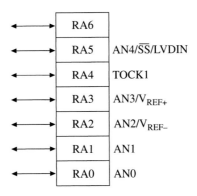

FIGURE 9-2 PIC18F452 PORTA

TABLE 9-1 PORTA Functions

Name	Bit#	Buffer	Function
RA0/AN0	bit0	TTL	Input/output or analog input.
RA1/AN1	bit1	TTL	Input/output or analog input.
RA2/AN2/V_{REF-}	bit2	TTL	Input/output or analog input or V_{REF-}.
RA3/AN3/V_{REF+}	bit3	TTL	Input/output or analog input or V_{REF+}.
RA4/T0CKI	bit4	ST	Input/output or external clock input for Timer0. Output is open drain type.
RA5/$\overline{SS}$/AN4/LVDIN	bit5	TTL	Input/output or slave select input for synchronous serial port or analog input, or low voltage detect input.
OSC2/CLKO/RA6	bit6	TTL	OSC2 or clock output or I/O pin.

Legend: TTL = TTL input, ST = Schmitt Trigger input

To set up these ports or individual pins, each I/O port is associated with the following three Special Function Registers (SFRs), and these SFRs need to be programmed based on an application.

- PORT: This register functions as a latch or a buffer. It enables the MPU to read from an input peripheral such as a keyboard or write to an output peripheral such as LEDs.
- TRIS: This is a data direction register. The logic level of the bits in this register determines the I/O function of an individual pin of the I/O port. Writing logic 0 to a bit in TRIS sets up the corresponding pin of the PORT as an output and logic 1 sets up the pin as an input.
- LAT: This is an output latch similar to PORT. This is used when the port is intended to function as a bidirectional port.

Another example is PORTB. Table 9-2 lists various functions of each pin of PORTB. Figure 9-3 shows the internal block diagram of PORTB in the simplified form, including three internal D flip-flops (or latches) labeled as Data Latch (Output Port), TRIS Latch (Data Direction Port), and Latch for Input Port. The I/O pin is external and available to the user. The data bit on the data bus is latched at Q when the processor asserts the WR signal by writing to the port, but the data bit is available at the I/O pin only if the tri-state buffer is enabled by the TRIS latch. When we write 0 into the TRIS latch, the tri-state buffer is enabled and the port functions as an output port. If we write 1 into the TRIS latch, the tri-state buffer is disabled, and PORTB functions as an input port.

TABLE 9-2   PORTB Functions

Name	Bit#	Buffer	Function
RB0/INT0	bit0	TTL/ST[(1)]	Input/output pin or external interrupt input0. Internal software program-mable weak pull-up.
RB1/INT1	bit1	TTL/ST[(1)]	Input/output pin or external interrupt input1. Internal software program-mable weak pull-up.
RB2/INT2	bit2	TTL/ST[(1)]	Input/output pin or external interrupt input2. Internal software program-mable weak pull-up.
RB3/CCP2[(3)]	bit3	TTL/ST[(4)]	Input/output pin or Capture2 input/Compare2 output/PWM output when CCP2MX configuration bit is enabled. Internal software program-mable weak pull-up.
RB4	bit4	TTL	Input/output pin (with interrupt-on-change). Internal software program-mable weak pull-up.
RB5/PGM[(5)]	bit5	TTL/ST[(2)]	Input/output pin (with interrupt-on-change). Internal software program-mable weak pull-up. Low voltage ICSP enable pin.
RB6/PGC	bit6	TTL/ST[(2)]	Input/output pin (with interrupt-on-change). Internal software program-mable weak pull-up. Serial programming clock.
RB7/PGD	bit7	TTL/ST[(2)]	Input/output pin (with interrupt-on-change). Internal software program-mable weak pull-up. Serial programming data.

Legend: TTL = TTL input, ST = Schmitt Trigger input

Note 1: This buffer is a Schmitt Trigger input when configured as the external interrupt.
   2: This buffer is a Schmitt Trigger input when used in Serial Programming mode.
   3: A device configuration bit selects which I/O pin the CCP2 pin is multiplexed on.
   4: This buffer is a Schmitt Trigger input when configured as the CCP2 input.
   5: Low Voltage ICSP Programming (LVP) is enabled by default, which disables the RB5 I/O function. LVP must be disabled to enable RB5 as an I/O pin and allow maximum compatibility to the other 28-pin and 40-pin mid-range devices.

EXAMPLE 9.1

Write instructions to set up PORTB as an input port and PORTC as an output port.

Solution

The port is set up as an input port by placing 1's and as an output port by placing 0's in the corresponding TRIS registers. These instructions can be written either by moving appropriate bytes in W register and placing them in TRIS registers or by using Set (SETF) and Clear (CLRF) instructions as shown.

Instructions:	SETF	TRISB	;Load all 1's in TRISB register to set up PORTB as ;an input port
	CLRF	TRISC	;Load all 0's in TRISC register to set up PORTC as ;an output port

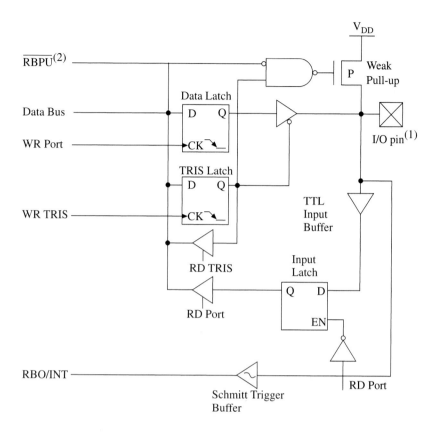

FIGURE 9-3   PORTB Internal Block Diagrams

Note 1: I/O pins have diode protection to V_{DD} and V_{SS}.

Note 1: I/O pins have diode protection to V_{DD} and V_{SS}.
   2: To enable weak pull-ups, set the appropriate TRIS bit(s) and clear the $\overline{RBPU}$ bit (Option_REG<7>).

EXAMPLE 9.2

Write instructions to set up pins RB7–RB4 of PORTB as inputs and RB3–RB0 pins as outputs.

Solution

The byte to set up upper half of PORTB as input and lower half as output is : 1 1 1 1 0 0 0 0.

Instructions:	MOVLW	0xF0	;Load 1 1 1 1 0 0 0 0 in W register
	MOVWF	TRISB	;Set up upper half of PORTB as input and lower
			;half as output

9.2 INTERFACING OUTPUT PERIPHERALS

The most commonly used output peripherals in embedded systems are LEDs, seven-segment LEDs, and LCDs. The simplest is LED that is used to indicate conditions such as power-on, emergency circuit break, battery charging, and normal operation. The LED is turned on when the anode voltage is higher than the cathode voltage by its forward bias voltage and is illuminated when there is sufficient current flow, typically, 10 to 15 mA in a regular LED and 5 to 6 mA in low power LED. A resistor is connected in series with the LED to limit the current.

Figure 9-4 shows two ways of connecting LEDs to I/O ports. In Figure 9-4a, LED cathodes are grounded and driven by the I/O port. In this case, logic 1 turns on the LED and the current is supplied by the chip; this process of supplying current is called current sourcing. In Figure 9-4b, LED cathodes are connected to the I/O port and driven by the power supply. In this case, logic 0 turns on the LED and the current flows towards the chip. In this case the current is received by the chip; thus, it is called current sinking. In general, a semiconductor chip has higher capacity to sink the current than to source it. If many peripherals are connected to I/O ports with heavy current load, the circuit in Figure 9-4b is preferred to that of in Figure 9-4a.

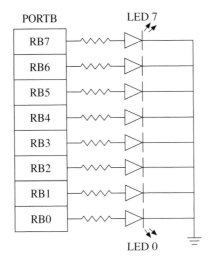
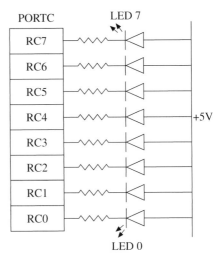

FIGURE 9-4 (a)   Interfacing LEDs with   **FIGURE 9-4 (b)**   Interfacing LEDs with
                     Cathode Grounded   Anode Connected to +5V

EXAMPLE 9.3

Write instructions to set up PORTC in Figure 9-4 b as an output port and turn on every other LED starting with LED0.

continued

Solution

In Figure 9-4 b, LED anodes are connected to + 5V and cathodes are connected to PORTC. The forward biased voltage is around 1.2 V and the current requirement is between 10 mA to 15 mA for-proper visibility. Therefore, logic 0 (low 0.7 V) output from PORTC should turn on the LEDs. Each I/O pin of PORTC is capable of sinking 25 mA of current.

    The instructions for setting up PORTC as an output port are the same as in Example 9.1. To turn on every other LED we need a bit pattern 1 0 1 0 1 0 1 0. Instructions should load AA_H in W register and output the bit pattern (AA_H) to PORTC.

Instructions:	CLRF PORTC	;Clear PORTC
	CLRF TRISC	;Load all 0's in TRISC to set up PORTC as an
		;output port
	MOVLW 0xAA	;Load appropriate bit pattern in W register
	MOVWF PORTC	;Turn on corresponding LEDs

9.2.1 Writing to I/O Ports

To understand how the processor in the PIC controller writes data into an output port, we must examine the instruction MOVWF PORTC (Example 9.3), which is stored in program memory. When the processor reads this instruction, it places the address of PORTC ($F82_H$) on the address bus of the data memory, the decoding network of the memory decodes the address and selects the register $F82_H$. The processor also places the contents of W register on the data bus and activates the WR signal. The WR signal enables the data into the PORTC latch and its tri-state buffer is already enabled by sending logic low to the TRIS latch. The data byte AA_H is latched at the port and logic 0's turn on the corresponding LEDs.

9.2.2 Interfacing Seven-Segment LEDs as an Output

Seven-segment LEDs are often used to display Binary Coded Decimal (BCD) numbers and a few letters that can be displayed in its physical structure. A seven-segment LED is a group of eight LEDs physically mounted in the shape of the number eight plus a decimal point. Each one is called a "segment." They are labeled "a" through "g", as shown in Figure 9-5a, and the eighth LED is mounted to show a decimal point either on the left or the right as shown. In terms of electrical characteristics, there are two types of seven-segment LEDs: (1) common anode where all anodes are connected together (Figure 9-5b) and (2) common cathode where all cathodes are connected together (Figure 9-5c).

    To display a digit, appropriate segments should be turned on. For example, to display the digit 0, all segments except "g" and "dp" should be turned on; to display digit 7, three segments "a, b, and c" should be turned on. Logic levels required to turn on a segment is determined by whether it is a common anode or common cathode seven-segment LED. For example, to display 8 without the decimal point at the common-anode seven-segment LED we need 1 0 0 0 0 0 0 (80_H) and at the common-cathode, we need the complement of 80_H - 0 1 1 1 1 1 1 1 ($7F_H$) on the data lines.

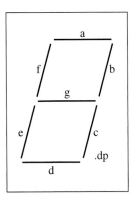

FIGURE 9-5 (a)   LED Physical Structure

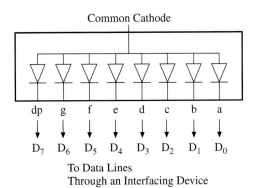

FIGURE 9-5 (b)   Logic Diagram - Common Anode LED

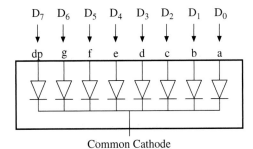

FIGURE 9-5 (c)   Logic Diagram - Common Cathode

EXAMPLE 9.4

A common anode seven-segment LED is connected to PORTB as shown in Figure 9-6a.and a BCD digit (0 to9) is stored in REG0. Write instructions to display the BCD number stored in data register REG0, at PORTB. Calculate the total current flow in the circuit when 8 is displayed if logic 0 on PORTB = 0.7 V, the voltage drop across the LED = 1.2 V, and the current limiting resistor is 330 ohms as shown.

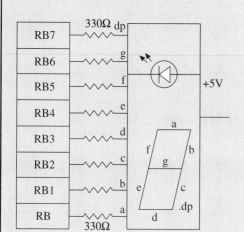

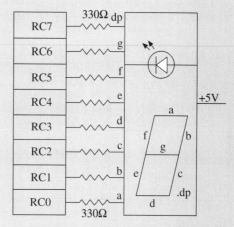

FIGURE 9-6 (a)   Interfacing Common-Anode Seven-Segment LED to PORTB

FIGURE 9-6 (b)   Interfacing Common-Anode Seven-Segment LED to PORTC

Solution

This is a common anode seven-segment LED; therefore, logic 0 is required to turn on a segment of the LED. We must find the LED codes for digits from 0 to 9 and store them sequentially as a table in program memory starting from the address CODEADDR as shown in the program. To display a digit stored in REG0, it should be copied into W register and added to the beginning address of the table. The result of the addition will have the address of the code for the digit in REG0. The following program illustrates the display of digit 5 stored in REG0.

Instructions:

```
    Title "BCD Digit Display"
    List p=18F452, f =inhx32
    #include <p18F452.inc>     ;This is a header file for PIC18F452

REG0        EQU    0x00        ;Define address of REG0

            ORG    0000
            GOTO   0X20
```

continued

```
                ORG    0x20              ;Begin assembly at memory location 000020H
    START:      CLRF   PORTB             ;Clear PORTB
                CLRF   TRISB             ;Set up PORTB as an output port
                MOVLW  5                 ;Digit to be displayed as an illustration
                MOVWF  REG0, 0
                MOVLW  UPPER CODEADDR    ;Copy upper 5 bits in Table Pointer of
                MOVWF  TBLPTRU           ; 21-bit address 00040H where code begins
                MOVLW  HIGH CODEADDR     ;Copy high 8 bit in Table Pointer
                MOVWF  TBLPTRH
                MOVLW  LOW CODEADDR      ;Copy low 8 bits in Table Pointer
                MOVWF  TBLPTRL
                MOVF   REG0, 0, 0        ;Get BCD number to be displayed
                ADDWF  TBLPTRL           ;Add BCD number to Table Pointer
                TBLRD*                   ;Read LED code from memory
                MOVFF  TABLAT, PORTB     ;Turn on LED

                ORG    0x40              ;Store LED code starting from location 0040H

    CODEADDR:   DB     0xc0, 0xF9, 0xA4, 0xB0, 0x99, 0x92; Code for digits 0 to 5
                DB     0x82, 0xF8, 0x80, 0x98          ; Code for digits 6 to 9
                END
```

Description:

The LED code for digits 0 to 9 is stored in program memory starting from the address labeled as CODEADDR 00040_H, which is a 21-bit address. We chose to store the LED codes starting from location 00040_H for convenience of calculation. After setting PORTB as an output port, the following six instructions copy the 21-bit address into the table pointer—upper, high, and low. The instruction MOVF REG0 copies the BCD byte into the W register which is added to the low address in the table pointer. In our example, we want to display digit 5. This number is added to the address in the table pointer. Now the address in the table pointer is $(00040 + 5)$ 00045_H pointing to the code 0x92. The instruction TABLRD* (Table Read) reads 0x92 into TABLAT (Table Latch) and the next instruction MOVFF outputs 0x92 to PORTB. This technique of finding code, stored sequentially in memory, is called table look-up.

Current Calculations:
The voltage drop across the resistor = 5V – 1.2V – 0.7V = 3.1V
Current flow in each LED segment = 3.1V/330 ohms ≈ 9.4 mA
Current flow for seven LED segments to display digit '8' = 65.8 mA

9.3 ILLUSTRATION: DISPLAYING BCD COUNTER AT SEVEN-SEGMENT LEDS

A BCD counter and seven-segment displays are commonly used in microprocessor-based systems. The programming techniques used for a counter can be used for designing many timing applications such as digital clocks, time delays, and periodic wave generations. The following illustration shows:

1. How to adjust a binary number as a BCD number for counting.
2. How to unpack a byte for a display.
3. How to get data from program memory by using the table look-up technique.

We have discussed some of these concepts in previous chapters, and in this illustration we will integrate them with hardware.

Applications

BCD numbers and seven-segment displays have numerous applications in our daily lives. Microprocessor/microcontrollers operate in the binary world, but we as human beings function in the decimal world. Therefore, the conversion of binary numbers into decimal numbers is essential. In our homes, we have many appliances and devices such as microwave ovens, dishwashers, digital clocks, and radios that display decimal numbers.

9.3.1 Problem Statement

Interface two common-anode seven-segment LEDs to PORTB and PORTC of the 18F microcontroller as shown in Figure 9-6a and b. Write instructions to design an up-counter counting from 00 to 59 at the interval of 100 ms and display the count at two seven-segment LEDs.

9.3.2 Problem Analysis

This problem requires the integration of hardware and software. We have already discussed the hardware aspect—how to interface seven-segment LEDs—in previous sections. The software problem must be divided into small segments and written as subroutines. The primary objective is to count in BCD from 00 to 59 every 100 ms and display the count. Therefore, we can divide the problem into three segments: (1) counting in BCD, (2) displaying a count, and (3) generating 100-millisecond delay. To display a BCD count, we must subdivide the task into two parts: (1) unpacking the BCD count and (2) finding the seven-segment code for each BCD digit. Thus we can subdivide the entire problem in five segments: (1) counting in BCD, (2) unpacking a byte, (3) getting the proper code for display by using the table look-up technique, (4) displaying a count, and (5) generating 100-millisecond delay.

Counting in BCD

As discussed in Chapter 6, a counter is a frequently used device, and we need an up-counter that counts from 00 to 59. Designing a software-based counter in a microprocessor-based system is very simple. In an up-counter, it involves clearing a register and incrementing it at a given interval up to the specified limit. However, the processor is a binary machine; it counts in binary. A 4-bit counter can count from 0 to F; however, after the number 1001 (9) the count goes to 1010 (A). In BCD we want the next number to be 10; the numbers from A to F are invalid. Therefore, when any binary number goes beyond nine, it needs to be adjusted by adding six as discussed in Chapter 6 or using a decimal adjust instruction.

The PIC18F family includes the instruction DW (Decimal Adjust W Register) that adjusts the binary count to the equivalent BCD count, and when this instruction is included in the program immediately after an arithmetic (or increment) operation, the processor adjusts the count to the proper value.

Unpacking a Byte and the Table Look-Up Technique

Two seven-segment LEDs are needed to display a byte, one for each group of four bits. However, each group of four bits must be separated or unpacked to find the code for each digit. For example, to display a number 39_H—called a packed number—the digits must be separated as 03 and 09, called unpacking. Once the number is unpacked, the seven-segment LED code can be found for each digit and displayed at two seven-segment LEDs. As discussed in Example 9.4, we store the seven-segment LED code in program memory in sequential order from 0 to 9, find the code for each digit by using the table look-up technique, and display the low-order digit at PORTC and the high-order digit at PORTB.

9.3.3 PROGRAM

```
;::::::::::::::::::::::::::::::::::::::::::::::::::::::::::::::::::::::::::::
;This program counts in BCD from 0 to 59 at the interval of 100 ms and    :
;displays the count at two seven-segment LEDs                             :
;::::::::::::::::::::::::::::::::::::::::::::::::::::::::::::::::::::::::::::

        Title "Illust9-3 BCD Digit Display"
        List p=18F452, f =inhx32
        #include <p18F452.inc>      ;Header file for PIC18F452

COUNTER     EQU     0x00            ;Register 00 is used as a BCD counter
REG1        EQU     0x01            ;Unpacked digits are saved in REG1 and REG2
REG2        EQU     0x02
TEMP        EQU     0x03            ;Temporary register for swapping nibbles
Count60     EQU     0x04            ;Register to hold last count for comparison
L1REG       EQU     0x10            ;Reg10 and REG11 are used to load delay
L2REG       EQU     0x11            ;counts


            ORG     00
            GOTO    MAIN

            ORG     0x20            ;Begin assembly at memory location 000020H
MAIN:       CLRF    PORTB           ;Initial readings at PORTB & PORTC = 00
            CLRF    PORTC
            CLRF    TRISB           ;Set up PORTB and PORTC as output ports
            CLRF    TRISC
REPEAT:     CLRF    COUNTER         ;BCD count begins at zero
```

continued

```
            MOVLW 0x60          ;The last count for comparison
            MOVWF COUNT60       ;Load count for comparison

START:      CALL  UNPACK        ;Unpack the count
            CALL  OUTLED        ;Display the count
            MOVLW D'200'        ;Multiplier count for delay
            MOVWF L2REG
            CALL  DELAY_100ms   ;Wait for 100 ms
            CALL  UPDATE        ;Go to next count
            CPFSEQ COUNT60      ;Is count = 60, if yes, skip
            GOTO  START         ;Go back to START to display
            GOTO REPEAT         ;Start again

UNPACK:     ;:::::::::::::::::::::::::::::::::::::::::::::::::::::::::::::::
            ;Function: This subroutine unpacks the BCD count in COUNTER   :
            ;          register that is used for counting and stores the  :
            ;          unpacked digits in REG1 and REG2                   :
            ;Input:    Packed BCD Count in COUNTER                        :
            ;Output:   Unpacked BCD in REG1 and REG2                      :
            ;:::::::::::::::::::::::::::::::::::::::::::::::::::::::::::::::

            MOVF  COUNTER,0,0   ;Get count in W register
            ANDLW 0x0F          ;Mask high-order nibble
            MOVWF REG1,0        ;Save low-order nibble in REG1
            MOVF  COUNTER,0     ;Get count in W again
            ANDLW 0xF0          ;Mask low-order nibble
            MOVWF REG2,0        ;Save high-order nibble in REG2
            SWAPF REG2,1,0      ;Swap high-order nibble for proper unpacking
            RETURN

OUTLED:
            ;:::::::::::::::::::::::::::::::::::::::::::::::::::::::::::::::
            ;Function: OUTLED gets the unpacked BCD digits from REG1 and  :
            ;          and REG2, gets seven-segment code by calling another :
            ;          subroutine GETCODE and displays BCD digits at PORTB :
            ;          and PORTC                                          :
            ;Input:    Unpacked BCD number in REG1 and REG2 Calls another :
            ;          subroutine GETCODE                                :
            ;:::::::::::::::::::::::::::::::::::::::::::::::::::::::::::::::

            MOVF  REG1,0,0      ;Get low-order BCD digit in W
            CALL  GETCODE       ;Find its seven-segment code
            MOVFF TABLAT, PORTC ;Display it at PORTC
            MOVF  REG2,0,0      ;Get high-order BCD
```

continued

```
                CALL   GETCODE        ;Get its code
                MOVFF  TABLAT, PORTB  ;Display at PORTB
                RETURN


GETCODE:        ;::::::::::::::::::::::::::::::::::::::::::::::::::::::::::::::::::::
                ;Function: This subroutine gets an unpacked digit from W        :
                ;          registerand looks up its seven-segment code and      :
                ;          stores it in TABLAT                                   :
                ;Input:    Unpacked digit in W                                  :
                ;Output:   Seven-segment LED code in TABLAT                     :
                ;::::::::::::::::::::::::::::::::::::::::::::::::::::::::::::::::::::

                MOVWF  TEMP
                MOVLW  UPPER LEDCODE  ;Copy upper 5 bits in Table Pointer of
                MOVWF  TBLPTRU        ;21_bit address of LEDCODE
                MOVLW  HIGH LEDCODE   ;Copy high 8 bit in Table Pointer
                MOVWF  TBLPTRH
                MOVLW  LOW LEDCODE    ;Copy low 8 bits in Table Pointer
                MOVWF  TBLPTRL
                MOVF   TEMP, 0, 0     ;Get BCD number to be displayed
                ADDWF  TBLPTRL        ;Add BCD number to Table Pointer
                TBLRD*                ;Read LED code from memory
                RETURN

LEDCODE:        DB     0xC0, 0xF9, 0xA4, 0xB0, 0x99, 0x92;Code for digits 0 to 5
                DB     0x82, 0xF8, 0x80, 0x98           ;Code for digits 6 to 9


DELAY_100ms  ;::::::::::::::::::::::::::::::::::::::::::::::::::::::::::::::::::::
                ;Function: Provides 100 ms delay Clock Frequency: 10 MHz        :
                ;Input:    A multiplier count in register L2REG                 :
                ;::::::::::::::::::::::::::::::::::::::::::::::::::::::::::::::::::::


LOOP2:          MOVLW  D'250'         ;Load decimal count in W
                MOVWF  L1REG          ;Set up Loop1 Register (L1REG) as a counter
LOOP1:          DECF   L1REG, 1       ;Decrement count in L1REG
                NOP                   ;Add instructions to increase delay
                NOP
                BNZ    LOOP1          ;Go back to LOOP1 if L1REG ≠ 0
                DECF   L2REG,1        ;Decrement Loop2 Register (L2REG)
                BNZ    LOOP2          ;Go back to load 250 in L1REG and start LOOP1
                RETURN
```

continued

```
UPDATE:        ;:::::::::::::::::::::::::::::::::::::::::::::::::::::::::::::::
               ;Function:   This subroutine increments count in COUNTER and     :
               ;           adjusts it as a: ;BCD count                          :
               ;Input:      BCD count in COUNTER                                 :
               ;Output:     Next adjusted BCD count in COUNTER                   :
               ;:::::::::::::::::::::::::::::::::::::::::::::::::::::::::::::::

               INCF  COUNTER,0,0     ;Next count
               DAW                   ;Adjust for BCD
               MOVWF COUNTER,0        ;Save count
               RETURN
               END
```

9.3.4 Program Description

This program includes five subroutines: UNPACK, OUTLED, GETCODE, DELAY_100ms, and UPDATE and uses COUNTER register as a counter that counts in BCD starting from 00. In the beginning, the directives (Equates) define the values of six data registers that are used in the program; the selection of these registers is arbitrary based on the user's choice. The main program clears PORTB and PORTC as a precaution against random values and initializes them as output ports using TRIS instructions. The remainder of the main program consists of calling the subroutines and repeating the process.

To explain how the program works, we will take a specific example. Let us assume that COUNTER has a byte 0 0 0 1 1 0 0 1 (19_H) and examine how the byte is displayed as 19_{BCD} at the seven-segment LEDs and it is adjusted to 20_{BCD} after the display.

1. UNPACK: The first subroutine called is UNPACK. The first instruction gets the byte 19_H in the WREG from COUNTER, the ANDLW 0x 0F instruction clears the high-order 4 bits (0 0 0 1) and stores 1 0 0 1 (9) in REG1. The next instruction MOVF gets 19_H in the WREG again, and the second AND instruction clears low-order 4 bits from 19_H and stores 10_H in REG2. The SWAP instruction changes 10_H into 01_H for proper unpacking. This can also be accomplished by using the instruction Rotate Right with No Carry (RRNCF) instruction four times. Before the RETURN instruction is executed, the contents of REG1 and REG2 will be 09_H and 01_H respectively.

2. OUTLED: This subroutine takes unpacked byte 09_H from REG1, gets the seven-segment code (98_H) in TABLAT (Table Latch) from another subroutine, GETCODE, and sends the code to PORTC which displays the low-order BCD digit 9. It repeats the same process for the unpacked byte 01; it gets the code $F9_H$ and displays 1 at PORTB.

3. GETCODE: This subroutine gets the input of unpacked BCD (09_H) from REG1 in the W register and saves it in TEMP register because the W register is used to copy the address of the code in TBLPTR (Table Pointer). The memory address of the LED-CODE is 21 bits wide; you can observe this address ($00007A_H$) in the list file. The next six instructions—two instructions at a time MOVLW and MOVF—copy 8-bit address segment into one of the three address registers of the table: Upper, High, and Low. The Upper segment has only five valid bits, the High segment holds 00 and the

Low segment holds the address $7A_H$. The next instruction gets the byte 09 from TEMP register into the W register and the ADDWF instruction adds 09 to $7A_H$, resulting into 83_H. The code for 09 is 98_H, which is stored in location 000083_H, and the instruction TABLRD* copies 98_H into the table latch (TABLAT). The entire process is repeated for the high-order digit 1.

4. DELAY: This subroutine provides an approximately 100-millisecond delay similar to the delay in Example 5.12 in Chapter 5, except that this is written as a subroutine (for the delay calculations, see Example 5.12.) This delay subroutine is called after the display of a count so that the user is able to see the count. This is a software delay based on the execution timing of the instructions, and it is generated using the nested loop (Loop-within-Loop) technique.

5. UPDATE: This subroutine increments the count in COUNTER after the delay, and the count is saved in the WREG. The next instruction DAW adjusts the count to the BCD value. In our example, the count 19_H goes to $1A_H$, and the DAW instruction adjusts that value to 20_H.

6. Checking the Last Count: The counter is expected to count and display the count up to 59_{BCD}, and when it is adjusted to the next count (60_{BCD}), the program should be terminated. After the UPDATE subroutine, the program goes back to the main program. The instruction CPFSEQ compares the count in the WREG with the count in register COUNT60. If the count is equal to 60_{BCD}, the program skips the next instruction GOTO and starts the program again by resetting the count to zero; otherwise, it returns to the label START to unpack the next count and continues until the count displays 59_{BCD}.

9.3.5 Program Execution and Troubleshooting

This program is a good illustration for troubleshooting using MPLAB simulator. This program is divided into five subroutines and we can check the output of each subroutine by setting breakpoints. The source code for this program is on the accompanying CD: Ch09\Illust9-3 BCD Counter.asm. You can execute and troubleshoot the program using the following procedure:

1. Open MPLAB IDE, create a new project as described in Section 4.5.1, add the source file Illust9-3 BCD Counter.asm, and build the project.

2. Select Debugger → Tool → MPLAB Sim. Resize the screen of the source program.

3. Select View → Watch. Add the following registers in the Watch window: WREG, TBLPTR, TABAT, PORTB, PORTC, REG0, REG1, and REG2. Adjust the screen to make all registers in the Window and the instructions in the source program (without comments) visible.

4. Select Debugger → Reset (F6).

5. Place the cursor on the first instruction (CLRF PORTB) of the source program and right-click. Select PC on Cursor and a green arrow should appear on the first instruction.

6. To set multiple breakpoints, double-click on each subroutine labels starting from CALL UNPACK, and red circles with the letter B will appear indicating the breakpoints (see Figure 9-7).

7. Run (F9) the program from the cursor, and it will stop at the first breakpoint (CALL UNPACK). Check all registers in the window; they should be cleared.

8. To begin counting from 19_{BCD} for an illustration, double-click on the value column of REG0 in the window, and insert 1 and 9 from the keyboard.

9. Run (F9) the program, and it will stop at the next breakpoint, CALL OUTLED. Check the registers in the Window. You should observe unpacked 19 as: REG1 = 09 and REG2 = 01. If you do not observe these values, it is an indication that the subroutine UNPACK has errors. To check those errors, reset the cursor back to CALL UNPACK and single-step each instruction and correct errors. Now you need to rebuild the project. Select Project and → Quick Build.

10. Run the program from the point where you left off and execute the subroutine OUTLED. Now you should observe seven-segment LED code $F9_H$ in PORTB representing digit 1, 98_H in PORTC representing digit 9, and the 21-bit memory address of the code $F9_H$ in the register TBLPTR. If you do not observe these values, you need to single-step the subroutine OUTLED.

11. The next subroutine is time delay: DELAY_100ms. This subroutine should not change any register values except WREG. After the execution, the program should return to the next subroutine UPDATE. If it does not return, the program must be caught in an infinite loop because of a wrong branch location. Check the subroutine using the single-step, correct where necessary, and rebuild the program.

12. Assuming the program returns to the UPDATE subroutine, you should observe 20_{BCD} after the execution of the UPDATE subroutine. If you observe 20_{BCD} after the execution of this subroutine, you can be assured that all subroutines are free of major bugs.

13. To check some remaining aspects, you should remove all breakpoints except the last one at GOTO START and run the program until you observe 30. After this observation, one more check should be performed when the program reaches 59. This is the completion of the simulation check; the rest is hardware troubleshooting if you build the circuit.

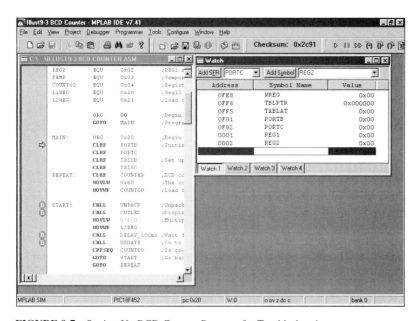

FIGURE 9-7   Setting Up BCD Counter Program for Troubleshooting

Postscript

In this illustration, two seven-segment LEDs are connected to two I/O ports; each seven-segment LED requires one port. In addition, the circuit draws excessive current (more than 65 mA per LED as calculated in Example 9.4) to light up all segments. This method is suitable if the system requires a few LEDs to indicate on/off conditions or two-digit seven-segment LEDs. However, many systems, such as clocks or instruments, require a display with four digits or more. To keep current consumption within the specified limits of the microcontroller, and use the minimum number of ports, the technique of time-multiplexing is often employed. It requires only two ports and can display eight digits. Another option is to use the LCD module to display multiple digits and ASCII characters. Both of these methods are described later in the chapter.

9.4 INTERFACING INPUT PERIPHERALS

The commonly used input peripherals are DIP switches, push-button keys, Hex and ASCII keyboards, and A/D converters. The circuit concept of an input from a DIP switch is simple: one side of the switch is tied high to a source (power supply) and the other side is grounded (Figure 9-8a) and the logic level changes as the position is switched. In a simple push-button key, the concept is the same except that the contact is momentary. To set up I/O ports of the PIC microcontroller as input ports, we need to send logic 1's to the TRIS latch, which disables the output latch buffer (Figure 9-3) and enables the processor to read logic levels on I/O pins.

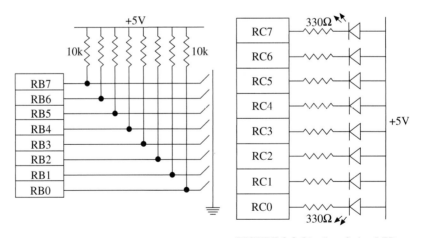

FIGURE 9-8 (a)   Interfacing DIP Switches   **FIGURE 9-8 (b)**   Interfacing LEDs

EXAMPLE 9.5

Write instructions to set up PORTB as input and PORTC as output. Read the switches connected to PORTB and display the switches that are grounded (on position) by turning on corresponding LEDs at PORTC.

Solution

In Figure 9-8a, if a switch is open, it reads high because it is connected to +5 V through a 10 k resistor. When a switch is closed (on), it is grounded; thus, it reads low. The LED anodes are connected to + 5V, and the cathodes are connected to PORTC (Figure 9-8b). Therefore, logic 0 (low 0.7 V) output from a PORTC pin should turn on an LED. Each I/O pin of PORTC is capable of sinking 25 mA of current.

The instructions for setting up ports are the same as in Example 9.1. To turn on the LEDs that correspond to the switches that are low, instructions should read PORTB and output the same reading to PORTC.

Instructions:	CLRF	PORTB	;Clear output data latches
	CLRF	PORTC	;Clear PORTC
	SETF	TRISB	;Load all 1's in TRISB register to set up ;PORTB as input
	CLRF	TRISC	;Load all 0's in TRISC to set up PORTC as an ;output
	MOVF	PORTB, W	;Read switches at PORTB
	MOVWF	PORTC	;Turn on corresponding LEDs

9.4.1 Reading from an I/O Port

To understand how the processor in the PIC controller reads data from an input port, we need to examine the instruction MOVF PORTB, W (Example 9.5) which is stored in program memory. When the processor reads this instruction, it places the address of PORTB (F81$_H$) on the address bus of the data memory. The decoding network of the memory decodes the address and selects the register F81$_H$. The processor follows that up with the control signal RD and enables the latch. The reading of the switches (logic highs and lows) is placed on the data bus, which is carried inside the processor and placed in the WREG. Meanwhile, the buffer is disabled (or the gate is closed) by deasserting the signal RD. The next instruction MOVWF PORTC writes the data into PORTC by using the similar process (described earlier in Section 9.2.1). The switch reading is latched in PORTC, and the LEDs corresponding to logic 0 are turned on.

9.4.2 Internal Pull-Up Resistor

In Figure 9-8a, we connected the pull-up resistors externally. However, PORTB can supply internal pull-up resistors through initialization. The internal connection of one of the I/O pins of PORTB is shown in Figure 9-9a. The FET connected to the I/O pin is turned off if the NAND gate driving the base is active low; thus the turned-off FET serves as a large pull-up resistor. To set up PORTB as an input port, we need to write logic 1 in the TRIS register that is connected to one of the inputs of the NAND gate. We can write to the other input signal RBPU (Pin RB Pull-Up) of the NAND gate through Bit7 of the Special Function Register INTCON2.

Figure 9-9b shows that Bit7 of INTCON2 is used to enable or disable $\overline{\text{RBPU}}$; other bits (B0 to B6) are used for various interrupt functions and not shown here. Therefore, rather than using external pull-up resistors, we can use internal pull-up resistors by writing logic 0 to Bit7 of INTCON2 register as follows: BCF INTCON2, 7, 0 (Clear Bit7 of INTCON2).

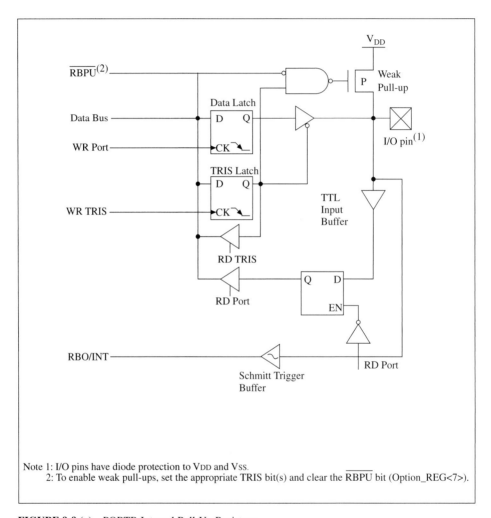

Note 1: I/O pins have diode protection to V$_{DD}$ and V$_{SS}$.
    2: To enable weak pull-ups, set the appropriate TRIS bit(s) and clear the $\overline{\text{RBPU}}$ bit (Option_REG<7>).

FIGURE 9-9 (a)   PORTB Internal Pull-Up Resistors

9.4.3 Interfacing Push-Button Keys

A group of push-button keys is another commonly used input device. There are various types of keys available including mechanical, plastic membrane, and capacitive. These keys are connected individually or they are arranged in a matrix format on a keypad. In this section, we will

B7	B6	B5	B4	B4	B3	B2	B1	BØ
$\overline{RBPU}$								

$\overline{RBPU}$ = PORTB pull-up resistor enable bit
   0 = Pull-up resistors are enabled
   1 = Pull-up resistors are disabled

FIGURE 9-9 (b)   Bit Specification of INTCON2 Register

focus on interfacing a group of individual mechanical keys, and discuss the interfacing of a matrix keypad later in the chapter.

The electrical connection of push-button keys is the same as the DIP switches—one side is grounded and the other side is connected to + 5 V through resistors, except that the contact is temporary (Figure 9-10a). In addition, when a mechanical switch is closed (or released), the metal contact bounces momentarily as shown in Figure 9-10b and can be read as multiple inputs by the processor. The reading of one contact as multiple inputs can be eliminated by a key-debounce technique, using either hardware or software.

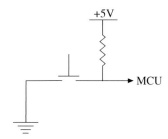

FIGURE 9-10 (a)   Push-Button Key Circuit

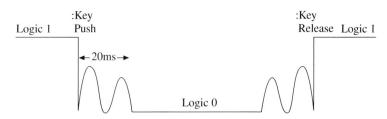

FIGURE 9-10 (b)   Push-Button Key Circuit

In the hardware technique, circuits are available that eliminate the bounce. Figure 9-11 shows two such circuits, which are based on the principles of generating a delay and switching the logic level at a certain threshold level. Figure 9-11a shows two NAND gates connected back to back, which is the equivalent of a S-R latch. The output of the S-R latch is a pulse without a bounce. Figure 9-11b shows another debounce circuit chip, MAX 6816, designed by MAXIM company. When a key is grounded, the chip debounces the key internally and provides a steady output, thus eliminating multiple output levels.

In the software technique, the program waits for 10 to 20 ms after it detects a switch closure and reads the switch closure again, and if it finds the reading still low, the program accepts and processes the reading. This technique is simpler and less expensive than the hardware technique and commonly used in microcontroller-based systems.

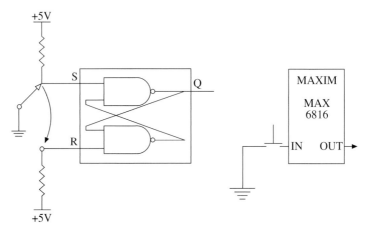

FIGURE 9-11 (a)   Back-to-Back Connected
NAND Gates – Same as
SR Flip-Flop

FIGURE 9-11 (b)   MAX 6816 — Debounce
Circuit by MAXIM

9.5 ILLUSTRATION: INTERFACING PUSH-BUTTON KEYS

This illustration shows the interfacing of a bank of push-button keys to a PIC microcontroller. When a key is pushed, the binary reading of the key is based on the position of the key in the bank. The software takes this reading and finds a way to recognize the position of the key using various techniques. This program shows how to establish the relationship between hardware and software and how to integrate hardware and software.

Applications

The application of push-button keys is a common occurrence in our daily lives. Generally, we see two types of switches in our homes: ones that make a permanent contact during the operation, such as a switch that turns on a light; and another group that makes a temporary contact to start or stop a process. Applications that require push-button keys include digital clocks, dish washers, telephone key pads, microwave ovens, computer keyboards, and television remotes. If we need more than eight keys, the contacts of these keys are generally arranged in a matrix format. The interfacing of matrix keys is illustrated later in the chapter.

9.5.1 Problem Statement

A bank of eight push-button keys K0 to K7 are connected as inputs to PORTB as shown in Figure 9-12. The pull-up resistors are internal to PORTB. Write a program to recognize a key pressed, debounce the key, and identify its location in the key bank with numbers from 0 to 7. The program should return the number representing the location in the WREG.

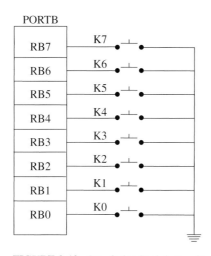

FIGURE 9-12   Interfacing Push-button Keys

9.5.2 Problem Analysis

Hardware

Figure 9-12 shows how keys are connected externally to PORTB. The internal pull-up resistors are not shown in the figure. In this problem, there are only two hardware set up requirements: PORTB should be set up as an input port, and its internal pull-up resistors should be enabled as discussed in Section 9.4.2.

Software

The software can be divided into the following three segments: (1) checking a key closure, (2) debouncing the key, and (3) encoding the key.

Checking a Key Closure

In the circuit shown in Figure 9-12, when a key is open, the logic level is one and when it is pressed, the logic level is zero. When all keys are open and if MPU reads PORTB, the reading will be FF_H, and when a key is pressed, the reading will be less than FF_H. Therefore, any binary reading less than FF_H indicates a key closure. In addition, we need to check for the situation where someone may press a key and hold it for a long time. Therefore, our program should first check whether all keys are open, as shown in Figure 9-13.

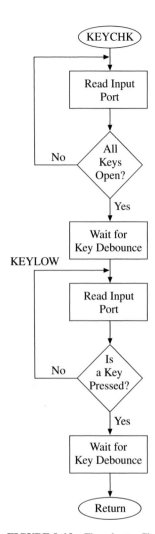

FIGURE 9-13   Flowchart – Checking a Key Closure

Debouncing the Key

Once we find a key closure, we need to debounce the key, as discussed in Section 9.4.3, to prevent the reading of multiple contacts. We will use the software technique of waiting for 20 ms and checking the key closure again. If the reading is still less than FF_H, we know that a key is pressed, and the key can be encoded.

Encoding the Key

Once a key closure is detected, the program needs to identify the key that is pressed. However, there is no direct relationship between the binary reading generated when a key is pressed and its position in the key bank. For example, when the key K1 is pressed, the binary reading is 1 1 1 1 1 1 0 1 (FD_H), and when K7 is pressed the reading is 0 1 1 1 1 1 1 1 ($7F_H$); therefore, we need to devise a technique to identify these keys as key number 1 and key number 7. One of the ways the key closure can be identified is by rotating right the binary reading and checking for a carry. For example, when the key K7 is pressed, the program will have to rotate the reading right seven times until it finds zero, and if the counter counts the number of rotations, the key can be encoded as number seven.

9.5.3 PROGRAM

```
;:::::::::::::::::::::::::::::::::::::::::::::::::::::::::::::::::::::::::::
;This program checks a key closure, debounces multiple key contacts,    :
;and encodes the key in binary digit that represents key's position     :
;:::::::::::::::::::::::::::::::::::::::::::::::::::::::::::::::::::::::::::


        Title "Reading Push Button Keys"
        List p=18F452, f =inhx32
        #include <p18F452.inc>        ;This is a header file that defines
                                      ;
SFR register values

KYSOPEN      EQU    0x00              ;Define data registers used
KYREG        EQU    0x01
COUNTER      EQU    0x02
L1REG        EQU    0x10              ;Registers for delay
L2REG        EQU    0x11


             ORG    00
             GOTO   START
```

continued

```
            ORG   0x20              ;Assemble at memory location 000020H
START:      SETF  TRISB             ;Set up PORTB as an input port
            BSF   INTCON2,7,0       ;Enable pull-up resistors
            SETF  KYSOPEN,0         ;Load FFH, reading when all keys are open


NEXT: CALL  KEYCHK                  ;Check a key closure
            CALL  KEYCODE           ;Encode the key number
            GOTO  NEXT              ;Wait


KEYCHK:     ;:::::::::::::::::::::::::::::::::::::::::::::::::::::::::::::::::::
            ;Function: KEYCHK checks first that all keys are open, then  :
            ;          checks a key closure, debounces the key, and saves:
            ;          the reading in KEYREG                              :
            ;Output: 8-bit reading of a key closure in KEYREG            :
            ;:::::::::::::::::::::::::::::::::::::::::::::::::::::::::::::::::::

            MOVF  PORTB,W,0         ;Read PORTB
            CPFSEQ KYSOPEN          ;Are all keys open?
            BRA   KEYCHK            ;If yes, go back to check again
            MOVLW D'40'             ;Load 40 multiplier in L2REG to multiply
            MOVWF L2REG             ;  500 µs delay in LOOP1
            CALL  DELAY_20ms        ;Wait 20 ms to debounce the key
KEYLOW:     MOVF  PORTB,W,0         ;Read PORTB to check key closure again
            MOVWF KYREG,0           ;Save the reading
            CPFSGT KYSOPEN          ;Is reading in W less than FFH
            BRA   KEYLOW            ;If no, go back and read again
            MOVLW D'40'             ;If yes, debounce the key - wait 20 ms
            MOVWF L2REG
            CALL DELAY_20ms
            MOVF PORTB,W,0          ;Read PORTB again
            CPFSEQ KEYREG           ;Is it same as before?
            BRA   KEYLOW            ;If no, it is a false reading
            RETURN                  ;If yes, return with a reading


KEYCODE:    ;:::::::::::::::::::::::::::::::::::::::::::::::::::::::::::::::::::
            ;Function: KEYCODE encodes the key and identifies the key    :
            ;          position                                          :
            ;Input: Key closure reading in KEYREG                        :
            ;Output: Encoded key position in W register                  :
            ;:::::::::::::::::::::::::::::::::::::::::::::::::::::::::::::::::::

            SETF  COUNTER           ;Set Counter to FFH
NEXTKEY:    INCF  COUNTER,1,0       ;Begin with 00 count after increment
            RRCF  KYREG,1,0         ;Rotate right input reading through
                                    ;Carry
```

continued

```
                BC      NEXTKEY         ;If Bit = 1, rotate again
                MOVF    COUNTER,0,0     ;Found zero, save in W
                RETURN


                ;:::::::::::::::::::::::::::::::::::::::::::::::::::::::::::::::
                ;Function: Provides 20 ms delay                              :
                ;Input: A multiplier count in register L2REG                 :
                ;:::::::::::::::::::::::::::::::::::::::::::::::::::::::::::::::

DELAY_20ms:
LOOP2:          MOVLW  D'250'   ;Load decimal count in W
                MOVWF  L1REG    ;Set up Loop1_Register as a counter
LOOP1:          DECF   L1REG, 1 ;Decrement count in L1REG
                NOP             ;Insructions added to get 500µs delay
                NOP
                BNZ    LOOP1    ;Go back to LOOP1 if L1REG ≠  0
                DECF   L2REG,1  ;Decrement L2REG
                BNZ    LOOP2    ;Go back to load 250 in L1REG and start LOOP1
                RETURN
                END
```

9.5.4 Program Description

The program consists of three subroutines: (1) KEYCHK, (2) KEYCODE, and (3) DELAY_
20ms. The KEYCHK checks whether a key is pressed, the KEYCODE finds which key is
pressed, and the DELAY_20ms provides a 20-millisecond delay to debounce a key.

In the equate block, the program defines the following three data registers: (1) KYSOPEN
that stores the byte FF_H representing the input reading when all keys are open, and it is used to
compare the input readings when a key is pressed; (2) KYREG is used to save the input reading when a key closure is found; and (3) COUNTER is set up to count the rotations of the input
reading and identify the key.

In the initialization block, PORTB is initialized as an input port, its pull-up resistors are
enabled by setting Bit7 in the INTCON2 register, and the KYSOPEN register is set to FF_H. The
main program consists of calling two subroutines:

- KEYCHK: This subroutine begins checking whether all keys are open as shown
  in the flowchart (Figure 9-13). This check prevents reading the same key if someone were to hold a key for a long time. After waiting for a debounce, the MPU
  reads PORTB again and continues to read until a key is pressed. When a key closure is found, the program saves the reading in KYREG, debounces the key, and
  reads the key closure again. The saved reading in KYREG is compared with the new
  reading. If they are equal, the program goes back to the main program to call the
  next subroutine KEYCODE; otherwise, the reading is viewed as a false alarm, and
  the program goes back to read PORTB again.
- KEYCODE: This subroutine begins with loading FF_H in the COUNTER register,
  and the next instruction INCF sets the COUNTER at 00. The rotate instruction

RRCF KYREG rotates the input reading through carry flag and stays in the loop until it finds logic 0 in the reading and for every rotation increments the counter. When the subroutine finds logic 0, it falls out of the loop, saves the counter reading in the W register, and goes back to the main program.

9.5.5 Program Execution and Troubleshooting

This program requires an input from PORTB, and the MPLAB simulator does not allow loading a byte in PORTB register in the Watch window. Therefore, to execute and troubleshoot this program, you need to add a data register that can accept any values as input readings; the program on the accompanying CD includes an additional register labeled as SimReg.

This program is divided into two major subroutines: KEYCHK and KEYCODE, and it also uses a 20-millisecond delay subroutine for debounce. You can check the output of each subroutine by setting breakpoints. The source code for this program is on the accompanying CD: Ch09\Illust9-5 Push Button Keys.asm. You can build the project using MPLAB IDE. Assuming the project is built without any assembly errors, and the MPLAB SIM is set up, you can execute and troubleshoot the program using the following procedure. To understand the loop structure, you will have to combine the processes of single-step and run.

Step 1: Select View → Watch. Add the following registers in the Watch window: WREG, INTCON2, PORTB, KYSOPEN, KYREG, COUNTER, STATUS, and SimReg. Adjust the screen to make all registers in the window and the instructions in the source program (without comments) visible.

Step 2: Select Debugger → Reset (F6). Place the cursor on the first instruction (SETF TRISB) of the source program and right-click. Select PC on the cursor and a green arrow should appear on the first instruction.

Step 3: Set breakpoints at subroutines KEYCHK, KEYCODE, and at the end of DELAY_ 20ms (See Figure 9-14).

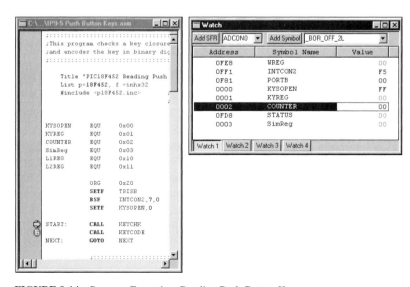

FIGURE 9-14   Program Execution: Reading Push-Button Keys

Step 4: Run (F9) the program from the cursor, and it will stop at the first breakpoint (CALL KEYCHK). Check registers KYSOPEN and INTCON2 for the following values: KYSOPEN = FFH, and Bit7 of INTCON2 should be set (some other bits are also set by the simulator).

Step 5: The KEYCHK subroutine first checks whether all keys are open. Therefore, to simulate the reading of all keys open, you should select Window screen, load FF_H in SimReg, and single-step the instructions. The program steps out of the loop and calls the delay routine.

Step 6: Run (F9) the program, and it will stop at the end of the delay subroutine.

Step 7: Now the subroutine is ready to check whether a key is pressed. Load $F7_H$ (1 1 1 1 0 1 1 1) in SimReg that represents the key K3 as low. Single-step the next few instructions, and the subroutine steps over the BRA instruction and calls the delay routine to debounce the key.

Step 8: Single-step the remaining instructions in the KEYCHK subroutine. At this point, we cannot really check whether the reading changes when PORTB is read after the debounce. You must load the same value ($F7_H$) in SimReg and compare. The values will be equal because you are loading the previous value. After this check, the program goes to the next subroutine, KEYCODE.

Step 9: Single-step the subroutine KEYCODE. You will see how COUNTER is set to 00 and bits are rotated in KEYREG. For every rotation, COUNTER is incremented and in the fourth rotation, when COUNTER = 3, the subroutine loads the count 3 representing the key K3 and returns to the main program.

Postscript

Applications and limitations of this illustrative program are as follows:

1. This program identifies the key pressed and provides that digit in W register at the end of the program. This digit can be used for many purposes. It can be displayed at a seven-segment LED port to indicate which key is pressed or it can be used to start a process such as turning on conveyer belt No. 3 or turning on a machine or appliance that is connected to Bit3 of an output port.

2. This program does not check if more than one key is pressed. The priority sequence is set up as K0 to K7. When two keys are pressed simultaneously, the key with a higher number is ignored.

3. When the number of keys required is larger than eight, a matrix keypad is used where a row and column are shorted when a key is pressed. A matrix keypad is a cost-effective way of connecting a large number of keys.

9.6 ILLUSTRATION: INTERFACING AN LCD (LIQUID CRYSTAL DISPLAY)

In Section 9.2 (Example 9.4) we illustrated the interfacing of seven-segment LEDs, which requires one I/O port per LED and consume an excessive current. In addition, the LEDs can display only limited characters. On the other hand, an LCD can display ASCII and graphic characters and consumes far less power. An LCD consists of crystal material sandwiched between two plates. When a low-frequency square wave pattern is applied to the plates, the internal crystals are arranged in the form of a dot matrix or segments that either block or pass the light and thus generate a display. The square wave pattern required to drive the plates is generated by a built-in driver. The interfacing of an LCD amounts to the interfacing of the driver. This illustration includes:

1. the hardware description of an LCD and its driver
2. the interfacing requirements
3. a program that displays a message on the LCD

Applications: LCDs are becoming very common. Examples include displays on wireless phones, watches, calculators, and instrument panels. Laptop computers always had LCDs, but now CRTs are being replaced by LCDs in PCs.

9.6.1 Problem Statement

1. Interface a 2-line × 20-Character LCD module with the built-in HD44780 controller to I/O ports of the PIC18452/4520 microcontroller.
2. Explain the requirements of the control signals.
3. Write a program to display the five ASCII characters representing time in hours and minutes and the colon separating the hours and minutes. These ASCII characters are stored in data registers starting at the label TIME. The display should be centered on the second line.

9.6.2 Problem Analysis

The problem analysis is divided into two major segments: hardware and software. The hardware segment describes the LCD with its built-in driver, data bus, control signals, and interfacing requirements. The software section consists of several subroutines that perform various tasks such as checking the readiness of the LCD to accept commands or data, sending a command or an ASCII character, and providing delays.

Hardware

The 20 × 2-line LCD can display two lines with 20 characters per line. The LCD has a display Data RAM that stores data in 8-bit character codes. Each register in data RAM has its own address that corresponds to the position on a line. Line 1 begins at the address 00 and goes up to 13_H and Line 2 begins at the address 40_H, and terminates at 53_H. The user can write to any address.

Figure 9-15 shows the 20-Character × 2-Line LCD with the built-in driver (Hitachi - HD44780) that includes an 8-bit data bus (DB7-DB0), three control signals (RS – Register Select, R/W – Read/Write, and E – Enable), and three power connections (V_{DD}, V_{SS}, and V_O) to connect power, ground, and a variable resistor to change the brightness. The LCD can be interfaced either in the 8-bit mode whereby all 8 data lines are connected for data transfer or in 4-bit mode where only 4 data lines (DB7-DB4 or DB3-DB0) are connected for data transfer. In the 4-bit mode, 2 transfers per character or instruction are needed. Figure 9-15 shows the interface of the 8-bit mode.

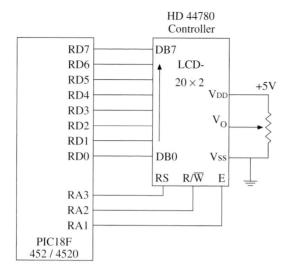

FIGURE 9-15 (a)   Interfacing LCD

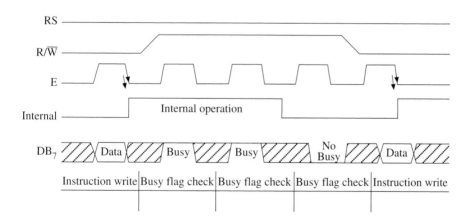

FIGURE 9-15 (b)   Timing Diagram

The driver HD44780 has two 8-bit internal registers: (1) Instruction Register (IR), and (2) Data Register (DR). The microcontroller must write into the IR to set up the LCD parameters for a desired operation and into the DR to display ASCII characters. The list of instructions that are needed to set up the LCD parameters such as Clear Display and Cursor Shift is shown in Table 9-3.

TABLE 9-3   Instructions

Instruction	RS	R/W	DB7	DB6	DB5	DB4	DB3	DB2	DB1	DB0	Description	Execution Time (max) (when f_{cp} or f_{osc} is 270 kHz)
					Code							
Clear display	0	0	0	0	0	0	0	0	0	1	Clears entire display and sets DDRAM address 0 in address counter.	1.52 ms
Return home	0	0	0	0	0	0	0	0	1	—	Sets DDRAM address 0 in address counter. Also returns display from being shifted to original position. DDRAM contents remain unchanged.	
Entry mode set	0	0	0	0	0	0	0	1	I/D	S	Sets cursor move direction and specifies display shift. These operations are performed during data write and read.	37 µs
Display on/off control	0	0	0	0	0	0	1	D	C	B	Sets entire display (D) on/off, cursor on/off (C), and blinking of cursor position character (B).	37 µs
Cursor or display shift	0	0	0	0	0	1	S/C	R/L	—	—	Moves cursor and shifts display without changing DDRAM contents.	37 µs
Function set	0	0	0	0	1	DL	N	F	—	—	Sets interface data length (DL), number of display lines (N), and character font (F).	37 µs
Set CGRAM address	0	0	0	1	ACG	ACG	ACG	ACG	ACG	ACG	Sets CGRAM address. CGRAM data is sent and received after this setting.	37 µs
Set DDRAM address	0	0	1	ADD	ADD	ADD	ADD	ADD	ADD	ADD	Sets DDRAM address. DDRAM data is sent and received after this setting.	37 µs
Read busy flag & address	0	1	BF	AC	AC	AC	AC	AC	AC	AC	Reads busy flag (BF) indicating internal operation is being performed and reads address counter contents.	0 µs
Write data to CG or DDRAM	1	0	Write data								Writes data into DDRAM or CGRAM.	37 µs $t_{ADD} = 4$ µs*
Read data from CG or DDRAM	1	1	Read data								Reads data from DDRAM or CGRAM.	37 µs $t_{ADD} = 4$ µs*

I/D = 1: Increment
I/D = 0: Decrement
S = 1: Accompanies display shift
S/C = 1: Display shift
S/C = 0: Cursor move
R/L = 1: Shift to the right

R/L = 0: Shift to the left
DL = 1: 8 bits, DL = 0: 4 bits
N = 1: 2 lines, N = 0: 1 line
F = 1: 5 × 10 dots, F = 0: 5 × 8 dots
BF = 1: Internally operating
BF = 0: Instructions acceptable

DDRAM: Display data RAM
CGRAM: Character generator RAM
ACG: CGRAM address
ADD: DDRAM address (corresponds to cursor address)
AC: Address counter used for both DD and CGRAM addresses

Execution time changes when frequency changes
Example:
When f_{cp} or f_{osc} is 250 kHz,

$$37 \,\mu s \times \frac{270}{250} = 40 \,\mu s$$

Note: — indicates no effect.

* After execution of the CGRAM/DDRAM data write or read instruction, the RAM address counter is incremented or decremented by 1. The RAM address counter is updated after the busy flag turns off. In Figure 10, t_{ADD} is the time elapsed after the busy flag turns off until the address counter is updated.

LCD Operation. To display a message on the LCD, the LCD should first be set up by sending appropriate initialization instructions to the IR. When the MPU writes an instruction into the IR or a character into the DR, the controller sets the data line DB7 high as a flag indicating that the controller is busy completing the internal operation, and that no additional information should be sent until the data line is reset. When the controller completes the operation, it resets the data line DB7. Therefore, the MPU should first check whether DB7 is low before sending an instruction or a data byte, except for the first two instructions. After the power-up, the DB7 cannot be checked for the first two initialization instructions.

LCD Interfacing. Figure 9-15 shows that the eight lines of PORTD are connected to the data lines DB7-DB0 of the LCD controller, and three lines, RA1, RA2, and RA3, from PORTA are used as control signals to read from and write into the LCD module. RA3 is connected to Register Select (RS) pin, RA2 to R/W pin, and RA1 to Enable (E) pin. The IR is selected when RS is low and the data register (DR) is selected when RS is high. The MPU asserts RS low to write instructions and high to write data as shown in Table 9-4. The MPU reads from the controller by asserting R/W high and writes into the controller by asserting it low. To latch either an instruction or a data byte, the MPU asserts E pin first high and then low.

TABLE 9-4

RS	R/$\overline{W}$	OPERATION
0	0	IR write as an internal operation (display clear, etc)
0	1	Read busy flag (DB7) and address counter (DB0 to DB6)
1	0	DR write as an internal operation (DR to DDRAM or CGRAM)
1	1	DR read as an internal operation (DDRAM or CGRAM to DR)

Software

To write into the LCD, the software should perform the following functions:

1. Send the initial instructions (commands) before it checks DB7 to identify whether it is operating in the 8-bit or 4-bit mode.
2. Check the data line DB7 and continue to check until it goes low.
3. Send instructions (command codes) to the IR to set up the LCD parameters such as the number of display lines, size of the characters, and the cursor status.
4. Send data to display the message.

The above tasks can be written as subroutines, as illustrated in the next section.

9.6.3 PROGRAM

```
;:::::::::::::::::::::::::::::::::::::::::::::::::::::::::::::::::::::::::
;This program displays time on the LCD.  The time is stored as ten     :
;ASCII characters in data registers labeled as TIME. The program       :
;initializes the LCD, checks the data line DB7.  When DB7 is low, it    :
;sends out instructions and data for display.                          :
;:::::::::::::::::::::::::::::::::::::::::::::::::::::::::::::::::::::::::


        Title "Illust9-6 Interfacing LCD"
        List p=18F452, f =inhx32
        #include <p18F452.inc>  ;This is a header file that defines
                                ;  SFR register values


LREG1       EQU   0x01          ;Delay register
DATAPORT    EQU   PORTD         ;Define PORTD as data port
TEMP        EQU   0x10          ;Temporary register to store values
DELAY_REG   EQU   0x11          ;Register that holds delay count
MSG_COUNTR  EQU   0x12          ;Register that holds number of display characters
CURSOR      EQU   0xC7          ;8th Location - 2nd line of DD RAM to start message
COUNT       EQU   D'50'         ;Delay count for 100µs
TIME        SET   0x30          ;Data registers where time characters are stored


#DEFINE     DB7   PORTD, 7      ;DB7 is defined to check flag
#DEFINE     E     PORTA, 1      ;Enable line
#DEFINE     RW    PORTA, 2      ;Read/Write line - Read high and Write low
#DEFINE     RS    PORTA, 3      ;Register select - low for IR and high for DR


            ORG   00            ;Reset memory location
            GOTO  0x20          ;Main program begins at 000020H


            ORG   0x20          ;Assemble at memory location 000020H
MAIN:       CALL  LCD_SETUP     ;Initialize LCD parameters
            MOVLW CURSOR        ;Starting location of the message
            CALL  LCD_CMD       ;Place cursor at desired location
            MOVLW D'05'         ;Number of characters to display
            MOVWF MSG_COUNTR    ;Load # in counter
            LFSR  FSR0, TIME    ;Set up data pointer
    NEXT:   MOVF  POSTINC0,W    ;Get character in W and increment pointer
            CALL  LCD_DATA      ;Send display character to LCD
            DECF  MSG_COUNTR,1  ;Decrement counter
            BNZ   NEXT          ;Is counter = 0? If not, get next chracter
    HERE:   BRA   HERE          ;Done - stay here
```

continued

```
LCD_SETUP:   ;::::::::::::::::::::::::::::::::::::::::::::::::::::::::::::::;
             ;Function:  This subroutine sends out initial instructions, :
             ;           provides sufficient delay for the LCD to get     :
             ;           ready, and again sends out various instructions :
             ;           to set up the LCD parameters such as the number :
             ;           of lines, size of the character, and the cursor :
             ;           increment.                                       :
             ;Calls:     Subroutines DELAY, LCD_OUT, AND LCD_CMD          :
             ;::::::::::::::::::::::::::::::::::::::::::::::::::::::::::::::::

             MOVLW  D'15'        ;Count for 1.5 ms delay
             MOVWF  DELAY_REG    ;Load count in Delay Register
             CALL   DELAY        ;Wait for 1.5 ms
             MOVLW  B'00110000'  ;First instruction to LCD
             MOVWF  TEMP         ;  to set up 8-bit mode interface
             CALL   LCD_OUT      ;Send instruction
             MOVLW  D'45'        ;Count for 4.5 ms delay
             MOVWF  DELAY_REG
             CALL   DELAY        ;Wait 4.5 ms
             MOVF   TEMP,W       ;Get same instruction from TEMP
             CALL   LCD_OUT      ;Send it out again
             MOVLW  D'01'        ;Count for 100 micro-sec delay
             MOVWF  DELAY_REG
             CALL   DELAY        ;Wait for 100 micro-sec
             MOVF   TEMP,W       ;Get same instruction from TEMP
             CALL   LCD_OUT      ;Send it out again
             MOVLW  B'00111000'  ;Code (38H) - 8 bits, 2 lines, and
             CALL   LCD_CMD      ;   5x7 dots
             MOVLW  B'00001000'  ;Code (08H) - Turn off display and
             CALL   LCD_CMD      ;  cursor
             MOVLW  B'00000001'  ;Code (01H) - Clear display
             CALL   LCD_CMD
             MOVLW  B'00000110'  ;Code (06H) - Entry mode, shift and
             CALL   LCD_CMD      ;  increment cursor
             MOVLW  B'00001100'  ;Code (0CH) - Turn on display
             CALL   LCD_CMD      ;  without cursor blinking
             RETURN


LCD_CMD:     ;:::::::::::::::::::::::::::::::::::::::::::::::::::::::::::::::::
             ;Function:  This subroutine checks DB7 by calling the CHK_FLAG  :
             ;           subroutine, and when the DB7 goes low, this subroutine:
             ;           accesses the instruction register and sends out a   :
             ;           command.                                            :
             ;Input:     Command code in W register                         :
             ;Calls:     Subroutine CHK_FLAG                                :
             ;:::::::::::::::::::::::::::::::::::::::::::::::::::::::::: ::::: :
```

continued

```
          MOVWF TEMP          ;Save command code in W register
          CALL  CHK_FLAG      ;Check if DB7 is low
LCD_OUT:  BCF   RS            ;Select Instruction Register (IR)
          BCF   RW            ;Set /Write low
          BSF   E             ;Set Enable high
          MOVF  TEMP,W        ;Get command code in W register
          MOVWF DATAPORT      ;Send command
          BCF   E             ;Assert Enable low to latch command in IR
          RETURN


LCD_DATA: ;::::::::::::::::::::::::::::::::::::::::::::::::::::::::::::::::::::::
          ;Function:  This subroutine checks DB7 by calling the CHK_FLAG    :
          ;           subroutine, and when the DB7 goes low, this subroutine :
          ;           accesses the data register and sends out an ASCII code.:
          ;Input:     ASCII code in W register                              :
          ;Calls:     Subroutine CHK_FLAG                                    :
          ;::::::::::::::::::::::::::::::::::::::::::::::::::::::::::::::::::::::

          MOVWF TEMP          ;Save ASCII code in W register
          CALL  CHK_FLAG      ;Check if DB7 is low
          BSF   RS            ;Select Data Register (DR)
          BCF   RW            ;Set Write low
          BSF   E             ;Set Enable high
          MOVF  TEMP,W        ;Grt ASCII code in W register
          MOVWF DATAPORT      ;Send ASCII code
          BCF   E             ;Assert Enable low to latch code
          RETURN


CHK_FLAG: ;::::::::::::::::::::::::::::::::::::::::::::::::::::::::::::::::::::::
          ;Function:  This subroutine initializes PORTD as an input port    :
          ;           and checks DB7 to verify whether the LCD              :
          ;           is busy in completing the previous operation.  It     :
          ;           continues to check the DB7 until it goes low          :
          ;           indicating that the LCD is free to accept the next    :
          ;           byte. It reinitializes PORTD as an output port and    :
          ;           returns.                                              :
          ;::::::::::::::::::::::::::::::::::::::::::::::::::::::::::::::::::::::

          SETF  TRISD         ;Set data port PORTD as input
          BCF   RS            ;Assert RS low to select Instruction Register
          BSF   RW            ;Set R/W high to read DB7 flag
READ:     BSF   E             ;Set Enable high
          NOP                 ;Stretch E pulse with a small delay
          BCF   E             ;Set Enable low
          MOVF  DATAPORT, W
          BTFSC DB7           ;Check if LCD busy
```

continued

```
             BRA    READ        ;If yes, go back and check again
             CLRF   TRISD       ;Set data port PORTD as output
             RETURN

DELAY:
       ;::::::::::::::::::::::::::::::::::::::::::::::::::::::::::::::::
       ;Function:  This subroutine provides multiples of 100 micro-sec  :
       ;           delay.  The count in DELAY_REG determines the        :
       ;           delay multiples. The clock is assumed to be 10 MHz.  :
       ;           Refer Section 5.4 for delay calculations             :
       ;Input:     Count in DELAY_REG                                   :
       ;::::::::::::::::::::::::::::::::::::::::::::::::::::::::::::::::


       LOOP: MOVLW  COUNT       ;Delay count for 100 micro-second
             MOVWF  LREG1
       LOOP1:DECF   LREG1,1     ;Decrement count
             NOP                ;Instruction to increase delay
             NOP
             BNZ    LOOP1       ;Is count = 0? If not, repeat LOOP1
             DECF   DELAY_REG   ;Decrement delay multiple
             BNZ    LOOP        ;Is multiple = 0? If not, repeat.
             RETURN
             END
```

9.6.4 Program Description

This program consists of five subroutines: (1) LCD_SETUP, (2) LCD_CMD, (3) LCD_DATA, (4) CHK_FLAG, and (5) DELAY. The LCD_SETUP initializes the desired parameters of the LCD, the LCD_CMD sends a command to the IR, the LCD_DATA sends data (ASCII character) to the DR, the CHK_FLAG checks the DB7 until it goes low, and the DELAY provides the necessary delay.

After the power up, the MAIN program calls the LCD_SETUP subroutine that sends the instruction to the LCD by calling the LCD_OUT and then sets up the 8-bit interface mode. The LCD_OUT is a part of the subroutine LCD_CMD, which does not check the DB7. This is repeated several times with the recommended delays until the LCD is ready to accept commands and data. After the first two instructions and appropriate delays, the LCD can set DB7 high when it is busy completing an internal operation. The LCD_SETUP then calls LCD_CMD to send instructions to set up the size of the characters, the number of lines, cursor increment, and to turn on the display. Before every instruction is sent out, the subroutine confirms the readiness of the LCD by checking the flag DB7.

After the setup, the MAIN program supplies the cursor position by specifying the address on the LCD line and loads the counter with the number of the ASCII characters to be displayed. We assume here that the ten ASCII characters to be displayed are: TIME XX:XX, and that they have been stored in data (file) registers of the PIC18F microcontroller starting with the label TIME by some other program. The MAIN sets up the FSR0 as a pointer to the data registers where ASCII characters are stored, gets these characters one at a time, and sends them to the LCD by calling the subroutine LCD_DATA until the counter goes to zero.

LCD_CMD and LCD_DATA. These two subroutines are in many ways similar except that the LCD_CMD writes into the IR and the LCD_DATA writes into the DR. The primary function of these subroutines is to assert the control signals RS to select the register, and R/W to read from or write into and toggle the E signal to latch the information.

CHK_FLAG. This subroutine first initializes the data port PORTD as an input port and checks the data line DB7. The LCD controller uses bit DB7 as a flag. When the LCD receives a command or a data byte, the controller sets DB7 and then resets it when the internal operation is complete. This subroutine continues to check DB7 until it is reset and then reinitializes the PORTD as the output port.

9.7 ILLUSTRATION: INTERFACING A MATRIX KEYBOARD

In Section 9.5 we illustrated the interfacing of eight push-button keys to one I/O port. When the number of keys to be interfaced increases beyond eight keys, the use of a matrix keypad (also called keyboard when the number of keys is large) is a cost-effective approach. In a matrix keypad, the keys are connected in rows and columns. When a key is pressed, the corresponding row and column make contacts and provide logic 0, and the software recognizes the key pressed and encodes the key. The following illustration includes an interfacing of 4 × 4 keypad to PORTB.

Applications: A keypad (or a keyboard) is a commonly used input device in many embedded systems that are used in homes and laboratories. Examples include microwave ovens, remote controls, telephones, copying machines, and training boards. A keyboard is almost always an integral part of a programmable system such as a PC.

9.7.1 Problem Statement

1. Interface a 4 × 4 (16) Hex keypad to PORTB of PIC18F452 microcontroller.
2. Write a program to recognize a key pressed and encode the key in its binary value from 0000 to 1111.

9.7.2 Problem Analysis

The problem analysis includes two segments: hardware and software. The hardware segment describes how a key contact is made in a matrix keypad, and the software identifies a key closure and encodes the key in its binary value.

Hardware

Figure 9-16 shows the interfacing of a 4 × 4 matrix keypad organized in the row and column format—four rows and four columns, whose keys are identified as 0 to F. Four rows are connected to the lower half of PORTB (RB0–RB3) and four columns are connected to the upper half of PORTB (RB4–RB7), and the internal pull-up resistors are turned on by clearing $\overline{RBPU}$ in the INTRCON2 <7> register. When a key is pressed, it makes a contact with the corresponding row and column; otherwise, there is no contact between a row and a column. For example, when key 0 on the pad is pressed, RB3 and RB4 are connected, and when key 3 is pushed, RB3 and RB7 are connected.

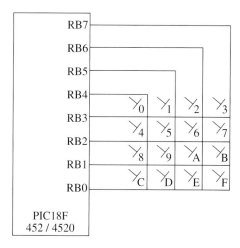

FIGURE 9-16   Interfacing a Matrix Keypad

Software

To recognize and encode the key pressed, the concept is to ground all the columns by sending zeros and check each key in a row for logic 0. The key that is pressed will read as logic 0. This can be accomplished by setting the column section of PORTB as output and the row section as input. The program should perform the following steps.

1. Check for a key closure.
   To check for a key closure, the program reads the input port and debounces the key. If a key is closed, the input reading will be less than all 1s.
2. Identify the key.
   When a key is closed, the input reading for four keys is the same. For example, when any of keys 0 to 3 is pressed, the input reading on the lines RB3–RB0 is the same: 0 1 1 1. To identify the key, the program grounds one column at a time and checks all the rows in that column. Once a key is identified, it is encoded based on its position in the column.

The above tasks are written as a subroutine in the next section. It is assumed here that when a key is pressed, some other program will call this subroutine.

9.7.3 PROGRAM

```
;::::::::::::::::::::::::::::::::::::::::::::::::::::::::::::::::::::::::
;This program is written as a subroutine.  It checks a key closure,   :
;debounces multiple key contacts, and encodes the key in binary digit :
;that represents the key's position                                   :
;::::::::::::::::::::::::::::::::::::::::::::::::::::::::::::::::::::::::


        Title "Illust9-7 Matrix Keypad"
        List p=18F452, f =inhx32
        #include <p18F452.inc>        ;This is a header file



KYSOPEN    EQU    0x00              ;Define data registers used
COUNTER    EQU    0x01
L1REG      EQU    0x10              ;Registers for delay
L2REG      EQU    0x11


           ORG    0x20              ;Assemble at memory location 000020H
KEYPAD:
SETUP:     MOVLW 0x0F
           MOVWF TRISB              ;Enable RB7-RB4 as output and RB3-RB0 as input
           BSF   INTCON2,7,0        ;Enable pull-up resistors
           MOVLW 0xFF               ;Load FFH, reading when all keys are open
           MOVWF KYSOPEN

KEYCHK:  ;:::::::::::::::::::::::::::::::::::::::::::::::::::::::::::::::::::
         ;Function: KEYCHK checks first that all keys are open, then     :
         ;          checks a key closure, and debounces the key.         :
         ;:::::::::::::::::::::::::::::::::::::::::::::::::::::::::::::::::::

           MOVF  PORTB,0,0   ;Read PORTB
           CPFSGT KYSOPEN    ;Are all keys open?
           RETURN            ;If yes, go back to check again
           MOVLW D'40'       ;Load 40 multiplier in L2REG to multiply
           MOVWF L2REG       ;  500 µs delay in LOOP1
           CALL  DELAY_20ms  ;Wait 20 ms to debounce the key

KEYCODE: ;:::::::::::::::::::::::::::::::::::::::::::::::::::::::::::::::::::
         ;Function: KEYCODE encodes the key and identify the key position:
         ;Output:   Encoded key position in W register                   :
         ;:::::::::::::::::::::::::::::::::::::::::::::::::::::::::::::::::::

    COLRB4: MOVLW 0xFF        ;Get ready to scan Column RB4
```

continued

```
          IORWF PORTB, F      ;All other keys should be 1s
  GNDB4:  BCF   PORTB, 4      ;Ground Column - RB4
  KEYB40: BTFSC PORTB, 0      ;Check RB0, if = 0, find code
          BRA   KEYB41        ;If RB0 = 1, check next key
          MOVLW 0x0C          ;Code for Key 'C"
          RETURN
  KEYB41: BTFSC PORTB, 1      ;Check RB1, if = 0, find code
          BRA   KEYB42        ;If RB1 = 1, check next key
          MOVLW 0x8           ;Code for key '8'
          RETURN
  KEYB42: BTFSC PORTB, 2      ;Check RB2, if = 0, find code
          BRA   KEYB43        ;If RB2 = 1, check next key
          MOVLW 0x4           ;Code for key '4'
          RETURN
  KEYB43: BTFSC PORTB, 3      ;Check RB3, if = 0, find code
          BRA   COLRB4        ;If RB3 = 1, go to next column
          MOVLW 0x0           ;Code for key '0'
          RETURN
  COLRB5: MOVLW 0xFF          ;Get ready to scan Column RB5
          IORWF PORTB, F      ;All other keys should be 1s
  GNDB5:  BCF   PORTB, 5      ;Ground Column RB5
          BTFSC PORTB, 0      ;Check RB0, if = 0, find code
          BRA   KEYB51
          MOVLW 0x0D
          RETURN
  KEYB51: BTFSC PORTB, 1      ;Check RB1, if = 0, find code
          BRA   KEYB52
          MOVLW0x9
          RETURN
  KEYB52: BTFSC PORTB, 2      ;Check RB2, if = 0, find code
          BRA   KEYB53
          MOVLW 0x9
          RETURN
  KEYB53: BTFSC PORTB, 3      ;Check RB3, if = 0, find code
          BRA   COLRB5        ;If RB3 = 1, go to next column
          MOVLW 0x1
          RETURN
  COLRB6: MOVLW 0xFF          ;Get ready to scan Column RB6
          IORWF PORTB, F      ;All other keys should be 1s
  GNDB6:  BCF   PORTB, 6      ;Ground Column RB6
          BTFSC PORTB, 0      ;Check RB0, if = 0, find code
          BRA   KEYB61
          MOVLW 0x0E
          RETURN
```

continued

```
  KEYB61:  BTFSC PORTB, 1    ;Check RB1, if = 0, find code
           BRA    KEYB62
           MOVLW 0xA
           RETURN
  KEYB62:  BTFSC PORTB, 2    ;Check RB2, if = 0, find code
           BRA    KEYB63
           MOVLW 0x6
           RETURN
  KEYB63:  BTFSC PORTB, 3    ;Check RB3, if = 0, find code
           BRA    COLRB7     ;If RB3 = 1, go to next column
           MOVLW 0x2
           RETURN
  COLRB7:  MOVLW 0xFF        ;Get ready to scan Column RB7
           IORWF PORTB, F    ;All other keys should be 1s
           BCF    PORTB, 7   ;Ground Column RB7
  KEY70:   BTFSC PORTB, 0    ;Check RB0, if = 0, find code
           BRA    KEYB71
           MOVLW 0x0F
           RETURN
  KEYB71:  BTFSC PORTB, 1    ;Check RB1, if = 0, find code
           BRA    KEYB72
           MOVLW 0xB
           RETURN
  KEYB72:  BTFSC PORTB, 2    ;Check RB2, if = 0, find code
           BRA    KEYB73
           MOVLW 0x7
           RETURN
  KEYB73:  BTFSC PORTB, 3    ;Check RB3, if = 0, find code
           RETURN
           MOVLW 0x3
           RETURN


DELAY_20ms: ;:::::::::::::::::::::::::::::::::::::::::::::::::::::::::::::::::::::
           ; Function: Provides 20 ms delay
           ;Input: A multiplier count in register L2REG             :
           ;:::::::::::::::::::::::::::::::::::::::::::::::::::::::::::::::::::::

  LOOP2:   MOVLW D'250'      ;Load decimal count in W
           MOVWF L1REG       ;Set up Loop1_Register as a counter
  LOOP1:   DECF  L1REG, 1    ;Decrement count in L1REG
           NOP
           NOP
           BNZ    LOOP1      ;Go back to LOOP1 if L1REG  =/ 0
           DECF  L2REG,1     ;Decrement L2REG
           BNZ    LOOP2      ;Go back to load 250 in L1REG and start LOOP1
           RETURN
           END
```

9.7.4 Program Description

The program includes two subroutines: KEYPAD and DELAY_20ms. The KEYPAD is divided into three sections: SETUP, KEYCHK, and KEYCODE. The instructions in the SETUP section initialize pins RB0–RB3 as inputs and pins RB4–RB7 as outputs, enable the pull-up resistors, and load FF_H in the register KYSOPEN, which is used for comparison when all the keys are open. The KEYCHK sections read PORTB and compare the reading with FF_H. If the reading is less than FF_H, it indicates that a key is pressed. The program debounces the key by waiting for 20 ms and goes to the next section KEYCODE. This section identifies which key is pressed as follows:

1. The IORWF instruction logically ORs the PORTB reading with FF_H to set up all column pins at logic 1; it will not affect the row in which a key is pressed.
2. The instruction BCF PORTB, 4 grounds the column RB4, and the instruction BTFSC checks the row RB0. If the reading is low, the program skips the next branch instruction, loads the binary code for key "C" in the W register, and returns to the calling program. The code is "C" because we chose to label the key as "C". We could have chosen to label the keys in a different sequence.

   If the reading is high, the key at RB0 must be open, and the program jumps to the next row RB1 at the label KEYB41 (Column 4 and Row 1 at RB1). If the key at the row RB1 is high, the program continues to check the next two rows RB2 and RB3 using the same process as when checking the row RB0.
3. After the completion of checking four rows in Step 2, the program arrives at the label COLRB5 and repeats Step 1 and Step 2 for the column RB5. The program sets the columns at logic 1, grounds the column RB5, and checks all the rows RB0–RB3 for a low key.
4. The key checking process continues by grounding the next two columns RB6 and RB7 one at a time. When a low key is found, the program loads the binary code of the key and returns to the calling program.

9.8 ILLUSTRATION: INTERFACING SEVEN SEGMENT LEDS – TIME MULTIPLEX SCANNING TECHNIQUE

Seven-segment LED displays are commonly used in various applications such as household appliances, laboratory instruments, and industrial embedded systems. As discussed in Section 9.3, we need one I/O port per LED, and the power consumption is excessive when more than two LEDs are needed for a display. To minimize the power consumption and the number of ports, the time multiplexed scanning technique is used whereby eight LEDs can be interfaced using two I/O ports. One port is used to send the display bytes (codes) and the other port is used to turn on and off LEDs continuously in a fast enough speed so that the display appears stable to the human eye. The following illustration demonstrates the interfacing of four seven-segment LEDs.

9.8.1 Problem Statement

Interface four common anode seven-segment LEDs to the PIC18F452 microcontroller using PORTB and PORTC, and explain how the circuit works. Write instructions to display a four-digit number (or a four-character message) stored in data registers starting from REG20 (address 20_H).

9.8.2 Problem Analysis

The problem is based on the concept that a stable display can be generated by turning on each seven-segment LED for a short period and rotating the turn on/off sequence with appropriate speed. This is similar to creating a visible circle by holding a flashlight and rotating it in a circular motion. By keeping the LEDs on for a short period, you can reduce the average current. Each pin of PORTB can source and sink 25 mA of current, and PORTB can source total 100 mA current and sink 150 mA current.

Hardware

Figure 9-17 shows that the eight data lines of PORTB are connected to anodes of eight segments (a through g) of each LED. These are common cathode seven-segment LEDs, and each cathode is connected to PORTC (pins RC0-RC3) through a transistor, which is necessary to sink the total current of each LED. By sending logic 1 to the base of the transistor, each transistor can be turned on. The on transistor grounds the cathode of the LED and turns on that LED. After an appropriate delay, the LED is turned off by sending logic 0 to the base of the transistor.

Software

The seven-segment LED codes necessary to display four digits or characters are stored sequentially in data registers starting from REG20. The program can get the codes from the data registers by using the pointer (FSR – File Select Register) and send them out to the LED segments through PORTB. Each LED is turned on by switching on the transistor by logic 1 from the pins of PORTC, and after an appropriate delay, each LED is turned off. This turning on/off of each LED in a sequence is known as scanning.

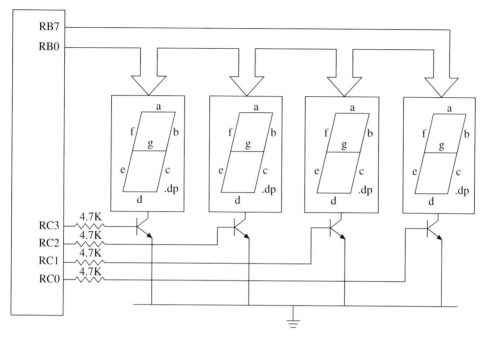

FIGURE 9-17   Interfacing Common Cathode Seven Segment LEDs Using Time Multiplex
Scanning Method

9.8.3 PROGRAM

```
        Title "Multiplex Seven Segment Display"
        List p=18F452, f =inhx32
        #include <p18F452.inc>        ;header file


COUNTR      EQU    0x00               ;Define registers that are used
DIGIT_ON    EQU    0x01
DIGIT_OFF   EQU    0x02
L1REG       EQU    0x10               ;Registers for delay count
L2REG       EQU    0x11
CODEREG     EQU    0x20


        ORG    00
        GOTO   START
```

continued

```
                ORG   0x20            ;Assemble at memory location 000020H
   START:       CLRF  PORTB           ;Initial reading at PORTB = 00
                CLRF  TRISB           ;Set up PORTB and PORTC as output ports
                CLRF  TRISC
   REPEAT:      MOVLW 4               ;Byte for four digits
                MOVWF COUNTR, 0       ;Set up counter
                CLRF  DIGIT_OFF       ;All 0s to turn off display
                MOVLW B'00000001'     ;Code to turn on first LED
                MOVWF DIGIT_ON        ;Load the code in DIGIT_ON
                LFSR  FSR0, CODEREG   ;Pointer for code
   ANODE_CODE:  MOVFF POSTINC0,PORTB  ;Output code and increment pointer
                MOVFF DIGIT_ON,PORTC  ;Turn on LED
                MOVLW 2               ;Count for 1 ms delay
                MOVWF L2REG
                CALL  DELAY_1MS       ;Call delay
                MOVFF DIGIT_OFF,PORTC ;Turn off LEDs
                RLNCF DIGIT_ON,1      ;Get ready to turn next LED
                DECF  COUNTR,1        ;Next count
                BNZ   ANODE_CODE      ;If counter ≠ 0, send next code
                BRA   REPEAT          ;Start again


   DELAY_1MS:

                ;:::::::::::::::::::::::::::::::::::::::::::::::::::::::::::::::::::
                ;Function: Provides 1 ms delay                                  :
                ;Input: A multiplier count in register L2REG                    :
                ;:::::::::::::::::::::::::::::::::::::::::::::::::::::::::::::::::::


   LOOP2:       MOVLW D'250'          ;Load decimal count in W
                MOVWF L1REG           ;Set up Loop1_Register as a counter
   LOOP1:       DECF  L1REG, 1        ;Decrement count in L1REG
                NOP
                NOP
                BNZ   LOOP1           ;Go back to LOOP1 if L1REG  ≠ 0
                DECF  L2REG,1         ;Decrement L2REG
                BNZ   LOOP2           ;Go back to load 250 in L1REG and start LOOP1
                RETURN
                END
```

9.8.4 Program Description

The initialization instructions: (1) set up PORTB and PORTC as output ports by sending 0s to TRISB and TRISC registers, (2) load 4 in the counter to count four rotations, (3) clear DIGIT_OFF register to send 0s to turn off the display, (4) load the byte 00000001 in the DIGIT_ON register, and (5) set up the pointer FSR0 for the data registers where the display code for the LEDs is stored. The instruction MOVFF POSTINC0, PORTB gets the code from the first data register (REG20), sends out to PORTB, and increments the pointer to point to the next code. The next

MOVFF instruction outputs the code 00000001 to PORTC, and the pin RC0 turns on the transistor and, in turn, the LED0 (Figure 9-17). The program loads the count 2 in the L2REG to get 1 ms delay, calls the delay subroutine, and after the delay, the byte in the DIGIT_OFF register turns off the display. The RLNCF instruction rotates the bits left in the DIGIT_ON register resulting the byte into 00000010, which will turn on the next LED1.

The program continues to send codes from the data registers and turn on/off each LED in a sequence until the counter becomes zero. When the counter is zero, the program goes back to reinitialize the counter, the pointer, and the code for PORTC. Then it continues to the next display cycle.

9.8.5 Program Execution

To execute this program, the codes for common cathode seven-segment LED in four data registers REG20, REG21, REG22, and REG23 must be loaded. For example, to display the year "2006," the following code is necessary:

CODEREG:	REG20: 0x7D	;Digit 6
	REG21: 0x3F	;Digit 0
	REG22: 0x3F	;Digit 0
	REG23: 0x5B	;Digit 2

Using MPLAB IDE

You should perform the following steps to execute the program using the MPLAB IDE.

Step 1: Assemble the program and build the project. Click on View Watch, and select the following registers to watch on the screen: COUNTR, FSR0, PORTB, PORTC, and DIGIT_ON.

Step 2: Select View again and click on File Registers. This opens up the screen of data registers. Load the above codes in REG20-REG23. You should have three screens on the monitor: Source File, Watch, and File (Data) Registers.

Step 3: Select Debugger → Tools → MPLAB Sim. Place the cursor on the first instruction of the assembly program and click on animate symbol or (Tools → Animate). Caution: For animation, the delay count is too long. Change the count in the delay subroutine from 250 to a smaller number between 5 and 10 and click on animate.

Step 4: As the instructions are being executed, the arrow moves along the side of each instruction in the source file. In the Watch window, the contents of FSR0 changes from 20 to 24 as the pointer points to the data register in each iteration. For each reading in the FSR0, the corresponding seven-segment LED code is displayed at PORTB. Both bytes in PORTC and DIGIT_ON change in the following sequence: 1 – 2 – 4 – 8 turning on each LED for a selected delay. You can also observe changes in the delay registers REG10 and REG11 and the count in the COUNTR register.

Using PIC18 Simulator IDE

The PIC18 Simulator IDE includes four seven-segment LEDs, and these LEDs can be used in simulation by performing the following steps (for detailed instructions see Appendix F):

Step 1: Open the PIC18 Simulator IDE by double-clicking on the symbol. To assemble the program: → Tools → Assembler → File → Open and select the Source file: Illust9-8 Multiplex Seven Segment LED from the Chapter 09 folder on the CD.

Step 2: On the Assembler Screen: → Tools → Assemble. If there are any errors, correct the errors and assemble the program again. After assembling the program: → Tools → Assemble and Load. The simulator loads the program in the program memory. Close the Assembler screen.

Step 3: Go to PIC18 Simulator IDE screen and → Tools → 7-Segment LED Display Panel. The simulator displays four seven-segment LEDs. Let us label the LEDs from the rightmost LED as LED0-LED3.

Step 4: To connect the LED segments (a through h) to PORTB: → Setup LED0 (the rightmost LED) → Space in front of the element 'a.' Pin selection screen is opened. → PORTB → 0 → Select. The simulator connects the segment element 'a' to RB0 of PORTB. Connect all elements (a through h) to pins RB0 to RB7 by repeating the same process.

Step 5: To connect the cathode of LED0 to pin RC0 of PORTC, click on the space that shows one of the following options: Always Enabled or Disabled or one of the PORTS. Connect pin RC0 from the Select Pin screen.

Step 6: Repeat Step 4 to connect the remaining LEDs (elements a through g) and Step 5 to connect cathodes of LED1 to RC1, LED2 to RC2, and LED3 to RC3.

Step 7: To load the seven-segment codes in registers REG20-REG23, you must begin the simulation in the Step mode.

On PIC18 Simulator IDE Screen: → Rate → Step by Step → Simulation → Start

Go to GPRs and → on register address 20 and load the code 7D. Load codes 3F, 3F, and 5B in registers 21–23, respectively.

Step 8: To begin simulation → Rate → Ultimate (No Refresh). You should observe digits 2006 on LED3 to LED0— one at a time.

SUMMARY

Microcontrollers communicate with peripherals, such as switches and LEDs, through I/O ports, a third major component of the microcontroller device. Peripherals are divided into two groups: input peripherals, including switches, push-button keys, and keyboards; and output peripherals, including LEDs, seven-segment LEDs, LCDs, and printers. These peripherals are interfaced with the microcontroller through I/O ports. The basic concepts in interfacing I/O ports and peripherals and their applications discussed in this chapter are summarized as follows:

1. I/O ports are buffers and latches on the MCU chip that can be assigned binary addresses by decoding the address bus, and the MPU accesses I/O ports through these addresses.

2. Peripherals are connected to the MCU through these I/O ports.

3. Peripherals are divided into two groups: input and output. Input peripherals such as switches and keyboards provide binary data as input to the MCU and output peripherals such as LEDs and seven-segment LEDs receive data from the MCU, generally for display.

4. To read (receive) binary data from an input peripheral, the MPU places the address of the input port on the address bus, enables the input port by asserting the RD signal, and reads data using the data bus.

5. To write (send) binary data to an output peripheral, the MPU places the address of the output port on the address bus, places the data on the data bus, and asserts the WR signal to enable the output port.

6. The PIC18 family of microcontrollers has a varied number of I/O ports depending on the version. They are identified with letters such as PORTA and PORTB.

7. Each I/O port is associated with a data direction register (called TRIS). By writing logic 1 in the TRIS register, the corresponding I/O pin is set up as input and by writing logic 0, I/O pin is set up as output. I/O ports are multiplexed, meaning in addition to their I/O functions, they are designed to perform additional functions. These functions are set up by writing to associated SFRs (Special Function Registers).

8. If PORTB is set up as an input port, its internal circuitry can provide pull-up registers by enabling Bit7 in the INTCON2 register.

9. A LED can be connected to an I/O port in two ways: (1) an anode is connected to an I/O pin and the cathode is grounded. This configuration is called current sourcing (supplied) by the I/O port, or (2) the cathode is connected to an I/O port and the anode is connected to a power supply. This configuration is called current sinking.

10. A seven-segment LED consists of eight LEDs physically mounted to display digits and some letters and are available in two types—common cathode and common anode. The code required to display a digit is based on how data lines from an I/O port are connected to the LEDs and the type of the LED.

11. Switches and push-button keys are connected as input peripherals to I/O ports. In the case of a push-button key, when the key is pushed it makes multiple contacts before it settles into a steady state. To avoid reading these multiple contacts as different readings, a debounce technique is used. The key can be debounced either by using a hardware chip such as an S-R latch or using a software 10 to 20 ms delay.

12. Liquid Crystal Display (LCD) is a commonly used output device when alphanumeric and graphic characters are required to display in an application. An LCD is interfaced with an I/O port using the HD44780 controller. The LCD can be interfaced either in the 4-bit mode or the 8-bit mode. In the 8-bit mode, eight data lines and three control signals are connected to the controller; thus the MCU must use two I/O ports. In the 4-bit mode, only four data lines are used; thus, the LCD can be interfaced using one I/O port. In the 4-bit mode, the program must send the same character twice.

13. In a matrix keyboard, the keys are arranged in rows and columns, and when a key is pressed, the corresponding row and column are connected. The matrix keyboard is a cost-effective way of interfacing a large number of keys; a maximum of 64 keys can be interfaced using two 8-bit I/O ports. The software program senses a key pressed, debounces the key, and encodes it.

14. To interface more than two seven-segment LEDs, the time multiplexed technique is generally used. In this technique, data are sent to all LEDs simultaneously, but the LEDs are turned on one at a time in a sequence with such a frequency that the display appears stable. The techniques requires only two I/O ports, and reduces the power consumption considerably.

QUESTIONS AND ASSIGNMENTS

Assume that data registers used in the following questions are in the Access bank:

9.1   Write instructions to initialize I/O pins of PORTB as follows: RB7 as an input and RB6 as an output. Read RB7, and if it is grounded, turn on the LED connected to pin RB6 (Figure 9-18a); otherwise, turn off the LED and continue.

9.2   Write instructions to initialize alternate I/O pins as inputs and outputs of PORTB as shown in Figure 9-18b and enable pull up resistors. Read the input pins, and, if they are on (grounded), turn on the corresponding LEDs.

9.3   In Figure 9-18a, assume the switch connected to RB7 is an emergency switch. Read RB7, and if it is on (grounded), flash the LED connected to RB6 at every 100 ms interval (assume that DELAY_100ms subroutine is available).

9.4   Write instructions to initialize PORTB as follows: RB0-RB3 as outputs and RB4-RB7 as inputs (Figure 9-19a). Enable pull-up resistors. Read the input switches and turn on the corresponding LEDs for the switches that are on (grounded).

9.5   Repeat Question 9.4 for Figure 9-19b.

9.6   Write instructions to initialize PORTB shown in Figure 9-19b. Read the input switches, and if switch S3 is closed (grounded), flash LD3 every 100 ms and ignore the other switches; otherwise, turn on/off LD0 to LD2 if the corresponding switches are on.

9.7   Write instructions to initialize pins RB7 as an input and RB0 as an output of PORTB in Figure 9-20a. Switch SW7 is used to turn on the solid-state relay (SSR0) that starts the conveyer belt. Read the switch and continue to read it until it is closed. Once it is closed, turn on SSR0.

9.8   Assuming that PORTB is properly initialized, write instructions to read two switches connected to pins RB7 and RB6, and if both of them are on, turn on the solid-state relay (SSR) in Figure 9-20b; otherwise continue to monitor both switches.

9.9   Refer to Figure 9-6b. Write instructions to display the letter "H".

9.10   In Figure 9-6a and Figure 9-6b flash letters "FF" continuously every 100 ms assuming that the subroutine DELAY_100ms is available.

9.11   Modify the program in Section 9.6.4 to display the following message on line 1 of the LCD: Temperature 68.9° F

9.12   In Figure 9-16, label the keys starting from the row RB0. The keys C,D,E, and F will be labeled as 0, 1, 2, and 3 respectively, and label the remaining keys in the same sequence. Rewrite the program to ground the rows one at a time and check the columns for a key closure and encode the keys.

9.13   In Q. 9.12, keys can be checked for a closure from 0 to F in a sequence. Rewrite the program to set up a counter starting from 0 and update the count as the keys are being checked and encode the keys. (Hint: the count value when a key is checked is the binary code for the key).

9.14   Modify Figure 9-17 to connect six seven-segment LEDs and rewrite the program to display six digit date: month day year –stored in six data registers starting from the register DATE. (Hint: Use table look-up technique).

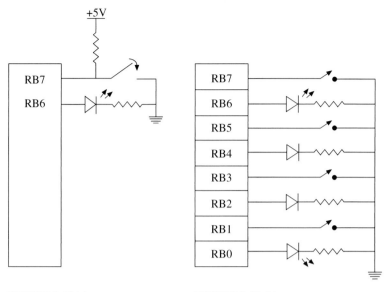

FIGURE 9-18 (a)   **FIGURE 9-18 (b)**

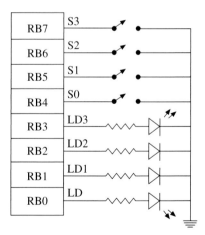

FIGURE 9-19 (a)   **FIGURE 9-19 (b)**

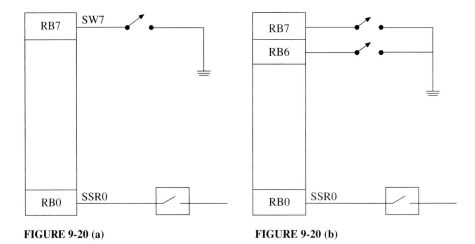

FIGURE 9-20 (a)   **FIGURE 9-20 (b)**

Troubleshooting

Simulation and Lab Exercises

The following assignment refer to Section 9.3 Illust9-3 Displaying BCD counter.

9.1 SE   Assemble the program Illust9-3 Displaying BCD Counter (see Assembly Programs on the enclosed CD) using PIC18 Simulator IDE. Connect two-seven segment LEDs to PORTB and PORTC. (Click on Tools and select 7-Segment Displays Panel and make connections two LEDs as explained in Appendix F). Execute the program and observe the display.

9.2 SE   Hardware Troubleshooting: When the following counts are displayed, digit 7 appears as digit 1, digit 6 appears as lowercase "b", but digit 0 is displayed properly. Explain the possible errors in the hardware wiring.

CHAPTER **10**

INTERRUPTS

OVERVIEW

The interrupt is one of the processes of communication or data transfer between an MPU and peripherals or devices. In microcontroller systems, these peripherals are divided into two groups: (1) devices external to the MCU, like a keyboard, and (2) internal or on-chip devices such as a timer. The interrupt is described as a process whereby a peripheral can request the processor to stop the ongoing execution process, attend to the new request, and then go back to the previous process where the MPU left off. This process is initiated by an external device asynchronously, meaning it can happen any time during the operation of the system.

In processor-based systems, there are two types of interrupts: hardware and software. Hardware interrupts are initiated by devices external to the processor, such as a keyboard or a timer. Software interrupts are instructions to direct the processor in an unusual situation, such as when dividing by zero. Most microcontrollers do not have software interrupts; therefore, our focus will be on hardware interrupts.

The hardware interrupts are divided into two groups: non-maskable and maskable. A non-maskable interrupt cannot be disabled and must be attended to by the MPU when it occurs. A maskable interrupt can be enabled or disabled by writing instructions. The basic concepts in the interrupt involve how to set up and implement the process. In many ways, the maskable interrupt is like a telephone communication in a home, which can be disabled by taking it off the hook.

The interrupt process is initiated by writing appropriate instructions in a program or enabling bits in appropriate registers; this is similar to having a dial tone for a telephone. To implement the process, we need to set up ways to interrupt the MPU, ways that the request can be serviced, and ways that the MPU can get back to the place where it was interrupted. In addition, we need to deal with issues of priority, including questions such as "What if the MPU receives more than one request at the same time?" or "What if an additional request comes in when the MPU is in the middle of attending a request?" This chapter focuses on these issues and on "Reset" as a special type of interrupt.

Next, we discuss the specific maskable interrupt scheme in the PIC18 microcontroller family in which the interrupts are divided into two main groups: high priority and low priority. Special function registers (SFRs) that are associated with the interrupt process are listed. Finally, an illustration that involves multiple interrupting sources with high- and low-priority are discussed. The illustration includes how to set up and implement the interrupt process using the PIC18F452/4520 microcontroller. In addition, the chapter includes the discussion of various types of resets in PIC18 microcontrollers.

OBJECTIVES

- Explain the interrupt process in microprocessor-based and microcontroller-based embedded systems and the difference between maskable and non-maskable interrupts.
- Explain the steps in enabling and implementing the interrupt process.
- Explain what reset is and how it can be viewed as a special type of interrupt.
- List PIC18 external and internal interrupt sources and associated special function registers (SFRs), and how the sources can be set up as either high- or low-priority interrupts.
- Explain the difference between a subroutine and an interrupt service routine (ISR), and instructions RETFIE and RETFIE FAST.
- Write instructions to enable the interrupt process for both high and low priorities, and implement these interrupts with ISRs.
- List various types of resets in PIC18 microcontrollers.

10.1 BASIC CONCEPTS IN INTERRUPTS

An interrupt is a communication process set up in a microprocessor or microcontroller in which an internal or external device requests the MPU to stop the processing, the MPU acknowledges the request, attends to the request, and goes back to the processing where it was interrupted. An interrupt is an asynchronous event that can occur any time. An interrupt process is an efficient way of implementing multitasking processes. For example, when you are using a computer, you expect it to print a file on a printer, receive keyboard inputs, copy a file, and keep track of updating the clock every second—and do all these things almost simultaneously. If there were no interrupt process, you would have to wait to type until the printing was done. Interrupts allow the keyboard to request that the MPU cease printing, accept the keystrokes you are typing, or update the clock and then go back to printing. This section describes how the interrupt process is set up during the design stage of an MPU, what the interrupting sources are, and how the process is implemented.

10.1.1 What is an Interrupt?

An interrupt is a process in which an external or internal (on-chip) device asks the MPU to stop whatever it is doing, attend to the request, and the MPU returns to the previous task. For example, while preparing a meal in the kitchen, you may be interrupted by many external and internal events as follows:

1. A phone rings, and you answer the phone after the completion of a task at hand. It is your neighbor asking you to move your car that is blocking his car. You hang up the phone, move the car, and go back to your cooking. In this situation, there is some chance that someone else may interrupt you by a second phone call before you go out the door ask you to do something else. Or another choice you have when you get the first phone call is not to hang up the phone, keep the phone busy, go out first and move your car, then come back and hang up the phone, and go back to your cooking.

2. You hear a buzzer from the washer in the laundry room indicating the end of washing cycle and you stop, take the clothes out of the washer, place them in the dryer, and go back to your cooking.

3. You smell something burning, turn off the stove, and stop cooking in frustration.

The above three incidents of interrupting the cooking process are similar to the interrupt process in a microprocessor or a microcontroller. Based on these analogies, we can describe the interrupt process in technical terms in the next section.

10.1.2 Types of Interrupts

Figure 10-1 shows various types of interrupts that are used in microprocessors or microcontrollers. They are divided into two main groups: hardware and software interrupts. The hardware interrupts are initiated by various devices or peripherals external to the MPU. Our first two examples described in the previous section fit into this category. The software interrupt is similar to our third example in the cooking scenario where something goes wrong and we need to stop cooking or start again.

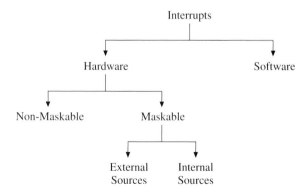

FIGURE 10-1   Interrupt Types

Some examples of software interrupts are as follows. In executing programs, we encounter situations such as divide by zero, for which the answer is indeterminate, or an overflow of stack where we run out of stack memory to save register information. In these cases, the MPU must stop processing and start again. These interrupts are also known as "traps" or "exceptions." Microprocessors that are used for general-purpose computing (such as processors used in personal computers) are generally designed to handle these types of interrupts. Most microcontrollers do not have software interrupts. Therefore, our focus here is on hardware interrupts.

As shown in Figure 10-1, hardware interrupts are divided into two groups: maskable and non-maskable. The non-maskable interrupts are generally used in emergency situations, such as a power failure whereby the MPU must respond immediately to a non-maskable interrupt. In the household, a smoke alarm can be viewed as a non-maskable interrupt. On the other hand, maskable interrupts can be masked, meaning they can be disabled. You can write instructions to disable maskable interrupts, and the MPU can not be interrupted. In our kitchen analogy, the telephone is a maskable interrupt in that you can disable it by not putting the receiver back on the hook. In general, maskable interrupts are disabled when the power is turned on, and you need to enable them by writing instructions.

The maskable interrupts are again divided into two groups: external interrupts and internal interrupts. In microprocessor or microcontroller design, certain pins are assigned to accept a signal from outside peripherals that can generate an interrupt request to the MPU. These are called external interrupts. For example, in the case of a microcontroller, any outside peripherals such as a keyboard or a switch can be connected to interrupt the MPU. In addition, there are many internal devices inside the MCU such as timers, A/D converter, and serial I/O modules that can send interrupt requests to the MPU. In our kitchen analogy, a buzzer from a washer can be considered an internal interrupt source.

Now let us consider an example of a simple switch that is connected to the interrupt pin of a MCU as shown in Figure 10-2. This is a switch that, when pressed, should cause the MPU to turn off the conveyer belt driven by one of the output ports and then to flash a light. Let us also assume that you have written the instructions to enable the interrupt of this microcontroller, and the MPU is running the conveyer belt and various other processes. Now the question is: how does the MPU recognize when the switch is pressed and how does it respond? This is discussed in the following section.

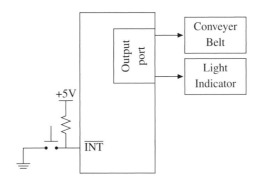

FIGURE 10-2   An Interrupt Circuit

10.1.3 MPU Response to Interrupts

When the interrupt process is enabled, the MPU checks the interrupt request line or a flag indicating the status of the interrupt request just before the end of the execution of each instruction. If the interrupt request is present, the MPU completes the execution of the instruction, saves the address of the program counter on the stack (which is the address of the next instruction), and stops the execution. Some MPUs are designed to save the contents of the accumulator, flag register, and registers in the MPU.

Now we need to get the process restarted to attend to the request. The nature of the request is generally asking the MPU to go to a set of instructions that accomplish a specified task, such as updating a clock. Let us review what the MPU does normally. When the power is turned on, it begins executing the instructions in sequence written in memory. Even if the program execution is redirected by a branch instruction, it begins the execution again in a sequence from the redirected memory location. The critical concept here is that the MPU is only capable of executing instructions stored in memory. Therefore, to restart the execution after an interrupt, the processor must perform the following two steps:

1. Find ways to get to the memory location where the code for the interrupt request is written; this is called an interrupt service routine (**ISR**) explained later.
2. Once the request is accomplished, the MPU should find its way back to the instruction where it was interrupted.

The first step is accomplished by interrupt vectors, and the second step is accomplished by the specific RETURN instruction in the ISR.

10.1.4 Interrupt Vectors

There are various ways to direct the processor to the memory location where the interrupt request is accomplished. They are as follows:

1. Defined Memory Location: When a processor or MCU is designed, a specific memory address is assigned to interrupt requests, called the interrupt vector. If the processor can be interrupted by multiple sources, the MPU identifies the source of the interrupt request either through software instructions or checking the flag assigned to a particular source. The PIC18 family of microcontrollers has two such defined memory locations called high-priority and low-priority vectors illustrated later in the chapter.
2. Defined Vector Location: In some microprocessors and microcontrollers, specific memory locations are assigned to store the vector addresses of the ISRs. When the MPU is interrupted and directed to these memory locations, it finds an address of the ISR. Motorola microcontrollers and microprocessors use this option.
3. Using External Hardware: The third method is providing a binary number (or a part of the memory address) through external hardware using an interrupt acknowledge signal. Earlier 8-bit processors such as 8085, Z80, and Intel processors used in personal computers employ this option.

10.1.5 Interrupt Service Routine (ISR)

These are a set of instructions, stored in memory and written to accomplish the task requested by the interrupting source. These instructions are written as a subroutine called interrupt service routine (ISR), and the only difference between a subroutine and an ISR is the RETURN instruction. The ISR must be terminated with a special RETURN instruction designed for interrupts. This RETURN instruction recovers the status of various registers that was stored on the stack, finds the address on the stack where the processor was interrupted, and redirects the program execution to that address.

10.1.6 Interrupt Priorities

Generally, there are multiple sources that send interrupt requests to the MPU, and some of the requests can come at the same time. Another situation is when one request is being served, another request can come during that time. The requests need to be prioritized. In our kitchen analogy, what should you do if the telephone and the buzzer from the washer ring at the same time? What if when you are going out the door to move the car, your cell phone rings? Should you answer the second phone when you are talking with someone on the first phone? In microcontrollers and microprocessors, there are many schemes for prioritizing interrupts. Some are based on hardware and some use software.

10.1.7 Summary of the Interrupt Process

There are many concepts involved in setting up the interrupt process. Let us review the interrupt process for maskable hardware interrupts in microcontrollers.

1. When power is turned on, maskable interrupts are disabled. They should be enabled either by writing instructions or by activating certain bits in appropriate registers.
2. An external source such as a keyboard or an internal source such as a timer must send a request for attention by asserting an electrical signal to the MPU.
3. The MPU checks the interrupt signal just before the end of each instruction.
4. If the interrupt is active, the MPU completes the execution of the instruction, saves the address of the next instruction on the stack, and disables the interrupt. Some MPUs also save the contents of microprocessor registers on the stack and stop the execution.
5. The MPU is directed to a specific memory location called an interrupt vector. In that memory location the MPU finds the beginning of the ISR or finds the address of an ISR.
6. Once the MPU locates the ISR, it begins the execution of the instructions in the ISR, and the last instruction in the ISR should be "Return from Interrupt."
7. When the MPU executes the last instruction, it goes to the stack, retrieves the address that is saved on the stack in Step 4 above, and returns to continue the execution where it was interrupted. If the contents of microprocessor registers were saved on the stack at the time of interrupt, the MPU also retrieves the contents in appropriate registers. In some processors, the instruction "Return from Interrupt" enables the interrupt process; in others, instructions must be written to enable the interrupt.
8. If multiple interrupt requests arrive at the same time or an additional interrupt request arrives during the servicing of an interrupt, the priority scheme designed in the processor determines how these requests are serviced.

10.1.8 Reset as a Special Purpose Interrupt

Reset is an external signal that enables the processor to begin execution or interrupts the processor if the processor is executing instructions. Generally, there are at least two types of resets: one is power-on reset and the other is manual reset. When the reset signal is activated, it establishes or reestablishes the initial conditions of the processor and directs the processor to a specific starting memory location. It does not save any register contents on the stack, unlike the interrupt; however, it can be viewed as a special-purpose interrupt.

10.2 PIC18 INTERRUPTS

PIC18 microcontroller family of devices has multiple sources that can send interrupt requests, and the number of sources varies among different family members. However, the interrupt process is the same. The PIC18 family of microcontrollers does not have any non-maskable or software interrupts; all interrupts are maskable. The priority scheme is divided into two groups—high priority and low priority—and multiple Special Function Registers (SFRs) are used to implement the interrupt process, as discussed in the following sections.

10.2.1 PIC18 Interrupt Sources

In PIC18 microcontrollers, the interrupt sources are divided into two groups: external sources and internal peripheral sources on the MCU chip

External Sources

PORTB pins are used as interrupt pins as follows:

- RB0/INT0, RB1/INT1, and RB2/INT2[1]: These three pins of PORTB are multiplexed, and external sources such as keyboards or switches can be connected to these pins as interrupting sources. These pins can be set up to recognize either rising or falling edge-triggered pulses.
- RB4-RB7: Change in logic levels of these pins of PORTB can be recognized as interrupts.

Internal Peripheral Sources

There are several internal devices on the PIC18 MCU that can send interrupt requests to the MPU. The number of devices varies among PIC18 MCUs. Typically, the interrupting devices include timers, the A/D converter, the serial I/O, EEPROM (write operation), and the low-voltage detection module.

To set up the interrupt process to these interrupt sources, the following SFRs are used:

- RCON: Register Control
- INTCON: Interrupt Control
- INTCON2:   Interrupt Control2
- INTCON3:   Interrupt Control3
- PIR1 and PIR2:   Peripheral Interrupt Register1 and 2

[1] Some versions of PIC18F controllers have additional interrupt pin RB3/INT3

- PIE1 and PIE2:   Peripheral Interrupt Enable1 and 2
- IPR1 and IPR2:   Interrupt Priority Register1 and 2

The RCON register sets up the global priority for all interrupts, the INTCON registers deal primarily with external interrupt resources (with a few exceptions), and the registers PIR, PIE, and IPR handle internal peripheral interrupts. To recognize the occurrence of an interrupt request, the MPU needs to check the following three bits for each interrupting source:

- The flag bit to indicate that an interrupt request is present
- The enable bit to redirect the program execution to the interrupt vector address
- The priority bit to select priority

These bits are defined in various SFRs. Even if there are multiple interrupting sources and many SFRs are needed to set up interrupts, the steps in implementing the interrupts are similar. Therefore, we will illustrate how the interrupt process is initialized and executed with a few examples.

10.2.2 Interrupt Priorities and RCON Register

In the PIC18, any interrupt source can be set up either as high or low priority. All high-priority interrupts are directed to the interrupt vector location 000008_H and all low-priority interrupts are directed to the interrupt vector location 000018_H. Furthermore, a high-priority interrupt can interrupt a low-priority interrupt in progress.

The interrupt priority feature is enabled by Bit7 (IPEN) in RCON (Reset Control) register as shown in Figure 10-3. Bits B0–B4 are used for various reset features explained later in Section 10.2.7.

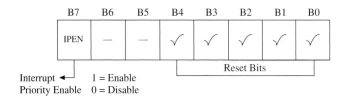

FIGURE 10-3   RCON Register

10.2.3 External Interrupts and INTCON Registers

Once the priority feature is enabled as illustrated in the previous section, the MPU needs to examine three bits as described before: (1) the priority bit to determine priority (high or low), (2) the flag bit to recognize an interrupt request, and (3) the enable bit to locate the interrupt vector.

Interrupt Control (INTCON) Registers: Figure 10-4 shows three registers with interrupt bit specifications primarily for external interrupt sources. Some bits are used for internal peripherals such as Timer0. Among these three registers, we should be able to find the necessary information to set up the interrupt process.

B7	B6	B5	B4	B3	B2	B1	BØ
GIE/GEIH	PEIE/GIEL	TMR0IE	INT0IE	RBIE	TMR0IF	INT0IF	RBIF

B7 **GIE/GEIH:** Global Interrupt Enable bit
<u>When IPEN = 0:</u>
1 = Enables all unmasked interrupts
0 = Disables all interrupts
<u>When IPEN = 1:</u>
1 = Enables all high priority interrupts
0 = Disables all high priority interrupts

B6 **PEIE/GIEL:** Peripheral Interrupt Enable bit
<u>When IPEN = 0:</u>
1 = Enables all unmasked peripheral interrupts
0 = Disables all peripheral interrupts
<u>When IPEN = 1:</u>
1 = Enables all low priority peripheral interrupts
0 = Disables all low priority peripheral interrupts

B5 **TMR0IE:** TMR0 Overflow Interrupt Enable bit
1 = Enables the TMR0 overflow interrupt
0 = Disables the TMR0 overflow interrupt

B4 **INT0IE:** INT0 External Interrupt Enable bit
1 = Enables the INT0 external interrupt
0 = Disables the INT0 external interrupt

B3 **RBIE:** RB Port Change Interrupt Enable bit
1 = Enables the RB port change interrupt
0 = Disables the RB port change interrupt

B2 **TMR0IF:** TMR0 Overflow Interrupt Flag bit
1 = TMR0 register has overflowed (must be cleared in software)
0 = TMR0 register did not overflow

B1 **INT0IF:** INT0 External Interrupt Flag bit
1 = The INT0 external interrupt occurred (must be cleared in software)
0 = The INT0 external interrupt did not occur

BØ **RBIF:** RB Port Change Interrupt Flag bit
1 = At least one of the RB7:RB4 pins changed state (must be cleared in software)
0 = None of the RB7:RB4 pins have changed state

Note: A mismatch condition will continue to set this bit. Reading PORTB
will end the mismatch condition and allow the bit to be cleared.

FIGURE 10-4 (a)   INTERCON Register

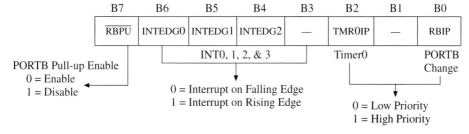

FIGURE 10-4 (b)   INTERCON2 Register

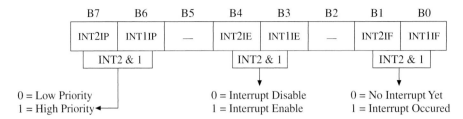

FIGURE 10-4 (c)   INTCON3 Register

10.2.4 Interrupt Service Routine (ISR)

The Interrupt Service Routine (ISR) attends to the request of an interrupting source, and the request generally does not require complex instruction code. The request can be as simple as updating a clock display after the elapsing of one second or reading a keyboard. This routine, as mentioned before, is similar to a subroutine except for the last instruction. In addition, we need to be aware of whether our code modifies the contents of registers used in the main-line program. The requirements of an ISR are as follows. It must:

- be terminated with the instruction RETFIE (Return from Interrupt).
- save register contents that may be affected by the code in the ISR.

Instruction RETFIE

When an interrupt occurs, the MPU completes the instruction that is being executed, disables global interrupt enable, and places the address from the program counter on the stack. This is the Return Address where the MPU should return after the completion of the ISR. The RETFIE instruction gets the address from the top of the stack and places it in the program counter, and enables the global interrupt enable bit (INTCON <7>). This instruction is similar to the RETURN instruction except for the last action of enabling the interrupt enable bit.

Syntax   RETFIE   [s]   ; s = 0 or 1. If s = 1, the MPU also retrieves the
                               ;contents of W, BSR, and STATUS registers (previously
                               ;saved) before enabling the global interrupt bit.

Saving Register Contents

The instructions in an ISR are likely to affect the contents of W and STATUS registers of the main program. Therefore, it is necessary to save the contents of W, STATUS, and BSR registers at the beginning of the routine and retrieve those contents before returning from the ISR. This is unnecessary if the ISR code does not affect the contents of registers used in the main-line program.

In high-priority interrupts, the contents of W, STATUS, and BSR registers are automatically saved into respective registers, called shadow registers, when an interrupt is accepted, and the instruction RETFIE 1 or RETFIE FAST in the ISR retrieves the contents of these registers as a part of the instruction.

In low-priority interrupts, these registers can be saved and retrieved by using software instructions as shown in Example 10.2.

EXAMPLE 10.1

Write instructions to set up INT1 pin as a high-priority interrupt input and set up a counter to count 10. Assume that a push button key (similar to Figure 10-2) is connected to the pin RB1/INT1. Write the ISR to count the number of interrupts up to 10 generated by the key and reset the counter again.

Solution

```
        Title "Ex10-1 Interrupts"
        List p=18F452, f =inhx32
        #include <p18F452.inc>      ;This is a header file

REG1        EQU   0x01              ;Define Register1 used as a
                                    ;counter
            ORG   0x00              ;Begin assembly
            GOTO  MAIN


            ORG   0x0008            ;High-priority interrupt vector
            GOTO  INT1_ISR          ;Go to ISR

MAIN:       BSF   RCON,   IPEN      ;Enable priority -RCON <7>
            BSF   INTCON, GIEH      ;Enable high-priority -
                                    ;INTCON <7>
            BCF   INTCON2,INTEDG1   ;Interrupt on falling edge-
                                    ;INTCON2 <5>
            BSF   INTCON3,INT1IP    ;Set high priority for INT1-
                                    ;INTCON3 <6>
            BSF   INTCON3,INT1IE    ;Enable Interrupt1- INTCON3 <3>
            MOVLW D'10'             ;Set up REG1 to count 10
            MOVWF REG1,0
HERE:       GOTO  HERE

            ORG   0x100
INT1_ISR:   BCF   INT1IF            ;Clear interrupt flag
            DECF  REG1,1,0          ;Next count
            BNZ   GOBACK            ;If counter ≠0, return to main
                                    ;program
            MOVLW D'10'             ;If counter =0, reset it to
                                    ;count 10
            MOVWF REG1,0
GOBACK:     RETFIE FAST             ;Retrieve W, STATUS, BSR and
                                    ;enable interrupt
            END                     ;bit and return to main program
```

Description:

The instructions are divided into two segments: initialization (MAIN) and the interrupt service routine (INT1_ISR).

continued

In MAIN segment, the first two instructions set up the overall interrupt process by enabling the priority feature and global interrupt. The next three instructions are specific to INT1. The INT1 is enabled in high-priority mode and can be interrupted by the falling edge of the incoming signal. The last three instructions initialize REG1 as a counter to count for 10 interrupts and stay in the loop for an interrupt to occur.

The INT1_ISR is an ISR. The MPU comes to this segment only when the key is pushed and an interrupt occurs. When an interrupt occurs, the INT1IF (INT1 Interrupt Flag) bit is set; therefore, the first instruction BCF INT1IF in this segment clears the flag so that subsequent interrupts can be recognized. If this flag is not cleared, before the completion of the ISR, thesame flag will be mistakenly recognized as a second interrupt. The function of the ISR is simple; it decrements the counter and returns to the main program where it was interrupted. When the counter becomes zero, the counter is again loaded with the count of 10 before returning to the main program. The last instruction RETFIE FAST finds the return address on the top of the stack and places the address in the program counter, retrieves the contents of W, BSR, and STATUS registers, and enables the interrupt enable bit. Thus, the program execution is redirected back to the GOTO instruction in the main program.

In the context of this example, the following points should be noted:

1. The instruction GOTO HERE creates a continuous loop, and it is used here to demonstrate how an interrupt works when the processor is busy executing instructions. In real embedded system, the processor will be executing other instructions.

2. The instruction RETFIE FAST can also be written as RETFIE 1; both formats are recognized by the MPLAB IDE assembler. If the instruction is not followed by the term FAST or the number 1, the instruction does not retrieve the contents of W, STATUS, or BSR register.

EXAMPLE 10.2

For a low-priority interrupt, write instructions to save W, STATUS, and BSR registers that may be modified in the ISR code.

Solution

```
        Title "Ex10-2 - Saving Interrupt Contexts"
        List p=18F452, f =inhx32
        #include <p18F452.inc>     ;This is a header file

STATUS_TEMP EQU   0x200            ;The addresses of these
WREG_TEMP   EQU   0x201            ;registers are chosen arbitrarily
BSR_TEMP    EQU   0x202
ISRCODE     EQU   0x100


            ORG   ISRCODE
```

continued

```
ISRLO:MOVFF STATUS, STATUS_TEMP   ;Save STATUS, W, and BSR in
      MOVWF WREG_TEMP             ;  temporary registers
      MOVFF BSR, BSR_TEMP
      ;
      ;ISR Code
      ;
      MOVFF BSR_TEMP, BSR         ;Retrieve information in BSR, W,
      MOVFW REG_TEMP, W           ;  and STATUS registers before
      MOVFF STATUS_TEMP,STATUS    ;  returning from ISR
      RETFIE                      ;Find return address on the
                                  ;top of stack, enable interrupts
                                  ;and return
```

Description:

In low-priority interrupts, W, STATUS, and BSR registers are not saved on the stack as a part of the interrupt process, and only the program counter is saved on the stack. Generally, the ISR code is likely to use these registers to attend to the request. Therefore, these registers must be saved in data registers at the beginning of the ISR and retrieved before the end of the ISR.

10.2.5 Internal Interrupts and Related Registers

As mentioned in a previous section, there are many internal devices in the PIC18 MCU such as timers and A/D converter that can interrupt the MPU. These devices are referred to as peripheral interrupts by Microchip. The interrupt process for internal devices is similar to that of the external interrupts in that three bits are associated with each interrupt: priority, interrupt request flag, and enable. The functions of various bits in these registers are similar to that of bits in INTCON registers discussed in the previous section. These registers are referred as follows:

- IPR: Interrupt Priority Register
- PIR: Peripheral Interrupt Request (Flag)
- PIE: Peripheral Interrupt Enable

There are many internal peripheral sources; therefore, multiple registers are needed to specify the bits for these sources. The number of internal peripherals differs within the PIC18 family members. Figures 10-5, 10-6, and 10-7 show the contents of these registers; however, all of these bits may not be relevant to some members of PIC18 microcontroller family.

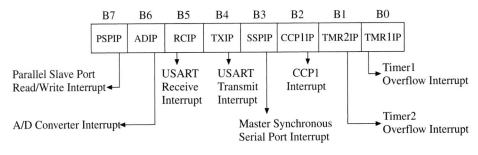

1 = High Priority
0 = Low Priority

FIGURE 10-5   IPR1 – Interrupt Priority Register1

B7	B6	B5	B4	B3	B2	B1	B0
PSPIF[1]	ADIF	RCIF	TXIF	SSPIF	CCP1IF	TMR2IF	TMR1IF

B7   **PSPIF[1]:** Parallel Slave Port Read/Write Interrupt Flag bit
    1 = A read or write operation has taken place (must be cleared in software)
    0 = No read or write has occurred
B6   **ADIF:** A/D Converter Interrupt Flag bit
    1 = An A/D conversion completed (must be cleared in software)
    0 = The A/D conversion is not complete
B5   **RCIF:** USART Receive Interrupt Flag bit
    1 = The USART receive buffer, RCREG, if full (cleared when RCREG is read)
    0 = The USART receive buffer is empty
B4   **TXIF:** USART Transmit Interrupt Flag bit
    1 = The USART transmit buffer, TXREG, is empty (cleared when TXREG is written)
    0 = The USART transmit buffer is full
B3   **SSPIF:** Master Synchronous Serial Port Interrupt Flag bit
    1 = The transmission/reception is complete (must be cleared in software)
    0 = Waiting to transmit/receive
B2   **CCP1IF:** CCP1 Interrupt Flag bit
    Capture mode:
    1 = A TMR1 register capture occurred (must be cleared in software)
    0 = No TMR1 register capture occurred
    Compare mode:
    1 = A TMR1 register compare match occurred (must be cleared in software)
    0 = No TMR1 register compare match occurred
    PWM mode:
    Unused in this mode
B1   **TMR2IF:** TMR2 to PR2 Match Interrupt Flag bit
    1 = TMR2 to PR2 match occurred (must be cleared in software)
    0 = No TMR2 to PR2 match occurred
B0   **TMR1IF:** TMR1 Overflow Interrupt Flag bit
    1 = TMR1 register overflowed (must be cleared in software)
    0 = TMR1 register did not overflow

    Note 1:   This bit is reserved on PCI18F2X2 devices; always maintain this bit clear.

FIGURE 10-6   PIR1 – Peripheral Interrupt Request Register1

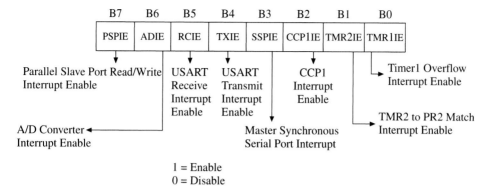

FIGURE 10-7   PIE1 – Peripheral Interrupt Enable Register1

10.2.6 Handling Multiple Interrupt Sources

As mentioned in a previous section, all interrupt requests in PIC18 microcontroller go to one of the two interrupt vectors: 00008_H (high priority) and 000018_H (low priority); however, multiple interrupt sources are directed to these vectors. Therefore, it is necessary to identify the interrupt source to direct the MPU to the appropriate ISR. In PIC18 microcontrollers, when multiple interrupt sources are enabled, the interrupting source is identified by checking the interrupt flags through software instructions as shown in Example 10.3.

EXAMPLE 10.3

Write instructions to identify the interrupting source when the following three internal sources are enabled as low-priority interrupts: A/D Converter, Timer1, and Timer2.

Solution

```
        Title "Ex10-3"
        List p=18F452, f =inhx32
        #include <p18F452.inc>            ;This is a header file
        ORG    0x00
        GOTO   MAIN

        ORG    0x0018       ;Low Priority Interrupt Vector
        BTFSC  PIR1,TMR1IF  ;Check Timer1 flag - skip if it is clear
        GOTO   TMR1_ISR     ;If Timer1 flag set, go to its ISR

        BTFSC  PIR1,TMR2IF  ;Check Timer2 flag - skip if it is clear
        GOTO   TMR2_ISR     ;If Timer2 flag set, go to its ISR
        BTFSC  PIR1,ADIF    ;Check A/D flag - skip if it is clear
        GOTO   ADC_ISR      ;If A/D flag set, go to its ISR
```

continued

```
        ORG    0x20     ;Begin MAIN program here
MAIN: ;
        ;
        ;
        ;
        ORG    0x100                ;Interrupt Service routines begin here
TMR1_ISR:RETFIE                     ;Timer1 Service Routine
TMR2_ISR:RETFIE                     ;Timer2 Service Routine
ADC_ISR:RETFIE                      ;A/D Converter Service Routine
        END
```

Description:

There are three interrupt sources enabled in this problem: Timer1, Timer2, and A/D converter, and when an interrupt occurs, the respective flag in the Peripheral Interrupt Register1 (PIR1) is set (see Figure 10-6). Bit0, Bit1, and Bit6 are used respectively for these interrupt sources. The MPLAB IDE defines these bits with labels; therefore, in writing instructions, we can use either labels or bit positions.

When an interrupt occurs, the MPU is directed to the location 00018_H; the instructions in this example are assembled at that location. The first instruction (BITFSC) tests Bit0 labeled as TIMR1IF (Timer1 Interrupt Flag), and if it is set, the MPU goes to the Timer1 ISR (TMR1_ISR); otherwise, it skips the GOTO instruction. The follow up instructions repeat the same process for Timer2 flag and A/D converter flag. Whenever the MPU finds the flag that is set, it goes to that ISR. The sequence in which these tests are performed gives Timer1 the highest priority among these interrupt sources. This priority can be changed by changing the sequence of checking the flags. This example shows only how to identify an interrupt source when an interrupt occurs and how to access the ISR of that interrupt; it does not show either the initialization of the interrupts or the codes for ISRs.

10.2.7 PIC18 Resets

In PIC18 microcontrollers, when the reset signal is activated, the MPU goes into a reset state, during which the initial conditions are established. When the MPU comes out of the reset state, its program counter is cleared to 00000, which is called the reset vector. After the reset, the MPU begins the execution of instructions from location 00000. PIC18 microcontrollers can be reset either by external source such as the push-button key shown in Figure 10-8(a) or when power is turned on or by various internal sources. All these resets are categorized in the following six categories.

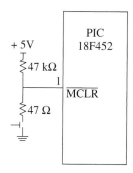

FIGURE 10-8 (a)   Manual Reset

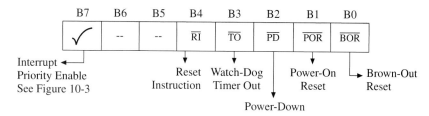

FIGURE 10-8 (b)   RCON (Reset Control) Register

1. External Manual Reset Key: The PIC18 has a pin called $\overline{\text{MCLR}}$ (Master Clear) as shown in Figure 10-8(a), and a push-button key is connected to the $\overline{\text{MCLR}}$ pin through a resistor divider. When the key is pushed manually, the $\overline{\text{MCLR}}$ signal is grounded and the MCU is reset. This key should be pushed to restart the program execution from the reset vector during an operation or to wake up the MCU from the sleep mode.

2. Power-on Reset (POR): When the power is turned on, the rise in power supply voltage (V_{DD}) is detected, and a reset signal is generated. When power-on reset occurs, bit B1 in RCON register (shown in Figure 10-8b) is cleared; otherwise, it stays high. If necessary, the additional time delay is provided by the Power-Up Timer (PWRT). The PWRT creates a time delay that allows V_{DD} to rise to an acceptable level, and PWRT can be enabled or disabled in the Configuration Register. After the power is turned on, a certain time delay is required in order for the crystal oscillator to stabilize. At the end of PWRT delay, the Oscillator Start-Up Timer (OST) provides the additional delay.

3. Watchdog Timer Reset (WDT): This timer resets the microcontroller when the timer runs out of time (or overflows) and clears bit B3 ($\overline{\text{TO}}$ – in Figure 10-8b) in RCON register. This timer is used to detect software bugs. When a set of instructions takes much longer than expected or is caught in an infinite loop, this timer will reset the MCU and the source of reset can be identified by checking bit B3 in RCON register.

4. Programmable Brown-Out Reset (BOR): If V_{DD} (power supply voltage) falls below a certain limit called BV_{DD} (typically, BV_{DD} is 4.2 V for a 5 V power supply), the MCU does not function properly. The BOR resets the MCU and clears bit B0 ($\overline{\text{BOR}}$) in RCON register, and the MCU stays in the reset state. When V_{DD} rises above BV_{DD} and if the Power-up Timer (PWRT) is enabled, the MCU goes to the reset vector after a

delay provided by PWRT. The $\overline{\text{BOR}}$ can be enabled or disabled by the specified bit in the configuration register.

5. RESET and SLEEP Instructions: The PIC18 instruction set includes the RESET instruction that resets the MCU. This is the software equivalent of hardware resetting $\overline{\text{MCLR}}$. When this instruction is executed, Bit4 ($\overline{\text{RI}}$) in RCON register is cleared. Similarly, when the MPU executes the SLEEP instruction, it clears Bit2 ($\overline{\text{PD}}$) in the RCON register and places the MCU in the power down mode. The manual reset key can wake up the MCU from the power down mode.

6. Stack Full or Underflow Reset: When the stack overflows or underflows, the MCU is reset.

10.3 ILLUSTRATION: IMPLEMENTATION OF INTERRUPT PROCESS IN PIC18 MICROCONTROLLER

This illustration shows the complete process of initializing and implementing the interrupts in PIC18 microcontrollers. The section includes high- and low-priority interrupts.

Applications: When multiple processes are monitored and controlled by a microcontroller in an embedded system, the interrupt process is essential for running the system efficiently. In a PC, many of the peripherals such as a keyboard and disk drives are interfaced using the interrupt; the processors used in a PC provide memory space for 256 interrupts. The processes in household appliances, monitored by microcontrollers, are interrupt driven. For example, if a dryer is on and someone opens the door, the system generates an interrupt to stop the dryer. The interrupts will be used frequently when we discuss various internal devices, such as A/D converters and timers in subsequent chapters.

10.3.1 Problem Statement

Write instructions to set up the interrupt process in PIC18 microcontroller as follows:

- Initialize the INT1 as a high-priority interrupt.
- Initialize Timer1 and Timer2 as low-priority interrupts.
- A push-button key is connected to INT1 through a latch to debounce the key. Write the interrupt service routine INT1_ISR to count the number of times the key is pushed, and when the count reaches 10, reset the counter and turn on the LED connected to pin RB7. This LED is turned off by the Timer1 ISR after 100 milliseconds.
- The instructions for Timer1 and Timer2 interrupt requests are already written and available as subroutines and can be called; timers are discussed in Chapter 11.

10.3.2 Problem Analysis

This illustration includes high- and low-priority interrupts as well as external and internal sources. A push-button is connected to INT1 (RB1) through a latch, and the D input to the latch is tied high, as shown in Figure 10-9. When the key is pushed, the clock input goes high, and the output Q goes high. That sends an interrupt request to INT1. This request (the output of the latch) must be reset; otherwise, after the completion of the ISR, it will be mistakenly interrupted as a second interrupt. Pin RB6 is connected to the reset pin of the latch. In addition, an

LED is connected to bit RB7 to indicate that the count has reached 10. Therefore, two bits RB7 and RB6 of PORTB must be set up as outputs.

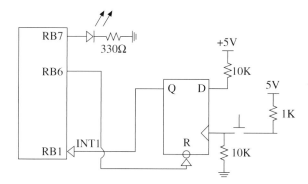

FIGURE 10-9   Interrupt Circuit

To set up the interrupt process, the following four bits should be enabled: priority scheme, global priority, source priority, and interrupt enable. The last two bits must be enabled for each interrupting source. The main program should also set up a counter to count a key being pushed ten times. For ISRs, the first task is to clear the flag set by the interrupt, and routines should be terminated with the instruction RETFIE. In case of high-priority interrupts, registers such as W and STATUS are saved when an interrupt occurs, and for low-priority interrupts these registers must be saved by using software instructions.

10.3.3 PROGRAM

```
;::::::::::::::::::::::::::::::::::::::::::::::::::::::::::::::::::::::
;The following program demonstrates how to set up high- and low-   :
;priority interrupts. INT1 is set up as a high-priority            :
;interrupt and Timer1 and Timer2 are set up as low-priority        :
;interrupts.                                                       :
;::::::::::::::::::::::::::::::::::::::::::::::::::::::::::::::::::::::

        Title "IP 10-3: Setting Up Interrupts"
        List p=18F452, f =inhx32
        #include <p18F452.inc>              ;This is a header file

REG1            EQU     0x01                ;Define Registers
STATUS_TEMP     EQU     0x100
WREG_TEMP       EQU     0x101
```

continued

```
                ORG     0x00
                GOTO    MAIN

                ORG     0x0008              ;High Priority Interrupt
                                            ;Vector
INTCHK:         GOTO    INT1_ISR            ;Go to ISR

                ORG     0x00018             ;Low-priority Interrupt
                                            ;Vector
TIMERCHK:       BTFSC   PIR1,TMR1IF         ;Check Timer1 flag -
                                            ;skip if it is clear
                GOTO    TMR1_ISR            ;If Timer1 flag set, go to
                                            ;its ISR
                BTFSC   PIR1,TMR2IF         ;Check Timer2 flag -
                                            ;skip if it is clear
                GOTO    TMR2_ISR            ;If Timer2 flag set, go to
                                            ;its ISR

MAIN:           MOVLW   B'00111111'         ;Bits to setup PORTB
                MOVWF   TRISB               ;Initialize Bits 6 & 7 of
                                            ;PORTB as output
                CLRF    INTCON3             ;Clear all INT flags
                CLRF    PIR1                ;Clear all internal
                                            ;peripheral flags
                BSF     RCON, IPEN          ;Enable priority -RCON <7>
                BSF     INTCON,GIEH         ;Enable global high-
                                            ;priority -INTCON <7>
                BSF     INTCON2,INTEDG1     ;Interrupt on rising edge-
                                            ;INTCON2 <5>
                BSF     INTCON3,INT1IP      ;Set high priority for
                                            ;INT1-INTCON3 <6>
                BSF     INTCON3,INT1IE      ;Enable Interrupt1-
                                            ;INTCON3 <3
                BSF     INTCON, GIEL        ;Enable global low-
                                            ;priority -INTCON <6>
                BCF     IPR1, TMR1IP        ;Set Timer1 as low-priority
                BSF     PIE1, TMR1IE        ;Enable Timer1 overflow
                                            ;interrupt
                BCF     IPR1, TMR2IP        ;Set Timer2 as low-priority
                BSF     PIE1, TMR2I         ;Enable Timer2 match
                                            ;interpret
                MOVLW   D'10'               ;Set up REG1 to count 10
                MOVWF   REG1,0
```

continued

```
HERE:          GOTO    HERE                    ;Wait here for an interrupt

               ORG     0x100
INT1_ISR:      ;:::::::::::::::::::::::::::::::::::::::::::::::::::::::
               ;This ISR counts the number of times the key is       :
counter to     ;pressed. When the count reaches 0, it resets the     :
               ;counter to 10 and turns on the LED which is turned   :
               ;off by Timer1 after 100 ms                           :
               ;:::::::::::::::::::::::::::::::::::::::::::::::::::::::
               BCF     INTCON3, INT1IF      ;Clear INT1 flag
               DECF    REG1,1,0             ;Count key press
               BNZ     GOBACK               ;If count ≠0, go back
               MOVLW   D'10'                ;Set counter again at 10
               MOVWF   REG1,0
               BSF     PORTB,7              ;Turn on LED
GOBACK:        RETFIE  FAST                 ;Retrieve registers
                                            ;and go back
TMR1_ISR:      ;:::::::::::::::::::::::::::::::::::::::::::::::::::::::
               ;TMR1_ISR and TMR2_ISR demonstrate how to save and   :
               ;retrieve STATUS and W register in low-priority       :
               ;interrupts                                           :
               ;:::::::::::::::::::::::::::::::::::::::::::::::::::::::
               MOVFF   STATUS,STATUS_TEMP ;Save registers
               MOVWF   WREG_TEMP
               BCF     PIR1, TMR1IF         ;Clear TMR1 flag
               CALL    TIMER1               ;Call service subroutine
               MOVF    WREG_TEMP, W         ;Retrieve registers
               MOVFF   STATUS_TEMP,STATUS
               RETFIE                       ;Go back to main
TMR2_ISR       MOVFF   STATUS,STATUS_TEMP ;Save registers
               MOVWF   WREG_TEMP
               BCF     PIR1,TMR2IF          ;Clear TMR2 flag
               CALL    TIMER2               ;Call service subroutine
               MOVF    WREG_TEMP, W         ;Retrieve registers
               MOVFF   STATUS_TEMP,STATUS
               RETFIE                       ;Go back to main
TIMER1:        ;Timer1 subroutine begins here
               ;
TIMER2:        ;Timer2 subroutine begins here
               END
```

10.3.4 Program Description

Before the assembly begins at location 0000, three registers are defined: one for the counter, and
the other two to save the contents of W and STATUS registers temporarily after an interrupt

occurs. The additional two ORG directives define the interrupt vectors: a high-priority interrupt vector at location 00008_H labeled as INTCHK and a low-priority interrupt vector location at 00018_H labeled as TIMERCHK. In this problem, there is only one high-priority interrupt INT1; therefore, the MPU is immediately directed to the ISR INT1_ISR. However, we have more than one low-priority interrupts; therefore, the MPU must first identify the interrupt source. The instructions at location TIMERCHK check the interrupt flags of Timer1 and Timer2 and determine the interrupting source. Once the source is identified, the MPU is directed to the respective ISR: either TMR1_ISR or TMR2_ISR.

The next group of instructions are identified as MAIN divided into three segments. This group of instructions perform various initialization tasks for INT1, Timer1, and Timer2. For INT1 the segment of instructions (1) sets up two bits of PORTB as outputs, (2) enables priority scheme and global priority, (3) sets up rising edge pulse as an input, (4) places INT1 in high-priority group, and (5) enables interrupt for INT1. The next set of instructions enables global low-priority interrupts, places Timer1 and Timer2 in low priority, and enables interrupts. The last segment of instructions sets up REG1 as a counter for ten and a continuous loop where the MPU waits for an interrupt.

The ISR, INT1_ISR, clears the interrupt flag and decrements the counter, and if the counter is not zero, the MPU returns to the main program. When the count becomes zero, the MPU sets the counter to 10 again and turns on the LED. The last instruction, RETFIE FAST, retrieves the information in W and STATUS registers, finds the address on the top of the stack, and goes back to that address.

The next two ISRs —TMR1_ISR and TMR2_ISR—are similar in nature. Because these are low-priority interrupts, the contents of W and STATUS registers are first saved in memory, and then the interrupt flag is cleared. Next, the MPU goes to the subroutine that completes the task of the interrupt request, returns to the ISR, retrieves the contents of W and STATUS registers, and returns to the main program. The subroutines Timer1 and Timer2 are identified by the labels for the assembler to assemble the program; however, the codes are not shown here. Timers will be introduced in the next chapter.

SUMMARY

The interrupt is an efficient communication process between the processor and the peripherals- external and on-chip devices. The interrupt process is asynchronous and initiated by peripherals. When an interrupt request is accepted by the MPU, it stops the processing (execution) of instructions, attends the interrupt request, and goes back to continue the processing where it was interrupted.

The interrupts are divided into two groups: hardware and software interrupts. In hardware interrupts, a signal is asserted by a peripheral to send an interrupt request. In software interrupts, the MPU can be interrupted by two ways: (1) when the MPU executes a software instruction to transfer the program execution to a specific location called an interrupt vector, and (2) when the MPU comes across an indeterminate result such as divide by zero, which is also known as a trap or an exception.

The hardware interrupts are again divided into two groups: non-maskable and maskable. A non-maskable interrupt cannot be disabled, and the MPU must attend to a non-maskable interrupt request when it is received. A maskable interrupt can be disabled or its servicing can be delayed. The basic concepts of the interrupt process are summarized in Section 10.1.7 and specific details about the PIC18 interrupts are as follows:

1. PIC18 microcontrollers include only maskable interrupts, and they are divided into two priority groups: high and low. Setting IPEN bit in RCON <7> register enables the priority scheme. Once the priority scheme is enabled, any interrupt source can be placed in either high or low priority.

2. The interrupt sources can be divided into two groups: external and internal. The external interrupt sources can be connected to three pins, RB0–RB2, of PORTB and any logic change in pins RB4–RB7 can generate an interrupt. There are many internal sources, such as timers and A/D converters, that can generate interrupt requests, but the number of sources differ among different PIC18 family members.

3. The high-priority interrupts are vectored to memory location 00008_H and low-priority interrupts are vectored to memory location 000018_H. If multiple sources in either category interrupt the MCU, they must be identified through software by checking the respective interrupt flags.

4. When the priority scheme is enabled, high-priority interrupts are enabled by setting the GIEH (INTCON <7>) global high-priority enable and the individual interrupt enable bit. For low-priority interrupts, three bits should be set: the GIEH (INTCON <7>), the GIEL (INTCON <6>) global interrupt enable low, and the individual interrupt enable bit.

5. When the priority scheme is disabled, the interrupts are compatible with the medium range PIC16 family of controllers in which interrupt sources are divided into two groups: core and peripheral. The core group includes four INT pin interrupts, a Timer0 overflow, and a change in PORTB pins <7-4>. The peripheral group includes the remaining interrupt sources.

   To enable the interrupts in the core group, two bits should be set: GIE <INTCON <7> and individual interrupt bits. In the peripheral group three bits should be set: GIE (INTCON <7>), PEIE (INTCON <6>, and the individual interrupt bit.

6. In high-priority interrupts, when an interrupt is recognized, the GIEH bit is cleared so that no more interrupts can be accepted, the address in the program counter and W and STATUS registers are pushed (saved) on the stack, and the MPU is directed to the interrupt vector 00008_H. In low-priority interrupts, the GIEL bit is cleared, only the address in the program counter is saved on the stack, and the MPU is directed to the interrupt vector 000018_H.

7. When the MPU arrives at one of the interrupt vectors, it is directed to the group of instructions called interrupt service routine (ISR) that attends to the interrupt request. The ISR must be terminated in the instruction RETFIE (Return from Interrupt).

8. In low-priority interrupt ISRs, the instruction RETFIE sets the GIEL bit and retrieves the return address from the stack, and the MPU returns to the instruction where it was interrupted. In high-priority interrupt ISRs, the return instruction used is RETFIE FAST. This instruction sets the GIEH bit, retrieves the contents of saved registers (W, STATUS, and BSR), retrieves the return address, and the MPU returns to the instruction where it was interrupted.

9. The Reset is a special type of an interrupt. In PIC18 microcontroller, it can be activated many ways: manually or when the power is turned on, or by various circuits within the MCU. When the "Reset" is asserted, the MPU establishes or reestablishes (when MPU is in the midst of an operation) the initial conditions of the registers and begins the execution from the memory location 0000.

QUESTIONS AND ASSIGNMENTS

10.1   What is an interrupt?

10.2   What is an interrupt service routine (ISR) and how does it differ from a subroutine?

10.3   What is an interrupt vector?

10.4   List the types of hardware interrupts and explain the difference between these types of interrupts.

10.5   What is a reset and how does the reset signal affect the MCU?

10.6   What is an interrupt priority and why is it necessary?

10.7   List the types of interrupts available in PIC18 microcontroller and how they are classified in terms of priorities.

10.8   In PIC18 microcontroller, list the interrupt pins that can accept external interrupt sources.

10.9   List internal peripheral sources that can send an interrupt request to the PIC18 MPU.

10.10   Assuming both priority interrupts are being enabled, how many bytes (or words) are available to write high-priority service routine?

10.11   Is there a memory limitation to write a low-priority service routine starting at the memory location 00018_H?

10.12   Assuming that the system is using only one high-priority interrupt source, specify the memory limitation on its service routine.

10.13   Write instructions to disable the interrupt priority scheme.

10.14   List the registers that are saved on the stack when an interrupt source with high-priority interrupts the MCU.

10.15   Write instructions to set up INT1 with high priority and INT2 with low priority and clear REG1 to set up as an up-counter and load REG2 with 60_H.

10.16   Write an ISR for INT1 in Question 10.15 to count up to 60_{BCD} and then clear the counter.

10.17   Write an ISR for INT2 in Question 10.15 to count in BCD from 60 to 0, and when the counter reaches zero, reset the counter to 60. Assume all registers are in the Access bank.

10.18 Write instructions to set up INT1 and Timer1 in high priority and Timer2, Timer3, and A/D converter in low priority.

10.19 In Question 10.18, when an interrupt occurs, write instructions to identify the interrupt source for high-priority interrupts.

10.20 In Question 10.18, when an interrupt occurs, write instructions to identify the interrupt source for low-priority interrupts with the A/D converter having higher priority than the timers.

CHAPTER 11

TIMERS

OVERVIEW

Embedded systems use counters and timers almost universally. Chapter 5 presented software time delays in which time is calculated by counting clock cycles needed to execute instructions. In this chapter, however, we focus on hardware timer/counter chips, which are more precise in accounting for time than the software time delays. A hardware timer/counter chip consists of a register that is incremented (or decremented) with every clock pulse. The register serves as a counter, and time is calculated by multiplying the count by the clock period.

The PIC18 family of devices has multiple 8-bit and 16-bit counter/timer modules, which are referred to as timers. The parameters of these timers are specified in their control registers. In addition, PIC18 microcontrollers include Capture, Compare, and Pulse Width Modulation (CCP) modules. As the name suggests, these modules are used to capture an event, compare two events, and modify on/off time of a pulse waveform (PWM). These modules are used in conjunction with timers to create time delays, measure pulse width, record an incoming pulse, generate pulse waveforms, and modify pulse on/off time. This chapter illustrates various applications of PIC18 timers and CCP modules either by checking an event flag or the interrupt generated by an event.

OBJECTIVES

- Explain what a counter/timer chip is and how time is calculated.
- List the PIC18 timers and their features.
- Explain how timers are used to generate time delays and waveforms.
- List the Capture, Compare, and PWM (CCP) modules and their functions, and explain how they are used in conjunction with timers.
- Explain the difference between the software approach (checking a flag) and the interrupt mode in various timer/CCP applications, and list the advantages of the interrupt mode.
- Write instructions to initialize timers and CCP modules.
- Write instructions to generate time delays, pulse waveforms, and PWM.
- Illustrate a timer application by designing a 12-hour clock.

11.1 BASIC CONCEPTS IN COUNTERS AND TIMERS

In digital systems, the concept of counting is as fundamental and important as the concept of amplification in analog systems. Similarly, the clock is an essential element of sequential digital circuits, and the events are directly related to the clock. In digital systems, we count the numbers in synchronization with the clock and convert the count to time. In this chapter, we focus on the basic concepts of hardware timer chips and how a change in a timer count relates to time.

11.1.1 What are Hardware Counters and Timers?

A counter is simply a register. It can be loaded with a binary count, and this count can be incremented or decremented at every clock pulse. Time is calculated by subtracting the beginning count from the end count and multiplying that difference by the clock period. For example, if a counter is loaded with 100_{10} and counts down to 00, and the clock frequency f is 2 MHz, the time is calculated as follows:

Clock Period = (1/f) = 1 / (2 × 10^6) = 0.5 μs;
Time = Difference in Counts × Clock Period = (100 − 0) × 0.5 μs = 50 μs.

This register can also be used as an event counter. For example, if the clock of this register is replaced by a signal coming from an event, such as a push button key, and whenever the key is pushed the count is incremented (or decremented) thus, we can obtain the number of times the key is pushed or the event has occurred.

In the following discussion, we have used the terms counter and timer synonymously. In embedded systems, there are numerous applications that include event counters or timers. Most microcontrollers include timer chips as a part of the MCU, and they are generally referred to as timers rather than counters.

A typical example of an event counter is a microcontroller that is monitoring a parking lot. Whenever a car enters the lot, the counter is incremented and when a car leaves, the counter is decremented. When the count reaches the maximum number of cars allotted for that lot, it can display a sign, "Parking Lot Full."

Similarly, we witness many timer applications such as digital clocks, traffic light controllers, timers in microwave ovens, or timing of dishwasher cycles.

11.1.2 Types of Counters and Timers and their Features

In microcontrollers, typically, the registers used for timers are either 8-bit or 16-bit in size. An 8-bit counter can count from 00 to FF_H (255_{10}) with the total of 256 counts (including zero) and a 16-bit counter can count 65536 counts (0–$FFFF_H$). The time is calculated by multiplying a count by the clock period. Microcontrollers generally include prescaler circuits that divide the clock frequency by certain number such as 2, 4, 8, or 16. The following types of counters are generally found in microcontrollers.

- Up-counter: This counter, at every clock cycle, counts up and when it reaches the maximum count, it is set to zero like an odometer in a car. This is called an overflow. When an overflow occurs, the counter can set a flag, and/or generate an interrupt. This counter can be readable and writable.
- Down-Counter: This counter, at every clock cycle, counts down and when it reaches the count zero, it may set a flag and/or generate an interrupt signal. This counter can be readable and writable. A count must be loaded in this type of counter to start the counter.
- Free-Running Counter: This counter runs continuously, and it is only readable. When it crosses its maximum count, it can set a flag and/or generate an interrupt.

11.1.3 Timer Applications

As mentioned before, timers are critical in operations of digital systems. There are numerous applications of timers such as time delay, pulse waveform generation, pulse width measurement, frequency measurement, and counting events.

Time Delay

In a writable timer register, we can write instructions to load a count for a given delay. If it is a down-counter, the program will set a flag or generate an interrupt when it reaches zero. If it is an up counter, the program will compare the count value until the timer reaches the specified count, and set a flag. If it is a free-running counter, the program needs to record the timer reading at the starting point, add the delay count to that reading, and keep comparing the readings until a match is found.

Pulse Wave Generation

To generate a periodic wave, we need an output port or a pin that is set as an output. If it is a square wave, the time delay for on-time and off-time is the same. Therefore, the program can load an appropriate count in the timer register (or find the required difference between two counts). When the counter becomes zero or a match is found in comparison, the program can output a high pulse (logic 1) to the output port (pin) and, in the next match, it can output a low (logic 0). The program will stay in a continuous loop.

Pulse Width or Frequency Measurement

To measure a pulse width, we need an input port (or input pin) that is set up to read a rising pulse and an immediate falling pulse. The program begins the counter on the rising pulse and stops the counter on the falling pulse. The difference between the two counts is the pulse width. To measure the frequency of the incoming pulse, the program should record the count difference between either two consecutive rising pulses or two falling pulses.

Timer in the Counter Mode

The timer register can be set up as a counter to count pulses representing events. In this mode, the counter is incremented on external incoming pulses, not on an internal clock.

11.1.4 Capture, Compare, and PWM (CCP) Modules

The Capture, Compare, and PWM (CCP) modules are commonly found in recent microcontrollers. The CCP modules are 16-bit (or two 8-bit) registers that are specially designed to perform three functions in conjunction with timers.

Capture

In the Capture mode, the associated CCP pin can be set as an input to record the arrival time of either a rising or falling pulse. When an edge of a pulse is sensed, the CCP module records the timer value and sets a flag or generates an interrupt.

Compare

In the Compare mode, the associated CCP pin is set as an output and a count is loaded in the CCP register. This count is compared with the timer register at every clock cycle, and when a match is found, the CCP pin can be driven low, high, or toggled. This module is commonly used to generate pulses or periodic waveforms.

Pulse Width Modulation (PWM)

In the Pulse Width Modulation (PWM) mode, the ratio of on time of a pulse to its period called duty cycle is modified. The period of a pulse is measured from rising (or falling) edge to the next rising (or falling) edge as shown in Figure 11-1; the period shown is 1000 μs resulting in a frequency of 1 kHz ($F = 1/T = 1/1000 \times 10^{-6} = 1$ kHz).

The term duty cycle is defined as the percentage ratio of on time of a pulse to its period. For example, Figure 11-1 shows the on time of the pulse is 400 μs and off time is 600 μs; thus, its duty cycle is 40 percent (400 μs / 1000 μs). The changing of the duty cycle of a pulse waveform is called PWM.

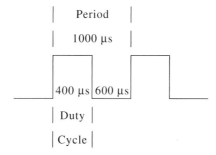

FIGURE 11-1 Pulse Waveform with 40 percent Duty Cycle

In the PWM mode, the associated CCP pin must be set be as an output and the necessary counts for a period and a duty cycle must be loaded in registers. The CCP module includes two registers—one to specify the period of the waveform and the other to specify the duty cycle. The timer register is first compared with the duty cycle register and, when a match is found, the output of the CCP pin is set low. Then the timer is compared with the period register and, when a match is found, the output pin is set high.

The PWM mode is commonly used in applications, such as modulation of the DC power for a lamp circuit to control the brightness of the lamp or controlling the speed of a DC motor.

11.2 PIC18 TIMERS

The PIC18 microcontroller family of devices has multiple timers: some have four timers and some have five timers. Timers are labeled as Timer0–Timer3 or Timer0–Timer4 and are divided into two groups: 8-bit timers and 16-bit timers. Timer0 can be set up as an 8-bit or 16-bit timer, Timer1 and Timer3 are 16-bit timers, and the Timer2 and Timer4 are 8-bit timers. Each timer is associated with its Special Function Register (SFR), labeled as T0CON through T4CON, and bits in these registers specify various parameters of these timers. As discussed in the previous section, all these timers generally function in the same fashion, with a few special features. We will discuss a few of these timers with examples and identify their special features.

11.2.1 Timer0

Timer0 is an up-counter that can be set up either as an 8-bit or 16-bit timer and has the following features (shown in Figure 11-2 (a) and (b)):

- It is readable and writable.
- It has eight options of prescale values (Bit2–Bit0).
- Its clock source can be internal (instruction cycle) or external (pin RA4/TOCK1).
- It can be set up on either rising edge or falling edge (Bit 4) when an external clock is used.
- It generates an interrupt or sets a flag when it overflows from FF_H to 00 in the 8-bit mode and from $FFFF_H$ to 0000 in the 16-bit mode.

All these above features can be set up by writing bits in T0CON register as shown in Figure 11-3.

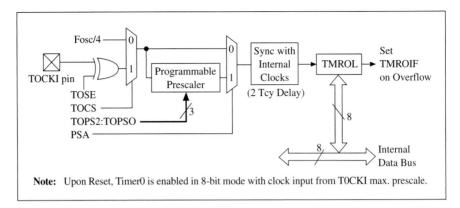

FIGURE 11-2 (a)   Timer0 Block Diagram In 8-Bit Mode

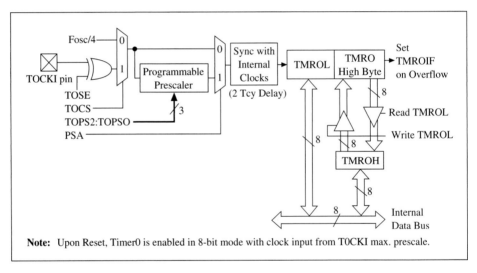

FIGURE 11-2 (b)   Timer0 Block Diagram In 16-Bit Mode

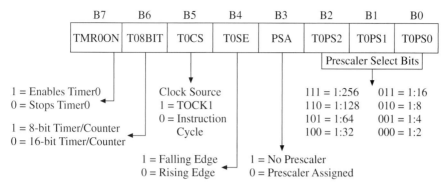

FIGURE 11-3   Timer0 Control Register (T0CON)

Timer0 as a Timer

To set up Timer0 as a timer, Bit5 in the T0CON register must be cleared to use the internal clock. At each instruction cycle (four clock cycles), the timer register is incremented. When the MPU writes into the timer register, the register does not increment during the following two instruction cycles. To obtain an accurate time delay, the count must be adjusted to account for these two instruction cycles.

Timer0 as a Counter

To set up Timer0 as a counter, Bit5 in the T0CON register must be set to 1 in order to use an external clock. In this mode, the input signal at PORTA—pin RA4/T0CK—is used as a clock. The Timer0 register can be incremented either at the rising or falling edge of pin RA4/T0CK1 determined by Bit4 (TOSE – Source Edge Select) in T0CON register. When Bit4 is 1 the register is incremented on the falling edge, and when Bit4 is 0 the register is incremented on the rising edge.

Prescaler

The prescaler enables the user to divide the frequency by a specified ratio. To use the prescaler feature Bit3 in the T0CON register must be cleared, and three bits, Bit0–Bit2, specify the scaler ratio from 1:2 to 1:256 as shown in Figure 11-3.

Interrupt

TMR0 sets a flag, which can be used as an interrupt, when the register overflows from FF_H to 00 in 8-bit mode and from $FFFF_H$ to 0000 in 16-bit mode. This overflow sets the TMR0IF (Timer0 Interrupt Flag) bit in the INTCON (Interrupt Control – Bit2) register (see Chapter 10, Figure 10-4a). This flag can be used two ways: a software loop can be set up to monitor the flag or to generate an interrupt. However, this flag must be cleared in the software to start the timer again. To use this flag to generate an interrupt, the interrupt bit TMR0IE (Timer0 Interrupt Enable–Bit5) in the INTCON register must be enabled. The disadvantage of checking the flag in a loop is that the processor will not be able to perform any other functions during that time.

16-Bit Mode

When TMR0 is set in the 16-bit mode, it uses two 8-bit registers, TMR0L and TMR0H, as shown in the block diagram in Figure 11-2b, which shows how the clock source is selected and the prescaler is used. It also shows a buffered register for TMR0H, which is used to latch a high-order byte when a low-order byte is read from TMR0L. This buffer avoids any mismatch between the high-order byte and the low-order byte that can occur when a low-order byte value of FE_H (or FF_H) is read. Without the buffer, the high-order byte will be incremented to the next number during the read cycles of the low-order byte.

EXAMPLE 11.1

Write instructions to initialize TMR0 as an 8-bit timer with the internal clock of 10 MHz. Calculate the maximum time delay available if the prescaler is 256.

Solution

When the timer runs on the internal clock, it is updated on every instruction cycle (not on MCU clock), which has four cycles. To calculate the total time, we need to calculate the clock period for an instruction cycle when the clock frequency is 10 MHz:

Clock period $T = 1 / f = 1 / 10 \ (10^6) = 0.1 \ \mu s$;
Instruction Cycle Clock Period $= 0.1 \ \mu s \times 4 = 0.4 \ \mu s$.

The prescaler divides the frequency, which is equivalent to multiplying the period by the prescaler. The largest count in an 8-bit counter is 256. Therefore, the maximum time delay is calculated as follows:

Time T_D = Instruction Cycle Clock $\times$ Prescaler $\times$ Count
$= 0.4 \ \mu s \times 256 \times 256 = 26.21$ ms.

The bits needed to set up the T0CON (Timer0 Control) are as follows in Figure 11-4:

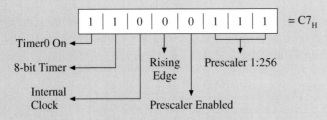

FIGURE 11-4   Control Word to Initialize Timer0

Instructions: MOVLW B'11000111'   ;Byte to enable T0CON register
                       MOVWF T0CON   ;Initialize Timer0

EXAMPLE 11.2

Write a program to set up Timer0 in the 16-bit mode to generate a high-priority interrupt every second with the clock frequency 10 MHz. Select an appropriate prescaler.

Solution

For a 10 MHz clock, each instruction cycles takes 0.4 µs (as calculated in Example 11.1). Therefore, to obtain a 1 second delay, the MPU needs 2,500,000 instruction cycles (count) as follows:

$$\text{Instruction Cycles} = \frac{1.0 \text{ s}}{0.4 \text{ µs}} = 2,500,000$$

If we select the prescaler of 1:128, we need the count 19531 (2,500,000 / 128 = 19531.25). This is an up counter; therefore, the timer counts up from a loaded number until $FFFF_H$ and subsequently rolls over to 0000 in one cycle. Therefore, the count should be $65,535 - 19531 = 46,004 = B3B4_H$. Now we need to load the count and start the counter. When an interrupt is generated, the interrupt service routine should load the count again and clear the flag.

11.2.2 PROGRAM

```
Title "Ex11-2 One Second Delay With Interrupt"
      List p=18F452, f =inhx32
      #include <p18F452.inc>        ;This is a header file


            ORG   00                ;Begin assembly
            GOTO  MAIN              ;Program begins at MAIN


            ORG   0x08
            GOTO  TMR0_ISR          ;Interrupt occurred - jump to service
                                    ;routine

MAIN:       CLRF  INTCON3           ;Disable all INT flags
            CLRF  PIR1              ;Clear all internal peripheral flags
            BSF   RCON,  IPEN       ;Enable priority - RCON <7>
            BSF   INTCON2,TMR0IP    ;Set Timer0 as high-priority
            MOVLW B'11100000'       ;Set Timer0: global interrupt, high
            IORWF INTCON,1          ;priority, overflow, interrupt flag
            MOVLW B'10000110'       ;Enable Timer0: 16-bit, internal
                                    ;clock,
            MOVWF T0CON             ;prescaler - 1:128

DELAY_1s:   MOVLW 0xB3              ;High count of B3B4_H
            MOVWF TMR0H             ;Load high count in Timer0
            MOVLW 0xB4              ;Low count of B3B4_H
```

continued

```
           MOVWF TMR0L              ;Load low count in Timer0
           BCF   INTCON, TMR0IF     ;Clear TIMR0 overflow flag – Start
                                    ;counter
HERE:      GOTO  HERE               ;Wait here for an interrupt

           ORG   0x100
TMR0_ISR:  MOVLW 0xB3               ;High count of B3B4_H
           MOVWF TMR0H              ;Load high count in Timer0
           MOVLW 0xB4               ;Low count of B3B4_H
           MOVWF TMR0L              ;Load low count in Timer0
           BCF   INTCON, TMR0IF     ;Clear TIMR0 overflow flag – Start
                                    ;counter
           RETFIE FAST              ;Return
           END
```

11.2.3 DESCRIPTION

The program is divided into three sections: MAIN, DELAY_1s, and an interrupt service routine TMR0_ISR. The MAIN section disables all the external and internal (peripheral) interrupt flags so that the initialization of the TIMR0 is not accidentally interrupted. Then it sets up the TIMR0 in the high priority mode by enabling global and peripheral interrupts and initializes various parameters of the TIMR0 (16-bit mode, prescaler) by writing to the T0CON register.

The next section DELAY_1s loads the count for a one second delay, starts the counter by clearing the interrupt flag, and waits in the loop. When the timer overflows from FFFF$_H$ to 0000, it generates an interrupt, and the MPU transfers the program to the high-priority interrupt vector location 0008$_H$ and, from that location, to the interrupt service routine TMR0_ISR. The interrupt service routine loads the count in TMR0, clears the interrupt flag, and returns to the main program.

11.2.4 Timer1

Timer1 is a 16-bit counter/timer with the following features (Figure 11-5):

- It is a 16-bit counter/timer with two 8-bit registers (TMR1H and TMR1L).
- Both registers are readable and writable.
- It has four options of prescale values (Bit5–Bit4).
- Its clock source (Bit1) can be internal (instruction cycle) or external (pin RC0/T13CK1 on the rising edge).
- It sets a flag that can be monitored in a software loop or can generate an interrupt when it overflows from FFFF$_H$ to 0000.

Timer1 Operation

This timer can be enabled or disabled by writing to Bit0 in the T1CON register and can operate in three modes: as a timer, synchronous counter, and asynchronous counter. When TMR1CS\ (Timer1 Clock Select–Bit1) is cleared, Timer1 operates as a timer and increments its register at

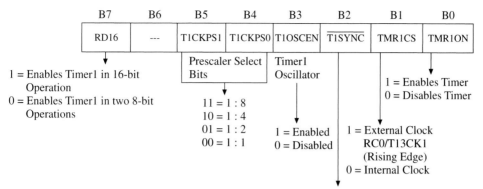

FIGURE 11-5   Timer1 Control Register (T1CON)

every instruction cycle. When TMR1CS is set to 1, Timer1 increments its register on every rising edge of the external clock input or Timer1 oscillator, if enabled. Setting Bit3 in T1CON register enables the timer1 oscillator (Figure 11-5), which is used for low-frequency operations up to 200 kHz.

Interrupt

When the register overflows from $FFFF_H$ to 0000, the timer generates an interrupt and sets Timer1 Interrupt Flag–Bit0 (TMR1IF) in register Peripheral Interrupt Register1 (PIR1) (see Chapter 10, Figure 10-6). The interrupt flag must be cleared in the software to enable the subsequent operations. However, the interrupt process can be disabled by clearing the TMR1IE (Timer1 Interrupt Enable – Bit0) in register PIE1 (Peripheral Interrupt Enable Register1 – see Chapter 10, Figure 10-7).

Resetting Timer1 using CCP Module

The PIC18 microcontroller includes CCP modules that can be used in conjunction with timers for various applications. When a CCP module associated with Timer1 is loaded with a 16-bit number and setup in the Compare mode, the count in Timer1 and the number in the CCP module are compared at every cycle. When a match is found, Timer1 is reset. This feature of Timer1 can be used for various applications, such as setting specific time delays and generating waveforms. This is further explained later in the chapter.

EXAMPLE 11.3

Write initialization instructions to set up Timer1: (1) in 16-bit counter mode with high-priority interrupt, (2) with the roll-over interrupt, (3) with the prescaler 1:1, and (4) to count external events, such as pulses arriving at RC0/T1CK1.

Solution

To meet the above specifications, the instructions are as follows:

```
BSF      RCON, IPEN      ;Enable priority interrupt
MOVLW    B'11000000'     ;Enable global and peripheral priority by
MOVWF    INTCON          ;setting Bits <7–6> in Interrupt Control Register
MOVLW    B'00000001'     ;Set up TMR1 interrupt as high priority by
MOVWF    IPR1            ;setting Bit0 in Interrupt Priority Register1
MOVWF    PIE1            ;Enable Timer1 overflow interrupt –Bit0 in PIE1
MOVLW    B'10000111'     ;Bit7 = 16-bit mode, Bit2 = No sync.with ext. clock
MOVWF    T1CON           ;Bit1 = External clock, and Bit0 = Enable TMR1
```

Description:

The first two groups of instructions set up interrupt priorities and enable the Timer1 interrupt. To initialize the counter mode, the external clock is required. Bit1 in the TCON register selects the external clock, and Bits <5-4> = 00 select 1:1 prescaler.

EXAMPLE 11.4

Write a subroutine to set up Timer1 in the 16-bit mode with the internal clock at 10 MHz to generate a 100 ms delay and provide multiples of 100 ms delay based on the count in REG1 (0x01). Select an appropriate scaler.

Solution

The initial calculations are similar to that of in Example 11.2. To obtain a 100 ms delay with the instruction cycle of 0.4 μs, the number of instruction cycles required is as follows:

Instruction Cycles: $100 \times 10^{-3} / 0.4 \times 10^{-6} = 250,000$.

If we select the prescaler value 1:8, the instruction cycles required = 250000 /8 = 31250

Therefore, the count to be loaded = 65,536 – 31250 = 34285 = $85ED_H$

11.2.5 PROGRAM

```
;::::::::::::::::::::::::::::::::::::::::::::::::::::::::::::::::::::
;Function:   This subroutine provides multiples of 100 ms delay    :
;            based on the value supplied in REG1 by the Calling    :
;            program.                                              :
;Input:      Count in REG1                                         :
;Output:     None                                                  :
;Registers modified: WREG and REG1                                 :
;::::::::::::::::::::::::::::::::::::::::::::::::::::::::::::::::::::

REG1    EQU    0x01                  ;Define Register1

TIMER1:

        MOVLW B'101100001'           ;Bit pattern for Timer1: Prescaler 1:8,
        MOVWF T1CON                  ;16-bit mode, Enable Timer1

DELAY100
        MOVLW 0x85                   ;High count of 85ED_H
        MOVWF TMR1H                  ;Load high count in Timer1 High Register
        MOVLW 0xED                   ;Low count of 85ED_H
        MOVWF TMR1L                  ;Load low count in Timer1 Low Register
        BCF   PIR1, TMR1IF           ;Clear TIMR1 overflow flag
LOOP:   BTFSS PIR1, TMR1IF           ;Test bit TMR1 flag - if set skip
        BRA   LOOP                   ;Keep in loop until TMR1 flag is set
        DECF  REG1,1,0               ;Decrement count
        BNZ   DELAY100               ;Is count = 0, if no repeat
        RETURN                       ;Go back to the calling program
        END
```

11.2.6 DESCRIPTION

The TIMER1 subroutine provides multiples of the 100 ms delay based on the count in REG1. For example, if REG1 holds the count of 5, this subroutine provides approximately 500 ms delay. The calling program should load a count in REG1 and call this program.

The first two instructions of the TIMER1 subroutine load the appropriate bit pattern in the T1CON register and specify all the timer parameters. The next group of instructions starting from the label DELAY100 loads the 16-bit count ($85ED_H$) in the Timer1 registers (High and Low), clears any residual timer overflow, and starts the timer. The instructions in the LOOP segment keep on checking whether the timer flag is set again. When it is set, the count is decremented and the program goes back to DELAY100. This is repeated until the counter becomes zero, and then the program goes back to the calling program.

continued

The actual delay in this subroutine is slightly more than 100 ms because the timer itself provides 100 ms delay, but there are 7 instructions outside the timer loop, and the execution time of these instructions is not accounted for in our delay calculations.

11.2.7 Timer2

Timer2 is an 8-bit timer with an 8-bit period register. Figure 11-6 shows the bit specifications of the Timer2 Control Register (T2CON). The timer includes the following features:

- Both registers (timer and period register) are readable and writable.
- It has three options of prescale values (Bit1–Bit0).
- It has 16 options of postscale values (Bit6–Bit3).
- It generates an interrupt when the TMR2 value matches that of the PR2.

The Synchronous Serial Port module (SSP) can use the output of this timer to generate a clock shift.

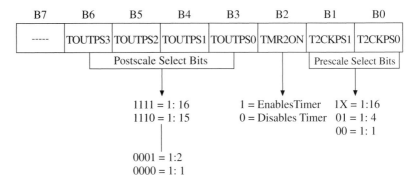

FIGURE 11-6   Timer2 Control Register (T2CON)

Timer2 Operation

The block diagram of the Timer2 module is shown in Figure 11-7. It shows that the module includes two 8-bit registers: a timer register, TMR2 and a period register, PR2. An 8-bit number is loaded in the PR2 and the timer is turned on, which is incremented at every instruction cycle. When the count in the timer register and the PR register match, an output pulse is generated and the timer register is reset to 00. The output pulse goes through a postscaler that divides the frequency by the scale factor and generates an interrupt. The interrupt sets the TMR2IF (Timer2 Interrupt Flag) in the PIR1 (Peripheral Interrupt Register1 – Bit1).

Timer2 can be used to generate periodic interrupt pulses or for Pulse Width Modulation (PWM) in conjunction with a CCP module.

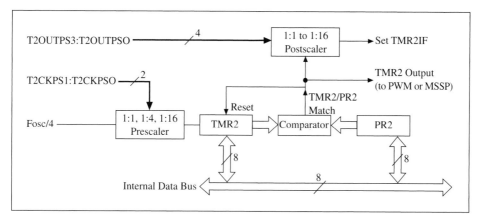

FIGURE 11-7 Timer2 Block Diagram

EXAMPLE 11.5

Write instructions to generate an interrupt every millisecond using Timer2 if the MCU clock frequency is 10 MHz.

Solution

We need to reset the Timer1 every millisecond by loading an appropriate count in register PR2. When the count in the timer register and the PR2 match, the EQ signal will reset Timer2 to zero in one instruction cycle. Therefore, the count for the PR2 is calculated as follows:

Time T_D = Instruction Clock Cycle × Prescaler × Postscaler × (Count + 1);
Count + 1 = T_D / (Instruction Clock Cycle × Prescaler × Postscaler); and
Count = 1×10^{-3} / ($0.4 \times 10^{-6} \times 4 \times 16$) — 1 = 39 – 1 = 38.

The prescaler and the postscaler values are selected (somewhat arbitrarily) to obtain a reasonable count with the smallest fraction, and one is subtracted from the count because Timer2 takes one instruction cycle to reset the timer.

Instructions:

```
        Title "Ex11-5 Generating an Interrupt Every Millisecond"
        List p=18F452, f =inhx32
        #include <p18F452.inc>          ;This is a header file

            ORG     0x00
            GOTO    TIMER2

            ORG     0x08
            GOTO    ISR_1ms             ;Go to clear the flag and start timer
                                        ;again
```

continued

```
TIMER2:      MOVLW  D'38'                  ;Count for PR2 to generate 1 ms
                                           ;interrupt
             MOVWF  PR2                    ;Load count in PR2
             BSF    RCON,   IPEN           ;Enable priority - RCON <7>
             BSF    INTCON,GIEH            ;Enable global high-priority -
                                           ;INTCON <7>

             BSF    IPR1, TMR2IP           ;Set Timer2 as high-priority
             MOVLW  B'01111101'            ;Set postscaler = 16,prescaler = 4,
             MOVWF  T2CON                  ;and turn on Timer2

             BSF    PIE1, TMR2IE           ;Enable Timer2 overflow interrupt
             BCF    PIR1, TMR2IF           ;Clear flag to begin

ISR_1ms      BCF    PIR1, TMR2IF           ;TMR2 and PR2 match occurred - clear
                                           ;flag

             RETFIE
```

Description:

The first two instructions load 38_{10} in the PR2 register, and the next four instructions enable priority mode, global and peripheral interrupts, and set up Timer2 in high priority. The byte loaded into the T2CON register sets up the prescaler and postscaler values and enables Timer2. The last two instructions are critical to start the process. When a match occurs between the timer register and the PR2, the TMR2IF (Timer2 Interrupt Flag) is set and needs to be cleared in the interrupt service routine.

11.2.8 Timer3

Timer3 is many ways similar to Timer1; therefore, we will discuss Timer3 very briefly, focusing on its control register, T3CON. We need to examine this register in detail because its bit specifications are essential to set up the CCP modules, which are discussed in Section 11.3.

Timer3 is a 16-bit counter/timer with two 8-bit registers, TMR3H and TMR3L, and can operate as a timer, a synchronous counter, or an asynchronous counter. It has the following features:

- An internal or external clock select option
- An interrupt-on-overflow from $FFFF_H$ to 0000
- A reset from CCP module trigger
- A prescale select option

The control register T3CON is shown in Figure 11-8

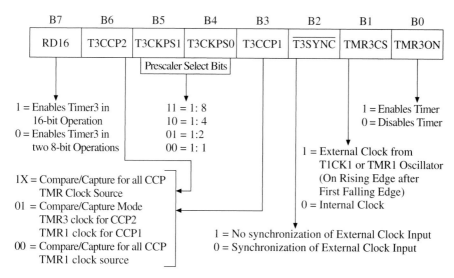

FIGURE 11-8 Timer3 Control Register (T3CON)

The bit specifications are self-explanatory, but the combination of bits B6 and B3 need some additional explanation, as follows:

1X = For 10 and 11, all CCPs are set up in the Compare/Capture mode with Timer3 as the clock source (for all CCPs); therefore, Timer3 must be enabled.

01 = CCP2 is set up in the Compare/Capture mode with Timer3 as the clock source; CCP1 is set up in the Compare/Capture mode with Timer1 as the clock source. Appropriate timers must be enabled.

00 = All CCPs are set up in the Compare/Capture mode with Timer1 as the clock source (for all CCPs); therefore, Timer1 must be enabled.

11.3 CCP (CAPTURE, COMPARE, AND PWM) MODULES

The PIC18 microcontroller family includes one or more CCP (Capture, Compare, and PWM) modules, and these modules include 16-bit registers. These registers are associated with timers and are used in recording timer readings for a specified event or between two events. The operations of all CCP modules are identical, except for the special event triggers in CCP1 and CCP2. The following description refers primarily to CCP1 module.

Each CCP module is comprised of two 8-bit registers: CCPPR1L (low-order byte) and CCPR1H (high-order byte). The modules can operate as a 16-bit Capture register, 16-bit Compare register, or duty-cycle PWM register. Timer1 and Timer3 are used as clock resources for Capture and Compare registers and Timer2 and Timer4 are used as clock sources for PWM registers. The bit specification to set up CCP modules is shown in Figure 11-9, and Bit6 and Bit3 in the T3CON register determine the assignment of a particular timer to a module (See Figure 11-8).

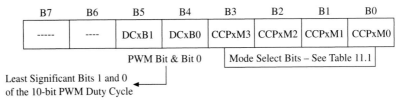

FIGURE 11-9   Capture, Compare, and PWM Control Register (CCP1CON)

TABLE 11-1   Definitions of Mode Select Bits M3–M0 in CCP1CON

M3-M0	Mode Description	M3-M0	Mode Description
0000	Capture/Compare/PWM disabled (Resets CCP Module)	1000	Compare mode – Initialize CCP pin Low – on match pin goes High (CCPIF is set)
0001	Reserved	1001	Compare mode – Initialize CCP pin High – on match pin goes Low (CCPIF is set)
0010	Compare Mode –Toggle output on match (CCPxIF bit is set)	1010	Compare mode – Generates software interrupt on Compare match (CCPIF is set)
0011	Reserved	1011	Compare mode – Triggers special event (CCPIF bit is set)
0100	Capture mode – Every falling edge	11xx	PWM Mode
0101	Capture mode – Every rising edge		
0110	Capture mode – Every 4th rising edge		
0111	Capture mode – Every 16th rising edge		

11.3.1 CCP in the Capture Mode

In Capture mode, the CCPR1 register captures the 16-bit value of Timer1 (or Timer3) when an event occurs on pin RC2/CCP1. An event is defined by four bits, Bit3–Bit0, in the control register CCP1CON (Figure 11-9). When a capture occurs, the interrupt request flag bit CCP1IF, which is Bit2 in PIR1 (Peripheral Interrupt Request Register1) (Figure 10-6), is set and must be cleared for the next operation. To capture an event, the following must take place:

1. Set up pin RC2/CCP1 of PORTC as the input.
2. Initialize Timer1 in the timer mode or the synchronized counter mode by writing to T1CON register.
3. Initialize CCP1 by writing to the CCP1CON register.
4. Clear the CCP1IF flag to continue the next operation when a capture occurs.
5. Clear the CCP1IE (Interrupt Enable Bit2 in Peripheral Interrupt Enable1 –PIE1) and the CCP1IF (Interrupt Flag in PIR1) to avoid a false interrupt when the capture mode is changed.

The above steps are also applicable to other CCP modules and associated timers.

EXAMPLE 11.6

Write a program to measure the period of an incoming pulse waveform; the source is connected to pin RC2 of PORTC, and the microcontroller is running at a clock frequency of 10 MHz. Use the CCP1 module to capture the incoming waveform with Timer1 as the clock source of the CCP1 module. The number of cycles will be the period of measurement.

Solution

To measure the period of an incoming pulse waveform, we need to capture two consecutive rising or falling pulses and record the timer values at those events. The difference between these two readings of the timer is the period, measured in the number of cycles of the incoming waveform.

The program should include the following steps:

1. Setting up pin RC2/CCP1 as an input.
2. Setting up the CCP1 module in capture mode to record two consecutive rising pulses.

Using Timer1 as a clock source for the CCP1 module.

11.3.2 PROGRAM

```
        Title "Ex11-6 Measuring Period of an Incoming Pulse Waveform"
        List p=18F452, f =inhx32
        #include <p18F452.inc>        ;This is a header file


REG0    EQU     0x00            ;Define data registers
REG1    EQU     0x01
REG2    EQU     0x02
REG3    EQU     0x03


        ORG     0x00
        GOTO    MAIN


        ORG     0x20
MAIN:   BSF     TRISC,CCP1      ;Set up CCP1 (RC2) pin as input
        MOVLW   B'10000001'     ;Select Timer1 as clock for
                                ;CCP1-Bits<6-3> 00
        MOVWF   T3CON,0         ;Timer3 on in 16-bit mode
        MOVLW   B'10000001'     ;Enable Timer1 in 16-bit mode, set
                                ;prescaler =1,
        MOVWF   T1CON,0         ;use instruction cycle clock
        MOVLW   B'00000101'     ;Set CCP1 to capture on every rising
                                ;edge
        MOVWF   CCP1CON,0
        BCF     PIE1,CCP1IE,0   ;Disable CCP1 capture interrupt
                                ;as a precaution
        CALL    PULSE
        MOVFF   CCPR1L,REG0     ;Save Timer1 at arrival of first
                                ;rising edge
```

continued

```
        MOVFF      CCPR1H,REG1

        CALL       PULSE
        CLRF       CCP1CON     ;Disable further captures
        MOVFF      CCPR1L,REG2 ;Save Timer1 at arrival of second rising
                               ;edge
        MOVFF      CCPR1H,REG3

        CALL       RESULT      ;Get period in clock cycles
HERE:   GOTO       HERE


PULSE:  BCF        PIR1,CCP1IF,0 ;Clear the CCP1IF flag
CHECK   BTFSS      PIR1,CCP1IF,0 ;Is there a rising edge?
        GOTO       CHECK       ;If not, wait in loop
        RETURN

RESULT: MOVF       REG0,W,0    ;Get first low reading in W
        SUBWF      REG2,1      ;Subtract W from second low reading -
                               ;Save in REG2
        MOVF       REG1,W      ;Get first high reading in W
        SUBWFB     REG3,1,0    ;Subtract W from second high reading -
                               ;Save in REG3

        RETURN
        END
```

11.3.3 PROGRAM DESCRIPTION

The program is written in three sections: a MAIN routine and two subroutines. The MAIN routine initializes the CCP1, using Timer1 as a clock source. The subroutine PULSE records the timer readings when pulses arrive. Finally, the subroutine RESULT determines the difference between the two readings to find the period. The program could be written in one section, but, conceptually, it is easier to follow when tasks are subdivided into modules.

The MAIN section sets pin RC2 as an input and selects Timer1 as the clock source for the CCP1 module by writing into the T3CON register. It may seem somewhat confusing to use the T3CON register for Timer1. However, timers for the CCP modules are selected by bits <6-3> in the T3CON register; this is a part of original design of the PIC18 microcontroller. The MAIN routine disables the CCP1 capture interrupt as a precaution against accidental interrupts during the subroutine PULSE.

The PULSE subroutine clears the flag CCPIF and waits in a loop for the arrival of a rising pulse. When the pulse arrives, it captures the timer readings in the CCPR1 registers and returns to the MAIN program, which saves the reading in REG0 and REG1. The next rising pulse is captured in the same fashion. The MAIN section saves all these readings in registers REG0–REG3 and calls the subroutine RESULT. This subroutine loads the low-order reading of the first pulse in the WREG and subtracts it from the low-order reading of the second pulse. It repeats the same process for high-order readings with the instruction SUBWFB that accounts for the borrow, if the borrow is generated by the first subtraction of the low-order reading. The result is saved in registers REG2 and REG3.

11.3.4 CCP in the Compare Mode

In the Compare mode, the 16-bit value loaded by the user in CCPR1 (or other initialized CCPRx) is constantly compared with the TMR1 (or TMR3) register when the timers are running in either timer mode or synchronized counter mode. When a match occurs, the pin RC2/CCP2 on PORTC is driven high, low, or toggled based on mode select bits in the CCP1CON (Bit3–Bit0 in CCP1 control register), and the interrupt flag bit CCP1IF is set.

To set up CCP1 in compare mode, the following steps are necessary.

1. Set up pin RC2/CCP1 of PORTC as output.
2. Initialize Timer1 in the timer mode or the synchronized counter mode by writing to the T1CON register.
3. Initialize CCP1 by writing to the CCP1CON register.
4. Clear the flag CCP1IF, which is set when a compare occurs, and must be cleared to continue to the next operation.
5. For a special event trigger, an internal hardware trigger is generated that can be used to initiate an action.
6. The special event trigger output resets Timer1.

The above steps are also applicable to other CCP modules and associated timers.

EXAMPLE 11.7

Write instructions to generate a trigger as a special event every 10 ms that can be used to initiate an A/D conversion if the crystal frequency is 10 MHz.

Solution

To generate an interrupt every 10 ms using a 10 MHz clock frequency, the timer needs a count equal to 25,000 (10 ms / 0.4 μs) without a prescaler; the count can be reduced to 6250 (186A$_H$) with the prescaler set to a ratio of 1:4. To generate a special event trigger, we need to use the CCP2 module in the compare mode. The instructions are as follows:

```
START:    MOVLW    B'10000001'     ;Bits <6 –3> 00 select Timer1 as clock source
                                   ;for CCP2
          MOVWF    T3CON           ;Initialize Timer3
          MOVLW    0x6A            ;Load count in CCPR1H and CCPR1L for 10 ms
                                   interrupt
          MOVWF    CCPR1L
          MOVLW    0x18
          MOVWF    CCPR1H
          CLRF     TMR1L           ;Clear Timer1 to begin counting from 0
          CLRF     TMR1H
          MOVLW    B'11000000'     ;Enable global and peripheral interrupts
          MOVWF    INTCON
          BSF      PIE2, CCP2IE    ;Enable CCP2 interrupt
```

continued

```
        MOVLW    B'00001011'    ;Initialize CCP2 in Compare mode
        MOVWF    CCP2CON
        MOVLW    B'10100001'    ;Initialize TMR1 – 16-bit mode and 1:4 prescaler
        MOVWF    T1CON
```

Description:

The first instruction initializes Bit6 and Bit3 of the T3CON register as 00 in order to select Timer1 as a resource timer for the CCP2 module and enable Timer3 in 16-bit mode. Because we are using Timer1 as a resource counter, the count ($186A_H$) is loaded in the CCPR1 register. When the counter register matches the count, an internal interrupt is generated that initializes an A/D conversion if the A/D module is enabled (See Chapter 12). To generate an interrupt, we need to enable global, peripheral, and CCP2 interrupts. Finally, the last instruction starts Timer1 with the 1:4 prescaler.

11.3.5 CCP in the Pulse Width Modulation (PWM) Mode

A CCP module can be set up to output a pulse waveform for a given frequency and a duty cycle.

CCP Module Operation

Timer2 (and Timer4) are used in conjunction with the CCP modules to generate PWM. The CCP modules use a 10-bit number to specify the duty cycle. For example, when the CCP1 module is used, high-order 8 bits are loaded in register CCPR1L and low-order two bits are loaded in the CCP1CON (CCP1 Control) register (Bit5 and Bit4). The PWM period is specified by writing into the PR2 register. When TMR2 is equal to PR2, the following three events occur in the next increment cycle:

- TMR2 is cleared.
- Pin RC2/CCP1 of PORTC is set to high.
- The PWM duty-cycle byte is latched from CCPR1L into CCPR1H.

When CCPR1H and TMR2 match again for the specified duty cycle, the CCP1 pin is cleared. To set up CCP1 in the PWM mode, the following steps are necessary:

1. Set up pin RC2/CCP1 of PORTC as the output.
2. Set up the PWM period by writing to the PR2 register.
3. Set the PWM duty cycle by writing to the CCPR1L register and Bit5–Bit4 of the CCP1CON register.
4. Set the TMR2 prescale value and initialize Timer2 in timer mode by writing to the T2CON register.
5. Enable the CCP1 module in PWM mode.
6. Initialize the CCP1 by writing to the CCP1CON register.

The above steps are also applicable to other CCP modules and Timer4.

EXAMPLE 11.8

Calculate the PR2 count to generate a 10 kHz (F_W) pulse waveform if the PIC18 oscillator frequency F_{OC} is 10 MHz.

Solution

The period for 10 kHz Tw = 1/FW = $1/10 \times 10^3$ = 0.1 ms

Now we need to calculate how many instructions (count) we need in order to generate a time of 0.1 ms.

Time is calculated as follows:

Tw = System Clock Period × 4 cycles × Prescale × (Count + 1).
If we convert all time units in their frequency and keep the Count on one side of the equation, then:

$$\frac{1}{Fw} = \frac{4 \times Prescale \times Count + 1}{Foc} \text{ therefore, Count} + 1 = \frac{Foc}{Fw \times 4 \times Prescale} = \frac{10 \times 10^6}{10 \times 10^3 \times 4 \times Prescale}$$

If we assume the prescale is 1, the count for PR2 = 250 − 1 = 249

EXAMPLE 11.9

Write instructions to set up the CCP1 in PWM mode to generate a pulse waveform at 10 kHz with a 40 percent duty cycle if the PIC18 microcontroller oscillator frequency = 10 MHz.

Solution

The value for PR2 (249) has been already calculated in the previous example, and the count for 40 percent duty cycle = 250 × 0.4 = 100. The instructions to set up the CCP1 in PWM mode are as follows.

```
BCF       TRISC, CCP1      ;Set up RC2/CCP1 of PORTC as output
MOVLW     D'249'           ;Load the count for 10 kHz frequency in PR2
MOVWF     PR2,0
MOVLW     D'100'           ;Load duty cycle count (100) in CCPR1L
MOVWF     CCPR1L,0
MOVWF     CCPR1H,0
MOVLW     B'1000001'       ;Reset Bit6 and Bit3 to 00 to select Timer2 as resource timer
MOVWF     T3CON, 0         ;for CCP1 module
CLRF      TMR2,0           ;Start counting from zero
MOVLW     B'00000100'      ;Set prescaler to 1 and Bit2 to turn on Timer2
MOVWF     T2CON, 0
MOVLW     B'00001100'      ;Set CCP1M3 – CCP1M0 to 1100 for PWM mode
MOVWF     CCP1CON, 0       ;Enable CCP1 in PWM mode
```

continued

> **Description:**
>
> The instructions follow the steps outlined earlier. The first instruction initializes pin RC2/CCP1 of the PORTC as an output, followed by the loading of the appropriate count in PR2 to generate 10 kHz and duty-cycle count in CCPR1L to have 40 percent duty cycle. The bits <6–3> in the T3CON register select Timer2 as a resource timer for all the CCP modules. We will be using the CCP1 module. The CLRF instruction clears Timer2, which enables it to begin counting from 0, and the next instruction initializes Timer2. The last instruction sets up the CCP1 in PWM mode.

11.4 ILLUSTRATION: GENERATING A PERIODIC WAVEFORM USING AN INTERRUPT

In the previous three sections, we examined timer operations and their interactions with the CCP modules. For various applications such as time delay, pulse width measurement, and frequency measurement, we have two options: (1) set up a software loop and wait in the loop until the timer sets up a flag, or (2) use the interrupt approach. If we use the software loop approach, the processor is kept busy in executing the loop and will be unable to perform any other operation. On the other hand, if we use the interrupt approach, the processor can perform other functions and attend to the timer/CCP function when an interrupt is generated. This following illustrative program (Section 11.4.3) generates a square wave using the interrupt approach.

11.4.1 Problem Statement

Write a program to generate a 1 kHz square wave at pin RC2/CCP1 of the PORTC using the interrupt method and Timer1, assuming a crystal oscillator frequency Fosc of 10 MHz.

11.4.2 Problem Analysis

To generate a 1 kHz square wave at pin RC2, the following steps should be performed:

1. Set up the RC2/CCP1 pin as an output.
2. Set up interrupt priorities, enable global priorities, and enable the CCP1 interrupt.
3. Initialize Timer1 as a base timer for CCP1.
4. Configure the CCP1 pin to toggle when a match occurs.
5. Calculate the instruction cycle count needed for a 1 kHz waveform on-time and off-time and add the count to the timer register.

Count for 1 kHz = 1 ms /0.4 μs = 2500 assuming the prescaler is 1:1. Therefore, the count needed for the pulse to be high as well as low is 1250 (04E2$_H$).

The above steps can be translated into PIC18 assembly language as follows:

11.4.3 PROGRAM

```
HICOUNT    EQU      0x04          ;High-order count
LOCOUNT    EQU      0xE2          ;Low-order count


           ORG      00
           GOTO     START         ;Begin at START


           ORG      0x08
           GOTO     ISR_PULSE     ;Interrupt occurred - go to service
                                  ;routine

START:     BCF      TRISC, CCP1   ;Set up RC2/CCP1 as an output

Block1     MOVLW    B'11000000'   ;Enable global and peripheral
                                  ;interrupts
           MOVWF    INTCON
           BSF      IPR1, CCP1IP  ;Set CCP1 high priority
           BSF      PIE1,CCP1IE   ;Enable CCP1 interrupt

Block2     MOVLW    B'10000001'   ;Select Timer1 for CCP1 -
                                  ;Bits <6-3> 00
           MOVWF    T3CON
           MOVLW    B'00000010'   ;Toggle on match
           MOVWF    CCP1CON

Block3     CLRF     CCPR1L        ;Clear CCP registers
           CLRF     CCPR1H
           CLRF     TMR1L         ;Clear Timer1 registers
           CLRF     TMR1H

Block4     BSF      PORTC, CCP1   ;Set RC2/CCP1 high
           BCF      PIR1, CCP1IF  ;Clear interrupt flag
           MOVLW    B'10000001'   ;Turn Timer1 on
           MOVWF    T1CON
HERE:      GOTO     HERE          ;Wait here until an interrupt occurs

ISR_PULSE  BTFSS    PIR1, CCP1IF  ;Check if it is CCP1 interrupt
           RETFIE                 ;If no, return
           BCF      PIR1, CCP1IF  ;Clear interrupt flag
           MOVLW    LOCOUNT       ;Add low match count to  CCPR1L
           ADDWF    CCPR1L, 1
           MOVLW    HICOUNT       ;Add high match count to CCPR1H
           ADDWFC   CCPR1H, 1
           RETFIE
```

11.4.4 PROGRAM DESCRIPTION

The program above follows the steps outlined in the Problem Analysis section. It begins by setting the RC2/CCP1 pin of PORTC as an output, and follows by setting interrupts in Block1. In Block2, we need to use the T3CON (Timer3 control register) to select Timer1 as a base timer for CCP1; Bit6 and Bit3 in the T3CON register select Timer1 for CCP1.

Instructions in Block3 clear both the CCPR1 and the Timer registers, enabling them to begin at 00, and the instructions in Block4 get ready to start the timer by clearing the interrupt flag, setting the CCP1 pin high, and starting Timer1.

At the beginning, the timer registers and CCP registers are cleared, and the match generates an interrupt that directs the program to the interrupt service routine ISR_PULSE. The first instruction in the ISR_PULSE checks the CCP1IF to verify that the interrupt came from CCP1. If it is a false alarm, the program returns to the place where it was interrupted. If CCP1IF is set, however, the MPU executes the interrupt routine, The ISR clears the interrupt flag, adds the match count for a 1 kHz waveform to the CCP1PR registers, and returns to the main program.

11.5 ILLUSTRATION: DESIGNING A 12-HOUR CLOCK

Designing a clock is a typical application of a timer. Recalling our Time and Temperature project from Chapter 1, we now explore designing a clock that can display time at an output device, such as LCD or seven-segment LEDs. The following illustration uses Timer0 to generate an interrupt every second, and the interrupt is used to design a clock.

11.5.1 Problem Statement

Set up Timer0 in the high-priority interrupt mode to generate an interrupt every second, given a clock frequency of 10 MHz. Use the interrupt to design a 12-hour clock that can record hours, minutes, and seconds as well as store those values in ASCII characters separated by a colon in a data table.

11.5.2 Problem Analysis

Example 11.2 illustrates setting up Timer0 in the high-priority interrupt mode to generate an interrupt every second. We can use that example to initialize Timer0 in the interrupt mode, but we will have to modify the instructions in the interrupt service routine to accommodate the clock requirements.

To design a clock, we need three registers to keep track of hours, minutes, and seconds. We can clear all these registers at the beginning and start counting the interrupts. At each interrupt, the program should increment the count in the register that stores the seconds, adjust the count for BCD values, and continue counting until 60. When the count reaches 60, the program should clear the "second" register, increment the "minute" register, and start counting again. When the "minute" register reaches the count of 60, the program should increment the "hour" register, clear the "minute" and "second" registers. It should also continue to count until the "hour" register reaches the count of 13. Now the program should load the count 1 in the "hour" register, clear the other two registers, and repeat the process. The interrupt service routine,

TMR0_ISR, in Example 11.2 reloads the count in Timer0 registers for a one-second delay and starts the counter again. Now the ISR should be modified to include the upgrading of the three registers discussed in this section.

So far the register values are in binary, representing two packed BCD digits. The conversion of binary digits (a packed byte) into ASCII characters is illustrated in Chapter 8, Section 8.4. In this problem, the algorithm will be even simpler than that of Section 8.4 because the digits are already in BCD, limited to 0 to 9. They can be unpacked and converted into ASCII characters by adding 30_H.

11.5.3 PROGRAM

```
;:::::::::::::::::::::::::::::::::::::::::::::::::::::::::::::::::::::::
;This is 12-hour clock program, designed to keep track of hours, minutes,  :
;and seconds,and stores the decimal values of these variables as ASCII     :
;characters in ;data registers. Timer0 is used to generate an interrupt    :
;every second, and the interrupt service routine updates the time that is  :
;adjusted for BCD values, and equivalent ASCII characters of these BCD     :
;values in data registers.                                                 :
;:::::::::::::::::::::::::::::::::::::::::::::::::::::::::::::::::::::::


            Title "Illust11-5: Designing a Clock"
            List p=18F452, f =inhx32
            #include <p18F452.inc>  ;This is a header file


SECONDS     RES   1                 ;Reserve one memory location each
MINUTES     RES   1
HOURS       RES   1

TEMP        EQU   0x00              ;Temporary register used in BCDASCII routine
BUFFER0     EQU   0x01              ;Registers to save BCD values
BUFFER1     EQU   0x02
TIME0       SET   0x10              ;Save ASCII characters for Hours, Minutes,
                                    ;and Seconds

            ORG   00
            GOTO  MAIN

            ORG   0x08
            GOTO  CLK_ISR           ;Go to interrupt service routine
                                    ;to attend an interrupt from Timer0
```

continued

```
              ORG   0x20
MAIN:         CLRF  SECONDS            ;Clear registers
              CLRF  MINUTES
              CLRF  HOURS

              CLRF  INTCON3            ;Disable all INT flags
              CLRF  PIR1               ;Clear all internal peripheral flags
              BSF   RCON, IPEN         ;Enable priority - RCON <7>
              BSF   INTCON2,TMR0IP     ;Set Timer0 as high-priority
              MOVLW B'11100000'        ;Set Timer0:global interrupt, high
              IORWF INTCON,1           ;priority, overflow, interrupt flag
              MOVLW B'10000110'        ;Enable Timer0: 16-bit, internal clock,
              MOVWF T0CON              ;prescaler- 1:128

DELAY_1s:     ;Sets up Timer0 to generate an interrupt every second

              MOVLW 0xB3               ;High count of B3B4H
              MOVWF TMR0H              ;Load high count in Timer0
              MOVLW 0xB4               ;Low count of B3B4H
              MOVWF TMR0L              ;Load low count in Timer0
              BCF   INTCON, TMR0IF     ;Clear TIMR0 overflow flag - Start
                                       ;counter
TIME:         MOVF  HOURS,W            ;Converts Hours, Minutes, and Seconds
              CALL  BCDASCII           ;in ASCII characters and saves them in
                                       ;a table
              MOVFF BUFFER1,TIME0      ;Save Hours in ASCII
              MOVFF BUFFER0,TIME0+1
              MOVLW 0x3A               ;ASCII for colon
              MOVWF TIME0+2
              MOVF  MINUTES,W
              CALL  BCDASCII           ;ASCII for minutes
              MOVFF BUFFER1,TIME0+3
              MOVFF BUFFER0,TIME0+4
              MOVLW 0x3A               ;ASCII for colon
              MOVWF TIME0+5
              MOVF  SECONDS,W          ;ASCII for seconds
              CALL  BCDASCII
              MOVFF BUFFER1,TIME0+6
              MOVFF BUFFER0,TIME0+7
              BRA   TIME

BCDASCII:     ;:::::::::::::::::::::::::::::::::::::::::::::::::::::::::::::::::::
              ;Function:  This subroutine converts BCD values from 0 to 9    :
              ;           in ASCII characters.                               :
              ;Input:     A packed BCD digit in register TEMP                :
              ;Output:    An equivalent ASCII value in WREG                  :
              ;:::::::::::::::::::::::::::::::::::::::::::::::::::::::::::::::::::
```

continued

```
          MOVWF TEMP                ;Converts Hours, Minutes, and Seconds
          ANDLW 0x0F                ;in two ASCII characters each and saves
          ADDLW 0x30                ;them ADDLW 0x30 in BUFFER0 and BUFFER1
          MOVWF BUFFER0
          SWAPF TEMP,W
          ANDLW 0x0F
          ADDLW 0x30
          MOVWF BUFFER1
          RETURN

CLK_ISR:  ;::::::::::::::::::::::::::::::::::::::::::::::::::::::::::::::::
          ;Function:  This is an interrupt service routine to update    :
          ;           time. It loads count in TMR0 for one-second delay :
          ;           and updates the values of seconds, minutes, and   :
          ;           hours and adjust them for BCD values.             :
          ;Input:     Time values in WREG                               :
          ;Output:    Updated BCD values of time in WREG                :
          ;::::::::::::::::::::::::::::::::::::::::::::::::::::::::::::::::

          MOVLW 0xB3                ;High count of B3B4H
          MOVWF TMR0H               ;Load high count in Timer0
          MOVLW 0xB4                ;Low count of B3B4H
          MOVWF TMR0L               ;Load low count in Timer0
          BCF   INTCON, TMR0IF      ;Clear TIMR0 overflow flag - Start
                                    ;counter

CLK_UPDATE:
          INCF  SECONDS,W           ;Update seconds
          DAW                       ;Adjust for BCD
          MOVWF SECONDS
          SUBLW 0x60                ;Is it 60 seconds?
          BTFSS STATUS, Z           ;If yes, go to minutes
          BRA   GOBACK

          CLRF  SECONDS             ;Clear seconds
          INCF  MINUTES,W           ;Update minutes
          DAW                       ;Adjust for BCD
          MOVWF MINUTES
          SUBLW 0x60                ;Is it 60 minutes?
          BTFSS STATUS, Z           ;If yes, go to hours
          BRA   GOBACK

          CLRF  MINUTES             ;Clear minutes
```

continued

```
            INCF   HOURS,W           ;Update hours
            DAW                      ;Adjust for BCD
            MOVWF  HOURS
            SUBLW 0x13               ;Is it the 13th hour?
            BTFSS STATUS, Z          ;If yes, load count 1 in hours
            BRA    GOBACK
            MOVLW  1                 ;Begin at 1o clock
            MOVWF  HOURS
GOBACK:     RETFIE    FAST           ;Return
            END
```

11.5.4 PROGRAM DESCRIPTION

The program is divided primarily into three major sections: (1) MAIN initializes the interrupt process and timer and saves the BCD time values, (2) BCDASCII converts BCD digits into ASCII characters, and (3) the CLK_ISR starts the timer again and updates the clock values. The MAIN segment begins by clearing three registers, HOURS, MINUTES, and SECONDS, and the clock begins at 00:00:00. There is no provision in this illustration to set the time. The next subsegment, DELAY_1s, loads the 16-bit count in TMR0H and TMR0L for one-second delay and starts the timer as in Example 11.2. The next subsegment, TIME, converts the time values into ASCII codes by calling the subroutine BCDASCII and saves them in a sequence for display on an LCD (not a part of this program). The CLK_ISR interrupt service routine loads the timer again with the count for one-second delay and increments the SECONDS register. It checks whether the count in the register has reached 60. It does so by subtracting the count from 60 and examining the zero flag. If the flag is set, it skips the next branch instruction and goes to increment the MINUTES register and clear the SECONDS register; otherwise, the program jumps to the end of the service routine and returns to the TIME segment to update the clock. It repeats a similar process for minutes and hours.

In this program, the clock begins at 00. If we are interested in setting time, we will have to build push-button keys and adjust the time values either through interrupts or a program that checks a key pressed, as explained in Time and Temperature Project design (Chapter 14).

SUMMARY

In recent years, counter/timer circuits have become an integral part of most microcontrollers. A timer is an 8-bit or 16-bit register that is incremented (or decremented) at a given clock cycle, and the count is converted into time by multiplying the count by the clock period. The PIC18 microcontroller family includes multiple 8-bit and 16-bit timers; the number of timers varies based on the version. Each timer is associated with a control register that is used to specify various timer parameters, such as internal or external clocks, prescalers, and rising or falling edges. In addition, the PIC18 family includes Capture, Compare, and PWM (CCP) modules that are used in conjunction with timers in various applications such as time delay, wave generation, frequency measurement, and PWM. This chapter discussed various aspects of the timers, CCP modules, and their applications. The important features of these timers are summarized below. The number of timers varies from three to five (TMR0 to TMR4), and some of the following comments may not be applicable to a particular version.

1. Timers are classified in two categories: 8-bit and 16-bit. Timer0, Timer1, and Timer3 are 16-bit timers, and Timer2 and Timer4 are 8-bit timers.

2. Each timer is associated with a control register and bits in these control registers, which specify various operating parameters of the timers.

3. The clock source of a timer, either internal or external, can be selected using its control register. Similarly, a prescaler, used to divide the clock frequency, can be specified.

4. Timer circuits can function as timers or counters (asynchronously or synchronously). When an internal clock source is used, the timer functions as a timer; when an external clock source is used, the timer functions as an event counter.

5. When a timer overflows from its largest count ($FFFF_H$ for a16-bit and FF_H for an 8-bit) to 00, it sets a flag or generates an interrupt (if enabled). The flag can be monitored in a software loop or used by the interrupt process, and the flag must be cleared to restart the timer.

6. PIC18 microcontrollers include multiple CCP modules, and the number of modules depends on the particular version of the PIC18 family. These modules are, for the most part, identical and are configured in the same manner.

7. These modules have three operational modes: Capture, Compare, and Pulse Width Modulation (PWM).

8. Timer1 and Timer3 are used as base timers in Capture and Compare modes, and Timer2 and Timer4 are used as base timers in the PWM mode.

9. In the Capture mode, when the selected edge of the incoming signal is detected, the 16-bit reading in the base timer is captured in the CCPR register pair, the interrupt flag is set, and an interrupt is generated if it is enabled.

10. In the Compare mode, the 16-bit reading in the CCPR is compared with the reading in the base timer at every cycle. When a match is found, the CCP pin can be toggled, set high, or set low. Additionally, the interrupt flag is set and an interrupt is generated if enabled.

11. The PWM mode is used primarily to generate a waveform with a given duty cycle. In PWM mode, the period of the waveform for Timer2 and the CCP1 register is specified by writing into the PR2 register. The duty cycle is specified by writing a 10-bit word, the eight MSB into the CCPR1L register and the two LSB in the CCP1CON register (Bits <5–4>). When the count in PR2 matches the base timer, three events occur: (1) the timer is cleared, (2) the CCP1 pin is set high, and (3) the PWM duty cycle is reloaded from the CCPR1L register to the CCPR1H register. When the duty-cycle count in CCPR1L register matches the timer, the CCP1 pin is pulled low. These features are also applicable to Timer4 (if available).

QUESTIONS AND ASSIGNMENTS

11.1 A free-running 16-bit timer has a clock frequency 5 MHz. The counter register in the timer is incremented every clock cycle. When the counter reaches $FFFF_H$, it rolls over to 0000 and continues to count. If the timer reading at the beginning of the event is $1FF8_H$ and at the end of the event is 3380_H, calculate the time delay between the two events.

11.2   In Q.11.1, the first reading is $F138_H$ and the second reading is $3F58_H$. Calculate the time delay.

11.3   In a pulse waveform, the on-time is 150 µs and the off-time is 300 µs. Calculate the duty cycle of the waveform.

11.4   Calculate the frequency of the waveform in Q.11.3.

11.5   In a pulse waveform, the on-time is 300 µs and the off-time is 100 µs. Calculate the duty cycle of the waveform.

11.6   What is the duty cycle of a square wave?

11.7   Calculate the maximum delay that can be obtained when Timer0 is running in 16-bit mode and the MCU clock frequency is 10 MHz.

11.8   Calculate the time delay when Timer0 is loaded with the count of $676B_H$, the instruction cycle is 0.1 µs, and the prescaler value is 128.

11.9   Write a subroutine to generate a 0.5 second delay using Timer0, assuming a clock frequency of 40 MHz. The program should check the overflow flag by setting a loop to generate the delay.

11.10   Write a subroutine to generate a 1.0 second delay using Timer0 with the clock frequency at 10 MHz.

11.11   Write a program to trigger a special event to start a conversion in the A/D module every second. Use Timer1 and CCP2 in the compare mode.

11.12   Write instructions to measure the pulse width of the input source connected to pin RC2 if the system clock frequency is 10 MHz. Use the CCP1 module in capture mode and Timer1 as a clock source for the module.

11.13   Write instructions to set up CCP1 in the PWM mode to generate a 1 kHz pulse waveform with a 50 percent duty cycle, given a system clock frequency of 10 MHz.

11.14   Use instructions in Question 11.10 to build a five-minute egg timer. Output minutes and seconds in BCD to PORTB and PORTC. When the timer reaches five minutes, reset the timer. Divide the program in small modules and write them as subroutines.

11.15   In Q.11.14, when the timer reaches five minutes, flash the LEDs on and off, simulating a buzzer (see Simulation Exercise 11.2SE).

SIMULATION EXERCISES

11.1 SE Modify the program in Section 11.5 – Illustration: Designing 12-Hour Clock to update Hours, Minutes, and Seconds in BCD and save the values starting from register 10_H. Assemble and execute the program using MPLAB IDE and display the time registers on the Watch screen.

11.2 SE Assemble and execute the program in Question 11.15 using PIC18 Simulator IDE. Display minutes and seconds at four seven-segment LEDs. At the end of five minutes, flash LEDs that can be connected to PORTE.

CHAPTER **12**

DATA CONVERTERS

OVERVIEW

Some embedded systems are designed to process real-world physical quantities such as temperature, pressure, sound, and light. These are nonelectrical quantities and must be converted into electrical signals by transducers (see Section 1.5). Chapter 1 introduced the Time and Temperature embedded system that includes a temperature sensor. The sensor converts temperature into an analog electrical signal that is continuous in nature. On the other hand, microcontrollers (or microprocessors) are binary machines, and process binary electrical signals that are discrete. Therefore, analog signals must be converted into equivalent digital binary signals if analog signals need to be processed by microcontrollers. The electronic circuit that converts an analog signal into a digital signal is called an analog-to-digital (A/D) converter. Similarly, a digital signals need to be translated into analog signals to drive physical devices, such as a pair of headphones in a CD music system. The electronic circuit that translates a digital signal into an analog signal is called a digital-to-analog (D/A) converter.

This chapter explains the need for data conversion and basic concepts in the operation of an A/D converter. It lists the various methods of designing A/D converters and focuses on the circuits inside an A/D converter that uses the successive approximation method. The PIC18 family of microcontrollers has an A/D converter module that is based on the successive approximation method. In this chapter, we will discuss interfacing a temperature sensor with the A/D converter module and the conversion of an analog temperature reading into BCD values. The chapter also discusses the basic concepts of a D/A converter and the interfacing of a D/A converter to PIC18 I/O ports.

OBJECTIVES

- Explain the need for the data conversion between analog and digital signals and the basic concepts in data converters.
- Define the terms: resolution, LSB, and MSB, and describe the full-scale output and relationship among these three terms.
- List various methods of designing A/D converters and explain concepts in the successive approximation technique.
- Explain the function of the reference voltages and their relationship to the output of the A/D converter.
- List various control registers of the PIC18 A/D converter module, explain the functions of individual bits, and demonstrate how to initialize a conversion.
- Demonstrate the interfacing of a transducer to the A/D converter module and write instructions to read an analog signal and convert it into its digital value.
- Explain the circuit concepts in designing a D/A converter.
- Demonstrate the interfacing of a D/A converter to PIC18 microcontroller and write instructions to generate waveforms.

12.1 DATA CONVERTERS: BASIC CONCEPTS

Analog signals are continuous, with infinite values in a given range. On the other hand, digital signals have discrete values, such as on and off, 0 and 1, or +5 V, and 0.7 V. Examples of analog systems include a clock face with hands or a voltmeter that gives its readout using a needle. In contrast, a digital clock displays time from one discrete value to the next. The same is true of a digital voltmeter. If the world around us is analog, why translate analog signals into a digital format? An obvious answer is that processors can handle only binary (digital) signals. We do use analog signals in systems such as an audio amplifier, but fundamentally, analog systems have the following limitations:

- Analog signals pick up noise as they are being amplified.
- Analog signals are difficult to store.
- Analog systems are more expensive in relation to digital systems.

In digital systems: (1) noise can be reduced by cutting off the peak values of the discrete signals because noise rides on the top of the signal; (2) binary signals of 0s and 1s can be easily stored in memory; and (3), technology for fabricating digital systems has become so advanced that they can now be produced at low cost. Therefore, analog systems are becoming increasingly digital. Examples include music CDs, digital cameras, and image processing and printing systems. The major limitation of a digital system is how accurately it represents the analog signals after conversion (discussed in the next section).

The critical element in the widespread use of digital systems is the advancement in the conversion process of signals: from analog to digital signals (abbreviated as A/D, ADC, or A to D) and from digital to analog signals (abbreviated as D/A, DAC, or D to A). A simplified block diagram of a typical system is shown in Figure 12-1. The first part uses a transducer, such as microphones or video cameras, which converts a nonelectrical signal into an equivalent analog electrical signal. The second block is an A/D converter that converts the analog signal into equivalent binary bits, which are processed by the processor. The output of the processor is also binary and is converted into an equivalent analog signal by a D/A converter. This signal is converted back into nonelectrical analog signals by another transducer, such as speakers or video monitors. The block diagram in Figure 12-1 could equally well represent systems, such as speech synthesizers or time and temperature systems described in Chapter 1. In our time and temperature system, however, we do not need to convert the output of the processor back to analog signals because it reads temperature, displays the reading, and turns on and off the fan and heater. Our focus here will be on A/D and D/A conversion and their applications.

12.1.1 Analog-to-Digital (A/D, ADC or A-to-D) Conversion

A/D conversion is the process of converting a continuously varying signal, such as voltage or current, into discrete digital quantities that represent the magnitude of the signal compared to a standard or reference voltage. Figure 12-2a shows a simple hypothetical A/D converter circuit with one analog input signal and three digital output lines with eight possible values: 000 to 111. If we calibrate this A/D converter for a 0-to-1 V full-scale input signal, we can divide 1 V voltage into eight digital values with each digital value equal to 1/8 V. If we begin with 0 V for 000, the values of the remaining combinations are from 001 = 1/8 V to 111 = 7/8 V as shown in Figure 12-2b. The graph shows that the value of each binary combination stays constant for a period and then jumps to the next value; this is called the quantization process. From the graph, the following points can be summarized.

1. The maximum value this quantization process reaches is 7/8 V for a 1 V analog signal; there is an inherent error of 1/8 V. This is also equal to the value of the Least Significant Bit (LSB): 001. The resolution of a converter is defined in terms of the number of discrete values it can produce. This number is expressed in the number of bits used for the conversion. It is also defined as $1/2^n$ where n is number of bits.
2. The value of the Most Significant Bit (MSB)—100—is equal to 1/2 the voltage of the full-scale value of 1 V.
3. The value of the largest digital number 111 is equal to the full-scale value minus the value of the LSB (i.e., Full-scale 1 V – 1/8 V LSB = 7/8 V).
4. The quantization error can be reduced or the resolution can be improved by increasing the number of bits used for the conversion.

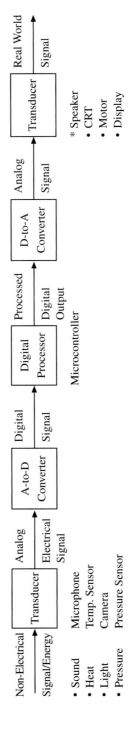

FIGURE 12-1   Embedded Systems: A-to-D and D-to-A Signal Conversion

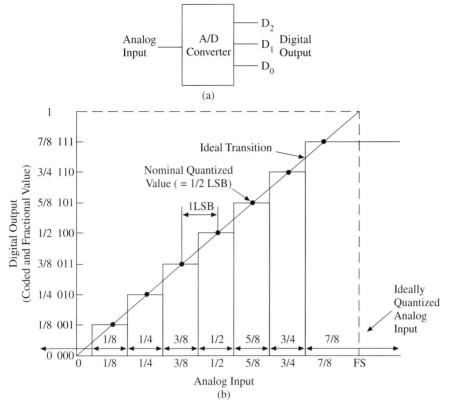

SOURCE: Analog Devices, Inc., *Integrated Circuit Converters, Data Acquisition Systems, and Analog Signal Conditioning Components* (Norwood, Mass.: Author, 1979), p. I–18.

FIGURE 12-2   A 3-Bit A/D Converter: (a) Block Diagram and (b) Analog Input versus Digital Output

---

EXAMPLE 12.1

Given a 0 to 5 V analog input signal and an 8-bit A/D converter, calculate the values of LSB and MSB, resolution, and the full-scale output.

Solution:

An 8-bit A/D converter has 256 (2^8) digital combinations from 00 to FF_H. Therefore,

1. The value of LSB 00000001 = 5 V / 256 = 19.53 mV
2. MSB = 10000000 = ½ of full scale value = 5 V / 2 = 2.5 V
3. Resolution is 8 bits or in terms of voltage = 5 V x (1 / 2^8) = 19.53 mV
4. Full-scale Output = Full-scale Analog Signal – LSB = 5 V – .01953 V ≈ 4.98 V

EXAMPLE 12.2

Calculate the voltage resolution for a 10-bit converter when the full-scale input voltage ranges from -5 V to $+5$ V

Solution:

1. The converter with the 10-bit resolution generates $2^{10} = 1024$ quantization levels
2. Voltage resolution is $(5 - (-5))/1024 = 10/1024 = 9.76$ mV

12.1.2 A/D Conversion Methods

There are various methods used for A/D conversion, and they can be classified primarily in four groups:

- Flash
- Integrator
- Successive Approximation
- Counter

As regards to the concepts used in the conversion, the above ADCs can be grouped in two categories: (1) comparing an analog signal with the internally generated equivalent signal—flash, counter, and successive approximation fall into this category—and (2) changing an analog signal into time or frequency, and comparing these new parameters to known values—integrator falls into this category. We focus here on the successive approximation method because the concepts underlying this method are helpful in using the A/D converter in the PIC18 family, but describe the remaining three briefly.

Flash

This converter uses multiple comparators in parallel as shown in Figure 12-3. A known signal is connected on one side of the comparator and the analog signal to be converted is connected to the other side of the comparator. If the analog signal is higher than the known signal, the output of the comparator goes to one. Otherwise, it remains zero. This is a high-speed, high-cost converter. The major drawback is its cost. To design an 8-bit converter, 255 comparators are needed.

Integrating

This type of A/D converter charges a capacitor for a given amount of time using the analog signal. It discharges back to zero with a known voltage and the counter provides the value of the unknown signal. This is a slow converter, but with a high accuracy and low noise. This converter is used in applications such as digital meters and monitoring systems where high accuracy is required.

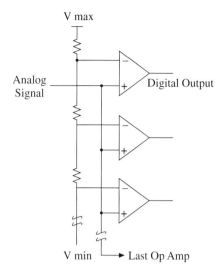

FIGURE 12-3 Flash Converter

Successive Approximation

This converter is more frequently used in industrial applications than any other type of converter. This converter offers an effective compromise among resolution, speed, and cost. It is also found in many microcontroller chips; the PIC18 microcontroller family includes this type of A/D converter.

This converter circuit includes a D/A converter and a comparator. An internal analog signal is generated by turning on successive bits (from MSB to LSB) in the D/A converter. This D/A signal and an analog signal are compared in the comparator until the voltages match (see details in the next section).

Counter

This is similar to the successive approximation converter circuit; it includes a counter, a D/A converter, and a comparator. The primary difference between the successive approximation circuit and this circuit is in how the internal analog signal is generated. In this circuit, the counter starting from zero feeds the signal to the D/A converter instead of turning on successive bits. When a match is found, the binary value in the counter represents the digital equivalent of the analog signal.

12.1.3 Successive Approximation A/D Converter Circuit

Figure 12-4 shows a block diagram of a successive approximation A/D converter circuit. It includes a comparator, a D/A converter, a successive approximation register (SAR), a control register, and an output register. As mentioned before, the technique used compares the analog signal V_{in} to the internally generated signal from the D/A converter until a match is found. Binary input to the D/A converter is provided by the successive approximation method as discussed below and, when a match is found, this binary input represents the equivalent digital value of the analog signal.

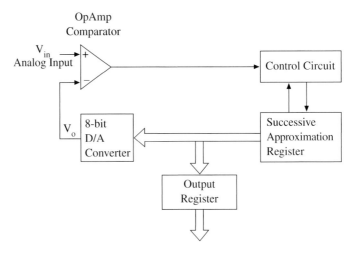

FIGURE 12-4   Block Diagram – Successive Approximation A/D Conversion

The successive approximation method of providing input to the D/A converter is similar to weighing an item less than 1 kg by the weights such as 1/2 kg, 1/4 kg, 1/8 kg, and 1/16 kg. The weighing procedure begins by putting the heaviest weight (1/2 kg) first. If the item is lighter than 1/2 kg, we will remove the 1/2 kg weight and add the next one (1/4 kg). If the item is heavier than 1/2 kg, we will continue to add other weights until the balance is tipped. The weight that tips the balance is removed and lighter weights are subsequently tried until the smallest weight is used. Similarly, the SAR begins with bit B7 as the first input to the DAC. If the comparator output goes low, indicating V_o is $> V_{in}$, bit B7 is turned off; otherwise, B7 is kept "on." Next, bit B6 is turned on. If the comparator indicates that $V_o < V_{in}$, bit B6 is added to the previous bit B7. This process is continued until the last bit B0. Now the 8-bit input to the DAC represents the digital equivalent of the analog signal V_{in}.

The circuit shown in Figure 12-4 can be fabricated on a chip as an A/D converter or implemented as a module inside a microcontroller. To use an A/D converter or the internal A/D module for data conversion, we need reference voltages in order to calibrate the input signal. An A/D converter should provide the necessary pins to connect these reference voltages (this is further explained later using the PIC18 microcontroller). The conversion process should include the steps listed below. The MPU should:

1. access the ADC to start a conversion;
2. wait in a loop until the conversion process (turning on and off successive bits of the DAC) is complete by checking the flag or responding to an interrupt; and
3. read the converted digital reading.

These steps are illustrated in reference to the A/D converter module in the PIC18 microcontroller (see Section 12.3).

12.1.4 Sample and Hold Circuit

If the input voltage to an A/D converter is variable, as with an AC waveform, the digital output is likely to be unreliable and unstable. Therefore, the varying voltage source is connected to the A/D converter through a sample and hold circuit as shown in Figure 12-5. Conceptually, the sample and hold circuit consists of a capacitor that presents high impedance to the input source with unity gain and a switch. When a switch is connected, it samples the input voltage; when the switch is open, it holds the sampled voltage by charging the capacitor. After the switch is open, the time to charge the capacitor and to let the output settle is called the acquisition time. After the completion of the acquisition time, a conversion can begin. During the conversion, each bit is turned on (or turned off if necessary—see Successive Approximation A/D converter), and the conversion time per bit is defined as T_{AD}.

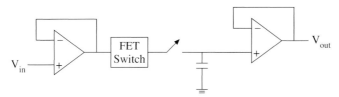

FIGURE 12-5   Sample and Hold Circuit

12.2 PIC18F4520 ANALOG-TO-DIGITAL (A/D) CONVERTER MODULE

The PIC18 microcontroller includes a 10-bit A/D converter with multiple channels. The earlier devices such as 18F452 had only two control registers; however, recent devices have three control registers: ADCON0 to ADCON2. We will focus here on a device that has three control registers: ADCON0, ADCON1, and ADCON2 and outline the differences between recent devices such as 18F4520 and the earlier devices such as 18F452. Figure 12-6 shows a simplified block diagram of a PIC18F4520 ADC module. The following functions must be specified through three control registers.

1. Select a channel: Figure 12-6 shows a 10-bit A/D converter with a multiplexer that can accept analog signals from 13 different channels.
2. Set up the I/O pins for analog signals from ports A, B, and E that are used as inputs for A/D conversion.
3. Set up pins RA2 and RA3 to connect external V_{REF} + and V_{REF} – if specified in the control register ADCON1.
4. Select an oscillator frequency divider through the control register ADCON2.
5. Select of an acquisition time through the control register ADCON2.

To specify the above five parameters, we now need to look at various control registers.

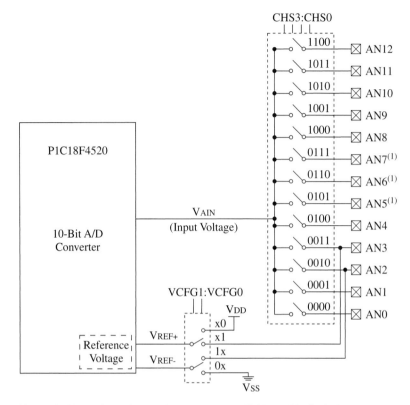

Note   1: Channels AN5 through AN7 are not available on 28-pin devices.
2: I/O pins have diode protection to VDD and VSS.

FIGURE 12-6   Block Diagram – P1C18F4520 A/D Module

12.2.1 A/D Control Register0 (ADCON0)

The primary function of the ADCON0 register is to select a channel for the input analog signal, start the conversion, and indicate the end of the conversion as shown in Figure 12-7. Bit1 is set to start the conversion, and at the end of the conversion this bit is reset.

12.2.2 A to D Control Register1 (ADCON1)

The A/D Control Register1 is primarily used to set up the I/O pins either for analog signal or for digital signals (see Table 12-1) and select V_{REF} voltages.

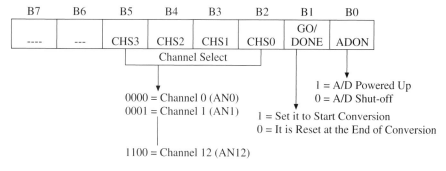

FIGURE 12-7   A/D Control Register0 (ADCON0)

TABLE 12-1   PCFG3 – PCFG0 <3 – 0>: A/D Port Configuration Control Bits

PCFG3: PCFG0	AN12	AN11	AN10	AN9	AN8	AN7[2]	AN6[2]	AN5[2]	AN4	AN3	AN2	AN1	AN0
0000[1]	A	A	A	A	A	A	A	A	A	A	A	A	A
0001	A	A	A	A	A	A	A	A	A	A	A	A	A
0010	A	A	A	A	A	A	A	A	A	A	A	A	A
0011	D	A	A	A	A	A	A	A	A	A	A	A	A
0100	D	D	A	A	A	A	A	A	A	A	A	A	A
0101	D	D	D	A	A	A	A	A	A	A	A	A	A
0110	D	D	D	D	A	A	A	A	A	A	A	A	A
0111[1]	D	D	D	D	D	A	A	A	A	A	A	A	A
1000	D	D	D	D	D	D	A	A	A	A	A	A	A
1001	D	D	D	D	D	D	D	A	A	A	A	A	A
1010	D	D	D	D	D	D	D	D	A	A	A	A	A
1011	D	D	D	D	D	D	D	D	D	A	A	A	A
1100	D	D	D	D	D	D	D	D	D	D	A	A	A
1101	D	D	D	D	D	D	D	D	D	D	D	A	A
1110	D	D	D	D	D	D	D	D	D	D	D	D	A
1111	D	D	D	D	D	D	D	D	D	D	D	D	D

A = Analog input   D = Digital I/O

Note 1: The POR value of the PCFG bits depends on the value of the PBADEN configuration bit. When PBADEN = 1, PCFG<3:0> = 0000; when PBADEN = 0, PCFG<3:0> = 0111.

 2: AN5 through AN7 are available only on 40/44-pin devices.

TABLE 12-2 ADCS2 – ADCS0 <2-0>: A/D Conversion Clock Select Bits

Bits <2–0>		Clock select
111	=	F_{RC} (clock derived from A/D RC oscillator)[1]
110	=	Fosc/64
101	=	Fosc/16
100	=	Fosc/4
011	=	F_{RC} (clock derived from A/D RC oscillator)[1]
010	=	Fosc/32
001	=	Fosc/8
000	=	Fosc/2

Note 1: If the A/D F_{RC} clock source is selected, a delay of one T_{CY} (instruction cycle) is added before the A/D clock starts. This allows the SLEEP instruction to be executed before starting a conversion.

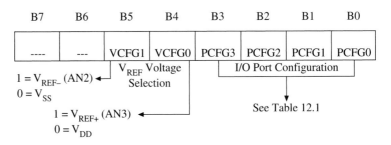

FIGURE 12-8 A/D Control Register1 (ADCON1)

12.2.3 A to D Control Register2 (ADCON2)

This register is used primarily to select acquisition time and clock frequency. In addition, it is also used to right or left justify the output reading. The output reading, after a conversion, is stored in 16-bit register ADRESH and ADRESL. However, this is a 10-bit A/D converter leaving six bit positions unused. Bit7 (ADFM) enables the user either to right justify or left justify the 16-bit reading leaving the unused positions as 0s.

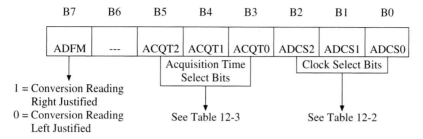

FIGURE 12-9   A/D Control Register2 (ADCON2)

TABLE 12-3   ACQT2 – ACQT0 <5-3>: A/D Acquisition Time Select Bits

Bits <5–3>	;	Acquisition Time
111	=	20 T_{AD}
110	=	16 T_{AD}
101	=	12 T_{AD}
100	=	8 T_{AD}
011	=	6 T_{AD}
010	=	4 T_{AD}
001	=	2 T_{AD}
000	=	0 T_{AD}

EXAMPLE 12.3

A 10 k potentiometer is connected to the PIC18F4520 A/D converter. Channel 4 (AN4/RA5) with the voltage range from + 5 V to 0 V as shown in Figure 12-10, the PIC18 controller has a clock frequency F_{OSC} = 10 MHz. Write instructions to initialize the A/D Converter module, start the conversion, and store the reading in REG10 and REG11. The instructions should meet the following specifications.

1. Set up channel 4 (AN4) to accept an analog signal.
2. Minimum sample time = 15 μs
3. Voltage reference: V_{REF} + = V_{DD} (Power Supply + 5 V) and V_{REF}- = V_{SS} (ground).
4. Conversion reading should be right justified.

continued

Solution:

To initialize the A/D converter module for the specified parameters, we need to select the clock period for the conversion and calculate an appropriate acquisition time. Then we should load these parameters in the three control registers: ADCON0–ADCON2.

1. If we use a frequency select ratio of $F_{OSC}/16$ and the crystal frequency is $F_{OSC} = 10$ MHz:
   Conversion time per bit = T = 1 / (10 MHz /16) = 1.6 µs

2. Specified acquisition time = 15 µs. Therefore, we need at least 12 T_{AD} (12 x 1.6 µs = 19.2 µs)

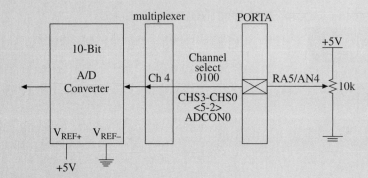

FIGURE 12-10   Interfacing a 10 k Pot

Instructions:

```
        ORG    00
        GOTO   START


        ORG    0x20
START: MOVLW   B'00010001'   ;Select channel AN4 and turn on A/D module
        MOVWF  ADCON0
        MOVLW  B'00001010'   ;Select VD_D and V_SS as reference voltages
        MOVWF  ADCON1        ;and set up RA5/AN4 for analog input
        MOVLW  B'10101101'   ;Conversion reading is right justified, T_AD = 12
        MOVWF  ADCON2        ;and conversion frequency = F_OSC / 16
        BSF    ADCON0,GO     ;Starts conversion
WAIT:  BTFSC   ADCON0, DONE  ;Is conversion complete? If not go to WAIT
        BRA    WAIT          ;If yes, skip this Branch instruction
        MOVFF  ADRESL, REG10 ;Save low-order reading in Register10
        MOVFF  ADRESH, REG11 ;Save high-order reading  in Register11
        END
```

continued

Description:

The first six instructions initialize the bits in the three control registers ADCON0, ADCON1, and ADCON2 to meet the specified parameters. The instruction BSF sets the Bit1—GO/DONE—to start a conversion. The next instruction BTFSC keeps checking Bit1, and the BRA instruction keeps the program in the loop until Bit1 (GO/DONE) is reset, which indicates the completion of the conversion. At the end of the conversion, the program saves the 10-bit conversion reading in registers ADRSL and ADRSH. The last two instructions save the reading in REG10 and REG11.

12.2.4 PIC18F452 A/D Converter Module

The PIC18F4520 is an enhanced version of the earlier device PIC18F452, and both are pin compatible. But programs written for these devices may not be compatible because of the differences between control registers. The differences between these devices are as follows:

1. The 40/44 pin 18F4520 device has thirteen channels with three control registers versus eight channels in the 18F452 with two control registers.
2. The 18F4520 includes additional features such as the selection of reference voltages and acquisition times.

Question 12.12 refers to the PIC18F452 microcontroller and provides additional information regarding its control registers

12.3 ILLUSTRATION: INTERFACING A TEMPERATURE SENSOR TO THE A/D CONVERTER MODULE

A temperature sensor is a transducer that converts temperature into an analog electrical signal. Nowadays, many temperature sensors are available as integrated circuits, and their outputs (voltage or current), in general, are linearly proportional to the temperature. However, the output voltage ranges of these transducers may not be ideally suited to the reference voltages of A/D converters, and may introduce inaccuracies during the conversion. Therefore, it is necessary to scale the output of a transducer to the range of reference voltages of an A/D converter. The scaling may require amplification or shifting of voltages at a different level.

The following example illustrates temperature readings taken with a sensor, shows how to scale the reading for the PIC18 A/D converter module, and discusses converting the reading into ASCII format.

12.3.1 Problem Statement

Interface the National Semiconductor LM34 temperature sensor to channel 0 (AN0) of the A/D converter module as shown in Figure 12-11. Assume that the output voltage of the LM34 for the temperature range from 0°F to 100°F is properly scaled to 0 V to +5 V. Write instructions to start a conversion, read the digital reading at the end of the conversion, and calculate the equivalent temperature reading in degrees Fahrenheit, convert it into BCD, and store the reading in ASCII code to the accuracy of one decimal point. The expected range of temperatures is 0°F to 99.9°F.

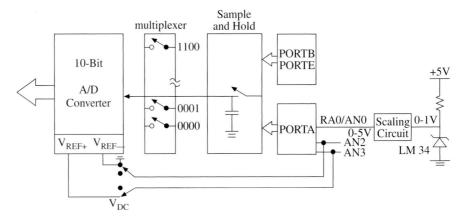

FIGURE 12-11   Interfacing LM34 – Temperature Sensor to PIC18 A/D Conversion Module

12.3.2 Problem Analysis

There are two parts to this problem: hardware and software. In the hardware part we need to look at the characteristics of the LM34 and the A/D converter module of the PIC18F4520 microcontroller. In the previous section, we discussed various features of the A/D converter module. Now we need to examine how to interface the LM34 using the scaling circuit with the A/D converter module.

Temperature Transducer LM34

This transducer is a three-terminal integrated circuit device that can operate in the +5 V to +30 V power supply range (see Appendix I for data sheets). It outputs 10 mV/ °F linearly over the temperature range of –50°F to 300°F with 0 V at 0°F. In this problem, we are interested in the temperature range from 0°F to 99.9°F; therefore, the expected output voltage range of the LM34 is 0 V to 1 V (rounded off to 100°F).

Scaling Circuit

If the output voltage range of the transducer is 0 V to 1 V for the given temperature range, we have two options to get the full dynamic range of the A/D conversion: (1) we can connect $+V_{REF}$ to +1 V, or (2) scale the output voltage +1 V to the voltage of the power supply +5 V as shown in Figure 12-11 (the op amp calculations are shown in Appendix C). This scaling enables us to connect PIC18 power supply V_{DD} as voltage reference $+V_{REF}$ and ground V_{SS} as $-V_{REF}$.

Temperature Calculations

The A/D converter in the PIC18 microcontroller has a 10-bit resolution. Therefore, for the temperature range 0°F to 100°F, the digital output is divided into 1023 steps (0 to $3FF_H$). As such, the digital value per degrees Fahrenheit is 10.23 ($1023/100 = 10.23_{10}$). To obtain the temperature reading from a digital reading of the A/D converter, the digital reading must be divided by the factor of 10.23.

Software Modules

To analyze the software requirements, let us examine the problem statement again. The program is expected to perform the following steps:

1. **Start a conversion and read the digital reading at the end of the conversion:** To start a conversion, the program should initialize the A/D converter module, start the conversion process, and wait until the flag is set, which indicates the end of the conversion. The program, then reads a 10-bit output of the A/D converter module. In Example 12.3, we have written code to meet similar requirements for Channel 4 – AN4. Here we can adapt the code from Example 12.3 and convert it into a subroutine.

2. **Calculate the equivalent temperature reading:** To calculate the temperature, the digital reading should be divided by 10.2 (ignoring the second decimal point) as explained in the previous section. The PIC18 microcontroller does not have a divide instruction. However, this calculation can be performed by multiplying the reading by 10 and dividing the product by 102; this process is known as fixed-point calculation. In this problem, the largest reading can be $3FF_H$. Therefore, we can adapt the subroutine as previously discussed in Section 8.5 – (Multiplication of 16-bit numbers) to multiply the readings by ten and the subroutine in Section 8.6 – (Division of 8-bit unsigned numbers) to divide the product by 102.

3. **Convert the result in BCD:** The temperature reading calculated in Step 2 will be in binary (Hex), and needs to be converted into BCD numbers. We can modify the subroutine in Section 8.2 (Binary to BCD conversion) to meet the requirements of this problem.

4. **Convert the BCD numbers in ASCII code:** We have a written a subroutine in Section 8.4 (Binary to ASCII code conversion). Now, however, we are dealing exclusively with BCD numbers that can be converted by adding 30_H.

12.3.3 PROGRAM

```
;:::::::::::::::::::::::::::::::::::::::::::::::::::::::::::::::::::::::::::
;The following program reads an analog temperature from LM34, a         :
;temperature;transducer,  by using the A/D converter module of the      :
;PIC18F4520.;;The temperature range is from 0°F to 99.9°F.;;The program  :
;converts the  binary ;reading in BCD digits and stores them in the ASCII :
;format.                                                                :
;:::::::::::::::::::::::::::::::::::::::::::::::::::::::::::::::::::::::::::
          Title "Illust12-3 Interfacing Temperature Sensor"
          List p=18F452, f =inhx32
          #include <p18F452.inc>         ;This is a header file

WORD1       SET    0x10                   ;Define registers for multiplication
WORD2       SET    0x12
RESULT      SET    0x14
DIVIDENDLO  EQU    0x20                   ;Define registers for division
DIVIDENDHI  EQU    0x21
QUOTIENT    EQU    0x22
```

continued

```
INTEGER     EQU     0x23
REMAINDR    EQU     0x24
DECIMAL     EQU     0x25
READING     EQU     0x26                    ;Register to get temperature
                                            ;reading for BCD
CFLAG       EQU     0                       ;Define bit position of carry flag

BCD0        EQU     0x30                    ;Define registers to save BCD digits
BCD1        EQU     0x31

DEGREES     SET     0x40                    ;Define registers to save ASCII
                                            ;characters for display


            ORG     00
            GOTO    MAIN

            ORG     0x20
MAIN:
ATOD_SETUP  MOVLW   B'00000001'             ;Select channel AN0 and turn on
                                            ;A/D module
            MOVWF   ADCON0
            MOVLW   B'00001110'             ;Select V_DD and V_SS as reference
                                            ;voltages
            MOVWF   ADCON1                  ;and set up RA0/AN0 for analog input
            MOVLW   B'10101101'             ;Conversion reading is right
                                            ;justified, T_AD = 12
            MOVWF   ADCON1                  ;and conversion frequency = F_osc/ 16

            CALL    CONVERT                 ;Initialize A/D module
            CALL    MULTIPLY10              ;Multiply temperature reading by 10
                                            ;and divide
            CALL    DIVIDE100               ;by 100 to adjust decimal point
            MOVFF   QUOTIENT,INTEGER        ;Save integer and remainder
            MOVFF   DIVIDENDLO,REMAINDR
            MOVFF   REMAINDR, WORD1         ;Adjust remainder to get decimal
                                            ;point by
            CLRF    WORD1+1                 ;multiplying by 10 and dividing
                                            ;by 100
            CALL    MULTIPLY10
            CALL    DIVIDE100
            MOVFF   QUOTIENT,DECIMAL        ;Save decimal point
            MOVF    INTEGER,W               ;Get integer and convert it in  BCD
            CALL    BINBCD

            MOVFF   BCD1, DEGREES           ;Save integer in registers
                                            ;for display
            MOVFF   BCD0, DEGREES+1
            MOVLW   0x2E                    ;Save ASCII character for
                                            ;decimal point
            MOVWF   DEGREES+2
```

continued

```
                MOVF    DECIMAL,W               ;Get decimal digit  and
                                                ;convert in BCD
                CALL    BINBCD
                MOVFF   BCD0, DEGREES+3         ;Save decimal digit in
                                                ;registers for display
                CLRF    DEGREES+4               ;Indicate end of display
                                                ;by null character
HERE            BRA     HERE

CONVERT:        ;:::::::::::::::::::::::::::::::::::::::::::::::::::::::::::::
                ;Function:  This subroutine starts the A/D           :
                ;           conversion and  waits in a loop until the :
                ;           conversion flag (DONE) is cleared. The   :
                ;           A/D module places the 10-bit converted   :
                ;           reading in registers ADRESH (High-byte)  :
                ;           and ADRESL (Low-byte).                    :
                ;Output:    10-bit digital reading equivalent of     :
                ;           the analog temperature in registers      :
                ;           WORD and WORD+1                           :
                ;:::::::::::::::::::::::::::::::::::::::::::::::::::::::::::::

                BSF     ADCON0,GO               ;Starts conversion
WAIT:           BTFSC   ADCON0, DONE            ;Is conversion complete?
                                                ;If not go to WAIT
                BRA     WAIT                    ;If yes, skip this
                                                ;Branch instruction
                MOVFF   ADRESL, WORD1           ;Save low-order reading
                                                ;for BCD conversion
                MOVFF   ADRESH, WORD1+1         ;Save high-order reading
                                                ;for BCD conversion
                RETURN

MULTIPLY10:     ;:::::::::::::::::::::::::::::::::::::::::::::::::::::::::::::
                ;Function:  This subroutine multiplies 16-bit        :
                ;           reading by decimal 10.                    :
                ;Input:     The 16-bit reading in registers          :
                ;           WORD1 and WORD1+1.                        :
                ;Output:    The 16-bit product in                    :
                ;           registers RESULT and RESULT+1.            :
                ;:::::::::::::::::::::::::::::::::::::::::::::::::::::::::::::

                MOVLW   0x0A                    ;8-bit Multiplier - decimal 10
                MOVWF   WORD2
                MOVF    WORD1,W                 ;Multiply low-order temperature
                                                ;reading by 10
                MULWF   WORD2
                MOVFF   PRODL, RESULT           ;Save result in RESULT and
                                                ;RESULT +1
                MOVFF   PRODH, RESULT+1

                MOVF    WORD1+1,W               ;Multiply high-order temperature
                                                ;reading by 10
                MULWF   WORD2
                MOVF    PRODL,W
                ADDWF   RESULT+1,1              ;Add previous product and
                                                ;save in RESULT+1
```

continued

```
              MOVF    PRODH,W
              ADDWFC  RESULT+2,1           ;Add previous product and
                                           ;save in RESULT+2
              RETURN

DIVIDE100:    ;:::::::::::::::::::::::::::::::::::::::::::::::::::::::::::::
              ;Function:    This subroutine divides 16-bit reading by     :
              ;             decimal 102.                                  :
              ;Input:       The 16-bit number in registers RESULT and     :
              ;             RESULT+1.                                      :
              ;Output:      The 8-bit quotient in register QUOTIENT and   :
              ;             the remainder in  DIVIDENDLO.                  :
              ;:::::::::::::::::::::::::::::::::::::::::::::::::::::::::::::

              MOVFF   RESULT, DIVIDENDLO   ;Get result after
                                           ;multiplication by 10
              MOVFF   RESULT+1, DIVIDENDHI
              CLRF    QUOTIENT             ;Register to save integer reading
AGAIN:        MOVLW   D'102'               ;Divisor
              INCF    QUOTIENT
              SUBWF   DIVIDENDLO,1         ;Begin division by subtraction
                                           ;low-order number
              BTFSC   STATUS, CFLAG        ;Is the remainder negative
              GOTO    AGAIN                ;If not, go back and
                                           ;subtract again
              DECF    DIVIDENDHI           ;Now remainder is negative,
                                           ;adjust

              BTFSC   STATUS, CFLAG
              GOTO    AGAIN
              DECF    QUOTIENT
              ADDWF   DIVIDENDLO,1
              RETURN

BINBCD:       ;:::::::::::::::::::::::::::::::::::::::::::::::::::::::::::::
              ;Function:    This subroutine converts an 8-bit reading     :
              ;             into two decimal BCD digits (0 to 9).         :
              ;Input:       An 8-bit number in W register.                :
              ;Output:      Two unpacked BCD digits in registers          :
              ;             BCD0 (low) and BCD1 (high).                    :
              :Calls another subroutine DIVIDE.                           :
              ;:::::::::::::::::::::::::::::::::::::::::::::::::::::::::::::

              CLRF    BCD0                 ;Clear registers to save BCD
                                           ;digits
              CLRF    BCD1

              MOVWF   READING              ;Get temperature reading
              LFSR    FSR0, BCD1           ;Pointer to BCD1
              MOVLW   D'10'                ;Divisor 10
```

continued

```
            RCALL    DIVIDE                    ;Find BCD1
            LFSR     FSR0, BCD0                ;Pointer to BUFF0 to save BCD0
            MOVFF    READING, BCD0             ;Remainder = BCD0
            MOVLW    0x30                      ;ASCII conversion
            ADDWF    BCD1,1                    ;Save ASCII readings in BCD1
            ADDWF    BCD0,1                    ;and BCD0
            RETURN

DIVIDE:     ;:::::::::::::::::::::::::::::::::::::::::::::::::::::::::::::::::
            ;Function: This subroutine divides an 8-bit number by 10     :
            ;Input:    An 8-bit number in register READING and a         :
            ;          pointer to BCD1                                   :
            ;Output:   Two unpacked BCD digits in registers BCD0 (low)   :
            ;          and BCD1 (high)                                   :
            ;:::::::::::::::::::::::::::::::::::::::::::::::::::::::::::::::::
            INCF     INDF0                     ;Begin counting number of
                                               ;subtractions
            SUBWF    READING, 1                ;Subtract power of ten from
                                               ;binary reading
            BTFSC    STATUS, CFLAG             ;If C flag =0, skip next
                                               ;instruction
            GOTO     DIVIDE                    ;If  C flag  =1, go back to next
                                               ;subtraction
            ADDWF    READING,1                 ;Adjust the temperature by
                                               ;adding divisor
            DECF     INDF0                     ;Adjust the number of subtraction
            RETURN
            END
```

12.3.4 Program Description

This program follows the same four steps described in Section 12.3.2. It consists primarily of calling various subroutines and supplying the necessary inputs to each subroutine. At the beginning, the MAIN program defines the registers that are used in various subroutines. The addresses of the registers are selected somewhat arbitrarily in this problem; our main concern is finding an easy-to-follow order of the file (data) registers during a simulation. Each row of the file registers in the View window begins with Hex addresses such as 10_H, 20_H, and 30_H; therefore, it is easy to observe the contents of various registers if their addresses begin at these locations. For example, in this problem, you can observe the final results in ASCII characters in the fifth row starting from the first column of address 40_H.

The MAIN program sets up all the parameters of the A/D converter module and calls the CONVERT subroutine, which starts the conversion process and waits until the end of the conversion. When the conversion is complete, the A/D module reads the result of the conversion and stores it in registers ADRESH and ADRESL, which is copied in registers WORD and WORD+1. The next subroutine multiplies the converted reading by ten, followed by the subroutine that divides the result by 102_{10} to find the equivalent temperature value. The integer value of the temperature is a quotient, saved in register INTEGER, and the remainder is saved in register REMAINDR, which is a decimal value. The program repeats the same process for the remainder and saves the value in register DECIMAL.

The next step is to convert the binary reading into BCD digits and find the ASCII characters. The problem states that the maximum temperature to be displayed in °F is 99.9_{10}. Therefore, the integer value will have two BCD digits and the decimal value will have only one BCD digit. The MAIN program loads the integer value (INTEGER) in the WREG and calls the BINBCD subroutine, which converts the integer in two BCD digits: BCD1 and BCD0. The range of the BCD digits is 0 to 9 and, therefore, they are converted into ASCII characters by just adding 30_H. The BCDBIN subroutine returns two ASCII characters to the main program, which are stored in registers DEGREES and DEGREES+1. To display the decimal point the program loads the byte $2E_H$, which is the ASCII value for the decimal point, and saves the byte in register DEGREES+2. For the decimal digit, the MAIN program loads the value of the decimal (DECIMAL) in the WREG and calls the BINBCD subroutine again, which returns two ASCII characters in BCD1 and BCD. The main program saves only BCD0 because the remainder is converted in one digit. Therefore, BCD1 will be always zero. The last instruction clears the register DEGREES+4 that indicates the null character. The main program stores a string of ASCII characters starting from the register DEGREES and terminates the string in the null character. Now an output device such as an LCD can display the string (a display is not a part of this program).

12.3.5 Program Execution and Troubleshooting

This program includes one main program and five subroutines. This program can be executed and verified one subroutine at a time by setting breakpoints at each subroutine in MPLAB IDE. You should assemble the program and add various registers to watch in the View window as shown in Figure 12-12 and perform the following steps:

1. Without any hardware, you will have to add ADCON register in Watch windows and clear the flag in the ADCON register and load a known binary value (less than $03FF_H$) in registers ARDESH and ARDESL to simulate a reading from the A/D converter module. Figure 12-12 shows that the temperature reading of $03FA_H$ is manually loaded in ARDESH and ARDESL. Set the breakpoint at CALL MULTIPLY10 and run the program, which results in the execution of the subroutine CONVERT. You should observe $03FA_H$ in registers 11_H and 10_H (WORD+1 and WORD). Continue to set breakpoints at each call instruction and observe the results.

2. The MULTIPLY10 multiplies $03FA_H$ by 10 and displays $27C4_H$ in registers 14_H and 15_H.

3. The next subroutine DIVIDE100 divides the reading by 10^2. Now you should single-step the next two MOVFF instructions. You should observe the contents of registers QUOTIENT = 63_H, (also copied in register INTEGER) and REMAINDR = 52_H and verify these values by independent calculation, using the calculator in the Accessories program of the PC. The byte 63_H in the QUOTIENT register is equal to 99_{10} in decimal and represents the temperature in whole number. The byte 52_H in the REMAINDR register is equal to 82_{10} and represents the fraction or the decimal digit after 99. It will be converted into 0.8 as explained in the next two steps.

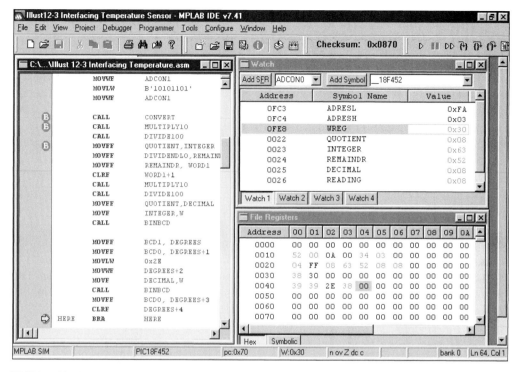

FIGURE 12-12   Program Execution – Interfacing a Temperature Sensor to A/D Converter

4. The same process (Steps 2 and 3) is repeated by the next two subroutines for the byte in the REMAINDR(52_H). This byte represents the decimal value saved in register DECIMAL (8_H).

5. Now the temperature reading is in Hex in two registers: INTEGER (52_H) and DECIMAL(8_H). Set the final breakpoint at the last instruction BRA HERE and run the program. The CALL BINBCD converts these readings in BCD digits 99_{BCD} and 8_{BCD} and stores them as ASCII characters in the register DEGREES in a sequence that also includes the ASCII decimal point ($2E_H$). You should observe the following ASCII strings starting from register 40_H: 39_H, 39_H, $2E_H$, 38_H, and 00_H

6. Troubleshooting: You should check the output of each subroutine, and when you do not observe the expected output, you should single-step that subroutine. You should also run the program with other values of the temperature readings.

7. Errors: The conversion process of an analog signal to digital signal has an inherent conversion error. The approximation in the division process will contribute an additional error. In this example, the expected reading is 99.5°F, but our result is 99.8°F. This error is caused because the divider 10.23 is rounded off to 10.2 (i.e., the loss of one decimal place). If you enter the temperature reading $03FF_H$, the conversion will be out of range, and this program does not check for an out-of-range temperature reading.

12.4 DIGITAL TO ANALOG (D/A, DAC, OR D-TO-A) CONVERSION

The D/A conversion is exactly the opposite process of the A/D conversion. The D/A conversion is the process of converting discrete signals, such as binary code representing a quantity, into voltage or current into discrete analog values that represent the magnitude of the input signal compared to a standard or reference voltage. The output of the DAC is discrete analog steps. However, by increasing the resolution (number of bits), the step size is reduced, and the output approximates a continuous analog signal. For example, music on MP3 players (and CDs) is stored in digital form—binary voltages of 0s and 1s—and DACs in MP players convert digital signals into analog signals to drive the headphones. To hear music for one second, approximately 44,000 binary voltages are converted into an analog signal. Another example of the D/A conversion is the display on the monitor of a computer where video signals in a computer are converted into analog form to display them on an analog monitor.

Figure 12-13a shows a simple hypothetical D/A converter circuit with three digital input lines with eight possible values: 000 to 111 and one output signal. The resolution of a DAC is defined in terms of bits—the same way as in ADC discussed in the previous section. Similarly, the values of LSB, MSB, and full-scale voltages are calculated the same way as in the ADC section. Figure 12-13b shows that the largest input signal 111 is equivalent of 7/8 of the full-scale analog value.

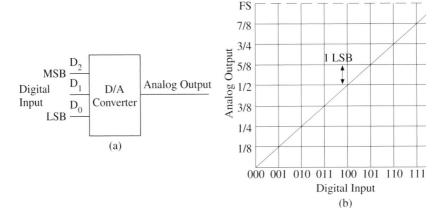

FIGURE 12-13 (a)   3-bit D/A Converter
FIGURE 12-13 (b)   Digital Input and Analog Output

12.4.1 D/A Converter Circuits

A D/A converter circuit can be designed using an operational amplifier and appropriate combination of resistors as shown in Figure 12-14a. The resistors connected to data bits are in binary weighted proportion, and each is twice the value of the previous one. Each input signal can be connected to the op amp by turning on its switch to the reference voltage that represents logic 1: if the switch is off, the input signal is logic 0. If the reference voltage is 1 V, and if all switches are connected, the output current can be calculated as follows:

$$I_O = I_T = I_1 + I_2 + I_3 = \frac{V_{REF}}{R_1} + \frac{V_{REF}}{R_2} + \frac{V_{REF}}{R_3} = \frac{V_{REF}}{1\ k}\left(\frac{1}{2} + \frac{1}{4} + \frac{1}{8}\right) = 0.875\ mA$$

The output voltage $V_O = -R_f I_T = -(1\ k) \times (0.875\ mA) = -0.875\ V = |\ 7/8\ V\ |$

These calculations match the value of the digital reading 111_2 in Figure 12-13b. Now we can extend this formula to any number of bits as follows:

$$I_O = \frac{V_{REF}}{R_{REF}}\left(\frac{A_1}{2} + \frac{A_2}{4} + \frac{A_3}{8} + \dots + \frac{A_n}{2^n}\right)$$

where R_{REF} is a reference resistor and A_1 to A_n can be either 0 or 1, A_1 is the MSB and A_n is the LSB. However, in a microprocessor, bit B7 (or D7) is the MSB, and in the above formula A1/2 will be replaced by B7/2; this is further explained in Example 12.4

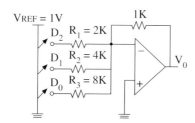

FIGURE 12-14 (a)   3-Bit D/A Converter Circuit

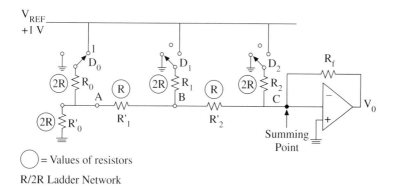

= Values of resistors

R/2R Ladder Network

FIGURE 12-14 (b)   R/2R Ladder Network for D/A Converter

12.4.2 D/A Converters as Integrated Circuits

We can build a D/A converter using the circuit shown in Figure 12-14a. However, it is too expensive to include precision resistors of different values for high-resolution D/A converters. The commercially available D/A converters on a chip use the same concept but use the R/2R ladder network. This ladder network uses resistors with two values, R and 2R, and provides resistance values in a binary weighted proportion as shown in Figure 12-14b.

D/A converters are available commercially as integrated circuits and can be classified in three categories: current output, voltage output, and multiplying type. The current output DAC provides I_O (discussed in the previous section) as the output signal. The voltage output D/A converts I_O into voltage internally by using an op amp and provides voltage as the output signal. In the multiplying DAC, output is the product of the input voltage and the reference source V_{REF}. Conceptually, there is not much difference between these three types.

The converter in Figure 12-15 is a microprocessor-compatible 8-bit converter, and has two control signals: $\overline{WR}$ (Write) and $\overline{CS}$ (Chip Select) in addition to 8-input data lines. In interfacing this converter with a microcontroller, no additional circuitry is required. To convert a binary code, the processor needs to place the code on the input data lines, access the converter by sending a low signal to $\overline{CS}$, and write the code (or input the code) by sending a low signal to the $\overline{WR}$ signal.

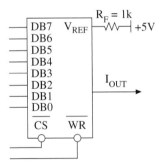

FIGURE 12-15   Microprocessor Compatible D/A Converter – Current Output

EXAMPLE 12.4

In Figure 12-15, calculate the full-scale reference current and the output current if the digital input is 1000 0001.

Solution:

   1.   Reference current = V_{REF} / R_{REF} = 5 V / 1 k = 5 mA

continued

2. To calculate the output current, we need to change the notation used earlier to minimize confusion. In the earlier formula, we used A_1 as the most significant bit and A_n was the least significant bit. In an 8-bit processor system, Bit7 is the most significant bit, and we labeled it as DB7 to follow the notation used for the data converter. Therefore, the output current I for digital input 1001 0001 is calculated as follows:

$$I_O = \frac{V_{REF}}{R_{REF}}\left(\frac{DB7}{2} + \frac{DB6}{4} + \frac{DB5}{8} + \ldots\ldots + \frac{DB0}{2^8}\right)$$

$$I_O = 5\text{ mA}\left(\frac{1}{2} + \frac{0}{4} + \frac{0}{8} + \frac{1}{16} + \frac{0}{32} + \frac{0}{64} + \frac{0}{128} + \frac{1}{256}\right) = 2.832\text{ mA}$$

12.5 ILLUSTRATION: GENERATING A RAMP WAVEFORM USING A D/A CONVERTER

A D/A converter is generally not designed as a peripheral resource on a microcontroller chip. The PIC18 microcontroller does not have a D/A converter on its chip. Therefore, we need to interface an external D/A converter to convert a digital signal into an analog signal. Typical applications of a D/A converter include digital waveform generation (ramp, triangle, and pulse), digital audio and video, and robotics. The following illustration demonstrates the generation of a ramp waveform using two I/O ports of the PIC18 controller.

12.5.1 Problem Statement

Write a subroutine to generate a ramp waveform at the output of the D/A converter AD558 (from Analog Devices, a manufacturer of data converters) shown in Figure 12-16. It is a voltage-output 8-bit DAC. The slope of the ramp should be variable based on the delay count provided by the caller.

12.5.2 Problem Analysis

The AD558 DAC requires ten signals from the microcontroller—8 bits for data input and 2 bits for the control signals. We can use 8 bits of PORTC for data input and the 2 bits RE0 and RE1 of PORTE for control signals. All these bits should be initialized as outputs.

To generate a ramp, the output should begin with zero and the waveform should be increased incrementally. This can be done by setting a register as an up-counter starting from zero and outputting that count. The output will begin with zero volts. As the counter is incremented and each count is outputted, the output waveform will increase in magnitude in a straight line with a slope depending upon the delay between counts. When the counter reaches the maximum value FF_H, it will be reset to 00 and start again. Therefore, the output will reach the maximum

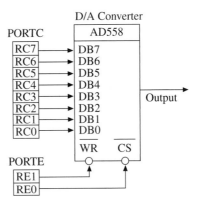

FIGURE 12-16   Interfacing a D/A Converter

value and start again from zero, generating a ramp waveform. The instructions are described in the following section.

12.5.3 PROGRAM

```
;:::::::::::::::::::::::::::::::::::::::::::::::::::::::::::::::::::::::::::::
;The following program generates ramp (saw tooth) waveform by outputting:
;the count to D/A Converter. It uses PORTC to output the count to the    :
;D/A and  two bits RE0 and RE1 to access and write into D/A.            :
;The slope of the waveform can be adjusted by changing the delay count. :
;:::::::::::::::::::::::::::::::::::::::::::::::::::::::::::::::::::::::::::::
        Title "Illust12-5 Generating Ramp Waveform Using DAC"
        List p=18F4520, f =inhx32
        #include <p18F452.inc>              ;This is a header file
REG10   EQU     0x10                        ;Delay register
COUNT   EQU     D'100'                      ;Delay count


        ORG     0x00
        GOTO    START


        ORG     0x20
START   CLRF    TRISC                       ;Initialize PORTC as output
        MOVLW   B'11111100'                 ;Set up RE0 and RE1 as output
        ANDWF   TRISE                       ;for control
        BCF     PORTE,  RE0                 ;Access  DAC
        BCF     PORTE,  RE1                 ;Write in  DAC and place it in
                                            ;transparent mode
        CLRF    PORTC                       ;Start with outputting 0 V to DAC
CONTINUE:INCF   PORTC                       ;Next output
        CALL    DELAY                       ;Wait for appropriate slope
        BRA     CONTINUE                    ;Continue
```

continued

```
;Function: This routine provides delay based on the count supplied to REG10

DELAY:    MOVLW    COUNT                    ;Load delay count in REG10
          MOVWF    REG10, 0
NEXT:     DECF     REG10,1                  ;Decrement count
          BNZ      NEXT                     ;If count is not zero, go
                                            ;back to decrement
          RETURN                            ;Go back to the caller
          END
```

12.5.4 Program Description

The CLRF TRISC instruction sets up PORTC as an output port by sending 0s to TRISC. The next instruction loads a byte with bits Bit0 and Bit1 as zeroes in WREG and logically ANDs with TRISE register. By ANDing the byte 11111100 with the TRISE register, the instruction sets only two bits as output bits without affecting the previous setting of other bits. The two BCF instructions reset Bit0 and Bit1 in PORTE, RE0 selects the DAC, and RE1 writes the input byte from PORTC in to the DAC. When both of these signals are low, the DAC is in the transparent mode, meaning any changes at the input will be sent to the output.

The CLRF PORTC instruction sends out zero to the DAC and the output of the DAC begins at zero voltage. The next instruction increments the count, and resends the count to PORTC. The subsequent counts are sent out after a delay generated by the DELAY subroutine. As the count increases, the output voltage of the DAC increases incrementally, generating a straight-line signal with a slope based on the delay. When the count reaches FF_H, the voltage output reaches its maximum value, and drops down to zero because the counter register rolls over from FF_H to 0.

SUMMARY

The real-world quantities such as temperature, pressure, sound, and light exist in the form of non-electrical energy. These quantities must be converted into electrical signals if they are to be monitored, processed, or used for control applications. The devices used for such energy conversion are called transducers. They convert these quantities into analog electrical signals, which must in turn be converted into digital signals. This chapter discussed the processes of data conversion and its applications in embedded systems using the PIC18F4520 microcontroller. The important concepts are as follows:

1. The analog signal is continuous in nature with infinite possibilities; the digital signal is discrete: logic 0 or logic 1.

2. The analog signal to be processed by the microcontroller must be converted into a digital signal, and the device that converts an analog signal into a digital signal is called an analog-to-digital, or A/D, converter.

3. To drive the real-world devices such as earphones or motors, the digital signal must be converted into an analog signal, and the device that converts a digital signal into an analog signal is called a digital-to-analog, or D/A, converter.

4. The conversion from an analog signal to a digital signal has an inherent quantization error equal to $1/2^n$, where n is the number of bits used in conversion. Increasing the number of bits for the conversion reduces this error. The term resolution is defined in terms of number of bits for conversion.

5. There are various methods of A/D conversion, but the successive approximation method is widely used. The PIC18 microcontroller family includes an A/D converter module that uses the successive approximation method.

6. The modules of recent PIC18 family devices have three control registers (ADCON0–ADCON2) that are used to select a channel and specify the following parameters: (1) reference voltages, (2) analog or digital nature of I/O ports used for conversion, (3) right or left justification for results, (4) acquisition time, and (5) the clock source. The earlier PIC18 family members such as the PIC18F452 have only two control signals.

7. When a temperature sensor (or a transducer) is connected as an input source to a PIC18 A/D converter channel, the output signal of the transducer may have to be amplified or scaled to make full use of the range of the reference voltages.

8. The PIC18 family of controllers does not include a D/A converter module as a peripheral device. Therefore, to interface a D/A converter, available I/O ports must be used.

QUESTIONS AND ASSIGNMENTS

12.1   Explain the functions of a transducer, A/D converter, and D/A converter.

12.2   Calculate the resolution of a 16-bit A/D converter.

12.3   Give an analog signal from 0 V to +10 V and a 12-bit A/D converter, and calculate the values of LSB, MSB, and the full-scale output voltage.

12.4   In Question 12.3, if the voltage range is –5 V to +5 V, calculate the values of MSB, the voltage resolution, and full-scale output.

12.5   If the reference voltages in the PIC18 A/D module are $V_{REF-} = 0$ V and $V_{REF+} = +5$ V, calculate the output voltages for the following digital values: 80_H, 200_H, and $03FA_H$.

12.6   In Question 12.5, explain the significance of the output voltage for 200_H.

12.7   If the reference voltages in the PIC18 A/D module are $V_{REF-} = -1$ V and $V_{REF+} = +5$ V, calculate the value of the voltage per bit.

12.8   In question 12.7, calculate the output voltages for the following digital values: 55_H, AA_H, and $03F0_H$.

12.9   Write initialization instructions to set up the PIC184520 A/D converter module to read two channels: AN0 and AN1, alternately, and meet the following requirements: (1) minimum sample time = 16μs if F = 20 MHz, (2) V_{DD} and V_{SS} are used as reference voltages, (3) results are right justified, and (4) frequency select ratio is $F_{OSC}/16$.

12.10   In Question 12.9, read the channels 16 times and find the average.

12.11   In Question 12.10, use the interrupts generated by the A/D converter module to read the channels 16 times instead of waiting in the loop.

12.12   Write initialization instructions to set up the PIC18452 A/D converter module to read channel AN0 and meet the following requirements: (1) minimum sample time = 16μs if F = 10 MHz, (2) V_{DD} and V_{SS} are used as reference voltages, (3) results are right justified, and (4) frequency select ratio is F_{OSC} / 16 (Refer to Figures 12.17 and 12.18 for the bit definitions of the control registers ADCON0 and ADCON1).

12.13 Set up Timer1 and the CCP2 module in the special event trigger mode, and read the temperature reading every second.

12.14 Write instructions to generate a 1-kHz square wave using the DAC circuit shown in Figure 12-16 if F_{OSC} = 10 MHz.

12.15 Write instructions to generate a triangular waveform using the DAC circuit shown in Figure 12-16 if F_{OSC} = 20 MHz. Calculate the frequency of this waveform.

SIMULATION EXERCISES

Using MPLAB IDE or PIC18 Simulator IDE

The following questions refer to Section 12.3 Illustration: Interfacing Temperature Sensor. Modify the program as required.

12.1 SE Assemble and execute the program for the output reading $03FA_H$ as described in Section 12.3.6, as well as for the reading 0000_H.

12.2 SE Execute the program for the output reading $03FF_H$, and explain the result.

12.3 SE Add instructions to meet the following requirement: when the temperature reading is beyond 100°F, store the ASCII code for "ERROR" in the same registers where the final reading was stored.

12.4 SE Display the output temperature at the LCD described in Chapter 9, Section 9.6.

12.5 SE In question 12.1SE, display the temperature as follows: add the word "TEMPERATURE" on line1 and display the reading on line 2 followed by the words "Degree F".

12.6 SE Select I/O ports to interface three common-anode seven-segment LEDs and display the output temperature.

CHAPTER 13

SERIAL I/O

OVERVIEW

The 8-bit microcontroller processes data in groups of 8 bits and transfers data to peripherals using the entire data bus. This is called parallel I/O or parallel communication. However, it is not always possible to use the entire data bus to transfer data to and from a peripheral or over long distance communications lines. Typical examples include: (1) data transferred to a serial tape or to the monitor of a computer, and (2) communication over telephone lines or via a fax machine. In this type of communication, generally two lines—one to receive and one to transmit—are required for data transfer and only one line is needed when the communication is one way at a time. This is called the serial communication whereby only one bit is transferred at a time over one line.

To transfer data serially, one bit at a time, the microcontroller must convert its parallel word into a stream of serial bits. This is called parallel-to-serial conversion. Similarly, to receive data serially, the microcontroller must receive one bit at a time and convert a stream of serial bits into a parallel word. This is called serial-to-parallel conversion. Over the years, many protocols have been developed that are suitable for given applications and accepted as standards. These are as follows: (1) EIA-232 (formerly known as RS-232) used primarily in PCs, (2) Serial Peripheral Interface (SPI): a synchronous serial interfacing protocol developed by Motorola for communication between a microcontroller and peripheral devices, and 3) Inter-Integrated Circuit (I^2C): a protocol developed by Philips for synchronous serial communication between a microcontroller and peripheral devices.

In addition to above three protocols, there are other protocols developed either for higher speed or specific applications. Some examples are as follows: (1) RS485 (EIA-485) is an extension of EIA-232, but transfers data with higher speed (100 kb/s to 10 Mb/s), (2) Universal Serial Bus (USB) has become very popular in connecting peripherals in PC systems, (3) Controller Area Network bus (CAN) and Local Interconnect Network (LIN) are designed for automotive applications in which the amount of information transferred is relatively small, (4) one-Wire bus protocol was developed by two companies—Maxim and Dallas (now one company)—in collaboration to control network systems.

In this chapter, we will focus on the first three protocols: EIA-232, SPI, and I^2C. The PIC18 family of microcontrollers includes two modules: the Addressable Universal Synchronous Asynchronous Receiver Transmitter (USART) module, which handles the EIA-232 protocol, and the

Master Synchronous Serial Port (MSSP) module, which implements serial communication in SPI and I^{2}C modes. The chapter also includes software modules that demonstrate protocols and examples that illustrate how to interface EEPROMs with PIC18 modules.

OBJECTIVES

- Explain the differences between serial I/O and parallel I/O. List advantages of serial I/O over parallel I/O.
- Explain differences between synchronous and asynchronous transmission. Define or explain the terms with examples: simplex, half-duplex and full duplex transmission, baud, check-sum, and parity check.
- Explain the EIA232 (formerly RS-232C) serial data communication protocol.
- Sketch a waveform of an 8-bit ASCII character that is transmitted in the asynchronous mode using the EIA-232 protocol.
- Identify the control registers and the functions of their specific bits to set up USARTs in the PIC18 microcontroller to receive and transmit data.
- Explain the Serial Peripheral Interface (SPI) and its signals and applications.
- Identify the control registers and the functions of their bits to implement the SPIprotocol. Write instructions to set up the SPI mode to transmit data from the PIC18 microcontroller to a serial device.
- Explain the I^{2}C protocol, and list its signals and applications.
- Demonstrate the interfacing of serial EEPROMs with the PIC18 MSSP module in the SPI and I^{2}C modes.

13.1 BASIC CONCEPTS IN SERIAL COMMUNICATION

In serial communication, we need to convert a parallel word into a stream of bits to send the data out over one line. The device that initiates the data transfer is called a transmitter, and the device that receives the data is called a receiver. This transmission process raises various issues: (1) how to inform the receiver of the Start and the End (Stop) of a transmission, (2) the type of transmission: one way or two ways, (3) simultaneous transmission (a telephone conversation) versus one at a time (a CB radio conversation), (4) the speed of transmission, (5) errors in transmission, and (6) protocols or standards in transmission. These questions are discussed in the following sections.

13.1.1 Synchronous versus Asynchronous Serial Communication

The serial communication between a receiver and a transmitter is classified primarily into two formats: synchronous and asynchronous. In the synchronous format, the receiver and the transmitter have the same clock. Figure 13-1a shows a typical synchronous transmission format that begins with sync characters followed by data. Until the 1980s, the synchronous format was used primarily in high-speed transmission. Since the advent of the microcontroller, the synchronous format is commonly used for the serial communication between the microcontroller and its peripherals.

The asynchronous format is shown in Figure 13-1b, and its typical application includes the communication between a PC and a modem at an agreed upon clock rate. As the term asynchronous suggests, the communication can happen any time. Therefore, the transmitter must let the receiver know when it begins to transmit and when it ends. Figure 13-1b shows that the "Start" and "Stop" bits are included as a part of the transmission. The transmitter begins by sending a logic low signal as a Start bit and terminates the transmission by sending a high signal as a Stop bit.

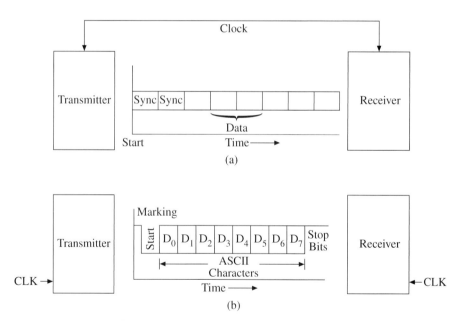

FIGURE 13-1 (a)   Synchronous Data Transfer

FIGURE 13-1 (b)   Asynchronous Data Transfer

13.1.2 Simplex versus Duplex Transmission

Serial communication is also classified as Simplex, Full Duplex, and Half Duplex transmissions. These terms refer to the direction and simultaneity of data flow.

Simplex: In this transmission the data flow in only one direction, such as from a PC to its peripheral, in the serial format.

Full Duplex: In this transmission, the data flow in both directions simultaneously. A telephone conversation or communication via a modem can be viewed as a full duplex transmission.

Half Duplex: In this transmission, the data flow in both directions, but only one direction at a time. A conversation over a CB radio is viewed as half duplex.

13.1.3 Rate of Transmission

In synchronous communication, both the receiver and the transmitter have the same clock, but in asynchronous communication, the transmission rate should match the reception rate. The rate of transmission is defined either in terms of bits per second or the unit called "baud"; sometimes they are equal, as explained in the next paragraph.

Baud and Bits Per Second (BPS)

In telecommunications and electronics, baud is a measure of the "signaling rate," which is the number of changes to the transmission medium per second in a modulated signal. It is named after Émile Baudot, the inventor of the Baudot code for telegraphy.

At slow speeds, only one bit of information is encoded in each electrical change. In this case, the baud and the bits per second are equal. At higher speeds, multiple bits are encoded in one electrical change. Therefore, data transmission rates are generally expressed in bits per second (bps). As an analogy, the baud is similar to counting vehicles passed through a toll gate and the bps is similar to the number of passengers passed through the same toll gate. If there is only one passenger per car, the number of cars and the passengers is the same.

13.1.4 Transmission Errors and Error Checks

During transmission, several errors can occur for various reasons, such as noise in the transmission, change in frequency, or limited bandwidth. Some errors can be checked through hardware configuration and some need to be checked through software. Errors that are commonly checked are as follows:

1. **Framing Error.** This error occurs when Start and Stop bits improperly frame a character, and the error is recognized when the Stop bit is zero which is expected to be one. A serial I/O chip, generally, provides a bit or a flag in its control register to check for this error.
2. **Overrun Error.** This error occurs when a new byte overwrites the earlier byte before the receiver completes the reading of the earlier byte. This error can also be verified in the control register of a serial I/O chip.
3. **Parity Error.** In some transmissions, the last bit is used as a parity check. This technique is based on the principle that the characters (data) are transmitted as either an even or odd number of 1s. The ASCII is a 7-bit alphanumeric code; the last bit, B7, of a byte is always zero, and in some systems, the last bit is used for parity information. For example, the ASCII letter "Y"— 0101 1001 (59_H)—has four 1s, which is even. If the transmission is set up with even parity, the letter "Y" is transmitted as 59_H. However, if the transmission is set up with odd parity, we need to send an odd number of 1s. Therefore, bit B7 is changed to 1 and the byte is sent as 1101 1001 ($D9_H$). The receiver

on the other side checks for the number of 1s: if the number is odd, it accepts the character by eliminating bit B7, and if the number of bits is even, it generates an error message.

4. **CheckSum.** The checksum method is used when blocks of data are transmitted. For example, when we download a Hex code file from a PC to a microcontroller trainer in our laboratory, the checksum method is generally used to check the accuracy of the transmission. In this method, all bytes to be transmitted are added (without a carry), and the 2's complement of the sum is calculated, which is sent as the last byte, called checksum. The receiver program adds all the bytes including the checksum (2's complement), and if the sum is zero, the receiver accepts the data. Otherwise, the receiver program generates an error message.

13.1.5 Standards and Protocols in Serial Data Transfer

Equipment needed for serial data transfer are designed and manufactured by various companies. Therefore, a common understanding must exist that guarantees compatibility among different types of equipment. For example, there is a common understanding among various manufacturers for the connections between a PC and its peripherals such as a monitor, printer, and scanner. This common understanding or guideline is known as a standard. If an operational procedure is defined, it is known as protocol. In many situations, these standards and protocols are defined by professional organizations, and in some situations, a company with a dominant product can establish these protocols and standards of compatibility. A standard may include various aspects such as assignments of pin connections, voltage levels, speed of transmission, format of transmission, and mechanical specifications. A protocol may include a certain sequence of steps or a procedure to follow. We will list here various standards and protocols commonly used in serial communication and discuss them later in the context of PIC18 serial communication modules.

1. **EIA-232 (formerly known as RS-232):** This protocol was developed in the 1960s and used primarily for data communication between a mainframe computer and a remote terminal. Now it is used for various types communication such as that between PCs, a PC and a scanner, a PC and telephone lines (over modems), and the microcontroller and the PC. The original specification includes twenty-five signals, and in PCs, generally, seven or nine signals are used. This protocol is discussed in Section 13.2.

2. **Serial Peripheral Interface (SPI):** This is a four-wire synchronous serial interfacing protocol (explained in Section 13.3) developed by Motorola for the communication between a microcontroller and its peripheral devices. This is a very simple protocol and is widely used in various microcontrollers. The microcontroller functions as a master and communicates with peripherals as slaves. This protocol can also be used for serial communication between two microcontrollers, one being a master and the other being a slave. This is a full-duplex protocol, and data bits are simultaneously transmitted and received. The MICROWIRE is another protocol, similar to SPI, developed by National Semiconductor. It is essentially a subset of the SPI.

3. **Inter-Integrated Circuit (I^2C):** This is a two-wire interface protocol (explained in Section 13.4) developed by Philips, used for synchronous serial communication between a microcontroller and integrated devices on a printed circuit board. Now it is also used

for communication between a microcontroller and its peripherals. The major advantage of this interface is that it uses only two lines (clock and data), which are bidirectional.

4. **One-wire (1-Wire®) Bus:** This is a one-wire bus used for network control and developed by the collaboration between two companies—Maxim Integrated Products and Dallas Semiconductor—after the merger, one company, Maxim. This bus was designed to lower cost and simplify interface design, and provides additional control, signaling, and power over a single wire.

5. **Controller Area Network Bus (CAN):** The controller area network protocol is designed for automotive applications where the amount of information transferred is relatively small. Its speed can be up to 1 Mb/s and it is designed to work with the concept of multiple masters.

6. **Local Interconnect Network (LIN):** This low-cost protocol is also used in automotive applications. It functions as an inexpensive subnetwork to CAN, allowing control over such items as mirrors, doors, and seats. This is a low-cost single-wire implementation and can run up to 20 kb/s. A typical LIN network consists of 1 LIN master and up to 16 LIN slaves with a possibility of connecting up to 64 different slaves. LIN and CAN are also used in applications in other industries.

In this chapter, we have discussed the first three protocols—EIA232, SPI, and I^{2}C—in the following sections.

13.2 EIA-232 AND PIC18 SERIAL COMMUNICATION MODULE USART

The PIC18 family of controllers includes two serial communication modules: the Addressable Universal Synchronous Asynchronous Receiver Transmitter (USART) and the Master Synchronous Serial Port (MSSP). We will discuss the USART module in this section and the MSSP module in the next section. In generic terminology, the USART is also known as the Serial Communications Interface (SCI), and the MSSP is known as the Serial Peripheral Interface (SPI). In this section, we focus on the EIA-232 communication standard and the USART module in the PIC18 family.

13.2.1 EIA-232 (RS-232) Serial I/O Standard

RS-232 (Revised Standard 232), now known as EIA-232, is a standard for serial binary data interconnection between data terminal equipment (DTE) such as a computer or a terminal, and data communication equipment (DCE) such as a modem or a router. The Electronics Industries Association (EIA) established the standard in the 1960s and, after various revisions, it came to be generally known as RS-232C.

Initially, the RS-232 standard was established for interconnecting equipment such as electromechanical teletypewriters, modems, terminals, and mainframe computers. The standard defines the electrical, mechanical, and functional characteristics for interfacing communication equipment. The equipment in the 1960s operated on high voltages, which were defined using negative true logic. The voltage range from –3 V to –25 V is defined as logic 1 or Mark and +3 V to + 25 V is defined as logic 0 or Space. The recommended connector type, known as the

DB25 subminiature, has twenty-five pins, and signals are divided into four groups: data, control, timing, and ground. All signals are defined in reference to DTE. For example, a DTE transmits on pin 2 and receives on pin 3; a DCE transmits on pin 3 and receives on pin 2. This standard, defined for equipment in the 1960s, is currently used for the serial communication in PCs; therefore, various modifications became necessary as follows:

1. Modern equipment, including the PC, uses positive TTL/CMOS logic levels, and +5 V power supply. The operating voltage levels in TTL/CMOS logic are less than 5 V. Therefore, these signals cannot be connected directly to EIA-232 interface circuitry. These signals need to be converted into negative logic and shifted to higher voltages by using line drivers and line receivers called transceivers. Figure 13-2 shows the minimum connections—transmit, receive, and ground—required to interface a DTE and a DCE. Nowadays, the transceivers are available as integrated circuits that can change the logic level and the voltages without having a power supply higher than +5 V. The necessary higher voltages are generated internally. Figure 13-2 shows that the transceiver changes TTL voltage +3.4 V to EIA-232 level –9 V and again back to the TTL level at the receiver side. Similarly, 0.2 V (logic 0) is changed to +9 V and back to 0.2 V at the receiver side.

2. The PC serial communication is primarily asynchronous. Therefore, many of the signals on the DB25 connector used in synchronous communication are unnecessary; only seven to nine signals are necessary in asynchronous communication. Eventually, the connector was changed to only nine pins, which is generally known as a DB9. Figure 13-3 shows the 25-pin and 9-pin connectors with the associated signals.

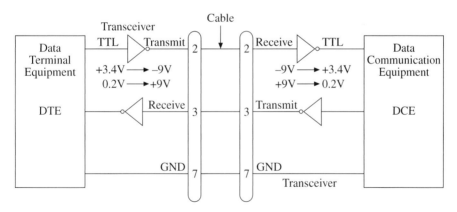

FIGURE 13-2   EIA 232 DTE and DCE Minimum Connection with MAX 232A Transceiver

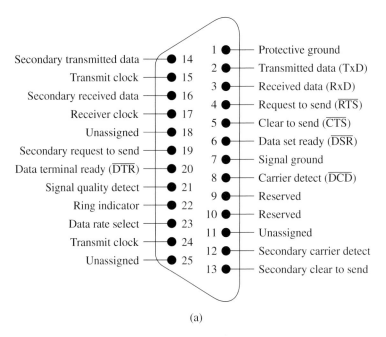

(a)

FIGURE 13-3 (a)   EIA-232 25-Pin DB Connector

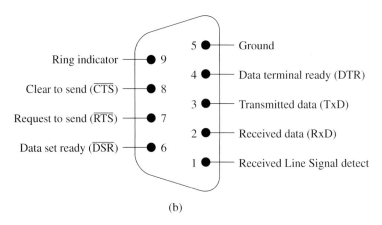

(b)

FIGURE 13-3 (b)   EIA-232 9-Pin DB Connector

EIA-232 Transmission Format (Framing) in the Asynchronous Mode

Typically, transmission consists of a Start bit, 7 or 8 data bits, optional parity bit, and 1 or 2 Stop bits. Figure 13-4 shows the transmission format of the ASCII character "Y" (59_H read backward from right to left). The transmission begins with the "Start" bit low (Logic 0) informing the receiver that this is the beginning of the transmission, and is followed by the bits of the ASCII character "Y" beginning with the LSB B0. It terminates with the "Stop" bit(s) high (Logic 1). The number of "Stop" bits used to be either 1, 1½, or 2, and these bits provide enough time delay for a slow receiver to get ready for the next transmission. Currently, 1 Stop bit is almost universal. In the serial communication terminology, the format shown in Figure 13-4 is called Framing; logic 1 is called Marking, and logic 0 is called Space.

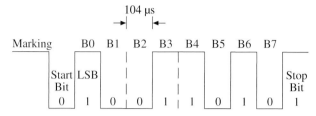

FIGURE 13-4   Framing: ASCII Character Y (59_H) – 9600 BAUD

In Figure 13-4 the bits are transmitted at 9600 bits per second. Therefore, the time per bit (bit time) is shown as 104 μs (1 / 9600 = 0.000104). If we assume 1 Stop bit, each 8-bit character needs 10 bits, and therefore, we can transmit 960 characters per second. In some transmission protocol ninth bit is added as a parity bit.

13.2.2 PIC18 USART Module

The PIC18 family of microcontrollers includes the USART module. The newer versions include an enhanced version of the USART module, called EUSART, which has additional features. We will focus primarily on the features of the USART module and when necessary identify the additional features of the EUSART module.

This serial USART module can be configured in the following modes:

- Asynchronous: full duplex
- Synchronous: Master half duplex
- Synchronous: Slave half duplex

This module is commonly used in PC serial communication in the asynchronous mode with the EIA-232 protocol. It is also used in the half duplex synchronous mode to communicate with peripherals, such as A/D and D/A converters, and with serial EEPROM. The registers used to set up and control the communication are:

- SPBRG: Baud Rate Generator
- TXSTA: Transmit Status and Control
- RCSTA: Receive Status and Control
- BAUDCON: Baud Rate Control

These are described in the following sections.

13.2.3 Data Transmission

Figure 13-5 shows a simplified block diagram of how serial bits are transmitted on pin RC6 of PORTC. Generally, two types of protocols – 8-bit and 9-bit –are used in serial transmission. The 9-bit protocol is used in multiprocessor communication where a PC as a host communicates with many microcontrollers. For a 9-bit transmission, the ninth bit must be placed in TX9D bit position of the TXSTA register (Figure 13-6) before writing 8 bits into the TXREG register. Once a data byte is written into the TXREG, bits are moved into the transmit shift register and clocked out onto the TX pin. Each byte (character) is preceded by a Start bit and followed by a Stop bit. The transmit shift register enables the MPU to write new data while the previous data bits are being transmitted. Data transmission is controlled by the TXSTA control register (Figure 13-6).

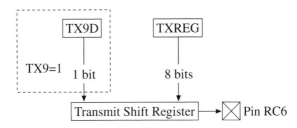

FIGURE 13-5   Simplified Block Diagram: USART Transmission on PIN RC6

B7	B6	B5	B4	B3	B2	B1	BØ
CSRC	TX9	TXEN	SYNC	SENDB	BRGH	TRMT	TX9D

B7 **CSRC:** Clock Source Select bit
Asynchronous mode:
Don't care.
Synchronous mode:
1 = Master mode (clock generated internally from BRG)
0 = Slave mode (clock from external source)

B6 **TX9:** 9-bit Transmit Enable bit
1 = Selects 9-bit transmission
0 = Selects 8-bit transmission

B5 **TXEN:** Transmit Enable bit
1 = Transmit enabled
0 = Transmit disabled
  Note: SREN/CREN overrides TXEN in Sync mode.

B4 **SYNC:** EUSART Mode Select bit
1 = Synchronous mode
0 = Asynchronous mode

B3 **SENDB:** Send Break Character bit
Asynchronous mode:
1 = Send Sync Break on next transmission (cleared by hardware upon completion)
0 = Sync Break transmission completed
Synchronous mode:
Don't care.

B2 **BRGH:** High Baud Rate Select bit
Asynchronous mode:
1 = High speed
0 = Low speed
Synchronous mode:
Unused in this mode.

B1 **TRMT:** Transmit Shift Register Status bit
1 = TSR empty
0 = TSR full

BØ **TX9D:** 9th bit of Transmit Data
Can be address/data bit or a parity bit.

FIGURE 13-6   Transmit Status and Control Register (TXSTA-PIC18F4520)—BitB3 is
                      Unimplemented in PIC18F452)

Transmit Status and Control (TXSTA)

This register is primarily used to select the mode of transmission, enable the transmission, and indicate the status of the transmit shift register as specified by the bits in Figure 13-6.

13.2.4 Data Reception

Figure 13-7 shows a simplified diagram of how serial bits are received on pin RC7 of PORTC. The USART can receive either 8 or 9 data bits. When the USART detects the Start bit, the receive shift register receives either 8 or 9 bits, one bit at a time. After checking the Stop bit, the USART moves data bits into the buffer and then into RCREG. For the 9-bit reception, the ninth bit is placed in RX9D-bit position of the RCSTA register. The receive shift register and the buffer enable the MPU to read data from the RCREG while receiving data in the receive shift register. The RCSTA control register controls data reception (Figure 13-8).

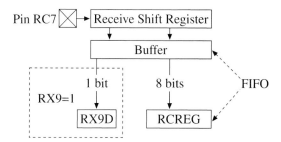

FIGURE 13-7   Simplified Block Diagram: USART Reception on Pin RC7

Receive Status and Control (RCSTA)

The bits in this register specify various parameters for the serial data reception as shown in Figure 13-8.

B7	B6	B5	B4	B3	B2	B1	B0
SPEN	RX9	SREN	CREN	ADDEN	FERR	OERR	RX9D

B7 **SPEN:** Serial Port Enable bit
1 = Serial port enabled (configures RX/DT and TX/CK pins as serial port pins)
0 = Serial port disabled (held in Reset)

B6 **RX9:** 9-bit Receive Enable bit
1 = Selects 9-bit reception
0 = Selects 8-bit reception

B5 **SREN:** Single Receive Enable bit
Asynchronous mode:
Don't care.
Synchronous mode – Master:
1 = Enables single receive
0 = Disables single receive
This bit is cleared after reception is complete.
Synchronous mode – Slave:
Don't care.

B4 **CREN:** Continuous Receive Enable bit
Asynchronous mode:
1 = Enables receiver
0 = Disables receiver
Synchronous mode:
1 = Enables continuous receive until enable bit CREN is cleared
(CREN overrides SREN)
0 = Disables continuous receive

B3 **ADDEN:** Address Detect Enable bit
Asynchronous mode 9-bit (RX9 = 1):
1 = Enables address detection, enables interrupt and loads the receive buffer
when RSR <8> is set
0 = Disables address detection, all bytes are received and ninth bit can be used
as parity bit
Asynchronous mode 9-bit (RX9 = 0):
Don't care

B2 **FERR:** Framing Error bit
1 = Framing error (can be updated by reading RCREG register and receiving next
valid byte)
0 = No framing error

B1 **OERR:** Overrun Error bit
1 = Overrun error (can be cleared by clearing bit CREN)
0 = No overrun error

B0 **RX9D:** 9th bit of Received Data
This can be address/data bit or a parity bit and must be calculated by
user firmware.

FIGURE 13-8   Receive Status and Control Register (RCSTA)

13.2.5 Baud Rate Control Register (BAUDCON)

This is the third control register, only available in EUSART. This is used to detect and calibrate baud rate automatically. It can also specify the 16-bit register for baud rate generation. Figure 13-9 shows the functions of the various bits in this control register.

B7	B6	B5	B4	B3	B2	B1	BØ
ABDOVF	RCIDL	—	SCKP	BRG16	—	WUE	ABDEN

B7   **ABDOVF:** Auto-Baud Acquisition Rollover Status bit
   1 = A BRG rollover has occurred during Auto-Baud Rate Detect mode
   (must be cleared in software)
   0 = No BRG rollover has occurred

B6   **RCIDL:** Receive Operation Idle Status bit
   1 = Receive operation is Idle
   0 = Receive operation is active

B5   **Unimplemented:** Read as '0'

B4   **SCKP:** Synchronous Clock Polarity Select bit
   Asynchronous mode:
   Unused in this mode.
   Synchronous mode:
   1 = Idle state for clock (CK) is a high level
   0 = Idle state for clock (CK) is a low level

B3   **BRG16:** 16-bit Baud Rate Register Enable bit
   1 = 16-bit Baud Rate Generator – SPBRGH and SPBRG
   0 = 8-bit Baud Rate Generator – SPBRG only (Compatible mode), SPBRGH
      value ignored

B2   **Unimplemented:** Read as '0'

B1   **WUE:** Wake-up Enable bit
   Asynchronous mode:
   1 = EUSART will continue to sample the RX pin – interrupt generated on falling
      edge; bit cleared in hardware on following rising edge
   0 = RX pin not monitored or rising edge detected
   Synchronous mode:
   Unused in this mode.

BØ   **ABDEN:** Auto-Baud Detect Enable bit
   Asynchronous mode:
   1 = Enable baud rate measurement on the next character. Requires reception
      of a Sync field (55h); cleared in hardware upon completion
   0 = Baud rate measurement disabled or completed
   Synchronous mode:
   Unused in this mode.

FIGURE 13-9   Baud Rate Control Register (BAUDCON—PIC18F4520 only)

13.2.6 Baud Rate Generator (SPBRG)

To set up the EUSART module for serial communication, Baud (the rate of data transfer) must be specified in the 16-bit register SPBRGH:SPBRG. Table 13-1 shows the configuration bits and the formulae to calculate baud.

TABLE 13-1 BAUD Rate Formulas

Configuration Bits			BRG/EUSART Mode	Baud Rate Formula
SYNC	BRG16	BRGH		
0	0	0	8-bit/Asynchronous	Fosc/[64 (n + 1)]
0	0	1	8-bit/Asynchronous	Fosc/[16 (n + 1)]
0	1	0	16-bit/Asynchronous	
0	1	1	16-bit/Asynchronous	Fosc/[4 (n + 1)]
1	0	x	8-bit/Synchronous	
1	1	x	16-bit/Synchronous	

Legend: x=Don't care, n=value of SPBRGH:SPBRG register pair

EXAMPLE 13.1

Calculate the byte for an 8-bit BRG to set up the EUSART for 9600 baud in the asynchronous mode when the clock frequency F_{OSC} = 10 MHz.

Solution:

A. If BRGH bit <2> in TXSTA register is set to 0, the formula given in Table 13.1 is:
Baud = F_{OSC} / [64 (n+1)] where n = number to be loaded in SPBRGH:SPBRG

$$9600 = \frac{10 \times 10^6}{64(n + 1)}; \text{ therefore, } n + 1 = \frac{10 \times 10^6}{9600 \times 64} = 16.276$$

Baud = n = 16.276 − 1 = 15.276 ≈ 15

B. If BRGH bit <2> in TXSTA register is set to 1, the formula given in Table 13.1 is:
Baud = F_{OSC} / [16 (n+1)] where n = number to be loaded in SPBRGH:SPBRG

$$9600 = \frac{10 \times 10^6}{16(n + 1)}; \text{ therefore, } n = \frac{10 \times 10^6}{9600 \times 16} - 1 = 64.1 ≈ 64$$

EXAMPLE 13.2

Write a subroutine to set up the EUSART module in the PIC18F4520 microcontroller to receive and transmit data at 9600 baud if the clock frequency $F_{OSC} = 10$ MHz.

Solution:

To set up the EUSART for the serial communication, the following pins and registers must be initialized:

1. Transmit pin RC6 on PORTC as an output.
2. Receive pin RC7 on PORTC as an input.
3. Transmit and Control Register (TXSTA) to enable transmission in the asynchronous mode to transmit 8 bits with low speed.
4. Receive and Control Register (RCSTA) to enable the serial port and 8-bit data reception without checking any errors.
5. Baud Rate Control Register (BUADCON) without autobaud detection and 8-bit BRGH register.
6. BRGH register to set up baud.

Instructions:

```
EUSART:
        BSF    TRISC, TX      ;Set RC7 and RC6 <7-6> as is
        BSF    TRISC, RX      ;The EUSART control changes input to output
                             ;as needed
        MOVLW B'00100000'     ;Enable transmit, 8 bits, low-speed baud,
        MOVWF TXSTA          ;asynchronous mode
        MOVLW B'00000000'     ;No baud rate detection, 8-bit baud
        MOVWF BAUDCON
        MOVLW D'15'           ;Byte for 9600 baud
        MOVWF SPBRG
        MOVLW B'10010000'     ;Enable serial port, 8 bits, receive enable,
        MOVWF RCSTA          ;No CHECK FOR framing or overrun error
        RETURN
```

Description:

The subroutine follows the six steps outlined above. It initializes the transmit (RC6) and receive (RC7) pins on PORTC, sets up the necessary parameters in the transmit, receive, baud control registers, and loads the byte as calculated in Example 13.1 for 9600 baud.

13.2.7 Data Transmission and Reception Methods: Checking Flags Versus Interrupts

There are two methods of implementing serial communication: (1) checking flags and (2) interrupts. The PIC18 family of microcontrollers uses the following bits in Peripheral Interrupt Request (Flag) Register1 (PIR1 – Chapter 10 Figure 10-6) to indicate the status of the transmission and reception. The description of the bits relevant to serial communication is repeated here from Figure 10-6 (Section 10.2.5) as follows.

- TXIF (Bit4): USART Transmit Interrupt Flag – this bit is 1 when TXREG (Transmit Register) is empty. When a byte is written in TXREG, this flag is set to 0 indicating that TXREG is full.
- RCIF (Bit5): USART Receive Interrupt Flag – this bit is 1 when RCREG (Receive Register) is full. When a byte is read from RCREG, this flag is set to 0 indicating that RCREG is empty.

These flags can be checked in software loops or used to interrupt the MPU. If we use the software loops, the MPU will be kept busy waiting in the loops and unavailable for the execution of any other instructions. If we set up the interrupt process, the reception and the transmission can be done in the background.

To set up the interrupt process, the PIC18 family provides two control registers as discussed in Chapter 10 and repeated here for easy reference: (1) the Interrupt Control Register (INTCON- Figure 10-4a) and (2) the Peripheral Interrupt Enable Register1 (PIE1- Figure 10-7). Bits used to set up the interrupt process are as follows:

INTCON:

- GIE/GIEH (Bit7): Global Interrupt Enable (H) – To enable the global interrupts, Bit7 in the INTCON register is set to 1; to disable all interrupts, the bit is set to 0. When the Interrupt Priority (IPEN Bit7) in the Reset Control Register (RCON) is set to 1, the priority levels are enabled.
- PEIE/ GIEL (Bit6): Peripheral Interrupt Enable bit – To enable peripheral interrupts, Bit6 in the INTCON register is set to 1; to disable peripheral interrupts, the bit is set to 0. When IPEN is 1, the low-priority peripheral interrupts are enabled.

PIE1:

- RCIE (Bit5): USART Receive Interrupt Enable – When this bit is 1, it enables the interrupt, and when this bit is 0, it disables the interrupt. TXIE (Bit4): USART Transmit Interrupt Enable – When this bit is 1, it enables the interrupt, and when this bit is 0, it disables the interrupt.
- TXIE (Bit4): USART Transmit Interrupt Enable – When this bit is 1, it enables the interrupt, and when this bit is 0, it disables the interrupt.

EXAMPLE 13.3

Write instructions to enable the USART interrupts.

Solution:

As discussed above, we need to set up Bit5 and Bit4 in the PIE1 register and Bit7 and Bit6 in the INTCON register. We have two options: we can either write a byte in these registers or set bits using symbolic labels for individual bits. To set or reset multiple bits, writing a byte in a control register is more efficient, but using symbols is easier to read and follow.

Instructions:

```
USARTINT:   BSF     PIE1, TXIE      ;Enable interrupt for transmitting data
            BSF     PIE1, RCIE      ;Enable interrupt for receiving data
            BSF     INTCON, PEIE    ;Enable peripheral interrupts
            BSF     INTCON, GIE     ;Enable global interrupts
```

EXAMPLE 13.4

Write a subroutine ECHO to receive a byte serially on pin RC7 and send that byte out on pin RC6 by checking the necessary flags. This is similar to echoing a character from the keyboard to the screen on a PC. Assume that the EUSART module in the PIC18F4520 has been initialized for serial communication as in Example 13.2.

Solution:

The Receive Interrupt Flag (RCIF) bit in the Peripheral Interrupt Request (PIR1) register is set when a new byte arrives and cleared when the byte is read. The subroutine should check the flag and stay in the loop until the flag is set, and once the flag is set, read the byte by copying it to a register. In this problem, we want to send the byte out to the screen. Therefore, we can save the byte in the W register.

Before we transmit the byte to the screen, we need to check the Transmit Interrupt Flag (TXIF) bit in the PIR1 to see whether the Transmit Register (TXREG) is empty. This flag is set when the TXREG is empty and cleared when a data byte is written.

continued

```
Instructions:

ECHO:
RECEIVE:    BTFSS    PIR1, RCIF    ;Check if new data byte arrived in RCREG
            GOTO     RECEIVE       ;If RCIF flag is not set, go back and wait
            MOVF     RCREG, W      ;If a new data byte has arrived, save in W
                                   ;register

TRANSMT:    BTFSS    PIR1, TXIF    ;Check if transmit register is empty
            GOTO     TRANSMT       ;If not, wait until it is empty
            MOVWF    TXREG         ;Write byte from W register into TXREG
            RETURN
```

13.3 SERIAL PERIPHERAL INTERFACE (SPI)

The SPI is a serial synchronous data exchange protocol between a master and a slave device. This is commonly used for high-speed serial communication between a microcontroller and its peripheral devices, such as EEPROMs, data converters, and display drivers. The basic concepts of the SPI are as follows:

- Four-wire interface. The four-wire interface includes four signals: clock, data in, data out, and slave select (also known as chip select).
- Synchronous protocol. The clock signal is provided and controlled by the master device, and data are sent with the clock pulses. Therefore, the clock can vary without disrupting the data flow.
- Master-Slave protocol. The master device controls the clock line (SCK), and can communicate with multiple slave devices. However, there is only one master in the SPI protocol.
- Data Exchange protocol. Each device (master and slave) has two data lines—one for input and one for output. Data are always exchanged between the devices; as data are being transmitted with the MSB first, new data are being received. A device cannot be only a transmitter or a receiver. For example, when the master transmits a data byte, it must read the incoming data byte from the slave before transmitting the next data byte, even if the received data may not have any use for a particular application.
- Master selects a slave. A slave device is selected by a signal from the master, and the signal is generally known as Chip Select ($\overline{CS}$) or Slave Select ($\overline{SS}$).
- Data Exchange and Clock. Data typically change during the rising or falling edge of the clock. The sampling and the data change take place on the opposite edge of the clock (this is further explained later in the context of the PIC18 module).

13.3.1 PIC18 Master Synchronous Serial Port (MSSP) Module

The MSSP module in the PIC18 family of microcontrollers is a serial interface used in communicating with other peripheral devices, and it can operate in one of two modes: (1) Serial

Peripheral Interface (SPI) or (2) Inter-Integrated Circuit (I^2C). In this section we will discuss the SPI mode, and in Section 13.4 we will discuss the I^2C mode.

The SPI module transmits and receives data simultaneously in the synchronous mode; it creates a loop as shown in Figure 13-10. Typically, it uses the following first three pins, and the fourth pin in the slave mode of operation:

- Serial Data Out (SDO): pin RC5/SDO on PORTC
- Serial Data In (SDI): pin RC4/SDI/SDA on PORTC
- Serial Clock (SCK): pin RC3/SCK/SCL on PORTC
- Slave Select (/SS): Available pin on a PORT. The $\overline{SS}$ pin is active low. It selects a slave device to communicate with and is generally used in configurations that call for multiple slaves. This pin can be any available pin on an I/O port.

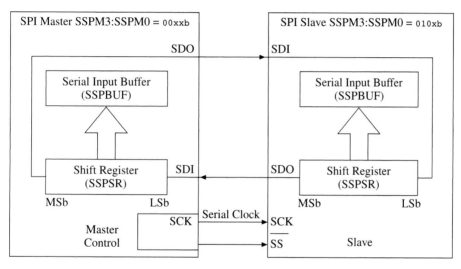

FIGURE 13-10   Master Synchronous Serial Port (MSSP) Module in SPI Mode – Master/Slave
Data Transfer

Figure 13-10 is a simplified block diagram of the MSSP module. The SSPSR is a shift register that is used to shift data out serially on the SDO pin and to receive data in on the SDI pin, and the SSPBUF is a buffer register that is used to write a byte to or read a byte from. For example, if a byte is to be sent from the master to the slave, the program must write the byte to the SSPBUF register of the master, and the module copies the byte into the shift register and shifts it out to the slave. At the same time, the master receives a byte from the slave. After the exchange of a byte between the master and the slave, the module copies the byte in the SSPBUF register, and the user program must read the SSPBUF register before transmitting the next byte. Because the Write and Read operations must be performed for each byte, some of the data bytes may not have any use in a given application. Therefore, the data transmission in the SPI mode can be divided in three categories: (1) transmission: the master sends data and receives dummy data from the slave, (2) transmission and reception: both exchange data, and (3) reception: the master receives data and sends dummy data to the slave.

13.3.2 Control of Data Exchange in MSSP Module

The data exchange operation in the SPI mode is controlled by two registers: (1) MSSP Control Register1 (SSPCON1) and (2) MSSP Status Register (SSPSTAT). The bit specifications of these registers are given in Figures 13-11 and 13-12.

B7	B6	B5	B4	B3	B2	B1	BØ
WCOL	SSPOV	SSPEN	CKP	SSPM3	SSPM2	SSPM1	SSPM0

B7   **WCOL:** Write Collision Detect bit (Transmit mode only)
     1 = The SSPBUF register is written while it is still transmitting the previous word (must be cleared in software)
     0 = No collision

B6   **SSPOV:** Receive Overflow Indicator bit
     SPI Slave mode:
     1 = A new byte is received while the SSPBUF register is still holding the previous data. In case of overflow, the data in SSPSR is lost. Overflow can only occur in Slave mode. The user must read the SSPBUF, even if only transmitting data, to avoid setting overflow (must be cleared in software).
     0 = No overflow

     Note: In Master mode, the overflow bit is not set since each new reception (and transmission) is initiated by writing to the SSPBUF register.

B5   **SSPEN:** Synchronous Serial Port Enable bit
     1 = Enables serial port and configures SCK, SDO, SDI and $\overline{SS}$ as serial port pins
     0 = Disables serial port and configures these pins as I/O port pins

     Note: When enabled, these pins must be properly configured as input or output.

B4   **CKP:** Clock Polarity Select bit
     1 = Idle state for clock is a high level
     0 = Idle state for clock is a low level

B3-BØ **SSPM3:SSPM0:** Synchronous Serial Port Mode Select bits
     0101 = SPI Slave mode, clock = SCK pin, $\overline{SS}$ pin control disabled, $\overline{SS}$ can be used as I/O pin
     0100 = SPI Slave mode, clock = SCK pin, $\overline{SS}$ pin control enabled
     0011 = SPI Master mode, clock = TMR2 output/2
     0010 = SPI Master mode, clock = Fosc/64
     0001 = SPI Master mode, clock = Fosc/16
     0000 = SPI Master mode, clock = Fosc/4

     Note: Bit combinations not specifically listed here are either reserved or implemented in I^{2}C mode only.

FIGURE 13-11   MSSP Control Register (SSPCON1): SPI Mode

B7	B6	B5	B4	B3	B2	B1	BØ
SMP	CKE	D/$\overline{\text{A}}$	P	S	R/$\overline{\text{W}}$	UA	BF

B7 **SMP:** Sample bit
    <u>SPI Master mode:</u>
    1 = Input data sampled at end of data output time
    0 = Input data sampled at middle of data output time
    <u>SPI Slave mode:</u>
    SMP must be cleared when SPI is used in Slave mode.

B6 **CKE:** SPI Clock Select bit
    1 = Transmit occurs on transition from active to Idle clock state
    0 = Transmit occurs on transition from Idle to active clock state

    Note: Polarity of clock state is set by the CKP bit (SSPCON1<4>).

B5 **D/$\overline{\text{A}}$:** Data/$\overline{\text{Address}}$ bit
    Used in I^{2}C mode only.

B4 **P:** Stop bit
    Used in I^{2}C mode only. This bit is cleared when the MSSP module is disabled,
    SSPEN is cleared.

B3 **S:** Start bit
    Used in I^{2}C mode only.

B2 **R/$\overline{\text{W}}$:** Read/$\overline{\text{Write}}$ Information bit
    Used in I^{2}C mode only.

B1 **UA:** Update Address bit
    Used in I^{2}C mode only.

BØ **BF:** Buffer Full Status bit (Receive mode only)
    1 = Receive complete, SSPBUF is full
    0 = Receive not complete, SSPBUF is empty

FIGURE 13-12   MSSP Status Register (SSPSTAT)

MSSP Control Register1 (SSPCON1)

This control register is used primarily to enable the pins for serial communication
<SSPEN – Bit5>, to select the clock polarity <CKP – Bit4> and the serial mode (such as
master or slave <Bit3 – Bit0>).

In this register, SSPOV <Bit6> is used to indicate an overflow in the Slave while receiving
data. An overflow occurs whenever the master writes a byte before the previous byte has been
read from the SSPBUF. If the SSPOV bit is set, it must be cleared by the user program. Figure
13-11 specifies the functions of each bit or combination of bits.

MSSP Status Register (SSPSTAT)—SPI Mode

This status register is used in both SPI and I^{2}C modes. In the SPI mode, this register is used to
specify when to sample the data (at the middle or at the end of the output line, <Bit7>), when
to transmit data (CKE – <Bit6>), and to indicate the status of the buffer register (BF–<Bit0>). The
remaining 5 bits of this register are used exclusively in the I^{2}C mode.

EXAMPLE 13.5

Write instructions to configure the PIC18F452/4520 MSSP module in the SPI master mode (Figure 13-13) with the following specifications.

1. The clock should idle high.
2. Given F_{OSC} = 10 MHz, the data transfer rate should be higher than 2 Mb/s.
3. The output should be sampled in the middle of the clock pulse.
4. The data should be sent on the rising edge.

After the configuration, write instructions to select the slave for communication and transfer a byte from the WREG to the slave.

Solution

We need to configure two control registers: SSPCON1 and SSPSTAT to meet the above specifications.

1. To configure SSPCON1 (Figure 13-11) the bits should be set as follows:
   - Bit5 (SSPEN) = 1 to enable the serial port and configure SCK (Serial Clock), SDO (Serial Data Output), SDI (Serial Data Input), and $\overline{SS}$ (Slave Select) as serial port pins.
   - Bit4 (CKP) = 1 to select clock polarity—idle state high.
   - Bit3-Bit0 = 0000 to place SPI in the master mode and set the data transfer rate equal to 2.5 Mbps (Clock = F_{OSC} /4)
   - Based on the above specifications, the byte in SSPCON1 = 00110000

2. To configure SSPSTAT (Figure 13-12) the bits should be set as follows:
   - Bit7 (SMP) = 0 to sample input data in the middle of the output time.
   - Bit 6 (CKE) = 1 to send data on the rising edge of SCK.
   - Bit0 (BF) = 0. This is Buffer Full flag for the receive mode, and 0 indicates that the SSPBUF is empty. The master should check this flag before sending a byte.
   - Therefore, the byte in SSPSTAT = 01000000

3. The other pins that should be initialized are as follows:
   - SDO (RC5) = 0. Serial Data Output as an output
   - SDI (RC4) = 1. Serial Data Input as an input
   - SCK (RC3) = 0. The Serial Clock is controlled by the master. Therefore, this pin should be output.
   - $\overline{SS}$ (RC0) = 0. The slave select signal is an output for the master, and in this example RC0 is chosen (somewhat arbitrarily, assuming that the pin is not used for any other function).
   - Therefore the byte in TRISC = 00010000

continued

Instructions:

#DEFINE	SS	PORTC,0	;Define PORTC pin 0 as Slave Select
SPI_MS	MOVLW	B'00010000'	;SDI should be input and remaining pins:
	MOVWF	TRISC	;SDO, SCK, and SS should be output
	MOVLW	B'00110000'	;Enable serial port and its pins, clock ;polarity
	MOVWF	SSPCON1	;idle high, master mode, and Fosc / 4
	MOVLW	B'01000000'	;Sample bit in middle, transition on ;rising
	MOVWF	SSPSTAT	;pulse
ACCESS:	BCF	SS	;Select slave device by making RC0 ;(SS) low
	MOVLW	0x55	;Arbitrary test byte
	MOVWF	SSPBUF	;Place byte in the buffer
CHK_BF	BTFSS	SSPSTAT, BF	;Check if buffer is full
	BRA	CHK_BF	;If yes, go back and check again
	MOVF	SSPBUF, W	;Read received byte and clear BF
	BSF	SS	;Disable slave device
	END		

Description:

The instructions in the segment SPI_MS configures the MSSP module in the SPI master mode as out-lined above. In the ACCESS segment, the first instruction asserts $\overline{SS}$ low and selects the Slave for serial communication. Next, the instructions load a test byte 55_H (any arbitrary 8-bit number) into the WREG and copies into the SSPBUF and checks the Buffer Full (BF) flag in the SSPSTAT register. The program stays in the loop until the BF flag is set. The MSSP module sets the BF flag when the data transfer is complete. Because this is a data exchange protocol, the Master also receives a byte, which is stored in SSPBUF. In our example, we have no use for this data byte, but it must be read to clear the flag. In the instruction MOVF SSPBUF, W reads the data in WREG and ignores it, and the last instruction deactivates the Slave.

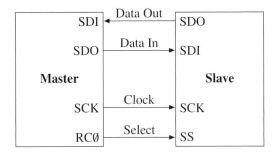

FIGURE 13-13   Master-Slave Data Transfer: SPI Mode

EXAMPLE 13.6

Write instructions to: (1) configure the PIC18F452/4520 MSSP module in the SPI slave mode with the same specifications as in Example 13.5, (2) receive the byte sent by the master in Example 13.5, and (3) save it in the OUTBUF with the address 20_H (Refer to Figure 13-13).

Solution

We need to configure two control registers: MSSP Control Register1 (SSPCON1) and MSSP Status Register and SSPSTAT to meet the above specifications.

1. To configure SSPCON1 (Figure 13-11) the bits should be set as follows:
   - Bit5 (SSPEN) and Bit4 = 1 the same as in the master.
   - Bit3-Bit0 = 0100 to place SPI in the slave mode with /SS control
   - The byte in SSPCON1 should be = 0011 0100

2. To configure SSPSTAT (Figure 13-12), the bits should be the same as before:
   - Byte in SSPSTAT = 0100 0000. data transfer on rising edge and sample data in the middle of the output time.

3. The other pins that should be initialized are as follows:
   - SDO (RC5) = 0. Serial Data Output as an output
   - SDI (RC4) = 1. Serial Data Input as an input
   - SCK (RC3) = 1. The Serial Clock is controlled by the master. Therefore, this pin should be an input.
   - $\overline{SS}$ (RC0) = 1. The slave select signal is an input for the slave.
   - The byte in TRISC should be = 0001 1001

continued

Instructions:

```
OUTBUF    EQU      0x20                  ;Define OUTBUF
SPI_SL:   MOVLW    B'00011001'           ;SDO should be output and remaining pins:
          MOVWF    TRISC                 ;SDI, SCK, and SS should be input
          MOVLW    B'00110100'           ;Enable serial port and its pins, clock polarity
          MOVWF    SSPCON1               ;idle high, and slave mode
          MOVLW    B'01000000'           ;Sample bit in middle, transition on rising
          MOVWF    SSPSTAT               ;pulse

          MOVFF    SSPBUF, OUTBUF        ;Save received byte in OUTBUF
          END
```

Description:

The instructions in the segment SPI_SL configure the MSSP module in the SPI slave mode as outlined above. The last instruction MOVFF saves the received byte in the OUTBUF register. In our example, the only function the slave is expected to perform is to save the received data.

13.3.3 SPI Applications

The SPI is a synchronous serial protocol and requires only four signals (pins). It is slower than the parallel mode, but many devices do not need high-speed communication. Because the SPI is simple and requires fewer signals, it has become a popular choice for interfacing peripherals in embedded systems. Many manufacturers have designed peripherals that are compatible with the SPI. These SPI compatible peripherals include converters (ADC and DAC), memories (EEPROM and Flash), sensors (temperature and pressure), and chips such as Real Time Clock (RTC), displays (LCD), and shift registers. The SPI operates in the master-slave configuration—one master and multiple slaves. Slaves can be connected in two different formats as shown in Figure 13-14a and b.

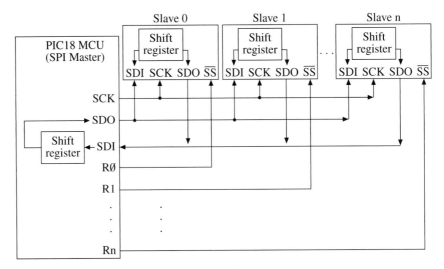

FIGURE 13-14 (a)   Master and Multiple Independent Slave Connections

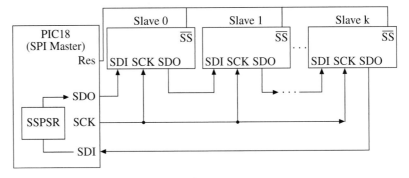

FIGURE 13-14 (b)   Master and Multiple Serial Slave Connections

Figure 13-14a shows one master with independent slaves. Each slave is supplied with a clock and connected with the same SDO and SDI signals, but each slave is accessed separately by using separate lines for $\overline{SS}$. These lines are shown as R0 to Rn where R represents an output line of an I/O port and n represents the number of output lines used for n slave devices. When a slave is not selected, it remains high, and the master can supply the data only to the selected slave. The configuration in Figure 13-14b is used when the same data should be supplied to all the slaves. Data are sent to k devices, from one device to the next.

EXAMPLE 13.7

Figure 13-15 shows a typical application of interfacing a shift register with a PIC18 microcontroller that is used to add an output port. Write instructions to display 55_H at the LEDs.

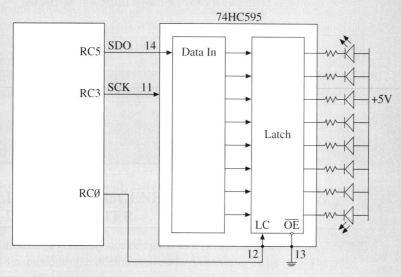

FIGURE 13-15   Interfacing 74HC – Shift Register in SPI Mode

Solution:

The 74HC595 includes a shift register and a latch. A byte can be clocked into the shift register and transferred into the latch by using the latch (LC) signal. As discussed before, all signals from PORTC should be output; RC0 is used to strobe the byte into the latch.

continued

Instructions:

RC0	EQU	0	
IOPORT:	BCF	TRISC, SCK	;Initialize SCK, SDO, and RC0 as ;outputs
	BCF	TRISC, SDO	
	BCF	TRISC, RC0	
	MOVLW	B'00100000'	;Enable SPI in master mode
	MOVWF	SSPCON1	;SSPEN = 1, CKP =0, <3-0> =0000
	MOVLW	B'11000000'	;CKE = 1, and SMP = 1
	MOVWF	SSPSTAT	
	BCF	PIR1, SSPIF	;Clear serial bit flag
	MOVLW	0x55	;Load byte for display
	MOVWF	SSPBUF	;Start shifting bits
CHK_IF:	BTFSS	PIR1,SSPIF	;Is shifting complete – flag =1? If not, ;wait
	BRA	CHK_IF	
	BCF	PORTC, RC0	;Strobe the pulse low and latch the byte
	BSF	PORTC, RC0	;Strobe the pulse high
	END		

Description:

The pins SCK (Clock), SDO (Serial Data Output), and RC0 are initialized as outputs. The SPI is configured in the master mode, and the input data bit is sampled at the end of data output line and transmitted on the rising pulse.

To start the transmission, the SSPIF flag must be cleared in the PIR1 register. The instructions load the byte to be displayed in WREG and copy into the SSPBUF register, at which point shifting begins. The program waits in the loop until the shifting is completed, which is indicated by setting the SSPIF flag in the PIR1 register. Once all the bits are shifted in the shift register, the byte is transferred into the buffer by asserting the signal RC0 low and then high.

13.4 THE INTER-INTEGRATED CIRCUIT (I²C) PROTOCOL

The I²C protocol was developed by Philips Inc. for data transfer between ICs on printed circuit boards, and now has gained considerable popularity. The features of the protocol are as follows:

- Two-wire Interface: This is a two-wire interface that has two open drain/collector lines, one for clock and one for data.
- Synchronous: Data are transferred with a synchronous clock initiated by the master device. The rate of data transfer is 100 kbps in the standard mode and 400 kbps in the fast mode.
- Master/Slave or Many Masters: The bus may have one master and many slaves, or multiple masters meaning multiple devices can initiate communication.
- SCL (Clock) and SDA (Data) lines: These lines are pulled high by resistors and stay high in the idle state. The clock controls the data transfer, and the data transfer is bidirectional. All changes in the SDA (data) line must occur when the SCL (clock) line is low.
- 8-bit Data Transfer: The master initiates all data transfers and transmits 8-bits starting with the MSb.
- Addressing: Two types of addresses are used: 7-bit or 10-bit. The SDA line is also used to send the addresses of the slaves.
- Framing: The data transfer from the master includes the following bits, Start (S), Address or Data (D), Stop (P), and the slave sends Acknowledge (A) or does not acknowledge ($\overline{A}$). A typical data transfer is shown in Figure 13-16a and b. Figure 13-16a shows the data transfer from the master to the slave, and acknowledge (A or $\overline{A}$) signal from the slave to the master. Figure 13-16b shows the master reading the slave, and the data transfer from the slave to the master.

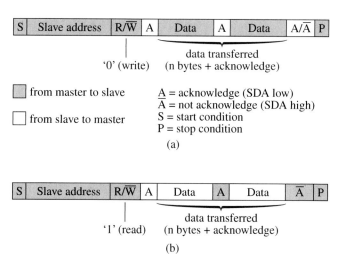

FIGURE 13-16   Data Transfer in I²C Mode

13.4.1 The PIC18 MSSP Module in the I²C Mode

The MSSP module in the I²C mode implements all the master and the slave functions and provides interrupts on Start and Stop bits in hardware to determine whether the bus is free when the multimasters are using the bus. The module supports both the 7-bit and the 10-bit addressing mode. It uses the following pins:

- SCL (Serial Clock): pin RC3/SCK/SCL on PORTC.
- SDA (Serial Data): pin RC4/SDI/SDA on PORTC.

These pins must be setup as inputs so that the I²C control circuitry can control the directions as necessary. These pins are used either for the SPI mode or the I²C mode but not for both in the same application.

The following control and status registers are used to configure the MSSP module in the I²C mode.

- SSPCON1 (MSSP Control Register1)
- SSPCON2 (MSSP Control Register2)
- SSPSTAT (MSSP Status Register)

The bits of the SSPCON1 and SSPSTAT registers are already described in Figures 13-11 and 13-12 in Section 13.3. However, when the MSSP module is configured in the I²C mode, some of the bits have different functions. Figures 13-17a, b and c show the definitions of these bits in these three registers.

B7	B6	B5	B4	B3	B2	B1	BØ
WCOL	SSPOV	SSPEN	CKP	SSPM3	SSPM2	SSPM1	SSPM0

B7   **WCOL:** Write Collision Detect bit
<u>In Master Transmit mode:</u>
1 = A write to the SSPBUF register was attempted while the I^{2}C conditions were not valid for a transmission to be started (must be cleared in software)
0 = No collision
<u>In Slave Transmit mode:</u>
1 = The SSPBUF register is written while it is still transmitting the previous word (must be cleared in software)
0 = No collision
<u>In Receive mode (Master or Slave modes):</u>
This is a "don't care" bit.

B6   **SSPOV:** Receive Overflow Indicator bit
<u>In Receive mode:</u>
1 = A byte is received while the SSPBUF register is still holding the previous byte (must be cleared in software)
0 = No overflow
<u>In Transmit mode:</u>
This is a "don't care" bit in Transmit mode.

B5   **SSPEN:** Synchronous Serial Port Enable bit
1 = Enables the serial port and configures the SDA and SCL pins as the serial port pins
0 = Disables serial port and configures these pins as I/O port pins
   Note: When enabled, the SDA and SCL pins must be properly configured as input or output.

B4   **CKP:** SCK Release Control bit
<u>In Slave mode:</u>
1 = Release clock
0 = Holds clock low (clock stretch), used to ensure data setup time
<u>In Master mode:</u>
Unused in this mode.

B3–BØ   **SSPM3:SSPM0:** Synchronous Serial Port Mode Select bits
1111 = I^{2}C Slave mode, 10-bit address with Start and Stop bit interrupts enabled
1110 = I^{2}C Slave mode, 7-bit address with Start and Stop bit interrupts enabled
1011 = I^{2}C Firmware Controlled Master mode (Slave Idle)
1000 = I^{2}C Master mode, clock = Fosc/(4 * (SSPADD + 1))
0111 = I^{2}C Slave mode, 10-bit address
0110 = I^{2}C Slave mode, 7-bit address
Bit combinations not specifically listed here are either reserved or implemented in SPI mode only.

FIGURE 13-17 (a)   MSSP Control Register1 (SSPCON1): I^{2}C Mode

B7	B6	B5	B4	B3	B2	B1	BØ
GCEN	ACKSTAT	ACKDT	ACKEN[1]	RCEN[1]	PEN[1]	RSEN[1]	SEN[1]

B7 **GCEN:** General Call Enable bit (Slave mode only)
1 = Enable interrupt when a general call address (0000h) is received in the SSPSR
0 = General call address disabled

B6 **ACKSTAT:** Acknowledge Status bit (Master Transmit mode only)
1 = Acknowledge was not received from slave
0 = Acknowledge was received from slave

B5 **ACKDT:** Acknowledge Data bit (Master Receive mode only)
1 = Not Acknowledge
0 = Acknowledge

> **Note:** Value that will be transmitted when the user initiates an Acknowledge sequence at the end of a receive.

B4 **ACKEN:** Acknowledge Sequence Enable bit (Master Receive mode only)[1]
1 = Initiate Acknowledge sequence on SDA and SCL pins and transmit ACKDT data bit. Automatically cleared by hardware.
0 = Acknowledge sequence Idle

B3 **RCEN:** Receive Enable bit (Master mode only)[1]
1 = Enables Receive mode for I^{2}C
0 = Receive Idle

B2 **PEN:** Stop Condition Enable bit (Master mode only)[1]
1 = Initiate Stop condition on SDA and SCL pins. Automatically cleared by hardware.
0 = Stop condition Idle

B1 **RSEN:** Repeated Start Condition Enable bit (Master mode only)[1]
1 = Initiate Repeated Start condition on SDA and SCL pins. Automatically cleared by hardware.
0 = Repeated Start condition Idle

BØ **SEN:** Start Condition Enable/Stretch Enable bit[1]
In Master mode:
1 = Initiate Start condition on SDA and SCL pins. Automatically cleared by hardware.
0 = Start condition Idle
In Slave mode:
1 = Clock stretching is enabled for both slave transmit and slave receive (stretch enabled)
0 = Clock stretching is disabled

> **Note 1:** For bits ACKEN, RCEN, PEN, RSEN, SEN: If the I^{2}C module is not in the Idle mode, these bits may not be set (no spooling) and the SSPBUF may not be written (or writes to the SSPBUF are disabled).

FIGURE 13-17 (b) MSSP Control Register 2 (SSPCON2): I^{2}C Mode

B7	B6	B5	B4	B3	B2	B1	BØ
SMP	CKE	D/$\overline{\text{A}}$	P	S	R/$\overline{\text{W}}$	UA	BF

B7   **SMP:** Slew Rate Control bit
      <u>In Master or Slave mode:</u>
      1 = Slew rate control disabled for standard speed mode (100 kHz and 1MHz)
      0 = Slew rate control enabled for high-speed mode (400 kHz)

B6   **CKE:** SMBus Select bit
      <u>In Master or Slave mode:</u>
      1 = Enable SMBus specific inputs
      0 = Disable SMBus specific inputs

B5   **D/$\overline{\text{A}}$:** Data/$\overline{\text{Address}}$ bit
      <u>In Master mode:</u>
      Reserved.
      <u>In Slave mode:</u>
      1 = Indicates that the last byte received or transmitted was data
      0 = Indicates that the last byte received or transmitted was address

B4   **P:** Stop bit
      1 = Indicates that a Stop bit has been detected last
      0 = Stop bit was not detected last

      Note: This bit is cleared on Reset and when SSPEN is cleared.

B3   **S:** Start bit
      1 = Indicates that a Start bit has been detected last
      0 = Start bit was not detected last

      Note: This bit is cleared on Reset and when SSPEN is cleared.

B2   **R/$\overline{\text{W}}$:** Read/$\overline{\text{Write}}$ Information bit (I^2C mode only)
      <u>In Slave mode:</u>
      1 = Read
      0 = Write

      Note: This bit holds R/$\overline{\text{W}}$ bit information following the last address
                   match. This bit is only valid from the address match to the next
                   Start bit, Stop bit or not $\overline{\text{ACK}}$ bit.
      <u>In Master mode:</u>
      1 = Transmit is in progress
      0 = Transmit is not in progress

      Note: ORing this bit with SEN, RSEN, PEN, RCEN, or ACKEN will
                   indicate if the MSSP is in Active mode.

B1   **UA:** Update Address bit (10-bit Slave mode only)
      1 = Indicates that the user needs to update the address in the SSPADD register
      0 = Address does not need to be updated

BØ   **BF:** Buffer Full Status bit
      <u>In Transmit mode:</u>
      1 = SSPBUF is full
      0 = SSPBUF is empty
      <u>In Receive mode:</u>
      1 = SSPBUF is full (does not include the $\overline{\text{ACK}}$ and Stop bits)
      0 = SSPBUF is empty (does not include the $\overline{\text{ACK}}$ and Stop bits)

FIGURE 13-17 (c)   MSSP Status Register (SSPSTAT): I^2C Mode

The other registers that are necessary to implement the I²C mode are shown in Figure 13-18, a simplified block diagram of the MSSP module in the I²C mode. These registers are as follows:

- SSPADD (MSSP Address Register): This is a register that holds the slave device address when the SSP is in slave mode. In master mode, it holds the baud rate generator reload value, which is only seven bits (without the MSb).
- SSPSR (Serial Receive/Transmit Buffer Register): This is a shift register to transmit and receive bytes.
- SSPBUF (Buffer Register): The MPU writes bytes to this register for transmission and reads bytes for reception.

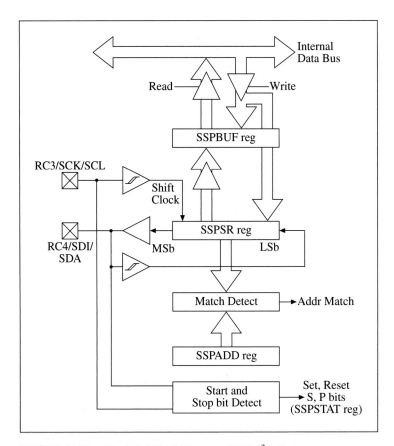

FIGURE 13-18   Simplified Block Diagram: MSSP I²C Mode

EXAMPLE 13.8

Write a subroutine to initialize the MSSP module in the I^2C master mode with the transmission rate of 100 kHz when clock frequency F_{osc} = 10 MHz

Solution

The count to set up the transmission Baud 100 kHz = $(10 \times 10^6 / 4) / 100 \times 10^3) - 1 = 24$

Instructions:

I2C:	MOVLW	B'00011000'	;Set up SDA and SCL PORTC as inputs
	IORWF	TRISC, F	
	MOVLW	D'24'	;Count to set up rate = 100 kHz
	MOVWF	SSPADD	;Initialize baud
	BCF	SSPSTAT,6	;Select I^2C input and not SME bus
	BCF	SSPSTAT,7	;Select normal slew rate
	MOVLW	B'00101000'	;No error detection, enable serial port – SDA ;& SCL pins
	MOVWF	SSPCON1	;<3-0> 1000 - I^2C master mode
	RETURN		

Description:

In the I^2C mode, the data pin (RC4/SDA) and the clock pin (RC3/SCL) should be set up as inputs so that the MSSP module control circuitry can change the I/O functions of these pins as necessary. The IORWF instruction performs the logical OR function of bits <4–3> with the TRISC register so that the status of the remaining bits in PORTC will not be affected. The remaining instructions configure the MSSP module in the I^2C mode as specified in the problem.

EXAMPLE 13.9

Write a subroutine to check whether the I^2C bus is idle prior to beginning an event, such as sending a byte on the bus. If the bus is idle, load the ASCII character 'A' in WREG and send it out on the bus from the Master using another subroutine.

Solution

The bus is busy if any of the following signals are active: Start, Repeated Start, Stop, Receive, and Acknowledge. The status of these conditions is indicated by five bits <4–0> in the SSPCON2 register. In addition, we need to check the status of the R/ /W signal <Bit2> in the SSPSTAT register.

continued

```
          Instructions:

          TRANSMT:    CALL       BUS_CHK         ;Check if I²C bus is idle
                      MOVLW      0x41            ;Load ASCII 'A' in W
                      CALL       SEND_BYTE       ;Send 'A' on the bus

          HERE: BRA HERE

          BUS_CHK:    MOVF       SSPCON2, WREG   ;Copy SSPCON2 in W
                      ANDLW      B'00011111'     ;Is any of bits <4–0> 1?
                      BNZ        BUS_CHK         ;If yes, go back and read again
          CHK_WR:     BTFSC      SSPSTAT, R_W    ;Check if transmission in progress
                      BRA        CHK_WR          ;If busy, go back and check again
                      RETURN


          SEND_BYTE:  MOVWF      SSPBUF          ;Write byte in Buffer
          CHK_BF:     BTFSC      SSPSTAT, BF     ;Is BF flag cleared?
                      BRA        CHK_BF          ;If not, go back and check
                      RETURN
                      END
```

Description:

The main program consists of calling two subroutines and loading a byte in the WREG for sending it out on the bus. The first subroutine BUS_CHK checks whether the bus is idle. The subroutine copies the status of the bits from the SSPCON2 register in the WREG and checks bits <4–0> whether they are logic 0 by ANDing with 1s. If all five conditions are idle, the AND instruction sets the zero flag. If the zero flag is not set, the subroutine stays in the loop until the flag is set. The next instruction checks the R/W bit <2> in the SSPSTAT register. For the Master mode, if this bit is 1, it indicates that the transmission is in progress; if it is zero, it indicates the absence of the transmission. The subroutine stays in the loop (CHK_WR) until this bit is cleared.

The next subroutine SEND_BYTE loads the ASCII character 'A' from the WREG into the buffer SSBUF. The shift register in the MSSP module gets the byte from the buffer and shifts this out serially. At the end of the transmission, the BF (Bit0) in the SSPSTAT register is cleared. The subroutine stays in the loop (CHK_BF) until the BF bit is cleared and returns to main program.

13.4.2 Steps in Master Mode Transmission

After initializing the MSSP in the I²C mode, we need to follow the data format as shown in Figure 13-16 to transmit data on the bus—Start, Address or Data, Stop, and Acknowledge. Based on the above discussion and the examples, we can summarize the steps necessary to set up the transmission in the I²C mode as follows:

- Send Start condition by enabling SEN—bit <0> in the SSPCON2 register.
- Write the slave address in the SSPBUF register for transmission.
- The MSSP module shifts out the byte on the SDA pin.
- The master receives the ACK bit from the slave, and the status of the bit <6> in the SSPCON2 register is modified. It is cleared when the master receives an acknowledgment, and set to 1 when there is no acknowledgment.

- The MSSP module generates an interrupt at the end of the ninth bit and sets up the SSPIF flag.
- Write the data byte to be transmitted in the SSPBUF register.
- The MSSP module shifts out the byte on the SDA pin.
- The master receives the ACK bit from the slave, and the status of the bit <6> in the SSPCON2 register is modified accordingly.
- The MSSP module generates an interrupt at the end of the ninth bit and sets up the SSPIF flag.
- Send Stop condition by enabling PEN—bit <2> in the SSPCON2 register.
- At the end of the Stop condition the module generates an interrupt.
- RSEN—bit <1> in the SSPCON2 register is set to send repeated Start conditions.

These steps are illustrated in Section 13.6, Interfacing EEPROM using the I^{2}C mode.

13.5 ILLUSTRATION: INTERFACING SERIAL EEPROM TO THE PIC18 MSSP MODULE IN THE SPI MODE

In many low-cost applications where the speed of processing is relatively unimportant and the number of pins available on a microcontroller is limited, a serial interface is an ideal solution to connect peripherals. In embedded systems, a microcontroller with limited memory can expand its memory size by adding a serial EEPROM. The advantages of serial EEPROM include features such as small size, low number of I/O pin requirements, byte level flexibility, low power consumption, and low cost. The primary benefit of using the MSSP module is that the signal timings are handled by the hardware rather than software. This enables the system to continue program execution while serial communication is handled in the background.

This illustration demonstrates the interfacing of the Microchip 25XXX serial EEPROM series using the PIC18 MSSP module in the SPI mode.

13.5.1 Problem Statement

Interface Microchip 25XX040 Serial EEPPROM to the PIC18F452/4520 MSSP module in the SPI mode. Explain the serial communication process between the memory chip and the microcontroller module, and write instructions to transfer 16 bytes from the microcontroller data registers to the EEPROM.

13.5.2 Problem Analysis

The problem analysis can be divided into two major segments: hardware and software. In the hardware segment, we need to examine the features of EEPROM and the MSSP module. We have already discussed the SPI mode of the MSSP module in Section 13.3. Here our primary focus is on EEPROM and its requirements.

Electrically Erasable PROM (EEPROM)

The Microchip 25XX040 is an 8-pin device with 4 Kb memory organized as 512×8. The address range for 512 bytes is 000 to $1FF_H$, and the size of the page is 16 bytes. The data transfer between memory and the microcontroller MSSP module takes place via a Serial Peripheral Interface™ (SPI™) bus. The bus requires only three signals (Figure 13-19): Serial Clock (SCK), Serial Input (SI), and Serial Output (SO). The memory chip is accessed through Chip Select ($\overline{CS}$) signal. Its $\overline{HOLD}$ signal is used to stop the data transfer, and the Write Protect ($\overline{WP}$) can prevent from writing to the chip. The $\overline{WP}$ and $\overline{HOLD}$ signals are tied high to disable these functions. The power supply V_{CC} range varies based on the version of the chip; the V_{CC} range for 25XX040 is 2.5 V–5.5 V.

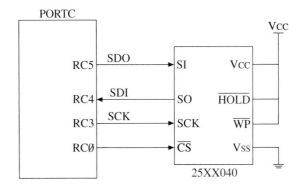

FIGURE 13-19 Interfacing Serial EEPROM with PIC18F MSSP Module in SPI Mode

The memory chip has an 8-bit instruction register with six instructions (commands) as shown in Table 13-2. When the $\overline{CS}$ is low, the data bits are clocked in on the rising edge of the clock (SCK) assuming the HOLD and WP are inactive (high). The chip includes the Read Status Register (RDSR), and the bits of this register are defined in Figure 13-20.

TABLE 13-2 25XX040 SPI™ Bus Serial EEPROM

INSTRUCTION SET		
Instruction Name	**Instruction Format**	**Description**
READ	0000 A8011	Read data from memory array beginning at selected address
WRITE	0000 A8010	Write data to memory array beginning at selected address
WRDI	0000 0100	Reset the write enable latch (disable write operations)
WREN	0000 0110	Set the write enable latch (enable write operations)
RDSR	0000 0101	Read Status regiser
WRSR	0000 0001	Write Status register

Note: A8 is the 9th address bit necessary to fully address 512 bytes.

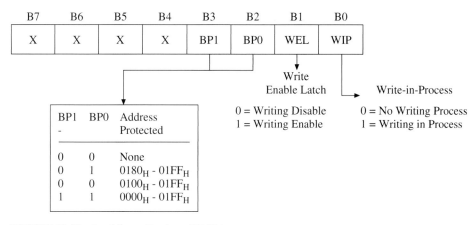

B7	B6	B5	B4	B3	B2	B1	B0
X	X	X	X	BP1	BP0	WEL	WIP

Write Enable Latch

Write-in-Process

0 = Writing Disable
1 = Writing Enable

0 = No Writing Process
1 = Writing in Process

BP1 -	BP0	Address Protected
0	0	None
0	1	0180_H - $01FF_H$
0	0	0100_H - $01FF_H$
1	1	0000_H - $01FF_H$

FIGURE 13-20   Read Status Register (RDSR)

Write Operation

To write into the memory, the following steps are required:

1. Enable the latch by sending the Write Enable (WREN) instruction. Assert $\overline{CS}$ low, send the WREN instruction, and disable the $\overline{CS}$ by setting it high. Figure 13-21a shows the timing diagram. The MSSP module asserts the $\overline{CS}$ low, and the 8 bits (06_H) of the latch-enable command are shifted in the memory chip on the SI line synchronized with the clock.

2. Read the Status register to check that the Write-in-Process (WIP) is no longer active. This is confirmed if the WIP bit is low; otherwise the program must stay in the loop until the WIP bit goes low. Figure 13-21b shows that the command to read status (05_H) is on the SI line, and data from the status register is sent out on the SO line.

3. Send a byte (or a page, which = 16 bytes) to EEPROM. Assert $\overline{CS}$ low, and send the WRITE instruction, followed by the address and data (Figure 13-21c). 16 bytes can be written before deactivating the $\overline{CS}$ signal.

4. During the write operation, the status register RDSR can be read to check any of the status bits. When the write cycle is completed, the Write Enable latch is reset.

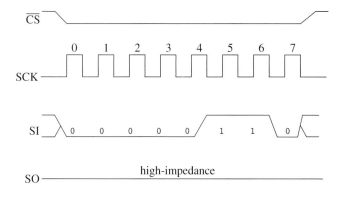

FIGURE 13-21 (a)   Write Enable (WREN) Command

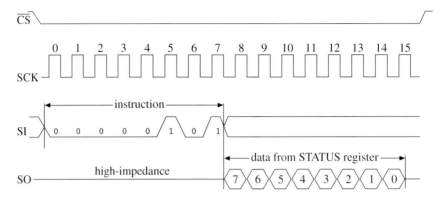

FIGURE 13-21 (b)   Read Status Register (RDSR) Command to Check Write-in-Process (WIP) Bit

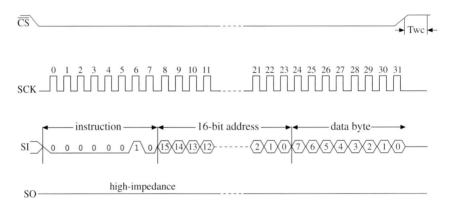

FIGURE 13-21 (c)   Write Command to Send Address and Data

Read Operation

To read from memory the following steps are required:

1. Access the memory chip. Assert $\overline{CS}$ low.
2. Send Read Instruction followed by the address.
3. Data are shifted out over the SO pin, and the operation continues until the $\overline{CS}$ is asserted high.

The timing diagram of a read operation is the same as in Figure 13-21c except that the command byte is 03_H.

13.5.3 PROGRAM

```
        Title "Illust 13-5 Interfacing Serial EEPROM"
        List p=18F452
        #include <p18F452.inc>  ;This is a header file


;:::::::::::::::::::::::::::::::::::::::::::::::::::::::::::::::::::::::::::::
;This program copies 16 bytes of data (1 page) from data registers with   :
;the starting address 50H into EEPROM with the starting address 0010H     :
;EEPROM used is 25LC040 - 256x8 - address range - 000 - 1FFH              :
;:::::::::::::::::::::::::::::::::::::::::::::::::::::::::::::::::::::::::::::

CSPROM      EQU   0                 ;RC0 is used as chip select to
                                    ;access EEPROM
OUTREG      EQU   0x20              ;Register used to transmit a byte
INREG       EQU   0x21              ;Register used to receive a byte
COUNTER     EQU   0x22              ;Number of bytes loaded here
HI_ADDR     EQU   0x00              ;EEPROM starting address 0010H
LO_ADDR     EQU   0x10
DATA_REG    EQU   0x50              ;Bytes to be copied are stored
                                    ;from 50H to 5FH


RD_CMD      EQU   B'00000011'       ;Read data from memory beginning at
                                    ;selected address
WR_CMD      EQU   B'00000010'       ;Write data to memory beginning at
                                    ;selected address
WRDI_CMD    EQU   B'00000100'       ;Disable latch
WREN_CMD    EQU   B'00000110'       ;Enable latch
RD_STATUS   EQU   B'00000101'       ;Read status register
WIP         EQU   0                 ;Write-in progress bit in status
                                    ;register


            ORG   00
            GOTO  MAIN


            ORG   0x20

MAIN:       RCALL SETUP             ;Initialize MSSP in SPI mode
            RCALL WR_ENABLE         ;Enable latch
            RCALL CHK_STATUS        ;Check whether write-in process is on
            LFSR  FSR0,DATA_REG     ;Set up memory pointer for data register
            RCALL PAGE_WR           ;Transfer 16 bytes (one page)
            RCALL WR_DSABLE         ;Disable latch
HERE        BRA   HERE

SETUP:      ;:::::::::::::::::::::::::::::::::::::::::::::::::::::::::::::::
            ;This subroutine initializes MMSP module in the SPI mode    :
            ;and PORTC pins as inputs and outputs as necessary. It      :
            ;disables global and  peripheral interrupts and sets bits   :
            ;in SSPCON1 and SSPSTAT registers to enable serial          :
            ;communication in the SPI mode                              :
            ;:::::::::::::::::::::::::::::::::::::::::::::::::::::::::::::::
```

continued

```
            CLRF    PORTC               ;Clear any previous settings
            CLRF    PIE1                ;Disable peripheral interrupts
            CLRF    INTCON              ;Disable all interrupts
            CLRF    SSPCON1             ;Clear SSPCON1 and SSPSTAT
            CLRF    SSPSTAT
            BSF     PORTC,CSPROM        ;Disable memory access

            MOVLW B'00010000'           ;All bits of PORTC are
                                        ;outputs except
            MOVWF TRISC                 ;SDI <4>
            MOVLW B'00110000'           ;Enable synchronous serial
                                        ;communication in
            MOVWF SSPCON1               ;Master mode, Clock - idle state
                                        ;high
            MOVLW B'10000000'           ;Sample data at end of pulse,
                                        ;transmit data on
            MOVWF SSPSTAT               ;rising edge
            RETURN

WR_ENABLE:  ;:::::::::::::::::::::::::::::::::::::::::::::::::::::::::::
            ;Function: This subroutine enables the latch. It sends   :
            ;          WREN command by calling the subroutine OUT.    :
            ;:::::::::::::::::::::::::::::::::::::::::::::::::::::::::::

            BCF     PORTC, CSPROM       ;Access EEPROM by asserting RC0
            MOVLW WREN_CMD              ;Copy latch enable command in
                                        ;OUTREG
            MOVWF OUTREG
            RCALL OUT                   ;Send command to EEPROM
            BSF     PORTC, CSPROM       ;Disable EEPROM by asserting
                                        ;RC0 high

            RETURN

WR_DSABLE:  ;:::::::::::::::::::::::::::::::::::::::::::::::::::::::::::
            ;Function:  This subroutine disables the latch by sending :
            ;           WRDI command. Calls OUT subroutine.           :
            ;:::::::::::::::::::::::::::   :::::::::::::::::::::::::::::::

            BCF     PORTC, CSPROM       ;Access EEPROM by asserting RC0 low
            MOVLW WRDI_CMD              ;Copy latch disable command in
                                        ;OUTREG
            MOVWF OUTREG
            RCALL OUT                   ;Send command to EEPROM
            BSF     PORTC, CSPROM       ;Disable EEPROM by asserting
                                        ;RC0 high

            RETURN
```

continued

```
PAGE_WR:    ;:::::::::::::::::::::::::::::::::::::::::::::::::::::::::
            ;Function: This subroutine copies 16 bytes starting from  :
            ;          data RAM (50H)into EEPROM starting from 0010H. :
            ;Input:    FSR0 as memory pointer to the starting data    :
            ;          register (50H) Calls OUT subroutine.           :
            ;:::::::::::::::::::::::::::::::::::::::::::::::::::::::::
            BCF    PORTC, CSPROM      ;Access EEPROM by asserting RC0 low
            MOVLW  WR_CMD             ;Copy write command in OUTREG
            MOVWF  OUTREG
            RCALL  OUT                ;Send write command
            MOVLW  HI_ADDR            ;EEPPROM starting address 0010H
            MOVWF  OUTREG
            RCALL  OUT                ;Send high address
            MOVLW  LO_ADDR
            MOVWF  OUTREG
            RCALL  OUT                ;Send low address
            MOVLW  D'16'              ;Number of bytes
            MOVWF  COUNTER            ;Load in the counter
NEXT:       MOVFF  POSTINC0, OUTREG   ;Copy data in OUTREG, increment
                                      ;memory, and
            RCALL  OUT                ;send data byte
            DECF   COUNTER,F          ;Reduce the count by one
            BNZ    NEXT               ;Is COUNTER = 0, if not, go back to
                                      ;copy next data
            BSF    PORTC, CSPROM      ;Deselect EEPROM by asserting RC0 high
            RETURN


CHK_STATUS: ;:::::::::::::::::::::::::::::::::::::::::::::::::::::::::::
            ;Function: This subroutine checks the (Write-in Process   :
            ;          (WIP) bit It reads the status register and     :
            ;          checks WIP in SSPBUF It stays in the loop until:
            ;          the WIP bit is cleared. Calls OUT subroutine.  :
            ;:::::::::::::::::::::::::::::::::::::::::::::::::::::::::::

            BCF    PORTC, CSPROM
            MOVLW  RD_STATUS          ;Read status
            MOVWF  OUTREG
            RCALL  OUT
            MOVLW  00                 ;Send a dummy Byte
            MOVWF  OUTREG
            RCALL  OUT
            BSF    PORTC, CSPROM
            BTFSC  SSPBUF,WIP         ;Is WIP cleared, if not continue
                                      ;checking
            BRA    CHK_STATUS
            BSF    PORTC, CSPROM
            RETURN
```

continued

```
OUT:         ;:::::::::::::::::::::::::::::::::::::::::::::::::::::::::::
             ;Function: This subroutine sends a byte to EEPROM by      :
             ;          loading it in SSPBUF register and checks the   :
             ;          Buffer Full (BF) flag in SSPSTAT register. The :
             ;          subroutine stays in the loop until the byte is :
             ;          received and the flag is set.                  :
             ;Input:    A byte in W register                           :
             ;:::::::::::::::::::::::::::::::::::::::::::::::::::::::::::

             MOVF OUTREG, W            ;Get data byte or command
             MOVWF SSPBUF              ;Load in SSPBUF
CHK_FLAG     BTFSS SSPSTAT, BF        ;Is flag set?
             BRA   CHK_FLAG           ;If no, go back and check
             MOVF  SSPBUF, W
             MOVWF INREG               ;This is byte received to be ignored
             RETURN
             END
```

Description:

The MAIN program calls five subroutines in a sequence. They are: (1) SETUP, (2) WR_ENABLE, (3) CHK_STATUS, (4) PAGE_WR, and (5) WR_DSABLE. The flowchart of the program is shown in Figure 13-22. The flowchart shows the MAIN module and the program execution sequence of the WR_Enable and CHK_STATUS subroutines. All these subroutines call the OUT subroutine to transfer commands or data bytes.

The SETUP subroutine initializes the MSSP module in the SPI mode, and the WR_ENABLE subroutine enables the latch. The CHK_STATUS subroutine reads the status register inside the EEPPROM and checks the Write-In-Process (WIP) bit to confirm that the previously sent byte has been written. When the previously sent byte is written, this bit is cleared, and the next byte can be sent.

The MAIN initializes the FSR0 register as a memory pointer where data bytes are stored (50_H) and calls the PAGE_WR subroutine. This subroutine includes a counter with the count 16. This loop sends out 16 bytes until the counter becomes zero. The next subroutine WR_DSABLE disables the latch and terminates any further data transfer.

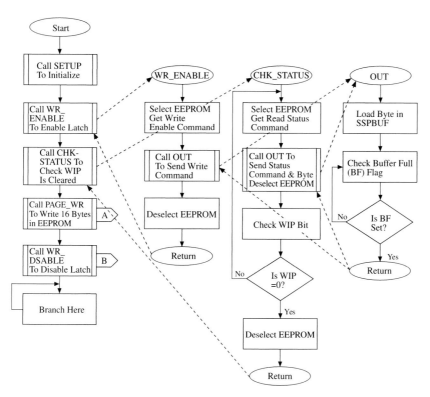

FIGURE 13-22 Flow Chart—Data Transfer from PIC18 Data (Memory) Registers to Serial EEPROM in the SPI Mode – Contd. Next Page

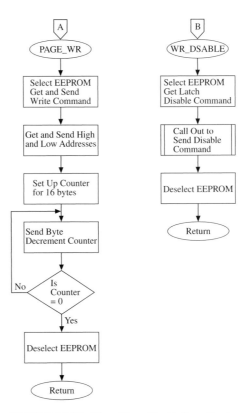

FIGURE 13-22   Flow Chart (Contd. from Previous Page)—Data Transfer from PIC18 Data (Memory) Registers to Serial EEPROM in the SPI Mode

13.6 ILLUSTRATION: INTERFACING SERIAL EEPROM TO THE PIC18 MSSP MODULE IN THE I²C MODE

As discussed in Section 13.4, the I²C is a two-wire interface consisting of SCL (Clock) and SDA (Data). It requires fewer pins than the SPI mode. Its advantages are similar to those of SPI mode and many devices, such as the I/O expander, D/A converter, and digital thermometers compatible with this bus, are available.

This illustration demonstrates the interfacing of Microchip 24XXX serial EEPROM series using the PIC18F452/4520 MSSP module in the I²C mode.

13.6.1 Problem Statement

Interface the Microchip 24LC024 Serial EEPPROM to the PIC18F452/4520 MSSP module in the I²C mode. Explain the serial communication process between the memory chip and the microcontroller module, and write instructions to transfer a byte from the microcontroller to the EEPROM.

13.6.2 Problem Analysis

As in the illustration of interfacing of EEPROM in the SPI mode, we will divide this problem into two segments: hardware and software. We have already discussed the I²C mode of the MSSP module in Section 13.4, and will discuss the requirements of the I²C compatible EEPROM in this section.

I²C Compatible Electrically Erasable PROM (EEPROM):

The Microchip 24LC024 is an 8-pin device with 2 Kb memory organized as 256×8. The address range for 256 bytes is 00 to FF_H, and the size of the page is 16 bytes. The data transfer between memory and the microcontroller MSSP module takes place via the I²C bus. The bus requires only two signals (Figure 13-23): Serial Clock (SCL) and Serial Data/Address (SDA). The power supply V_{CC} range varies based on the version of the chip; the V_{CC} range for 24LC024 is 2.5 V–5.5 V. It has three address lines A2–A0 that enable the user to address eight different EEPROM chips in the circuit. The WP is a Write Protection pin; it should be tied to V_{CC} to enable the write protection or grounded to disable the function. The SDA and SCL are open drain pins, and require pull-up resistors as shown in Figure 13-23. The SDA is bidirectional, and data flow in both directions, from the master to a slave and from a slave to the master. However, the master controls the clock and the direction of the data flow. The format of the control byte required to read and write to EEPROM is shown in Figure 13-24.

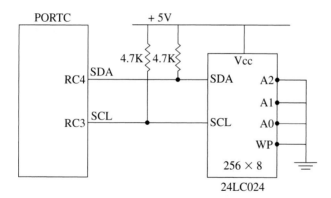

FIGURE 13-23   Interfacing Serial EEPROM with PIC18F MSSP Module in I2C Mode

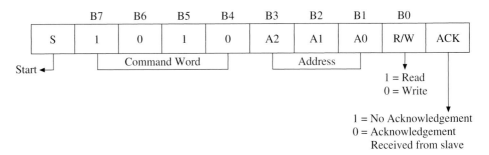

FIGURE 13-24   Control Byte Format

Write Operation

To write to the EEPROM, the following steps are required:

1. Send a Start bit. All I^2C commands must begin with the Start bit. Before sending the Start bit, the SSPIF flag in the PIR1 register must be cleared. To enable the Start condition, the bit SEN <Bit 0> in the SSPCON2 register is set, and the program should wait in the loop until the SSPIF flag is set.

2. Send the control byte. The code for the control byte is A0 $_H$ because (1) the code for the high-order nibble for this memory is A, (2) the address bits are 0 because they are grounded, and (3) the bit <0> is low for the write operation. To send the control byte, we follow the similar procedure as in Step 1: (1) clear the SSPIF flag, (2) load the control byte in SSPBUF, (3) wait in the loop until the SSPIF flag is set, and (4) check if the Acknowledge bit (Bit ACKSTAT <6>) is cleared.

   The EEPROM device must acknowledge by pulling the SDA line low for the ninth clock cycle.

3. Send the address. The 24LC024 EEPROM has only 256 bytes, and requires only one byte address. Memory chips larger than 2 Kb would require a 16-bit or 2-byte address. To send the address, we follow the same procedure as in Step 2.

4. Send the data byte. Once the address has been sent and acknowledged, the data byte can be sent by following the same procedure as in Step Two.

5. Send the Stop bit. The procedure to send the Stop is similar to that of the Start bit except that the bit set in the SSPCON2 register is PEN (Stop condition – Bit <2>).

Read Operation

To read from memory, the procedure is similar to the write operation, but somewhat more complex:

1. Send Start bit, followed by the write command and the address of the memory location to be read. These steps are identical to the write operation.

2. Send Restart bit. This is necessary to send the read command.

3. Send the read command from the master to the slave. Therefore, this is also a write operation.

4. Read the byte and send the Stop bit.

13.6.3 PROGRAM

```
;::::::::::::::::::::::::::::::::::::::::::::::::::::::::::::::::::::::::::::::
;This program sends the byte 55H to the EEPROM memory address 0010H        :
;The MSSP module of the PIC18F4520 microcontroller is initialized in       :
;I²C mode                                                                  :
;::::::::::::::::::::::::::::::::::::::::::::::::::::::::::::::::::::::::::::::


        Title " Illust 13-6 Interfacing Serial EEPROM in I²C Mode"
        List p=18F452
        #include <p18F452.inc>        ;This is a header file

OUTREG      EQU    0x20              ;Register used to send byte
ADDR        EQU    0x10              ;Address where byte is written
DATA_BYTE   EQU    0x55              ;Byte to be written in EEPROM

RD_CMD      EQU    B'10100000'       ;Command to read from EEPROM
WR_CMD      EQU    B'10100001'       ;Command to write to EEPROM

            ORG    00
            GOTO   MAIN

            ORG    0x20
MAIN:       RCALL  SETUP             ;Initialize MSSP in I2C mode
            RCALL  START_BIT         ;Send Start bit
            RCALL  TX_BYTE           ;Write a byte to EEPROM
            RCALL  STOP_BIT          ;Send Stop bit
WR_FAIL     NOP                      ;This location can be used to call
HERE        BRA    HERE              ;Error subroutine - otherwise, stay here

SETUP:      ;:::::::::::::::::::::::::::::::::::::::::::::::::::::::::::::::::
            ;Function:  The subroutine SETUP initializes SDA and SCL    :
            ;           (Bits <4-3>) of PORTC as inputs, clears control :
            ;           registers and flags, loads a baud rate count,   :
            ;           and places the MSSP module in I²C mode          :
            ;:::::::::::::::::::::::::::::::::::::::::::::::::::::::::::::::::

            MOVLW B'00011000'        ;Initialize bits <4-3> of PORTC
                                     ;as inputs
            IORWF TRISC
            CLRF   SSPCON2           ;Clear SSPCON2 and SSPSTAT
            CLRF   SSPSTAT
            BSF    SSPSTAT, SMP      ;Disable slew rate control
            BCF    PIR1, SSPIF       ;Clear SSP interrupt flag
            BCF    PIR2, BCLIF       ;Clear bus collision interrupt flag
            MOVLW B'00011000'        ;Load 24₁₀ in for 100 kHz baud
```

continued

```
            MOVWF SSPADD               ;See Example 13.8 for calculations
            MOVLW B'00101000'          ;Enable synchronous I²C master mode
            MOVWF SSPCON1
            RETURN

START_BIT:  ;:::::::::::::::::::::::::::::::::::::::::::::::::::::::::::::::::
            ;Function:  This subroutine clears the flag, and sets the     :
            ;           Start bit - SEN in SSPCON2 register and waits in :
            ;           the loop until the  flag  is set.                  :
            ;:::::::::::::::::::::::::::::::::::::::::::::::::::::::::::::::::

            BCF   PIR1, SSPIF          ;Clear SSP interrupt flag
            BSF   SSPCON2, SEN         ;Initiate Start - bit <0>
  CHK_SEN:  BTFSS PIR1, SSPIF          ;Is the flag set, if yes - return
            BRA   CHK_SEN              ;If not, go back to check again
            RETURN

STOP_BIT:   ;:::::::::::::::::::::::::::::::::::::::::::::::::::::::::::::::::
            ;Function:  This subroutine clears the flag, and sets the     :
            ;           Stop bit - PEN in SSPCON2 register and waits in   :
            ;           the loop until the : flag  is set.                 :
            ;:::::::::::::::::::::::::::::::::::::::::::::::::::::::::::::::::

            BCF   PIR1, SSPIF          ;Clear SSP interrupt flag
            BSF   SSPCON2, PEN         ;Initiate Stop - bit <2>
  CHK_PEN:  BTFSS PIR1, SSPIF          ;Is the flag set, if yes - return
            BRA   CHK_PEN              ;If not, go back to check again
            RETURN

WR_BYTE:    ;:::::::::::::::::::::::::::::::::::::::::::::::::::::::::::::::::
            ;Function:  This subroutine clears the flag, and loads the    :
            ;           byte in the SSPBUF register and waits in the loop :
            ;           until the flag  is set and checks acknowledge bit.:
            ;Input:     Receives the byte to be sent to EEPROM in OUTREG  :
            ;:::::::::::::::::::::::::::::::::::::::::::::::::::::::::::::::::


            BCF   PIR1, SSPIF          ;Clear SSP interrupt flag
            MOVF  OUTREG, W            ;Get byte from OUTREG
            MOVWF SSPBUF               ;Load byte in SSP buffer
  CHK_BYTE: BTFSS PIR1, SSPIF          ;Is the flag set, if yes - return
            BRA   CHK_BYTE             ;If not, go back to check again
            BTFSC SSPCON2,ACKSTAT      ;Is byte acknowledged? If yes, return
            BRA   WR_FAIL              ;If not, jump to failure routine
            RETURN
```

continued

```
TX_BYTE:     ;::::::::::::::::::::::::::::::::::::::::::::::::::::::::::::
             ;Function:  This subroutine loads: the write command, the   :
             ;                    address of memory location in EEPROM where a byte:
             ;                    should be stored, and the data byte in OUTREG –  :
             ;                    one at a time and calls the subroutine WR_BYTE    :
             ;                    to send these bytes to EEPROM                     :
             ;::::::::::::::::::::::::::::::::::::::::::::::::::::::::::::

  SEND_CMD:  MOVLW WR_CMD              ;Load write command in OUTREG
             MOVWF OUTREG
             RCALL WR_BYTE             ;Send write command to EEPROM

  SEND_ADDR: MOVLW ADDR               ;Load memory address in OUTREG
             MOVWF OUTREG
             RCALL WR_BYTE             ;Send address to EEPROM

  SEND_BYTE: MOVLW DATA_BYTE          ;Load data byte in OUTREG
             MOVWF OUTREG
             RCALL WR_BYTE             ;Send data byte to EEPROM
             RETURN
             END
```

Description:

This program includes the MAIN program and five subroutines: SETUP, START_BIT, STOP_BIT, WR_BYTE, and TX_BYTE. The MAIN program calls these subroutines in a sequence to store a byte from WREG to memory location 0010_H in EEPROM. The SETUP subroutine initializes the MSSP module in PIC18F452/4520 microcontroller in the I^2C mode.

In the problem analysis section, the sequence to send a byte to EEPROM is outlined; this program follows that sequence, and Figure 13-25 shows the flowchart of the program execution.

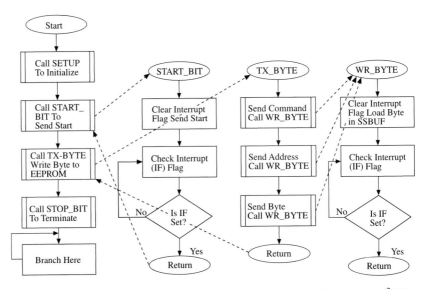

FIGURE 13-25   Flowchart: Data Transfer from PIC18 WREG to EEPROM in the I²C Mode

SUMMARY

The data communication between a microcontroller (MCU) and its peripherals can be divided into two major categories: parallel and serial. This chapter focused on serial communication in which one bit is transferred over one line. Therefore, to send out a parallel word from a microcontroller register, the word must be converted into serial format (stream of bits), called parallel to serial conversion. In serial reception, the MCU receives a stream of bits, and they must be converted into a word for the MCU to process

The serial communication between a receiver and a transmitter is classified into two formats: (1) asynchronous, which can take place any time, and (2) synchronous, which is controlled by the same frequency for the receiver and the transmitter. A variety of equipment for serial communication is manufactured by different manufacturers; therefore, there is a need of common understanding, called standards and protocols, among these manufacturers. This chapter focused on three frequently used protocols or standards: (1) EIA-232, (2) Serial Peripheral Interface (SPI), and (3) Inter-Integrated Circuit (I²C). Modules in the PIC18 microcontroller were used to illustrate these protocols. The important features of these protocols are summarized in the following section.

Serial Communication Interface (SCI) and EIA-232

1. EIA-232 (formerly known as RS-232) is a standard established for the asynchronous communication between Data Terminal Equipment (DTE) such as computers or terminals, and Data Communication Equipment (DCE) such as modems

2. EIA-232 standard for electrical signals protocol is defined using negative true logic: the voltage range –3 V to –25 V is defined as logic 1, and +3 V to + 25 V is defined as logic 0. It is not compatible with TTL logic levels. Transceivers are used to translate these voltage levels into TTL logic levels.

3. The asynchronous format of transmitting data is primarily character oriented. It includes a Start bit (low), 7 or 8 data bits starting with the LSB (D0), and one or two Stop bits (high); this is known as framing.

4. Serial communication is also classified in terms of the direction of flow: simplex, in which data flow in only one direction; full-duplex, in which data flow in both directions simultaneously; and half-duplex, in which data flow in both directions, but one direction at a time.

5. In serial transmission, errors are identified as: (1) a framing error when the Stop bit remains low, (2) an overrun error when a byte is transferred before the receiver completes the reading of the previously transmitted byte, and (3) a parity error when a data bit changes its logic level (as a result of noise in transmission).

6. The PIC18 family of controllers includes the module called USART (Universal Synchronous Asynchronous Receiver Transmitter); some versions have enhanced module called EUSART. Generically, this is also known as Serial Communication Interface (SCI).

7. The USART (or EUSART) specifies the parameters of the serial communication through three control registers: (1) the Transmit Status and Control Register (TXSTA), (2) the Receive Status and Control Register (RCSTA), and (3) the BAUD Rate Control Register (BAUDCON).

8. The USART can transmit and receive either 8-bit data or 9-bit data, and it uses two flags in the PIR1 register to indicate the status of the communication. Serial communication can be implemented either by monitoring the flags or through the interrupt by enabling the interrupt process.

Serial Peripheral Interface (SPI):

9. The SPI is a serial synchronous data exchange protocol between a master, such as a microcontroller, and slaves, such as data converters and serial EEPROMs.

10. The SPI uses four signals: Serial Data Output (SDO), Serial Data Input (SDI), Serial Clock (SCK), and Chip Select (CS), also known as Slave Select (SS).

11. The master controls the clock, and the user can set up the data exchange either on the rising or the falling edge of the clock.

12. The PIC18 family of controllers includes a module called Master Synchronous Serial Port (MSSP) that can be set up either in the SPI mode or the I^{2}C mode. In the SPI mode, 8 bits of data are synchronously transmitted and received simultaneously.

13. The MSSP module specifies and monitors the synchronous communication through two registers: the MSSP Control Register1 (SSPCON1) and the Status Register (SSPSTAT).

Inter-Integrated Circuit (I2C) Mode

14. The I^{2}C is a serial synchronous master-slave or multiple master protocols used in interfacing peripherals, such as data converters and serial EEPROMs, with a microcontroller.

15. It uses two signals: – Clock (SCL) and Serial Data/Address (SDA). These are open drain signals, and the master controls the clock.

16. The master initiates all data transfers, and it transfers 8 bits starting with the MSB. Framing includes a Start bit, 8-bit data, and an acknowledgement by the slave. The Stop bit terminates the data transfer either at the end of one byte or multiple bytes.

17. The MSSP module can be configured in the I^2C mode, and it can fully implement all master and slave functions. The module specifies and monitors the synchronous data transfer through three registers: MSSP Control Register1 (SSPCON1), MSSP Control Register2 (SSPCON2), and Status Register (SSPSTAT).

18. The mode can be implemented either by monitoring the flags or by generating interrupts.

QUESTIONS AND ASSIGNMENTS

13.1   Explain the difference between serial I/O and parallel I/O.

13.2   List advantages of serial I/O.

13.3   Explain the difference between asynchronous and synchronous data transmission.

13.4   Calculate the bit time when a fax machine transmits data at 1200 bps.

13.5   Classify a walkie-talkie conversation mode (simplex, half, or full duplex).

13.6   Sketch a waveform with timing when the character Z is being transmitted in the asynchronous mode with 56 k baud and one stop bit.

13.7   In Q.13.6, calculate the total time required to transmit 250 characters.

13.8   What is the Hex code used to transmit the 7-bit character "A" with odd parity?

13.9   What is a framing error and how is it recognized?

13.10   In Q.13.8, if the receiver receives the following bits, 0100 0001, is there a parity error?

13.11   When a block of data is being transmitted, what is the last byte sent in the check-sum method?

13.12   In EIA-232, identify the voltage range defined as logic 1.

13.13   Calculate the byte to be loaded in the SPBRG register to set the baud in the asynchronous mode = 9600 if F_{OSC} = 16 MHz and the BRGH bit is cleared in the TXSTA register.

13.14   Calculate the baud percent error in Q.13.13

13.15   In Q.13.13, calculate the byte value if BRGH is 1 (high speed) and the baud percent error.

13.16   Write a subroutine to initialize the EUSART in asynchronous mode to meet the following specifications: (1) 9600 baud with F_{OSC} = 16 MHz, (2) one Start and Stop bits and 8-bit data, and (3) enable transmit and receive.

13.17   Write instructions to send 68_H stored in data register 0020_H as ASCII characters assuming that the EUSART is initialized as in Q.13.16.

13.18   Write a subroutine to initialize the EUSART in asynchronous mode to meet the following specifications: (1) 9600 baud with F_{OSC} = 32 MHz, (2) one Start and Stop bits and 9 bits data, and (3) enable transmit and receive.

13.19   Write instructions to transmit 9-bit data assuming that the EUSART is initialized as in Q.13.18. 8 bits are stored in data register 20_H and the ninth bit is stored as B0 in register 21_H.

13.20   Identify the signals necessary to implement the SPI protocol.

13.21   Why the SPI protocol is referred to as data exchange rather than data transfer.

13.22   Identify the signals that are required in the I²C mode.

SIMULATION EXERCISES

13.1 SE Write a main program to call the subroutines EUSART (Example 13.2) and ECHO (Example 13.4) and terminate the program in a continuous loop. Assemble and load the program using the PIC18 Simulator IDE and execute the program as follows.

   a.   Single step the subroutine EUSART.

   b.   Click on Tools and open up the Hardware UART Simulation Interface. Click on Send Character and type the letter "A" in the dialog box.

   c.   Change the simulation rate to Fast or Extremely Fast and observe the screen UART Interface.

13.2 SE Rewrite the program in 13.1SE to send a string and execute the program using the Hardware Simulation Interface.

CHAPTER 14

DESIGNING EMBEDDED SYSTEMS

OVERVIEW

In Chapter 1, we began with an introduction to microcontrollers and microcontroller-based systems, called embedded systems. We examined the internal architecture of a microcontroller and the interdependency of the hardware and software, and emphasized the critical importance of integrating them when designing embedded systems. A block diagram of a Time and Temperature Monitoring System (TTMS) was introduced as an illustration of an embedded system.

Chapters 2 and 3 focused on a particular microcontroller family, the PIC18FXXX and PIC18FXXXX by Microchip, to illustrate Harvard architecture and its software model, and Chapter 4 examined programming as a problem-solving tool. Chapters 5 through 8 introduced the assembly language of the PIC18 family of microcontrollers with simple examples and built various illustrations—software components—of the block diagram. Chapters 9 through 13 illustrated various processes of data transfers—simple input/outputs, interrupts, and serial I/Os—using various I/O ports and communication modules of the PIC18 microcontroller family. We also discussed interfacing commonly used peripherals such as keys, temperature sensors, data converters, LEDs, seven-segment LEDs, and LCDs.

Now it is time to integrate hardware and software in a project—to bring together the various components we have built along the way and to examine how they come together. This chapter focuses on that integration and the tradeoffs between hardware and software by designing the TTMS introduced in Chapter 1. Before designing the system, the chapter discusses two topics: (1) the features of embedded systems and how they differ from other electronic systems, and (2) designing process of an embedded system. The chapter concludes by discussing various features of the PIC18 microcontroller family that are important in designing specialized embedded systems with various constraints, but were not discussed in previous chapters.

OBJECTIVES

- List the various features of embedded systems that differ from other electronic systems.
- List the steps in designing an embedded system.
- Explain why the hardware and software design decisions have to be made almost simultaneously in the early stage of a product development.
- Design the Time and Temperature Monitoring System (TTMS) to illustrate the design process in the embedded systems.
- List the features of the PIC18F microcontroller family that are necessary in designing specialized embedded systems.

14.1 FEATURES OF EMBEDDED SYSTEMS

Before we begin to design an embedded system, it is essential to examine how embedded systems differ from other electronic systems such as a personal computer. As discussed earlier, embedded systems are "embedded" inside a system controlling operations or as part of a large system performing a specific task, and are, therefore, not accessible to the user. Embedded systems are designed for specific dedicated tasks. They are constrained by space, memory size, power, and cost. They need protection schemes to prevent them from software failure, and some systems must operate within real-time constraints. In this section we will examine several features of embedded systems.

14.1.1 Specific Dedicated Purpose

Embedded systems are generally small and designed for a specific task, or a few tasks, and are generally designed using a microcontroller. There are at least twenty microcontrollers in an automobile; each one is dedicated to a specific task, for example, controlling the windows, checking seat belts, or setting the seat positions. Similarly, a large copying machine includes an embedded system that is designed to feed papers; in a PC, a microcontroller looks after the keyboard; in a digital clock, a microcontroller keeps track of time and displays the time. On the other hand, in contrast to embedded systems a PC is a general-purpose multitasking machine.

14.1.2 Space Constraints

Available space on a printed circuit board, generally known as "circuit real estate," is prime property. The circuit board must fit exactly within very tight quarters. For example, there is limited space available for a printed circuit board in modern cell phones, remote controls, iPods, or Walkmans. These portable devices are designed for higher and higher performance every year while they continue to grow smaller. Therefore, the number of pins on a microcontroller becomes a critical issue. Microchip has a series of microcontroller devices that range from 6 pins to 100 pins. The PIC10F200 microcontroller has 375 bytes of program memory and a total of 6 pins—4 I/O pins, a power, and a ground pin. On the other hand, PIC18F97J60 microcontroller has 100 pins and 128 K bytes of program memory.

14.1.3 Memory Constraints

One of the major advantages of using a microcontroller in an embedded system is that the processor, memory, and I/O ports are on one chip. However, integrating these features limits the size of the memory available to store software. The size of the program memory (ROM or flash) may vary from less than one kilobytes to 128 K bytes. In some microcontrollers it is possible to expand the size of memory by giving up some I/O ports, but that may increase the cost to an unacceptable level.

The size of the memory may dictate the selection of the language used to write programs: high-level or assembly language. If memory is limited, the designer may have to write programs in assembly language. In the PIC microcontroller families, memory size ranges from 375 bytes to 128 K bytes. At the present stage of memory technology, however, the size of memory is not really a major concern because it has become inexpensive. However, memory size becomes a concern if the number of pins and the board size are limited.

14.1.4 Power Constraints and the Trade-off Between Power and Frequency

In many systems, batteries with a limited lifetime power the onboard devices. Similarly, devices in space satellites are operated on solar power. In all these devices, power consumption must be minimized. Therefore, the sleep mode and the interrupts become very critical in designing these products. In the sleep mode, the system oscillator is turned off to minimize the power consumption. An interrupt is generated at the end of an event that can wake the MPU, which attends to a short interrupt service routine (ISR) and then goes back into the sleep mode again.

To minimize power consumption, the designer should also consider the tradeoff between power and frequency. As the frequency increases, the switching rate increases, which requires more power.

14.1.5 Cost Sensitivity

Electronics may be the only field where prices go down exponentially after the introduction of a product. In the early 1980s, the price of a PC was in the range of $1,000 to $10,000, and now, twenty years later, the price of a PC with much higher performance ranges from $300 to $1000. Cost is critical in consumer-oriented products, and saving a few cents per unit in manufacturing can add up to large savings for a company.

To minimize the cost of a product, the designer needs to consider many factors, such as the cost of the components, manufacturing, and service. A trade-off between hardware and software must be optimized. In the design stage, functions are partitioned between the hardware and the software. If the hardware (components) increases, then manufacturing cost increases proportionately and cost will be incurred during the entire product cycle. If more functions are assigned to software, the development cost increases, but is a one-time cost. The price of the Microchip microcontrollers, in 2006, ranges from less than fifty cents to more than six dollars.

14.1.6 Protection Against Software Failure

The implications of software failure in embedded systems are far more serious than systems such as a PC. Embedded systems are used in health-related systems such as heart monitoring, medical instruments, and pacemakers. Embedded systems are used in instruments where access to the device after the installation is extremely difficult, such as satellites. In a PC, when software fails, the system can be reset. In embedded systems, when the MPU is caught in a continuous loop, there is a need to reset the system automatically. Today, many microcontrollers include a watchdog timer inside the controller. It is based on the simple principle of resetting the watchdog timer before it runs out of its allotted time and then starts again. For example, if the watchdog timer is set for the maximum time of 20 ms to watch critical loops, and when a loop is executed within 20 ms, the watchdog timer is reset, and begins to monitor the next loop. Consider the analogy of a timed chess match in which players are limited to a certain amount of time to make a move. As soon as one player completes the move, the player resets the timer and the time of the second player begins. If the first player runs out of time, the player is penalized which is equivalent of resetting the play. If for some reason, for instance, a power glitch, the MPU is caught in an infinite loop and cannot execute the instructions within the allotted time, the watchdog timer resets the entire system, and the system starts again.

14.1.7 Real-Time Constraints

Some embedded systems are expected to operate under real-time constraints. These systems are grouped into two categories: (1) time-critical and (2) time-sensitive.

The time-critical embedded system must respond within a strict time limit, otherwise the consequences could be disastrous. For example, the antilock system on car brakes responds to the lack of traction in the tires in a timely fashion or within a strict time limit.

The time-sensitive embedded system also responds within a specified time limit. For example, a printer that comes with the specification of printing twelve pages per minute. However, in this case, if the system fails to print twelve pages per minute, the consequences are not serious.

In designing an embedded system, the designer should consider the above features and constraints on the embedded systems in:

- Selecting the microcontroller as a controlling device.
- Selecting the language to write software.
- Partitioning tasks between software and hardware to optimize the cost.

14.2 DESIGNING EMBEDDED SYSTEMS

In embedded systems, hardware and software are closely interlinked. A decision in hardware components may have considerable impact on the software, and vice versa. In designing embedded systems, hardware and software designs begin almost simultaneously, unlike the design processes of other computer systems. Designing embedded systems generally follows the following steps:

- Establishing system specifications
- Partitioning tasks between hardware and software

- Reevaluating partitioning decisions
- Designing hardware and software components
- Integrating hardware and software
- Testing and releasing to production
- Troubleshooting during production
- Upgrading and maintenance

Many of these steps are similar to those of any product design, except two steps, the step that calls for partitioning between hardware and software, and the step that requires integrating them. We will discuss these steps in the context of our Time and Temperature Monitoring System (TTMS) design project. Because this is a classroom project, and the focus of the textbook is on the PIC18 microcontroller family, we may not have freedom to make decisions, but we can discuss the issues in the context of the above steps. The project, even though it is simple, enables us to integrate the various concepts discussed throughout the text. However the issues related to the re-evaluation of various decisions, production, and maintenance are difficult to illustrate with examples from the project.

14.2.1 System Specifications

The objective of the TTMS project is to integrate various hardware and software concepts we discussed throughout the textbook in the context of PIC18 microcontroller family. The project is expected to monitor temperature and time, and display the time and temperature on an LCD. In addition, if the temperature falls below a certain limit, the system should turn on a heater, and if the temperature goes above a certain limit, it should turn on a fan. This project needs to be specified in terms of the following devices and parameters:

- Temperature range and limits
- Time format: 24-hour versus 12-hour clock
- Display: Devices for display
- Temperature input
- Power and frequency requirements
- User's ability to interrupt the system to set the time.
- Control of I/O peripherals such as the fan and the heater

Temperature

The temperature is an input to the TTMS. The temperature range to be monitored will have a significant impact in the selection process of the temperature sensor used in the project. The commonly used temperature sensors are one of four basic types: thermocouple, resistance temperature device (RTD), thermistor, and integrated silicone-based chips. The selection of the sensor depends on the temperature range, accuracy, linearity, ruggedness, and price. Table 14-1 shows some of the characteristics of these temperature sensors.

TABLE 14-1   Characteristics of Temperature Sensors

	Thermocouple	RTD	Thermistor	Integrated Silicon
Temperature Range	–270 to 1800°C	–250 to 900°C	–100 to 450°C	–55 to 150°C
Sensitivity	10s of μV/°C	0.00385 Ω/Ω/ °C (Platinum)	Several Ω/Ω/ °C	Based on technology that is –2mV/°C sensitive
Accuracy	±0.5°C	±0.01°C	±0.1°C	±1°C
Linearity	Requires at least a 4th order polynomial or equivalent look up table.	Requires at least a 2nd order polynomial or equivalent look up table.	Requires at least a 3rd order polynomial or equivalent look up table.	At best within ±1°C. No linearization required.
Excitation	None Required	Current Source	Voltage Source	Typically Supply Voltage
Form of Output	Voltage	Resistance	Resistance	Voltage, Current, or Digital
Typical Size	Bead diameter = 5 × wire diameter	0.25 × 0.25 in.	0.1 × 0.1 in.	From TO-18 Transistors to Plastic DIP
Price	$1 to $50	$25 to $1000	$2 to $10	$1 to $10

The TTMS is being designed to monitor indoor room temperature that may normally range from 40°F to 90°F. Considering the narrow temperature range to be monitored, the limited space on the printed circuit board, and the price, a silicone-based IC seems to be an ideal choice. We can extend the temperature range into the negative region as well, as beyond 120°F with the same temperature sensor. However, we may have to add a scaling circuit to adjust the output of the sensor to take the advantage of the full-scale range of the A/D converter.

The signal from the temperature sensor is an analog signal that must be converted into digital signal by an A/D converter. We need to select a microcontroller that includes an A/D converter as a peripheral module. The quantization error in the conversion from analog to digital depends on the resolution of the A/D converter module, and if a higher resolution is needed than what is available on the microcontroller, we may have to use a separate A/D converter chip with a higher resolution.

Time

Time is calculated by counting at a given interval (such as a clock period), which can be based on hardware timers or software delay loops. Hardware timers are more accurate than the software delay loops, and now multiple timers are available as peripherals on microcontroller chips. The decision to use hardware timers is an easy one.

Now the next question: should the clock be a 12-hour or 24-hour clock? This decision is based primarily on the customer preference and the market niche for which the system is designed. If the customer is a military unit, it may prefer 24-hour clock. However, this preference does not affect any major hardware decisions because this is done in software.

Display

This is an output, and the options to display time and temperature are seven-segment LEDs, matrix LEDs, and LCDs.

The seven-segment LEDs are least expensive, but limited in displaying various characters. The current required in each LED should be in the range of 10 ma to 15 ma for proper visibility. To minimize the power consumption, we will have to build a circuit to multiplex these LEDs and use two I/O ports, and in addition, our display will be limited in characters. If we limit our display primarily to digits, we may need six or seven digits to display time and temperature simultaneously. If we display time and temperature alternatively, we may need only four digits. This hardware decision affects the software decisions considerably.

The matrix LEDs are bright and very attractive displays. They would require a display driver chip or a driver would have to be written in software. In addition, they are more expensive than other options and consume more power.

The LCD display has become popular because of its ability to display alphanumeric and graphic characters even if the LCD is more expensive than the seven-segment LED display. We can use an 8-bit or 4-bit parallel interface depending on the number of pins available. If we have a limited number of pins available, we can use LCDs with a serial interface.

Power and Frequency

The integrated semiconductor chips such as microcontrollers are TTL logic compatible, and can run on a 5 V power supply. Alternately, the unit can be battery operated.

If we use a 5 V power supply, power consumption is not a critical issue. However, if the unit is to be marketed outside the United States, the power line voltages can be different. Therefore, the transformers used to step down the power line voltages will have to be selectable. A common practice is to step down the line voltage to unregulated 9 V and use a semiconductor regulator chip to supply 5 V DC.

If it is a battery-operated unit, the power consumption becomes a critical issue, and every effort should be made to minimize power consumption. One of the ways to minimize the power consumption is to use a lower clock frequency. In this system, the clock has to be updated every second, and there is ample time to complete all the operations well before one second elapses.

Two types of oscillator circuits are commonly used to generate frequency: an RC circuit and a crystal-driven circuit. The RC oscillator is generally used in time-insensitive applications, and is less expensive than the crystal-driven oscillator. In the RC oscillator circuit, the frequency is a function of temperature and power supply voltage in addition to the values of the RC components. From unit to unit there are likely to be some variations in frequency because of the tolerances in the component values. In the TTMS project the accuracy and the stability of the frequency is essential to maintain accurate timing. Therefore, we will use the crystal-driven circuit.

Specifications

Based on the above discussion, we can specify the various parameters as follows. Some of the specifications are somewhat arbitrary or simply the designer's choice.

- Temperature range: 0°F to 99°F
- Temperature sensor: semiconductor chip
- Clock: 12-hour
- Display: LCD
- Power supply: 5 V regulated and an additional negative voltage is necessary if an op amp scaling circuit is used to interface the temperature sensor to the microcontroller.
- Frequency: 10 MHz—crystal driven oscillator
- Clock setting: two switches

Figure 14-1 shows the block diagram of the above system.

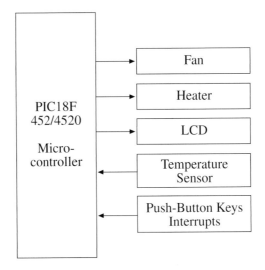

FIGURE 14-1   Block Diagram – Time and Temperature Monitoring Embedded System

14.2.2 Partition of Tasks Between Hardware and Software

Before we can begin to partition the tasks we must select a microcontroller. To select a microcontroller we need to examine the following items: (1) the number of I/O pins that are required, (2) the memory requirements, (3) cost, (4) availability of troubleshooting tools, (5) technical support from the manufacturer, (6) in-house capability and familiarity with the microcontroller being considered, and (7) availability of secondary sources.

I/O Pins Requirement

Based on the specifications outlined in the previous section, the TTMS project includes an LCD module, temperature sensor, fan, and a heater as output peripherals, and switches for interrupts. All of these devices require one pin each except the LCD. We can explore three options in interfacing the LCD: 8-bit interface, 4-bit interface, and serial LCD. Table 14-2 shows the pin count of these three options. The requirement ranges from fourteen pins to twenty-one pins.

TABLE 14-2   TTMS Project –Pin Requirements

System Components and Signals	8-bit LCD Interface/ Pin Requirements	4-bit LCD Interface/ Pin Requirements	Serial LCD Interface/ Pin Requirements
LCD – Output	8	4	1
Control Signals for LCD – Output	3	3	3
Temperature Sensor – Input	1	1	1
Fan and Heater – Output	2	2	2
Two Switches – Input	2	2	2
Power and Ground – Input	2	2	2
Clock Frequency - Input	2	2	2
Reset	1	1	1
Total Number of Pins Required	21	17	14

Memory Requirements

The size of the memory is determined by the size of the program. The size of the required memory is minimized if the programs are to be written in assembly language, especially for a small program. If we select a high-level language such as "C", the size of the memory requirement may increase depending on the "C" compiler and the size of the program.

The memory is divided primarily in two groups: R/W memory and Read-Only Memory (ROM). In order to choose between these options, we must answer two questions: (1) once the product is shipped, does the user need to modify any information to run the unit and (2) does the MPU need any memory for temporary storage during the execution of the program? If the user needs to provide any information—such as clock setting in TTMS, the R/W memory is needed to store this new information. The MPU also needs some memory for temporary storage if programs include subroutines or if memory registers are used to store information during the execution time. The TTMS unit requires input from the user to set the time, and that information must be stored in R/W memory. Similarly, the TTMS program is expected to include subroutines and the use of data registers. Therefore, we need both types of memory in our unit: ROM to store permanent programs and R/W to store the inputs from the user.

There are a number of ROM choices, including Masked ROM, one-time programmable (OTP), erasable programmable read-only memory (EPROM), Flash, or electrically erasable programmable read-only memory (EEPROM). The masked ROM becomes cost effective if the production quantities are in the thousands, but the product designer needs to depend on a memory manufacturer that may require a six- to twelve-week waiting period. If the production quantities are large, the product does not anticipate changes in near future, and the software is stable, the masked ROM is the most cost-effective choice. The advantage of using OTP over the masked ROM is that it can be programmed in-house, and it is somewhat cheaper than flash memory. However, one disadvantage is that it cannot be reprogrammed. The Flash memory has become popular because it can be programmed in-house and reprogrammed (1000 to 100 k erase/write cycles). The use of EPROM is almost obsolete because it needs to be taken out of the circuit to be reprogrammed and exposed to ultraviolet light to be erased. Today, many microcontrollers include on-chip flash memory and EEPROM. The advantage of EEPROM is that it can be easily reprogrammed a byte at a time in the field from a remote location. As a first estimate, based on the examination of similar systems, the TTMS may need less than 1 K of flash memory, fewer than 64 data registers, and no EEPROM.

Cost

The price of an 8-bit microcontroller in quantity ranges from less than fifty cents to more than six dollars in 2006. If a microcontroller meets the pin and memory requirements, we can select the least expensive microcontroller.

Troubleshooting Tools and Technical Support

This is also a very important consideration. In the product development process, tools such as a simulator, C-compiler, and an in-circuit debugger should be easily available for writing programs and troubleshooting software and hardware. If these tools are not available from the manufacturer, the designer will have to depend on the third-party tools.

In-house Capability

In-house capability is another very important consideration. If the company is designing products using Intel or Motorola microcontrollers, and most of the engineers are very familiar with those microcontrollers, there would have to be a compelling reason to switch to a new microcontroller. The company will have to incur the cost of retraining in-house personnel by switching to a new microcontroller.

Availability of Secondary Sources

When the unit is in production, there are always concerns about whether the selected microcontroller will be available in sufficient production quantities from the manufacturer. In a fast-changing technological environment, the manufacturer may discontinue the product unexpectedly for reasons such as decreasing demand or company takeover. In such circumstances, the availability of parts from second sources becomes important.

For the TTMS project we do not have much freedom to select a microcontroller because the textbook focuses on the Microchip 18F microcontroller. Given that, our objective here is to explain various options available in this microcontroller for different types of projects and product development. Among various family members of the 18F microcontroller family, we will focus primarily on the PIC 18F452 because the availability of trainers for this particular part and the 18F4520 because it is an updated part compatible with 18F452. The 18F4520 is recommended as a replacement for the 18F452 for new designs.

Partitioning Tasks Between Hardware and Software

This is a simple project, and tasks are almost divided between hardware and software as soon as the selection of the processor is done. In a large project, some of the examples of tradeoff between the hardware and the software are as follows:

1. Many products include a keyboard as an input device. When a key is pressed, it is necessary to verify which key is pressed and find its equivalent code. Software can perform this function or it can be assigned to a smaller microcontroller chip or a specially designed (called application-specific) hardware chip. The first option of performing this task in software is the slowest among all the options, but the least expensive. The last option of using a hardware chip provides the fastest response, enables the MPU to attend to other tasks, but it can be expensive. The option of using a smaller microcontroller is a compromise in terms of the cost and the response time.
2. Many products need inputs from multiple signals, and if all signals are present (the AND function), an action can be initiated. The software can check each input pin in a sequence and perform the AND function. However, this can also be done by using an AND gate that can provide much faster response than the software check.

In the TTMS project, modules inside the microcontroller perform many of the functions. One illustration of dividing tasks between hardware and software is the small chore of debouncing the interrupt key. A key can be debounced by building a hardware circuit as discussed in Chapter 10 or using a software delay. In this project we will use a software delay.

Now the next steps are re-evaluating the partitioning decisions and designing hardware and software modules. In a large project, these decisions must be re-evaluated before both teams—hardware and software—begin to work independently. Some decisions have to be made based on a team's past experience and its educated guesses. We will discuss the design aspects in the next two sections because these steps require an extensive discussion and specific details of the TTMS project. However, we can discuss the remaining steps in the design process in the following sections.

14.2.3 Integration of Hardware and Software and Testing

When the hardware and the software are developed separately, the earlier assumptions made by the design team may not hold true when both parts of the design are brought together. In most situations of developing a product, both hardware and software are tested separately to meet various specifications. Typically, in a classroom project, one team takes the responsibility of putting hardware together, and another team takes the responsibility of writing software.

Hardware testing is somewhat straightforward. In the TTMS system, one can check voltages with a meter or an oscilloscope at various points, for instance, at the output of the power supply, system clock frequency, power supply voltages, and the grounds of the every chip used on the board.

To test I/O pins, a simple code needs to be written. This testing is different from testing or troubleshooting analog systems. To troubleshoot an analog system, after checking the power and ground connections, one injects a signal at the input and traces that signal through various stages of the system. This method does not work in testing an MPU-driven system. For example, if you write an instruction to output some value to an LED port and nothing is lit up, then there is no way of checking the voltages at the I/O pins because the instruction is executed before voltages at the LEDs can be checked. You need to write a continuous loop to generate a steady pattern of output signals and then check the output pin voltages on the scope in relation to the output instruction. In classroom projects, typically, most students will download the entire software as soon as the hardware is built. However, it is highly recommended that you check every hardware component before the software is downloaded.

Software testing is generally done with a simulator, and each software module should be tested independently. When modules are combined, it is almost certain that the interaction of the independent modules will create several software errors called bugs. These errors must be corrected by checking the input and the output of each module. After all the errors are corrected, the software can be downloaded in the hardware target system.

Finally, hardware and software can be tested together. In our TTMS system, the first test is the power-on reset test. The LCD should display the time 00:00 and the room temperature. If the LCD is functioning properly, various elements of the system can be tested. The next test is of interrupt switches that you can observe on the LCD. Similarly, the system can be tested for the range of temperatures by cooling and heating the sensor within the limits. If some elements of the system are not functioning properly, you need to follow an iterative process of checking hardware and software. In designing embedded systems, the commonly used tools to troubleshoot hardware in conjunction with software are in-circuit emulators and in-circuit debuggers. The MPLAB ICD2, in-circuit debugger, is described in Appendix B.

14.2.4 Quality Control, Maintenance, and Upgrading

When a product is designed in a development lab, it may receive attention and the best selection of components When the unit goes to production, the components used tend to vary within a specified tolerance and the unit, which worked in the development lab, may not work for all instances on the production floor. In addition, what works electronically may not fit into the mechanical box that comes out of the production. At this stage the test will have to be the simple, "go/no go" type, and the quality control department will check whether sampled units meet the specifications. If any of these problems occur, some redesigning of the product may be necessary.

Maintenance and Upgrading

Maintaining and upgrading a product once it is shipped to the customer may affect the design process. When something goes wrong with a product in the field, generally, three methods can be employed to fix the problem: (1) the customer can ship the unit back, (2) some field personnel can visit the customer's site, or (3) online (remote) maintenance and upgrading.

The first two options do not affect the product design, but the third option may affect the design process. If it is possible to fix or upgrade the software code from remote location, the unit should be available online, and memory that is used in the product should be easily available for remote access. The code in EEPROM can be serially accessed and changed very easily. The code in flash memory can also be changed, but it should be built into the design process.

14.3 TTMS PROJECT DESIGN: HARDWARE

The design of the Time and Temperature Monitoring System (TTMS) is similar to that of the PICDEM™ 2 Plus demonstration board from Microchip. In this section, we will discuss the hardware aspect of the TTMS project and the software design in the next section. As discussed in the previous section, the hardware section is divided into the following components:

- Power supply and Oscillator (Clock)
- Temperature sensor as an input device
- Liquid Crystal Display (LCD) as an output device
- Switches to interrupt the MCU and set time
- Fan and heater as controlled output devices

These components are discussed in the following sections.

14.3.1 Power Supply and Oscillator (Clock)

The required voltage range for the PIC18F452/4520 microcontroller is 2 V to 5.5V; therefore, the TTMS can be designed as a battery-operated system. For our discussion, we will focus on a standard 5 V power supply as shown in Figure 14-2. This power supply can be plugged into a 120 V wall socket.

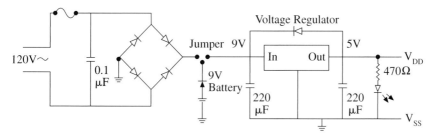

FIGURE 14-2   Regulated 5V-100mA Power Supply

Power Supply

The components required for this power supply are: (1) a wall transformer that converts 120 V into 9 V DC and provides 200 mA current, (2) a voltage regulator at 5 V with 1 A current capacity, (3) capacitors, (4) an LED, and (5) a 9 V battery.

- Wall Transformer: This transformer converts 120 V AC from the wall socket, using a full-wave rectifier, into 9 V unregulated DC with 120-cycle ripple.

- Voltage Regulator: This three-terminal integrated circuit provides a constant output voltage, even if the load current varies.
- Capacitors: These are large electrolytic capacitors. The input capacitor provides constant input voltage to the voltage regulator and filters out the 120-cycle ripple generated by 60-cycle AC voltage.
- LED: The circuit shows an LED with a 470-ohm resistor; this LED is turned on whenever the power supply is plugged in, indicating that the power is on.
- Nine-Volt Battery: The circuit also shows that the system can operate with a 9 V battery. The battery can be connected in place of the power supply, and the voltage regulator is protected by the indicated diode if the battery is connected accidently with the voltage reversed.

Oscillator

The PIC18F452 has eight and the 4520 has ten different oscillator modes. A mode can be programmed by setting appropriate logic levels of the bits FOSC3:FOSC0 in Configuration Register 1 H (discussed later in the chapter). These modes can be grouped in four categories: (1) using a crystal or a ceramic resonator, (2) RC oscillator circuit, (3) internal oscillator, and (4) external oscillator.

Figure 14-3a shows an oscillator circuit with a 10 MHz crystal connected to pins OSC1 and OSC2 with two 22-pico-farad capacitors. This is a quartz crystal with a fundamental frequency that behaves like a resonance circuit of a capacitor, inductor, and a resistor. An RC circuit can also be connected to generate the clock frequency as shown in Figure 14-3b, which is more cost effective but provides less stable frequency.

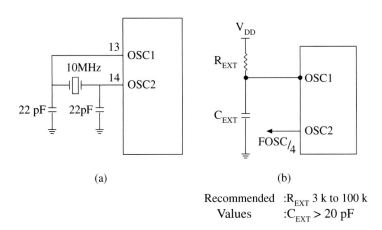

(a)   (b)

Recommended   :R_{EXT} 3 k to 100 k
Values   :C_{EXT} > 20 pF

FIGURE 14-3   Oscillator Circuits

The clock frequency directly affects the power consumption, and many special features are available in this microcontroller family to reduce the power consumption (discussed in Section 14.5).

Other Considerations: If the TTMS unit is to be marketed outside the United States, we may have to change the transformer ratio as well as the shape of the wall socket; many countries have 220–240 V as a standard AC input and use different types of wall sockets.

14.3.2 Temperature Sensor as an Input Device

The TTMS is to be designed for indoors, and the specified temperature range is 0°F to 99°F. Various types of temperature sensors are available as integrated semiconductor chips from different manufacturers; some are calibrated in Fahrenheit, some in Celsius, and some in Kelvin. For our project, a sensor with the voltage output proportional to Fahrenheit is an ideal choice. However, if the system is to be marketed outside the United States, the display should be converted to Centigrade. This conversion can be done in software, which does not affect hardware decisions.

In Chapter 12, Section 12.3.2, we discussed the National Semiconductor LM34 temperature sensor. This is a three-terminal device and its output changes linearly 10 mV per degree Fahrenheit starting with 0 volts at zero-degree temperature. The output voltage range of the LM34 will be 0 to 0.99 V for the temperature range of the system. The PIC18F4520 microcontroller includes an A/D converter module with 10-bit resolution and thirteen channels. We can use Channel AN0 (pin RA0) as shown in Figure 14-4. The reference voltages V_{REF+} and V_{REF-} can be the same as power supply voltage +5 V and ground. However, the input voltage from the temperature sensor to the A/D converter module ranges from 0 to 0.99 V. Therefore, we need to scale the input voltage range to match the entire range of the reference voltages of the A/D converter. This can be easily accomplished by using an operational amplifier with the gain of five as shown in Figure 14-4.

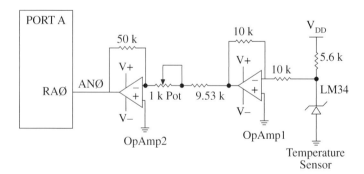

FIGURE 14-4   Connecting a Temperature Sensor to A/D Converter Input through a Sealing Circuit

The scaling circuit shown in Figure 14-4 has two precision operational amplifiers. The first operational amplifier (Op Amp1) functions as a unity gain amplifier. It changes the polarity of the input voltage range from positive to negative. The second Op Amp (Op Amp2) amplifies the input signal with the necessary gain (approximately five) and changes the polarity of the signal back to the positive voltage range. Therefore, the output of the scaling circuit ranges from

0 V to +4.99 V. The scaling circuit is shown here to demonstrate a method of interfacing a sensor to an MCU; otherwise, this circuit is unnecessary in this project because for such a limited range of temperature the 10-bit converter will provide the necessary resolution.

14.3.3 Liquid Crystal Display (LCD) as an Output

As discussed in the previous section, to display time and temperature in digits and view the display from a distance, seven-segment LEDs or matrix LEDs are more appropriate than LCD displays which are more expensive than LED displays. The primary reasons to use LCD displays in this project are: (1) any alpha numeric message can be displayed, (2) anyone designing an embedded system should be familiar with the interfacing of the LCD display, and (3) this display is generally available in trainers (such as Microchip PICDEMO™ 2 Plus) used in college laboratories.

The project specifies an LCD display, and the Hitachi LM032L (2-line x 20 Character) or equivalent is a commonly used display. This module can operate in three modes: (1) an 8-bit interface, (2) a 4-bit interface using high-order four data lines <7-4>, and (3) a 4-bit interface using low-order four data lines <3-0>. An 8-bit data interface requires less programming, but uses 11 (8 data lines and 3 control signals) I/O pins. A 4-bit data interface uses only 7 I/O pins, but requires more programming. This is a tradeoff between programming and I/O pins. In this project, we will use an 8-bit interface because of the availability of I/O pins.

Figure 14-5 shows the interfacing of the LCD module. The circuit uses eight pins of PORTD for LCD data lines, and three pins from PORTA for control signals. The operation of this circuit is discussed in Chapter 9 (Section 9.6).

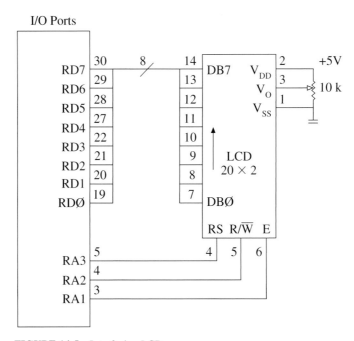

FIGURE 14-5 Interfacing LCD

14.3.4 Reset and Interrupts

As discussed in Chapter 10, a reset signal is used in a system to establish the initial conditions at start-up or to reestablish the initial conditions in the middle of an operation. The PIC18 family of controllers include multiple types of resets, but we are concerned with two types of reset: (1) Master Clear ($\overline{\text{MCLR}}$) reset and (2) Power-On Reset (POR). The question we need to answer is: Does the user need the $\overline{\text{MCLR}}$ reset? The user is expected to plug in the power cord and set the time. There is no reason for the user to reset the system. The $\overline{\text{MCLR}}$ can be tied to the power supply voltage V_{DD} as shown in Figure 14-6. When the user plugs the unit in the wall socket, the power-on reset will establish the initial conditions for the TTMS.

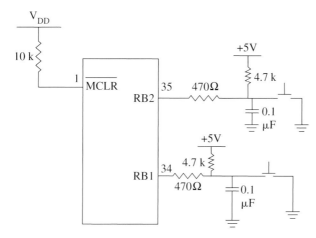

FIGURE 14-6   Reset and Interrupts

The next step is to set the time in hours and minutes. The only way to set the time is to provide input signals through a switch. A switch input can be used to interrupt the MCU or software can be written to poll the switch input. The Microchip PICDEM™ Plus trainer uses the polling method; we use the interrupt method. Figure 14-6 shows two push-button switches—one to set hours and the other to set minutes—connected to pins RB1 and RB2. These pins can accept interrupts from outside sources. The ISR (interrupt service routine) associated with each switch should increment the digits for hours and minutes.

14.3.5 Fan and Heater as Controlled Output Devices

The TTMS is expected to turn the fan and heater on and off at given temperatures. These devices are controlled by two output bits: RB7 controls the fan and RB6 controls the heater. The fan and the heater are high-voltage devices run on 120 V AC. These devices should not be connected directly to low-voltage digital systems, and their ground connections must be isolated. Figure 14-7 shows that the fan and heater are connected to bits RB7 and RB6, respectively, through the optically isolated triac drivers. Logic 0 at bit RB7 (and at RB6) turns on the LED of the optoisolator and the current flow in the LED, a minimum of 10 mA, turns on the triac driver, which, in turn, switches on the power triac 2N6071. The same explanation is applicable to the circuit that drives the heater. This is a high voltage circuit and should not be tried in a laboratory without additional safety measures and supervision. To verify the software, we do not need to build these circuits; the program logic can be verified by connecting LEDs to pins RB7 and RB6 as shown in Figure 14-8

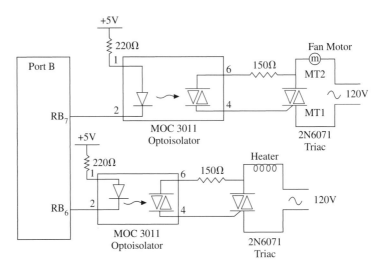

FIGURE 14-7   Interfacing of the Fan and the Heater

14.3.6 Complete TTMS System

Based on our discussion in the previous sections, Figure 14-8 shows the complete schematic of the TTMS system using the PIC18F452/4520 microcontroller. It is a 40-pin device, generally used in PIC18 available trainers. The TTMS uses only three ports: PORTA, PORTB, and PORTD. The eight data lines of the LCD are connected to PORTD, and three pins from PORTA are used for control signals. The temperature sensor, National Semiconductor LM34, is connected to channel AN0 (pin RA0). LEDs connected to pin RB7 and RB6 simulate the fan and the heater. To set the time, two switches—one to set hours and the second to set minutes—are connected to pins RB1 and RB2. The Master Clear $\overline{\text{MCLR}}$ is connected to the power supply through a 10 k resistor, and a 10 MHz crystal is connected to two pins—OSC1 and OSC2—to provide a 10 MHz clock to the system. The power supply that can be plugged into a 120 V AC wall socket supplies a 9 V DC input to the regulator which, in turn, provides a regulated 5 V DC to the system. The LED at the power supply indicates the on/off condition of the power supply.

When the user plugs the system in to 120V AC, the power supply turns on the LED, the power-on reset (POR) establishes the initial conditions of the system, and the LCD module displays time 00:00 AM and temperature 00.0°F. If the user does not set the time, the system begins the time at 00:00 AM and displays the room temperature. If the user pushes the hour switch, the system is interrupted, and the interrupt service routine (ISR) of that switch increments the hour digit at each contact of the switch until 12 PM and begins again at 1 PM. The user can set the minutes by pressing the minute switch, which also interrupts the system, and its ISR increments the minutes until 59 and restarts from 00.

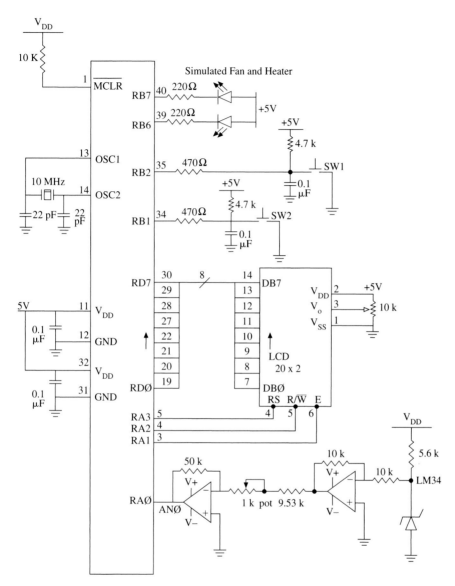

FIGURE 14-8   TTMS Circuit Schematic

14.4 TTMS PROJECT DESIGN – SOFTWARE

The software segment of the project involves the following tasks:

- Reading the room temperature
- Checking the reading against the set values of high and low to turn on/off the fan and the heater
- Displaying the temperature on line two of the LCD
- Setting a timer for a one-second delay
- Updating the clock every second and displaying the time on line 1 of the LCD

We can convert these tasks into independent modules, and write them as subroutines. We have written these modules in various earlier chapters, though we may have to modify these modules to meet the specifications of this project. Before we examine the details of each module, we need to discuss some critical issues regarding the use of interrupts versus polling flag bits.

This project includes two major tasks: monitoring the temperature and monitoring the clock. Both can be interrupt driven: when the A/D converter module completes a conversion, it can generate an interrupt; when the timer overflows, it too can generate an interrupt. In this scenario, the main program will consist primarily of instructions that call "display routines. The second option is to use an interrupt for one task and check the status flag for the other task. For example, the MPU monitors the temperature conversion loop, and when the timer overflows, it interrupts the MPU, the ISR updates the clock, and the MPU goes back to monitoring the temperature. In this option, we can also reverse the functions: the MPU waits for the timer to overflow and the A/D module generates an interrupt at the end of each conversion. The third option is to check flags for both tasks. The MPU waits in the timer loop and when the timer overflows, the MPU updates the clock and begins monitoring the temperature conversion loop. When the conversion is complete, the MPU updates the temperature on the display and goes back to the timer loop. In a large project, the third option, checking flags, can become problematic if the MPU is expected to perform many tasks. The interrupt-driven tasks keep the MPU free to perform other tasks. In this project, we can use any of the options; however, for an illustration, we focus on option two in which the MPU keeps checking the conversion flag until it is interrupted by a timer.

14.4.1 Project Flowchart and Interrupts

Based on our discussion in the previous section, if we choose option 2, the project software can be divided into two independent segments: (1) A main segment consists of subroutines that read temperature from the temperature sensor, convert analog readings into digital equivalents, and display the temperature at the LCD, and (2) A clock interrupt segment consists of interrupt service routines (ISRs) that update the clock when an interrupt is generated.

MAIN Segment

This segment consists of four modules as shown in Figure 14-9: (1) Initialization Module to initialize various I/O ports and modules such as the A/D converter, (2) Temperature Module to start an A/D conversion, read the temperature at the end of the conversion, convert the reading in a proper format, and turn on/off the fan and the heater (3) Display Module to display the time-and temperature at the LCD, and (4) Time Delay Module to provide delays to stabilize the LCD display.

Initialization Module

This module calls the SETUP subroutine to initialize various I/O pins either as inputs or outputs as well as the A/D converter module in the 18F4520 microcontroller. The SETUP subroutine initializes:

- Output pins: (1) all eight pins of PORTD that are used as data lines to the LCD; (2) the LCD control signals RA1 and RA2 of PORTA; (3) RB7 and RB6 pins of PORTB that control the fan and the heater.
- Input pins: RA0/AN0 for the analog signal from the temperature sensor and RB1 and RB2 for (as external interrupts from the switches) to set the time.
- A/D Module: It selects channel AN0, sets it up for analog input, and turns on the A/D module. Selects V_{DD} and V_{SS} as reference voltages, right justified display, and an appropriate conversion frequency F_{OC} and bit conversion time T_{AD}.

Temperature Module

This is one of the major modules illustrated in Chapter 12, Section 12.3. The module consists of multiple subroutines: (1) the A-TO-D subroutine to start a conversion and read the temperature at the end of the conversion; (2) the MULTIPLY10 and DIVIDE100 subroutines to convert the temperature reading into its equivalent integer and fraction; and (3) the BINBCD subroutine to convert the reading into BCD numbers and their ASCII equivalent. These readings are stored in data registers for the LCD module.

The module also checks the low and high temperature limits. If the temperature reading is higher than 80°F, the module calls the FANON subroutine to turn on the fan, and if the temperature reading is lower than 60°F, the module calls the HEATERON subroutine to turn on the heater. The flowchart shows two decision points for the fan on/off and for the heater on/off. At the outset, it looks redundant to ask the same question twice, but the subroutine FANON, illustrated in the next section, explains the rationale. Another issue is an attempt to turn on the fan even if the fan is already on. For example, when the temperature goes above 80°F, it is likely to stay above 80°F for a while. Rather than sending a signal to turn on the fan when it is already on, the subroutine needs to check whether it is already on; this can be accomplished by using a flag bit for on/off conditions. This flag concept is explained in Section 14.4.2.

Display Module

This is the LCD module illustrated in Chapter 9, Section 9.6. As explained in Chapter 9, this module initializes the LCD, sends the necessary commands, and displays the ASCII characters stored in data registers at TIME0, TIME, and DEGREE. When the TTMS system is turned on, it should display time as 00:00. Five ASCII characters for 00:00 are stored at location TIME0 that can be displayed by this module. The Temperature module saves temperature readings at data registers defined as DEGREE, and the ISR of the timer saves the time at data registers defined as TIME (explained in the ISR segment).

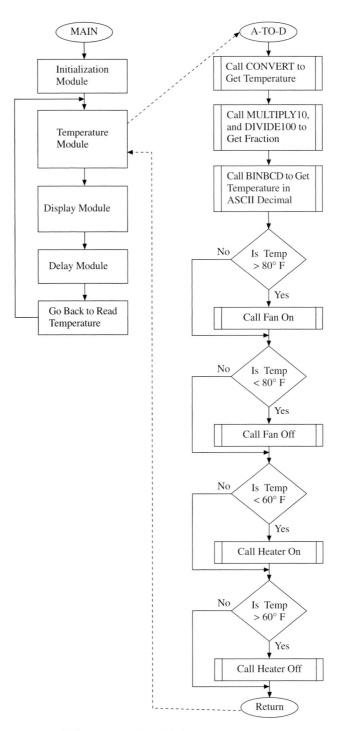

FIGURE 14-9 Flowchart TTMS Software

Delay Module

This subroutine provides multiples of a 20-millisecond delay, and the multiple is determined by the count in data register MULTIPLE. This delay is used between outputting consecutive displays and debouncing the interrupt keys.

Clock Interrupt Segment

This segment consists of three interrupts: one from Timer0, which is used for the clock., and two from the two external push-button switches connected to pins RB1 and RB2 to set time. The Timer0 interrupt is set up as a high-priority interrupt, and the interrupts generated by push-button switches are set up as low-priority interrupts.

Timer0 Interrupt

In Chapter 11, Section 11.5, the design of a 12-hour clock is illustrated. In this illustration, Timer0 is set up to generate an interrupt every second, the interrupt service routine CLK_ISR updates the time, and the BCDASCII subroutine converts hours, minutes, and seconds into equivalent ASCII characters that are saved in data registers labeled as TIME0 to TIME+7

Push-Button Interrupts

Two push-button keys are connected to pins RB1 and RB2 to set minutes and hours respectively. When the user pushes one of the keys, the key sends an interrupt signal, and the MPU stops the processing. Because it is a low-priority interrupt, the MPU jumps to memory location 00018_H and then to KEY_ISR. The flowchart of the KEY_ISR interrupt is shown in Figure 14-10.

The interrupt KEY_ISR first waits in a loop for 20-millisecond to debounce the key and then checks RB1 and if it is low, the service routine checks if minutes = 59. If yes, the ISR clears the minutes, increments hours, and returns to the place where it was interrupted; otherwise, it increments minutes only and returns. If RB1 is high, the KEY_ISR checks RB2 next, and if it is low, the KEY_ISR updates hours and returns to the places where the MPU was interrupted. If none of the keys is low, the KEY_ISR assumes it is a false alarm and returns to the place of the interrupt.

14.4.2 Subroutines – FANON and FANOFF

In the previous section, Figure 14-9 shows the flowchart of the TTMS software. The flowchart of the subroutine A-TO-D shows one decision point checking whether the temperature is higher than 80°F, followed by the second decision point checking whether the temperature is less than or equal to 80°F. This is somewhat unusual. If we do not ask the second question, the fan will be turned off, even if the fan has just been turned on by the previous subroutine because the temperature was higher than 80°F. The same logic is also applied to the process of running the heater.

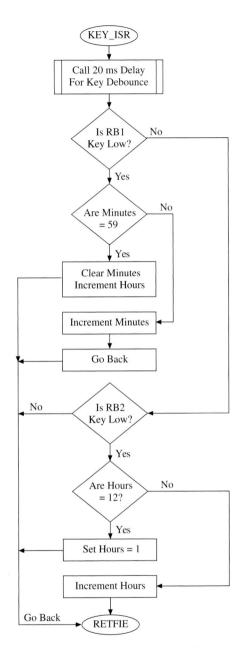

FIGURE 14-10 Flowchart for KEY_ISR

The following instructions check the temperature against the set limits and turn the fan on or off.

Instructions:

CHK_80	MOVLW	D'80'	; Load 80 in WREG
	CPFSLT	INTEGER	; Is temperature less than 80°F, if yes, do not turn ;on fan
	CALL	FANON	; Turn on fan
	CPFSGT	INTEGER	;Is temperature higher than 80°F, if yes, do not turn ;off fan
	CALL	FANOFF	;Turn off fan
	RETURN		;Go back
FANON:	BTFSS	FANFLAG, 0	;If Bit0 in FAN register =1, fan is already ;on – skip turning on
	BSF	PORTB, RB7	;Turn on fan
	BSF	FANFLAG, 0	;Set Bit0 in FAN register to indicate fan is on
	RETURN		
FANOFF:	BITFSC	FANFLAG, 0	;If Bit0 in FAN register =0, fan is already ;off – skip turning off
	BCF	PORTB, RB7	;Turn off fan
	BCF	FANFLAG, 0	;Reset Bit0 in FAN register to indicate fan is off
	RETURN		

Description: The first instruction loads the temperature limit value 80°F in W register and the next instruction CPFSLT (Compare F with W) compares the temperature value in the INTEGER register with the limit value in W register. If the temperature value is less than 80°F, it skips calling the subroutine FANON. If it is higher than 80°F, it calls the subroutine FANON to turn on the fan. The next instruction again compares the same values, and if the temperature reading in the INTEGER register is greater than 80°F, it skips calling the subroutine FANOFF. If it is less than 80°F, it calls the subroutine FANOFF to turn off the fan. An interesting question here is: what would happen if both temperatures – the limit value and the actual temperature – are equal? See the simulation exercise 14.2SE. When the subroutine FANON is called, the subroutine first checks Bit0 in the FANFLAG register, which is used as a flag register (not the STATUS register in the controller). The FANFLAG is any data register used to set Bit0 as a reminder when the fan is turned on and to reset Bit0 when the fan is turned off. This is like writing a reminder in your appointment book, and when the appointment is over, you erase that appointment. If the Bit0 is at logic 1, it means the fan is already on, and there is no need to turn it on again. If Bit0 is at logic 0, the subroutine turns on the fan and sets Bit0 in the FANFLAG register. The same logic is followed in the FANOFF register.

14.5 SPECIAL FEATURES AND CONFIGURATION REGISTERS

As discussed before, the design of embedded systems has many constraints, from limited space to minimum power consumption. The PIC18F microcontroller family includes many onboard features that assist designers in designing systems within these constraints. These special features are listed here with brief descriptions. Many of these features are available only in some families, and the configuration of these features is device specific.

14.5.1 Special Features

The special features of the PIC18F4520 family are as follows (some of the following features are not available in the PIC18F452):

- Watchdog Timer (WDT)
- Power-Down Mode (Sleep Mode)
- In-Circuit Serial Programming (ICSP)
- In-Circuit Debugger
- Fail-Safe Clock Monitor
- Two-Speed Start-up
- Code Protection
- ID Locations

These features are briefly described in the following paragraphs.

1. Watchdog Timer (WDT). We have already described the concept behind the watchdog timer in the context of software protection. The WDT is a circuit that will reset the microcontroller when the MPU is unable to complete the code execution in an allotted time. The WDT is enabled or disabled, or its time out interval is expanded by writing to CONFIG2H register – listed later.

2. Power-Down Mode (Sleep). This mode is used primarily to reduce power consumption, especially in battery-operated systems. When the MPU executes the SLEEP instruction, the controller stops the system clock and enters into the power-down mode. It can be awakened by different events such as resets, interrupts from various peripherals, and transmit and receive operations.

3. In-Circuit Serial Programming (ICSP). The PIC18 microcontrollers with flash memory can be programmed using the serial mode. This allows the manufacturers of embedded systems to manufacture boards without programs and install programs just before shipping the product. This enables the designer to install the most recent or custommade firmware in the products.

4. In-Circuit Debugger. This function enables the designer to debug the software while testing the system. To debug the software, the internal operations of the microcontroller should be transparent. The system under test uses two I/O pins (RB7 and RB6) to communicate serially with the outside world concerning the events internal to the microcontroller. With the appropriate software, these events can be checked or observed on a PC screen.

5. Fail-Safe Clock Monitor (FSCM). The microcontroller includes an internal RC oscillator (INTRC). If FSCM is enabled, the INTRC oscillator runs at all times to monitor

clocks to peripherals, and the FSCM allows the microcontroller to continue operation in the event of an external oscillator failure, by automatically switching the device clock to the internal oscillator clock.

6. Two-Speed Start-up. This feature minimizes the latency period (delay) from oscillator start-up to code execution by allowing the microcontroller to use Internal Oscillator (INTRC) as a clock source until the primary clock source is available.

7. Code Protection. The PIC18F microcontroller family provides code protection from piracy. The user program is divided into five blocks. Each block has three code protection bits associated with it: (1) Code-Protect bit, (2) Write-Protect bit, and (3) External Block Table Read bit.

8. ID Locations. Eight memory locations (200000_H – 200007_H) are designated as ID locations. The user can store checksum or other code identification numbers in these locations.

All the above features can be programmed by resetting bits in the configuration registers, which are discussed in the next section.

14.5.2 Configuration Registers

The PIC18F4520 microcontroller includes eleven configuration registers (shown in Table 14-3) that can be programmed to select various features described above; these registers are upward compatible with the registers in the PIC18F452. The last two registers in Table 14-3 show the device identification code, and these are read-only registers. The configuration registers are mapped starting at program memory location 300000_H, and can be accessed by using table-read (TBLRD) and table-write (TBLWT) instructions. The functions of various bits in these registers are listed in Appendix J.

TABLE 14-3   Configuration Bits and Device IDs

File Name		Bit 7	Bit 6	Bit 5	Bit 4	Bit 3	Bit 2	Bit 1	Bit 0	Default/ Unprogrammed Value
300001h	CONFIG1H	IESO	FCMEN			FOSC3	FOSC2	FOSC1	FOSC0	00-- 0111
300002h	CONFIG2L				BORV1	BORV0	BOREN1	BOREN0	PWRTEN	---1 1111
300003h	CONFIG2H				WDTPS3	WDTPS2	WDTPS1	WDTPS0	WDTEN	---1 1111
300005h	CONFIG3H	MCLRE					LPT1OSC	PBADEN	CCP2MX	1--- -011
300006h	CONFIG4L	DEBUG	XINST				LVP		STVREN	10-- -1-1
300008h	CONFIG5L					CP3[1]	CP2[1]	CP1	CP0	---- 1111
300009h	CONFIG5H	CPD	CPB							11-- ----
30000Ah	CONFIG6L					WRT3[1]	WRT2[1]	WRT1	WRT0	---- 1111
30000Bh	CONFIG6H	WRTD	WRTB	WRTC						111- ----
30000Ch	CONFIG7L					EBTR3[1]	EBTR2[1]	EBTR1	EBTR0	---- 1111
30000Dh	CONFIG7H		EBTRB							-1-- ----
3FFFFEh	DEVID1[1]	DEV2	DEV1	DEV0	REV4	REV3	REV2	REV1	REV0	xxxx xxxx[2]
3FFFFFh	DEVID2[1]	DEV10	DEV9	DEV8	DEV7	DEV6	DEV5	DEV4	DEV3	0000 1100

Legend: x = unknown, u = unchanged, = unimplemented, q = value depends on condition.
Shaded cells are unimplemented, read as 0 .
Note 1: Unimplemented in PIC18F2420/4420 devices; maintain this bit set.
   2: See Register 23-14 for DEVID1 values. DEVID registers are read-only and cannot be programmed by the user.

14.5.3 Programming Configuration Registers

To program a function or a special feature, the specific bits should be reset to 0 in the configuration registers. The configuration registers can be programmed by using the assembler directive __CONFIG (the assembly directive CONFIG preceded by two underscores), and the format is:

__CONFIG <register address>, <expression>

Example 1: __CONFIG   0x300009, 0x80

Description: Configure register 300009_H to protect the boot block code, but do not protect EEPROM code. This register is labeled as CONFIG5H and only two bits <7-6> labeled as CPD and CPB are used for code protection. The remaining bits <5-0> are unimplemented. When Bit7 and Bit6 are reset to 0, codes are protected. The address of this register is shown in Table 14-3, and bit specifications are described in Appendix J.

The addresses of the configuration registers and bit labels such as CPD and CPB are defined in the MPLAB IDE. In assembling the program using the MPLAB IDE, the Include File and INX32 should be specified as shown in previous chapters in assembling various programs. To assemble the pseudo command using the MPLAB IDE, the above example can be written by substituting the register address and bits by their respective labels. The alternative ways of writing the pseudo code in Example 1 are illustrated as follows:

Example 2: __CONFIG   CONFIG5H, 0x80
Example 3: __CONFIG   CONFIG5H, CPD_OFF_5H & _CPB_ON_5H
Example 4: __CONFIG   CONFIG15, _CPB_ON_5H

In Example 1, the address of the register CONFIG5H is included, and in Example 2, the address is substituted by the label CONFIG5H. In Example 3, bits <7-6> are specified with labels CPD and CPB, respectively, followed by the commands OFF or ON as necessary. In Example 4, the Bit6 – CPB – is turned on, meaning the bit is reset and the boot block code protection is on. If any bit is not included in the assembler command, the MPLAB IDE assumes the bit is set to 1 and is not to be programmed.

SUMMARY

Embedded systems differ considerably from traditional computer or programmable electronic systems. Embedded systems are built for specific dedicated tasks. Designers must consider the constraints of available board space, memory, power consumption, and cost sensitivity. Many of these systems need to be designed with real-time constraints and protection against software failure.

The process of designing embedded systems resembles the process used for other electronic systems, but with two exceptions: (1) partitioning of tasks between the hardware and software in the early stages of the product development, and (2) integration of the hardware and software.

The primary objective of this chapter is to illustrate the integration of the hardware and software concepts discussed throughout the text. This chapter illustrated various design steps and the integration of hardware and software by designing the TTMS. The design includes a complete schematic of the hardware system based on the PIC18F452/4520 microcontroller and its related software flowchart. The software design has drawn upon various program illustrations and examples throughout the text.

Finally, the chapter discusses the following various special features available in the PIC18F family of microcontrollers:

- Watchdog timer to prevent software failures

- Power-down (sleep) mode to minimize the power consumption

- In-circuit serial programming to install software after the hardware boards are built and to upgrade the boards online.

- In-circuit debugging to troubleshoot various hardware and software problems

- Program code protection to prevent the piracy of the installed software

- Configuration registers to configure various settings and operational modes

QUESTIONS AND ASSIGNMENTS

14.1   Explain how embedded systems differ from other electronic systems.

14.2   List various constraints under which embedded systems need to be designed.

14.3   Explain the concept of real-time constraints.

14.4   List the design steps that are required in the development of embedded systems that are different from designing other electronic systems.

14.5   The operating range of the temperature sensor LM34 is from -50°F to +300°F. The output voltage range of the sensor is -500 mV to +3000 mV, which is scaled to the range of 0 to 5 V as an input to the PIC18 A/D converter module with the 10-bit resolution. Calculate the input voltage to the A/D converter when the temperature is 0°F.

14.6   Calculate the temperature reading when the converted digital reading is 32_H.

14.7   Calculate the temperature reading when the converted digital reading is 200_H.

14.8   Calculate the digital Hex reading when the temperature is +100°F.

14.9   In Figure 14-2, explain the function of the diode connected to a battery source.

14.10   Write instructions to check whether a temperature reading is less than 60°F. If the temperature is less than 60°F, call the subroutine HEATERON. If the temperature is higher than 60°F, call the subroutine HEATEROFF.

14.11   Write the instructions for the subroutine HEATERON. The subroutine should check if the heater is already on, and it is already on, skip the step of turning on the heater. Use Bit1 of the data register TEMPFLAG to indicate whether the heater is on or off.

14.12   Write a 20-ms delay subroutine that can provide delays in multiples of 20 ms.

SIMULATION EXERCISES

14.1 SE Write a main program to define registers and initialize PORTB , and call the subroutine CHK_80 (Section 14.4.2). Execute the program in the animate mode to verify turning on/off of the fan and the heater at temperatures higher than the set limit and lower than the set limit.

14.2 SE Test the program in 14.1SE when the set limit and the temperature reading are equal. Explain your results.

APPENDIX

PIC18FXXX/XXXX
INSTRUCTION SET

APPENDIX A

This Appendix includes all the instructions of the PIC18FXXX family of microcontrollers in the alphabetical orders, but it does not include the extended set of the instructions of the 18FXXXX family. The instructions are explained with examples.

The following abbreviations and symbols are used in this Appendix:

Abbreviations and Symbols	Description
a	If a = 0, file (data) register resides in the Access Bank If a = 1, the bank of file register is specified by the BSR
b	Bit address 0-7 within 8-bit data register
BSR	Bank Select Register
d	Destination select bit. If d = 0 ; store result in WREG and if d = 1, store result in file (data) register
F	File or data register
k	Literal or constant: 8-, 12-, or 20-bit value
PC	Program Counter
s	Fast Call/Return mode select bit s = 0, save (or retrieve) WREG in execution of Call/Return s = 1, save (or retrieve) WREG, STATUS, and BSR
W or WREG	Working register (accumulator)
TBLPTR	21-bit Table Pointer
TABLAT	8-bit Table Latch
TOS	Top-of-Stack
N, OV, Z, DC, C	Flags: Sign, Overflow, Zero, Digit Carry, Carry/Borrow

ADDLW   Add Literal (8-bit) to   WREG   (1W/1Cy)

Syntax: ADDLW   8-bit   Status: All flags are affected

Description: Add operand (8-bit number) to W and place the result in W

Example: Assume the working register W holds the data byte $F2_H$. Add 37_H to W and identify the flags in the status register.

Instruction: ADDLW   0 x 37

$$
\begin{array}{lllllllll}
W & = & 1 & 1 & 1 & 1 & 0 & 0 & 1 & 0 & (F2_H) \\
+ & & & & & & & & & \\
\text{8-bit} & = & 0 & 0 & 1 & 1 & 0 & 1 & 1 & 1 & (37_H) \\
CY\,[1] & & 1 & 1 & 1 & & 1 & & & \\
\hline
W & = & 0 & 0 & 1 & 0 & 1 & 0 & 0 & 1 & (29_H)
\end{array}
$$

N OV Z DC C

Status: | - | - | - | - | 0 | 0 | 0 | 0 | 1 |

Flags: After the addition the result in the W register is 29_H with the carry flag set. The result (29_H) determines the status of the remaining flags. N flag is reset because Bit7 of 29_H is zero, and DC is reset because there is no carry generated from Bit3 to Bit4. Similarly, Z and OV flags are reset because the result is not zero and there is no overflow.

If these numbers are assumed to be unsigned, the carry flag is relevant, and the result is 1 29_H.

The sum generates a carry but no overflow. If these numbers are assumed to signed numbers, N and OV flags are relevant. $F2_H$ is a negative number and 37_H is a positive number. By adding a negative number to a positive number, the result will be smaller. In this example, the result 29_H is smaller than 37_H. Thus there is no overflow, and the sum is positive.

ADDWF   Add WREG to File (Data) Register   (1W/1Cy)

Syntax: ADDWF   F, d, a   Status: All flags are affected.

Description: Add WREG to a data register and place the result either in W if the destination d = 0 or in the data register if d = 1 (default). If a = 0, the data register is from the Access bank, and if a =1, BSR specifies the bank.

Example: Assume the working register W holds the data byte $F2_H$ and the data register REG1 (with the address 01_H) holds the byte 84_H. Add the data bytes, save the sum in data register, and identify the flags in the status register.

Instruction: ADDWF   0 x 01, 1,0

$$W = 1 \; 1 \; 1 \; 1 \quad 0 \; 0 \; 1 \; 0 \quad (F2_H)$$
$$+$$
$$(0x01) = 1 \; 0 \; 0 \; 0 \quad 0 \; 1 \; 0 \; 0 \quad (84_H)$$

CY [1]

$$\overline{\text{REG1} \quad 0 \; 1 \; 1 \; 1 \quad 0 \; 1 \; 1 \; 0 \quad (76_H)}$$

$$\text{N \; OV \; Z \; DC \; C}$$

Status: | - | - | - | 0 | 1 | 0 | 0 | 1 |

Flags:

After the addition the result 76_H is stored in data register REG1 and W register is unchanged. The result (76_H) determines the status of the remaining flags.

N flag is reset because Bit7 of 76_H is zero, DC is reset set because there is not a carry generated from Bit3 to Bit4, and Z is reset because the result is not zero. However, OV and C flags are set.

If these numbers are assumed to be signed numbers, $F2_H$ and 84_H are negative numbers. By adding two negative numbers, the sum should be negative. This sum is shown as positive because Bit7 = 0. A carry from Bit6 flows into Bit7, thus generating a positive number (76_H); therefore, there is an overflow.

If these numbers are assumed to be unsigned numbers, the addition generates a carry and the sum is 1 76_H.

ADDWFC   Add WREG to File (Data) Register and Carry   (1W/1Cy)

Syntax:

ADDWFC   F, d, a   Status: All flags are affected.

Description:

Add W register to data register and carry. This instruction is used for 16-bit addition. If d = 0, place the result in W and if d = 1, place the result in the data register. If a = 0, select the data register from Access bank and if a = 1, BSR (Bank Select Register) specifies the bank.

Example:

Assume that the working register W holds the data byte 72_H, the general purpose data register REG (REG =20_H) holds the byte $1F_H$, and the previous addition has set the carry flag. Add the bytes, save the result in the data register, and identify the flag status.

Instruction:

ADDWFC   0x20, 1, 0   Substitute REG = 0 x 20 or define REG with an equate

```
W    =   0  1  1  1    0  0  1  0    (72_H)
+
REG20 = 0  0  0  1    1  1  1  1    (1F_H)
Carry                          1
CY [0]   1  1  1  1    1  1  1
```
_____

```
REG20 = 1  0  0  1    0  0  1  0    (92_H)
                   N  OV  C  DC  Z
```

Status: | - | - | - | 1 | 1 | 0 | 1 | 0 |

Flags:

After the addition the result 92_H is stored in REG1 and the W register is unchanged. The result (92_H) determines the status of the remaining flags.

N flag is set because Bit7 of 92_H is one, DC is set because there is a carry generated from Bit3 to Bit4, and Z is reset because the result is not zero. However, OV flag is set.

If these numbers are assumed to signed numbers, 72_H and $1F_H$ are positive numbers. By adding two positive numbers, the sum should be positive. This sum is shown as negative because Bit B7 = 1. A carry from Bit6 flows into Bit7, thus generating a negative number (92_H); therefore, there is an overflow.

ANDLW   Logically AND Literal (8-bit) with WREG.   (1W/1Cy)

Syntax: ANDLW   8-bit   Status: N and Z flags are affected.

Description: AND the 8-bit number with the contents of the W register and place the result in the W register.

Example: Assume that the W register holds the data byte 78_H and AND 82_H with the contents of W.

Instruction: ANDLW   0 x 82

$$W \quad = \quad 0 \ 1 \ 1 \ 1 \quad 1 \ 0 \ 0 \ 0 \quad (78_H)$$

AND

$$\text{8-bit} \quad = \quad 1 \ 0 \ 0 \ 0 \quad 0 \ 0 \ 1 \ 0 \quad (82_H)$$

$$W \quad = \quad 0 \ 0 \ 0 \ 0 \quad 0 \ 0 \ 0 \ 0 \quad (00H)$$

N   Z

Status: | - | - | - | 0 | - | 1 | - | - |

Flags:	After the AND operation the result is 00_H, which sets the Z flag and resets the N flag.
ANDWF	Logically AND WREG with File (Data) Register (1W/1Cy)
Syntax:	ANDWF F, d, a Status: N and Z flags are affected.
Description:	AND the contents of the W register with the contents of data register. If d = 0, place the result in W and if d = 1, place the result in the data register. If a = 0, select file (data) register from the Access bank and if a = 1, use BSR.
Example:	Assume that the W register holds the data byte $F2_H$ and the general purpose file register REG20 (with the address 20_H) holds the byte $8F_H$. AND the bytes and save the result in the W register and identify the flag status.
Instruction:	ANDWF 0x20, 0, 0 (Substitute REG20 = 0 x 20 or define REG20 with an equate)

$$W \quad = 1 \ 1 \ 1 \ 1 \quad 0 \ 0 \ 1 \ 0 \quad (F2_H)$$

AND

$$REG = 1 \ 0 \ 0 \ 0 \quad 1 \ 1 \ 1 \ 1 \quad (8F_H)$$

$$W \quad = 1 \ 0 \ 0 \ 0 \quad 0 \ 0 \ 1 \ 0 \quad (82_H)$$

N   Z

Status: | - | - | - | 1 | - | 0 | - | - |

Flags:	After ANDing, the result 82_H is stored in the W register and REG20 is unchanged. The result (82_H) sets the N flag and resets the Z flag.

Branch (Relative): Conditional Relative Jump Instructions

The following instructions are relative conditional branch instructions based on four data flags (C, Z, N, and OV). There are two instructions per flag: one when a flag is set and the other when a flag is reset. The branch (jump) location is relative to the program counter.

BC	Branch if Carry (C = 1)	Syntax:	BC	n
BNC	Branch if No Carry (C = 0)		BNC	n
BZ	Branch if Zero (Z = 1)		BZ	n
BNZ	Branch if No Zero (Z = 0)		BNZ	n
BN	Branch if Negative (N = 1)		BN	n
BNN	Branch if No Negative (N = 0)		BNN	n
BOV	Branch if Overflow (OV = 1)		BOV	n
BNOV	Branch if No Overflow (OV = 0)		BNOV	n

Status: No flags are affected.   (1W/1-2Cy)

Cycles: When the MPU branches, it requires two cycles; otherwise, it takes one cycle.

Description: If the flag condition is true, redirect (jump) the program execution to memory location by "n" words from the program counter where "n" is an 8-bit signed number.

Example :   Assume that the branch instruction BC (Branch on Carry) is located at memory location 2000_H. Calculate the memory location where the program is redirected if $n = 05_H$ and $n = FA_H$.

Instruction 1:   BC   0x05   ; Stored at memory location 2000_H

If the carry flag is 1, the memory location where the program is redirected is calculated by adding:

1) 2 to the program counter because the PC will be incremented by 2 locations when the MPU reads the branch instruction.

2) (n x 2) because n is given in words (not bytes).

3) The memory location is calculated as follows:

$PC + 2 + n \times 2 = 2000_H + 2 + 05_H \times 2 = 200C_H$

If the carry flag is 0, the program goes to the next instruction located at memory location 2002_H, and the instruction is executed in one cycle.

Instruction 2:   BC   0xFA   ; Stored at memory location 2000_H

In this instruction, $n = FA_H$ is a negative number, and the program is redirected backward (backward jump). 2's complement of $FA_H = 06_H$. Therefore, the memory location where the program is redirected is as follows:

$PC + 2 + (-n \times 2) = 2000_H + 2 + (-06_H \times 2) = 2002_H - 0C_H = 1FF6_H$

BRA   Branch (Relative) Unconditionally   (1W/2Cy)

Syntax: BRA   n   Status: No flags are affected.

Description:	Branch unconditionally by n location relative to the program counter where n is signed number given in words (16 bits) between -1024_{10} and $+ 1023_{10}$
Example:	Assume that the branch instruction BRA is located at memory location 0100_H. Calculate the memory location where the program is redirected if n = 12_H and n = 86_H.
Instruction 1:	BRA 0x12 ; Stored at memory location 0100_H

The memory location where the program is redirected is calculated by adding:

1) 2 to the program counter because the PC will be incremented by 2 memory locations when the MPU reads the branch instruction.

2) (n x 2) because n is given in words (not bytes).

3) The memory location is calculated as follows:

$$PC + 2 + n \text{ x } 2 = 0100_H + 2 + 12_H \text{ x } 2 = 0126_H$$

Instruction 2:	BRA 0x86 ; Stored at memory location 0100_H

In this instruction, n = 86_H is a negative number, and the program is redirected backward (backward jump). 2's complement of $86_H = 7A_H$. Therefore, the memory location where the program is redirected is as follows:

$$PC + 2 + (-n \text{ x } 2) = 0100_H + 2 + (-7A_H \text{ x } 2) = 0102_H - F4_H = 000E_H$$

BCF	Bit Clear in File (Data) Register	(1W/1Cy)
Syntax:	BCF F, b, a Status: No flags are affected.	
Description:	Clear Bit B <0-7> in File (data) register. If a = 0, select data register from the Access bank and if a = 1, use BSR.	
Example:	Clear Bit7 in data register 15_H.	
Instruction:	BCF 0x15, 7, 0	
BSF	Bit Set in File (Data) Register	(1W/1Cy)
Syntax:	BSF F, b, a Status: No flags are affected.	
Description:	Set Bit B <0-7> in data register. If a = 0, select data register from the Access bank and if a = 1, use BSR.	
Example:	Clear Bit0 in data register 12_H.	
Instruction:	BSF 0x12, 0, 0	
BTFC	Bit Test in File (Data) Register, Skip if Clear	(1W/1-2Cy)
Syntax:	BTFC F, b, a Status: No flags are affected.	
Description:	Test Bit B <0-7> in data register. If bit is 0, (cleared), skip the next instruction. If a = 0, select data register from the Access bank and if a = 1, use BSR.	

Example:	Test Bit7 in data register 10_H, assuming the register holds the byte $7F_H$.
Instruction:	BTFC 0x10, 7, 0

This instruction tests Bit7 in data register 10_H, which holds the byte 0111 1111 ($7F_H$). This bit is zero; therefore, the MPU skips the next instruction. This instruction requires two cycles.

BTFSS   Bit Test in File (Data) Register; Skip if Set   (1W/1-2Cy)

Syntax:	BTFSS F, b, a Status: No flags are affected.
Description:	Test Bit B <0-7> in data register. If bit is 1 (set), skip the next instruction. If a = 0, select data register from the Access bank and if a = 1, use BSR (Bank Select Register).
Example:	Test Bit4 in data register 10_H, assuming the register holds the byte 69_H.
Instruction:	BTFSS 0x10, , 4, 0

This instruction tests Bit4 in data register 10_H, which holds the byte 0110 1001 (69_H). This bit is zero; therefore, the MPU does not skip the next instruction. This instruction requires one cycle.

BTG   Bit Toggle in File (Data) Register   (1W/1Cy)

Syntax:	BTG F, b, a Status: No flags are affected.
Description:	Toggle Bit B <0-7> in data register. If bit is 0, set it to 1 and if it is 1, reset to 0. If a = 0, select data register from the Access bank and if a = 1, use BSR (Bank Select Register).
Example:	Toggle Bit 7 in PORTB with the address 81_H, assuming PORTB holds the byte 78_H
Instruction:	BTG PORTB, 7, 0 ; PORTB must be defined using the equate or it ;should be substituted by the address 81_H.

This instruction toggles Bit7 (sets to 1) in PORTB that holds the byte 0111 1000 (78_H). After the execution of this instruction, PORTB holds the byte 1111 1000 ($F8_H$).

CALL   Call a Subroutine   (2W/2Cy)

Syntax:	Call Label, s Status: No flags are affected
Description:	This is a two-word instruction, and the MPU transfers the program execution to the address specified by the label within the entire memory range of 2 Mbyte. Before the transfer, the MPU pushes the program counter (PC +4), the return address, on the stack. If s = 1, three registers (W, STATUS, and BSR) are also pushed into their respective shadow registers. If s = 0, only the program counter is pushed on the stack.

Example: Assume that the following CALL instruction is written at location 0100_H that calls the subroutine located at location 120_H. Explain how the instruction is executed.

Instruction: CALL   0 x 120, 1

This is a two-word instruction stored at memory location 0100_H. When the MPU completes the reading of this instruction, the program counter (PC) holds the address of the next instruction (0104_H). When the MPU executes this instruction, it places the address 0104_H of the next instruction on the hardware stack, stores the contents of W, STATUS, and BSR registers in their respective shadow registers, places the address of the subroutine (0120_H) in the program counter, and transfers the program execution to location 0120_H.

CLRF   Clear File (Data) Register   (1W/1Cy)

Syntax: CLRF   F, a   Status: Z flag is set.

Description: Clear data register. If a = 0, select data register from the Access bank and if a = 1, use BSR.

Example: Clear data register 15_H, which is in the Access bank.

Instruction: CLRF 0x15, 0

CLRWDT   Clear Watchdog Timer   (1W/1Cy)

Syntax: CLRWDT   Status: No flags are affected.

Description: Reset the watchdog timer.

Instruction: CLRWDT.

This instruction clears the watchdog timer (WDT) and its postscaler, and sets Bit3 ($\overline{TO}$–Timer Out Flag) and Bit2 ($\overline{PD}$–Power Down Detection Flag) in RCON (Reset Control register), indicating the normal operation.

COMF   Complement File (Data) Register   (1W/1Cy)

Syntax: COMF   F, d, a   Status: N, Z flags are affected.

Description: Complement each bit in data register. If d = 0, save the result in W, and if d = 1, save the result back in data register (default). If a = 0, select data register from the Access bank and if a = 1, use BSR.

Example: Assume that the data register REG2 with the address 02_H holds the byte $7F_H$. Specify the result after the following instructions.

Instruction1:   COMF 0x02, 0   REG2 = 0111 1111

Result: W =1000 0000   REG2 = 0111 1111   N = 1, and Z = 0

Instruction2:   COMF 0x02, 1   REG2 = 0111 1111

Result: REG2 =1000 0000   N = 1, and Z = 0

CPFSEQ   Compare File (Data) Register with WREG; Skip if F = W   (1W/1-2Cy)

Syntax: CPFSEQ   F, a   Status: No flags are affected.

Description: Compare the contents of the file (data) register F with the contents of WREG. If they are equal, skip the next instruction. The comparison is performed by an unsigned subtraction of two bytes. If a = 0, select the file register from the Access bank, and if a = 1, use BSR.

Example: Assume that the WREG holds the data byte $F8_H$ and REG20 (with the address 20_H) also holds the byte $F8_H$.

Instruction: CPFSEQ   0 x 20, 0

The contents of WREG and the file register are equal; therefore, the MPU will skip the next instruction.

CPFSGT   Compare File (Data) Register with WREG; Skip if F > W   (1W/1-2Cy)

Syntax: CPFSGT   F, a   Status: No flags are affected.

Description: Compare the contents of the file (data) register F with the contents of the WREG. If (F) > (W), skip the next instruction. The comparison is performed by an unsigned subtraction of two bytes. If a = 0, select the file register from the Access bank, and if a = 1, use BSR.

Example: Assume that the WREG holds the data byte 98_H and REG30 (with the address 30_H) holds the byte $F2_H$.

Instruction: CPFSGT   0 x 30, 0

The byte in the file register > the byte in the WREG; therefore, the MPU will skip the next instruction.

CPFSLT   Compare File (Data) Register with WREG; Skip if (F) < (W)   (1W/1-2Cy)

Syntax: CPFSEQ   F, a   Status: No flags are affected.

Description: Compare the contents of the file (data) register F with the contents of the WREG. If (F) < (W), skip the next instruction. The comparison is performed by an unsigned subtraction of two bytes. If a = 0, select the file register from the Access bank, and if a = 1, use BSR.

Example: Assume that the WREG holds the data byte 47_H and REG10 (with the address 10_H) holds the byte $F8_H$.

Instruction: CPFSLT   0 x 10, 0

The byte in the file register > the byte in the WREG; therefore, the MPU will not skip the next instruction.

DAW   Decimal Adjust WREG   (1W/1Cy)

Syntax: DAW   Status: C flag is affected.

Description: Adjust the 8-bit value in the WREG resulting from the earlier addition of two vari-
                ables in the packed BCD format. The result will be in correct packed BCD
                format.

Example 1:   Add the following two packed BCD numbers—34 and 91—and adjust the sum
                for BCD numbers by using the instruction DAW.

Instructions: MOLW   0x34   ;Load 34_H (packed BCD) in WREG

                ADDLW   0x91   ;Add 91_H (packed BCD) and save the sum in WREG

                DAW   ;Adjust the sum to 34 + 91 = 1] 2 5

                The MPU adds these numbers in Hex as $34_H + 91_H = C\ 5_H$, but the instruction
                DAW adjusts the sum to the BCD value 25 in WREG and C = 1.

Example 2:   Add the following two packed BCD numbers—24_H and 58_H—and adjust the
                sum for BCD numbers by using the instruction DAW.

Instructions: MOLW   0x24   ;Load 24_H (packed BCD) in WREG

                ADDLW   0x58   ;Add 58_H (packed BCD) and save the sum in WREG

                DAW   ;Adjust the sum to 24 + 58 = 82

                The MPU adds these numbers in Hex as $24_H + 58_H = 7\ C_H$, but the instruction
                DAW adjusts the sum to the BCD value 82 in WREG and C = 0.

DECF   Decrement File (Data) Register by 1   (1W/1Cy)

Syntax: DECF   F, d, a   Status: All flags are affected.

Description: Decrement contents of data register by 1 and place the result either in W if the
                destination d = 0 or in data register if d = 1 (default). If a = 0, the register is
                from the Access bank, and if a =1, BSR specifies the bank.

Example: Assume that the data register REG1 (with the address 01_H) holds the byte 72_H.
                Decrement the byte and save the result in WREG. Identify the flags in the sta-
                tus register.

Instruction: DECF   0x 01, 0,0

                The decrement is a subtraction process. To subtract 1 by 2's complement
                method we need to take 2's complement of 1, which is FF_H, and add FF_H
                to 72_H as shown.

$(0x01) = 0\ 1\ 1\ 1\quad 0\ 0\ 1\ 0\ (72_H)$

$+$

$(-1)\quad = 1\ 1\ 1\ 1\quad 1\ 1\ 1\ 1\ (FF_H)$

; 2's complement of 1 is FF_H

CY [1]   1 1 1   1 1

REG1   0 1 1 1   0 0 0 1 (71_H)

N OV Z DC C

Status: | - | - | - | 0 | 0 | 0 | 1 | 1 |

Flags: After adding FF to 72 the result is 71 with the carry flag set. During the addition of FF, it also sets the Digital Carry because the addition of Bit3 generates a carry to Bit4 position.

DECFSZ   Decrement File (Data) Register; Skip if F = 0   (1W/1-2Cy)

Syntax: DECFSZ   F, d, a   Status: No flags are affected.

Description: Decrement the contents of file (data) register. If the result is zero, skip the next instruction. If d = 0, place the result in W and if d = 1, place the result in the file register. If a = 0, select file register from the Access bank and if a = 1, use BSR.

Example: Assume that the data register REG20 (with the address 20_H) holds the byte 05_H. Explain the following instructions.

Instruction: LOOP: DECFSZ 0x20, 1, 0

BRA   LOOP

The data register 20_H holds the byte and the instruction DECFSZ decrements the byte to 04_H and saves it in REG20 because d = 1. The branch instruction takes the execution back to decrement the register. This process is repeated five times, and when the byte goes to zero, the MPU skips the branch instruction and continues the program.

DCFSNZ   Decrement File (Data) Register; Skip if F ≠ 0   (1W/1-2Cy)

Syntax: DCFSNZ   F, d, a   Status: No flags are affected.

Description: Decrement the contents of file (data) register. If the result is not zero, skip the next instruction. If d = 0, place the result in W and if d = 1, place the result in the data register. If a = 0, select data register from the Access bank and if a = 1, use BSR.

Example: Assume that the data register REG20 (with the address 20_H) holds the byte 05_H. Explain the following instructions.

Instruction:	DCFSNZ	0x20, 1, 0
	BRA	STOP

The data register 20_H holds the byte 05_H and the instruction DECFSZ decrements the byte to 04_H and saves it in REG20 because d = 1. The result is not zero; therefore, the MPU skips the next instruction and continues the program. This branch instruction terminates the program by jumping to the label STOP.

GOTO	Unconditional Branch	(2W/2Cy)

Syntax: GOTO Label Status: No flags are affected.

Description: Branch unconditionally to any memory location within the entire 2 Mbyte memory range.

Example: Branch to the location labeled as START.

Instruction: GOTO START

INCF	Increment File (Data) Register by 1	(1W/1Cy)

Syntax: INCF F, d, a Status: All flags are affected.

Description: Increment the contents of the data register by 1 and place the result either in W if the destination d = 0 or in the data register if d = 1 (default). If a = 0, the data register is from the Access bank, and if a =1, BSR specifies the bank.

Example: Assume that the data register REG1 (with the address 01_H) holds the byte $7F_H$. Increment the byte and save the result in the data register. Identify the flags in the status register.

Instruction: INCF 0x 01, 1,0

$(0x01) =$ 0 1 1 1 1 1 1 1 $(7F_H)$

$+$

$(+1)$ $=$ 0 0 0 0 0 0 0 1 (01_H)

CY [0] 1 1 1 1 1 1 1

REG1 $=$ 1 0 0 0 0 0 0 0 (80_H)

N OV Z DC C

Status: | - | - | - | 1 | 1 | 0 | 1 | 0 |

Flags: After incrementing the byte $7F_H$ the result is 80_H. The sum does not generate a carry, but it sets three flags: N (Negative), OV (Overflow), and DC (Digital Carry). The overflow flag is set because by adding the positive numbers, the sum becomes negative; therefore, there must be an overflow. The N flag is set because the bit B7 is 1, and the DC flag is set because the sum of Bit3 generates a carry for Bit4.

INCFSZ   Increment File (Data) Register; Skip if F = 0   (1W/1-2Cy)

Syntax: INCFSZ   F, d, a   Status: No flags are affected.

Description: Increment the contents of file (data) register. If the result is zero, skip the next instruction. If d = 0, place the result in W and if d = 1, place the result in the data register. If a = 0, select the data register from the Access bank, and if a = 1, use BSR.

Example: Assume that the data register REG20 (with the address 20_H) holds the byte 05_H. Explain the following instructions.

Instruction: LOOP: INCFSZ   0x20, 1, 0

                              BRA   LOOP

The (data) register 20_H holds the byte 05_H and the instruction INCFSZ increments the byte to 06_H and saves it in the data register because d = 1. The branch instruction takes the execution back to increment the register. This process is repeated until the byte goes from FF to zero. The MPU then skips the branch instruction and continues the program.

INFSNZ   Increment File (Data) Register; Skip if F ≠ 0   (1W/1-2Cy)

Syntax: INFSNZ   F, d, a   Status: No flags are affected.

Description: Increment the contents of file (data) register. If the result is not zero, skip the next instruction. If d = 0, place the result in W and if d = 1, place the result in the data register. If a = 0, select data register from the Access bank and if a = 1, use BSR.

Example: Assume that the data register REG20 (with the address 20_H) holds the byte FA_H. Explain the following instructions.

Instruction: INFSNZ   0x20, 1, 0

               BRA   STOP

The file (data) register 20_H holds the byte FA_H and the instruction INFFSZ increments the byte to FB_H and saves it in the register because d = 1. The result is not zero; therefore, the MPU skips the next instruction and continues the program. This branch instruction terminates the program by jumping to the label STOP.

IORLW   Logically (Inclusive) OR Literal (8-bit number) with WREG   (1W/1Cy)

Syntax: IORLW   8-bit   Status: N and Z flags are affected.

Description: Inclusive OR the 8-bit number with the contents of the WREG and place the result in W register.

Example: Assume that the WREG holds the data byte 78_H. Write an instruction to inclusive OR 82_H with the contents of W.

Instruction: IORLW   0 x 82

W   = 0 1 1 1 1 0 0 0 (78_H)

OR

8-bit = 1 0 0 0 0 0 1 0 (82_H)

_____

W   = 1 1 1 1 1 0 1 0 (FA_H)

                       N   Z

Status: | - | - | - | 1 | - | 0 | - | - |

Flags: After the OR operation the result is FA_H, which sets the N flag and resets the Z flag.

IORWF   Logically (Inclusive) OR WREG with a File (Data) Register   (1W/1Cy)

Syntax: IORWF   F, d, a   Status: N and Z flags are affected.

Description: Inclusive OR the contents of the WREG with the contents of data register. If d = 0, place the result in W and if d = 1, place the result in the data register. If a = 0, select data register from the Access bank and if a = 1, use BSR.

Example: Assume that the working register W holds the data byte F2_H and the general purpose data register REG20 (with the address 20_H) holds the byte 8F_H. OR the bytes, save the result in W register, and identify the flag status.

Instruction: IORWF   0x20, 0, 0   (Substitute REG20 = 0 x 20 or define REG with an equate)

W   = 1 1 1 1 0 0 1 0 (F2_H)

OR

REG20 = 1 0 0 0 1 1 1 1 (8F_H)

_____

W   = 1 1 1 1 1 1 1 1 (FF_H)

                       N   Z

Status: | - | - | - | 1 | - | 0 | - | - |

Flags: After ORing the result, FF_H is stored in W register and REG20 is unchanged. The result (FF_H) sets the N flag and resets the Z flag.

LFSR   Load File Select Register (FSR)   (2W/2Cy)

Syntax: LFSR   F, k   Status: No flags are affected.

Description:	Load 12-bit literal <0-4095> in FSRH (High) and FSRL (Low). The MPU includes three FSR registers—FSR0, FSR1, and FSR2.
Example:	Write instructions to use FSR1 to point to the data register.
Instruction:	LFSR1, 0x0150

This instruction loads 01_H in high register FSR1H and 50_H in low register FSR1L.

MOVF   Move (Copy) File Register in WREG or Back in Same   (1W/1Cy)
Register

Syntax:	MOVF F, d, a Status: N and Z flags are affected.
Description:	Copy file (data) register in WREG if d = 0, and if d = 1, copy back into the same file register. If a = 0, select data register from the Access bank and if a = 1, use the BSR.
Example:	Assume that the data register REG20 (with the address 20_H) holds the byte 80_H. Copy the byte back in the same register to set the N flag.
Instruction:	MOVF 0x20, 1, 0 (Substitute REG20 = 0 x 20 or define REG20 with an equate)

The instruction copies the byte back in the same register 20_H

Flags: N = 1 and Z = 0

MOVFF   Move (Copy) from a File (Data) Register to Another File   (2W/2-3Cy)
(Data) Register

Syntax:	MOVFF Fs, Fd Status: No flags are affected.
Description:	Copy from one file (data) register called Fs (source register) to another file (data) register called Fd (destination register). These registers can be from anywhere in the data register space (4096 bytes). WREG can be one of the source or the destination register.
Example:	Assume that the data register REG20 (with the address 20_H) holds the byte $7F_H$. Copy the byte from REG20 into REG30 (with the address 30_H).
Instruction:	MOVFF REG20, REG30 ; REG20 and REG30 must be defined using ; Equates or as shown
Alternative:	MOVFF 0x20, 0x30

The instruction copies the byte from REG20 into REG30.

MOVLB   Move (Load) an 8-bit Literal in BSR (Bank Select   (1W/1Cy)
Register)

Syntax:	MOVLB k Status: No flags are affected.
Description:	Load an 8-bit number in the Bank Select Register. The least significant 4 bits <3-0> specify the bank, and the most significant 4 bits <7-4> are always set to 0.

Example: Load in the BSR to select the bank number 15.

Instruction: MOVLB   0x0F

The instruction loads $0F_H$ in the BSR register.

MOVLW   Move (Load) an 8-bit Literal in WREG   (1W/1Cy)

Syntax: MOVLW   k   Status: No flags are affected.

Description: Load an 8-bit number in WREG.

Example: Load $5F_H$ in WREG.

Instruction: MOVLW   0x5F

The instruction loads $5F_H$ in WREG.

MOVWF   Move (Copy) WREG into File (Data) Register.   (1W/1Cy)

Syntax: MOVWF   F, a   Status: No flags are affected.

Description: Copy contents of WREG into file (data) register. If a = 0, select the register from the Access bank and if a = 1, use the BSR.

Example: Assume that WREG holds the byte $5A_H$. Copy the byte from WREG into the data register REG20 (with the address 20_H).

Instruction: MOVWF   0x20, 0

The instruction copies the byte $5A_H$ into register 20_H.

MULLW   Multiply 8-bit Literal with WREG.   (1W/1Cy)

Syntax: MULLW   k   Status: No flags are affected.

Description: Multiply 8-bit unsigned number with the contents of WREG and save the 16-bit product in the PRODH:PRODL registers. This is a multiplication of two 8-bit unsigned numbers; the high byte of the product is saved in PRODH and the low-byte is saved in PRODL.

Example: Assume that WREG holds the byte $4F_H$. Multiply the byte with the unsigned number $9A_H$.

Instruction: MULLW   0x9A

The instruction multiplies $4F_H$ in WREG and the 8-bit literal $9A_H$ and saves the product $2F86_H$ in PRODH ($2F_H$) and PRODL ($9A_H$) registers.

MULWF   Multiply WREG and File (Data) Register.   (1W/1Cy)

Syntax: MULWF   F, a   Status: No flags are affected.

Description: Multiply 8-bit unsigned numbers in WREG and file (data) register and save the 16-bit product in the PRODH:PRODL registers. This is a multiplication of two 8-bit unsigned numbers; save the high byte of the product in PRODH and the low-byte in PRODL. If a = 0, select the register from the Access bank and if a = 1, use the BSR.

Example: Assume that the WREG holds the byte $C7_H$ and REG10 (with the address 10) holds the byte $A5_H$. Calculate the product of these two bytes.

Instruction: MULWF   0x10

The instruction multiplies $C7_H$ in WREG and $A5_H$ in REG10 and saves the product 8043_H in PRODH (80_H) and PRODL (43_H) registers.

NEGF   Take 2's Complement of File (Data) Register   (1W/1Cy)

Syntax: NEGF   F, a   Status: All flags are affected

Description: Take 2's complement of the byte in file (data) register and save the result in the same register.

Example: Assume that REG10 (with the address 10_H) holds the byte $C7_H$. Take 2's complement of the byte.

Instruction: NEGF   0x 10

REG10   = 1 1 0 0   0 1 1 1 ($C7_H$)

2's Complement   = 0 0 1 1   1 0 0 1 (39_H)

All flags are cleared based on the result 39_H.

                 N OV Z DC C

| - | - | - | 0 | 0 | 0 | 0 | 0 |

NOP   No Operation   (1W/1Cy)

Syntax: NOP   Status: No flags are affected.

Description: No operation. This instruction is used to provide additional delay or keep the MPU busy for four cycles.

POP   Discard Address From Top of the Stack   (1W/1Cy)

Syntax: POP   Status: No flags are affected.

Description: Discard the memory address from top of the stack (TOS) and move the stack pointer to the previous address.

Example: Explain what changes takes place after the execution of the following instruction.

Instruction: POP

Assume that the stack pointer is pointing to the 5th location on the stack. The POP instruction that discards the address is the 5th location and points to the 4th location.

PUSH   Store Memory Address on the Stack   (1W/1Cy)

Syntax: PUSH   Status: No flags are affected.

Description: Save the memory address PC + 2 on the stack and move the stack pointer by one stack location.

Example: Explain what is stored on the stack if the following PUSH instruction is written at location 000020_H.

Instruction: 000020   PUSH

The memory address PC + 2 (000022_H) is stored on the next available stack. Assume that the stack pointer is pointing to the 1st location on the stack. The PUSH instruction stores the address 000022_H on the 1st stack and the stack pointer is incremented to point to the 2nd location.

RCALL   Call a Subroutine   (1W/2Cy)

Syntax: RCALL n   Status: No flags are affected

Description: This is a relative call instruction followed by a 12-bit signed number that ranges from –1024 to + 1023. This instruction can transfer the program execution forward by 1023 memory locations and backward by 1024 locations. Before the transfer, the MPU pushes the program counter (PC +2), the return address, on the stack.

Example 1:   Assume that the following Relative Call instruction is written at location 0100_H with the displacement n = 50_H. Calculate the memory address where the program is redirected

Instruction: RCALL   0 x 50

This is a 1-word instruction stored at memory location 0100_H. When the MPU completes the reading of this instruction, the program counter (PC) holds the address of the next instruction (0102_H). When the MPU executes this instruction, it places the address 0102_H (the address of the next instruction) on the hardware stack and transfers the program execution to the location 050_H with the backward jump of $A7_H$ locations. In practice, this instruction is written with a label, such as RCALL NEXT, and the assembler calculates the code for the jump.

RESET   Reset   (1W/1Cy)

Syntax: RESET   Status: No flags are affected.

Description: Rest all registers and flags that are affected by $\overline{MCLR}$ Reset. This instruction performs a software reset equivalent to $\overline{MCLR}$ hardware reset.

RETFIE   Return from Interrupt   (1W/2Cy)

Syntax: RETFIE, s   Status: GIE/GIEH, PEIE/GIEL

Description: This instruction is written at the end of an interrupt service routine (ISR), and it redirects the program back to the next address where the program was interrupted. The MPU places the address from the TOS in the program counter, decrements the stack pointer, and enables interrupts by setting either the high or low priority global interrupt enable bit. The program execution begins at the address found on the top of the stack.

If s = 1, the MPU retrieves the contents of three registers: W, STATUS, and BSR from shadow registers WS, STATUSS, and BSRS. If s = 0, these registers are not affected. This instruction must be written at the end of the interrupt service routines.

RETLW   Return Literal to W   (1W/2Cy)

Syntax: RETLW k   Status: No flags are affected

Description: Load 8-bit literal in the WREG and load the program counter with the address from the TOS.

RETURN   Return from Subroutine   (1W/2Cy)

Syntax: RETURN, s   Status: No flags are affected

Description: This instruction is written at the end of a subroutine, and it redirects the program back to the address next to the call instruction. The MPU places the address from the TOS in the program counter, decrements the stack pointer, and transfers the program execution begining at the address found on the top of the stack.

If s = 1, the MPU retrieves the contents of three registers: W, STATUS, and BSR from shadow registers WS, STATUSS, and BSRS. If s = 0, these registers are not affected. This instruction must be written at the end of the subroutines.

RLCF   Rotate Left through Carry   (1W/1Cy)

Syntax: RLCF   F, d, a   Status: N, Z, and C flags are affected

Description: Rotate each bit of data (file) register to the left position including the carry. Bit7 is moved into the carry bit and the carry bit is moved into Bit0 position. If d = 0, save the result in W, and if d = 1, save the result back in data register (default). If a = 0, select data register from the Access bank and if a = 1, use BSR.

Example: Assume the data register REG20 (with the address 20_H) holds the byte (1 0 0 1 0 0 0 1) and the carry flag is reset. Explain the result after the execution of the following instruction.

Instruction: RLCF   0x20, 1, 0

Before the execution of the instruction RLCF, REG20 holds the byte $91_H.$ The instruction shifts each bit to the left position, Bit7 goes into the Carry flag, and the carry bit goes into Bit0 position; the result is 22_H with the Carry flag set. The MPU saves the result (22_H) back into REG20 because d = 1.

Flag Status: N = 0, Z = 0, C = 1

Before Instruction:

REG20 = 91$_H$

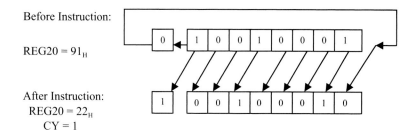

After Instruction:
REG20 = 22$_H$
CY = 1

RLNCF   Rotate Left Through No Carry   (1W/1Cy)

Syntax: RLNCF   F, d, a   Status: N and Z flags are affected

Description: Rotate each bit of data (file) register to the left position; place Bit7 into Bit0 position. If d = 0, save the result in W, and if d = 1, save the result back in data register (default). If a = 0, select data register from the Access bank and if a = 1, use BSR.

Example: Assume the data register REG20 (with the address 20$_H$) holds the byte (1 0 0 1 0 0 0 1) and the carry flag is reset. Explain the result after the execution of the following instruction.

Instruction: RLNCF   0x20, 0, 0

Before Instruction:

REG20 = 91$_H$

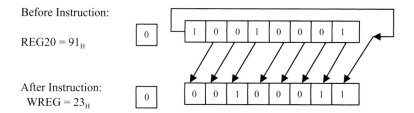

After Instruction:
WREG = 23$_H$

Before the execution of the instruction RLNCF, REG20 holds the byte 91$_H$. The instruction shifts each bit to the left position, Bit7 goes into Bit0 position; the result is 23$_H$. The MPU saves the result (23$_H$) in WREG because d = 0.

RRCF   Rotate Right Through Carry   (1W/1Cy)

Syntax: RRCF   F, d, a   Status: N, Z, and C flags are affected

Description: Rotate each bit of data (file) register to the right position including the carry. Bit0 is moved into the carry bit and the carry bit is moved into Bit7 position. If d = 0, save the result in W, and if d = 1, save the result back in data register (default). If a = 0, select data register from the Access bank and if a = 1, use BSR.

Example: Assume the data register REG30 (with the address 30$_H$) holds the byte (0 1 1 1 0 0 0 1) and the carry flag is reset. Explain the result after the execution of the following instruction.

Instruction: RRCF   0x30, 0, 0

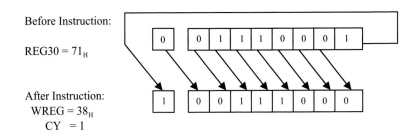

Before Instruction:

REG30 = 71_H

After Instruction:
  WREG = 38_H
  CY = 1

Before the execution of the instruction RRCF, REG30 holds the byte 71_H. The instruction shifts each bit to the right position, Bit0 goes into the Carry flag, and the carry bit goes into Bit7 position; the result is 38_H with the Carry flag set. The MPU saves the result (38_H) in WREG because d = 0.

Flag Status: N = 0, Z = 0, C = 1

RRNCF   Rotate Right Through No Carry   (1W/1Cy)

Syntax: RRNCF   F, d, a   Status: N and Z flags are affected

Description: Rotate each bit of data (file) register to the right position; place Bit0 into Bit7 position. If d = 0, save the result in W, and if d = 1, save the result back in data register (default). If a = 0, select data register from the Access bank and if a = 1, use BSR.

Example: Assume the data register REG30 (with the address 30H) holds the byte (0 1 1 1 0 0 0 1) and the carry flag is reset. Explain the result after the execution of the following instruction.

Instruction: RRNCF   0x30, 1, 0

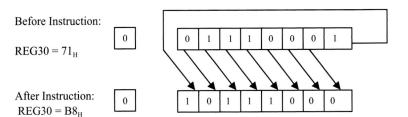

Before Instruction:

REG30 = 71_H

After Instruction:
  REG30 = $B8_H$

Before the execution of the instruction RRCF, REG30 holds the byte 71_H. The instruction shifts each bit to the right position and Bit0 goes into Bit7 position; the result is $B8_H$. The MPU saves the result ($B8_H$) back in REG30 because d = 1.

SETF   Set All Bits in File Register   (1W/1Cy)

Syntax: SETF   F, a   Status: No flags are affected.

Description: Set all bits in a data (file) register to 1 (= FF$_H$). If a = 0, select data register from the Access bank and if a = 1, use BSR.

Example: Set all bits in REG10 (with the address 10$_H$).

Instruction: SETF   0x10, 0

                 This instruction sets all eight bits in REG10 making the byte = FF$_H$.

SLEEP   Enter in Sleep Mode   (1W/1Cy)

Syntax: SLEEP   Status: Power down status bit PD is cleared.
                                              The time-out status bit TO is set.

Description: Place the MPU (processor) into the sleep mode. Clear watchdog timer and its postscaler and stop the oscillator.

SUBLW   Subtract WREG from Literal   (1W/1Cy)

Syntax: SUBLW   8-bit   Status: All flags are affected

Description: Subtract W from 8-bit literal and place the result in W

Example: Assume the working register W holds the data byte 92$_H$. Subtract W from the literal A7$_H$ and identify the flags in the status register.

Instruction: SUBLW   0 x A7

                 W   = 1 0 0 1   0 0 1 0 (92$_H$)

                 2's W   = 0 1 1 0   1 1 1 0 (6E$_H$)
                 +
                 8-bit   = 1 0 1 0   0 1 1 1 (A7$_H$)
                 CY [1]   1 1   1   1 1

                 _____

                 W   = 0 0 0 1   0 1 0 1 (15$_H$)
                                      N OV Z DC C

Status: | - | - | - | - | 0 | 0 | 0 | 1 | 1 |

                 This instruction subtracts the contents of WREG (92$_H$) from the 8-bit literal (A7$_H$) by 2's complement method. The MPU takes 2's complement of (92$_H$) and adds to (A7$_H$). The result is 15$_H$ with the Carry set as shown above.

Flags: In 2's complement addition, the sum 15$_H$ generates a carry and Bit3 generates a digit carry. Therefore, C and DC flags will be set.

SUBWF   Subtract WREG from File Register   (1W/1Cy)

Syntax: SUBWF   F, d, a   Status: All flags are affected.

Description: Subtract WREG from data (file) register and place the result either in W if the destination d = 0 or in data register if d = 1 (default). If a = 0, the register is from the Access bank, and if a =1, BSR specifies the bank.

Example: Assume the working register W holds the data byte 29_H and the data register 01_H holds the byte 72_H. Subtract the WREG from REG01, save the sum in data register, and identify the flags in the status register.

Instruction: SUBWF   0 x 01, 1,0

W   = 0 0 1 0   1 0 0 1 (29_H)

2's W = 1 1 0 1   0 1 1 1 ($D7_H$)
    +
(0x01) = 0 1 1 1   0 0 1 0 (72_H)

CY [1] 1 1 1   1 1
        _____

REG1   0 1 0 0   1 0 0 1 (49_H)
                    N OV Z DC C

Status: | - | - | - | 0 | 0 | 0 | 0 | 1 |

Flags: After the subtraction ($72_H - 29_H$) by 2's complement method, the result is 49_H and the carry flag is set as shown above. The result does not set any other flags.

SUBWFB   Subtract WREG and Borrow from File (Data) Register.   (1W/1Cy)

Syntax: SUBWFB   F, d, a   Status: All flags are affected.

Description: Subtract WREG and Borrow from the data (file) register. This instruction is used for 16-bit subtraction. If d = 0, place the result in W and if d = 1, place the result in data register. If a = 0, select data register from the Access bank and if a = 1, use BSR.

Example: Assume that the working register W holds the data byte $1F_H$ and the general purpose data register REG20 (with the address =20_H) holds the byte 72_H. Subtract W from REG20 with the Borrow flag set, save the result in the data register, and identify the flag status.

Instruction: SUBWFB   0x20, 1, 0   ; $72_H - 1F_H = 53_H$

$$
\begin{array}{llllllllll}
W & = & 0\ 0\ 0\ 1 & 1\ 1\ 1\ 1 & (1F_H) \\
2\text{'s W} & = & 1\ 1\ 1\ 0 & 0\ 0\ 0\ 1 & (E1_H) \\
& + & & & \\
REG20 & = & 0\ 1\ 1\ 1 & 0\ 0\ 1\ 0 & (72_H) \\
CY & [1] & 1\ 1 & & \\
\end{array}
$$

$$
\begin{array}{llllll}
REG20 & = & 0\ 1\ 0\ 1 & 0\ 0\ 1\ 1 & (53_H) \\
& & N\ \ OV\ \ Z\ \ DC\ \ C & & \\
\end{array}
$$

Status: | - | - | - | 0 | 0 | 0 | 0 | 1 |

Flags: This is a subtraction by 2's complement addition. The result is 53_H with the Borrow flag set. If C = 1 before the subtraction, the PIC controller cancels that borrow in the addition of 2's complement; therefore, the result is 53_H and not 52_H. The PIC performs 2's complement subtraction differently than other processors. The result, 53_H, resets N, OV, DC, and Z flags.

N flag is reset because Bit7 of 53_H is 0, DC is reset because there is no carry generated from Bit3 to Bit4, Z is reset because the result is not zero, and OV flag is reset because the subtraction of two positive numbers results in a smaller number.

SUBFWB   Subtract File (Data) Register from WREG with Borrow   (1W/1Cy)

Syntax: SUBFWB   F, d, a   Status: All flags are affected.

Description: Subtract data register and borrow from the WREG. This instruction is used for 16-bit subtraction. If d = 0, place the result in W and if d = 1, place the result in data register. If a = 0, select data register from Access bank and if a = 1, use BSR.

Example: Assume that the working register W holds the data byte 18_H, the general purpose data register REG20 (with the address $=20_H$) holds the byte 72_H, and the borrow flag (C=1) is set. Subtract REG20 from WREG, save the result in the data register, and identify the flag status.

Instruction: SUBFWB   0x20, 1, 0   ; $18_H - 72_H = A6_H$

Flags: This is a subtraction by 2's complement addition. The result is $A6_H$ with the Borrow flag reset. The result, $A6_H$, sets N and DC flags and resets OV and Z flags.

N flag is set because Bit7 of $A5_H$ is 1, DC is set because there is a carry generated from Bit3 to Bit4, Z is reset because the result is not zero, and OV flag is reset because the subtraction of two positive numbers results in a smaller number.

REG20   = 0 1 1 1   0 0 1 0 (72_H)

2's REG20 = 1 0 0 0   1 1 1 0 ($8E_H$)
                    +

WREG   = 0 0 0 1   1 0 0 0 (18_H)
    CY [0]   0 0 1 1

REG20   = 1 0 1 0   0 1 1 0   ($A6_H$)

                        N OV Z DC C

Status: | - | - | - | 1 | 0 | 0 | 1 | 0 |

SWAPF   Swap Nibbles   (1W/1Cy)

Syntax: SWAPF   F, d, a   Status: No flags are affected

Description: Exchange low-order four bits <3-0> (low-order nibble) with high-order four bits <7-4> (high-order nibble). If d = 0, place the result in W and if d = 1, place the result in data register. If a = 0, select the data register from the Access bank and if a = 1, use BSR.

Example: Assume that the general purpose data register REG20 (with the address =20_H) holds the byte 72_H. Exchange the nibbles and save the result in REG20.

Instruction: SWAPF   0x20, 1, 0

        This instruction swaps the nibbles, and the result, 27_H, is saved back in REG20.

TBLRD   Table Read   Status: No flags are affected.   (1W/2Cy)

Syntax: TBLRD*   : Table Read

        TBLRD*+   : Table Read and Increment Table Pointer

        TBLRD*-   : Table Read and Decrement Table Pointer

        TBLRD+*   : Increment Table Pointer First and Table Read

Description: Read (copy) the contents of program memory into the 8-bit register TABLAT. The 21-bit address of the program memory is stored in the register TBLPTR (Upper–TBLPTRU, High–TBLPTRH, and Low–TBLPTRL). The instruction copies the contents of program memory into TABLAT, and there are four options for the address in the TBLPTR as shown above: (1) no change in the TBLPTR, (2) increment the TBLPTR, (3) decrement the TBLPTR, and (4) increment the TBLPTR prior to the copy operation.

        This instruction is used to copy data from program memory into the 8-bit register TABLAT and the MOV instruction is used to copy from TABLAT to data registers (data memory).

Example: Identify the changes in various registers after the execution of the instruction
               TBLRD*+ if the 21-bit memory address is 008005_H and the memory location
               holds the byte $F2_H$.

Instruction: TBLRD*+   ; Before instruction: (008005_H) = $F2_H$

               After the execution of the instruction TBLRD*+, the register contents are:

               TABLAT = $F2_H$

               TBLPTR = 008006_H (TBLPTRU =00, TBLPTRH =80, TBLPTRL =06)

TBLWT   Table Write   Status: No flags are affected.   (1W/2Cy)

Syntax: TBLWT*   :Table Write

               TBLRD*+   :Table Write and Increment Table Pointer

               TBLRD*-   : Table Write and Decrement Table Pointer

               TBLRD+*   : Increment Table Pointer First and Table Write

Description: Write (copy) the contents of TABLAT into program memory of which the address
               is in the TBLPTR. The 21-bit address of the program memory is stored in the reg-
               ister TBLPTR (Upper–TBLPTRU, High–TBLPTRH, and Low–TBLPTRL). The
               instruction copies the contents of TABLAT into the program memory, and there
               are four options for the address in the TBLPTR as shown above: (1) no change in
               the TBLPTR, (2) increment the TBLPTR, (3) decrement the TBLPTR, and (4) incre-
               ment the TBLPTR prior to the copy operation.

Example: Identify the changes in various registers after the execution of the instruction
               TBLWT*- if the 21-bit memory address in the TBLPTR register is (018042_H) and
               the TABLAT holds the byte $A7_H$.

Instruction: TBLWT*-   ; Before instruction: (TABLAT) = $A7_H$

               After the execution of the instruction TBLWT*-, the register contents are:

               TABLAT = $A7_H$

               (018042_H) = $A7_H$

               TBLPTR = 018041_H (TBLPTRU =01, TBLPTRH =80, TBLPTRL =41)

TSTFSZ   Test File (Data) Register; Skip if F = 0   (1W/1-2Cy)

Syntax: TSTFSZ   F, a   Status: No flags are affected.

Description: Test the data (file) register, and if it is zero, skip the next instruction. If a = 0,
               select the data register from the Access bank, and if a = 1, use BSR.

XORLW   Logically Exclusive OR 8-bit Literal with WREG   (1W/1Cy)

Syntax: XORLW   8-bit   Status: N and Z flags are affected.

Description: Exclusive OR 8-bit literal with the contents of the WREG and save the result in
               WREG.

Example: Assume that the working register W holds the data byte F7$_H$. Exclusive OR the contents of WREG with the literal 78$_H$ and identify the flag status.

Instruction: XORLWF   0x78

W   = 1 1 1 1   0 1 1 1   (F7$_H$)
XOR
8-bit = 0 1 1 1   1 0 0 0   (78$_H$)

W   = 1 0 0 0   1 1 1 1   (8F$_H$)
                                N   Z

Status: | - | - | - | 1 | - | 0 | - | - |

Flags: After Exclusive ORing F7$_H$ and 78$_H$, the result 8F$_H$ is stored in W register. The result (8F$_H$) sets the N flag and resets Z flag.

XORWF   Logically Exclusive OR WREG with File (Data) Register   (1W/1Cy)

Syntax: XORWF   F, d, a   Status: N and Z flags are affected.

Description: Exclusive OR the contents of the WREG with the contents of data register. If d = 0, place the result in W and if d = 1, place the result in the data register. If a = 0, select File Register from Access Bank and if a = 1, use BSR.

Example: Assume that the working register W holds the data byte F6$_H$ and the general purpose file register REG20 (with the address =20$_H$) holds the byte 8F$_H$. Exclusive OR the bytes, save the result in W register, and identify the flag status.

Instruction: XORWF   0x20, 0, 0   (Substitute REG = 0 x 20 or define REG with an equate)

W   = 1 1 1 1   0 1 1 0   (F6$_H$)
XOR
REG20 = 1 0 0 0   1 1 1 1   (8F$_H$)

W   = 0 1 1 1   1 0 0 1   (79$_H$)
                                N   Z

Status: | - | - | - | 0 | - | 0 | - | - |

Flags: After Exclusive ORing F6$_H$ and 8F$_H$, the result 79$_H$ is stored in W register and REG20 is unchanged. The result (79$_H$) resets both the N and Z flags.

IN-CIRCUIT EMULATOR (ICE), IN-CIRCUIT DEBUGGER (ICD), AND MICROCHIP MPLAB® ICD 2

Throughout the text we used MPLAB IDE to write, assemble, execute, and debug programs (software). To execute and debug programs we focused exclusively on the MPLAB simulator. However, working software in a simulator is one part of an embedded system. Another part is hardware, known as a target system, and software should work with the target system. We need to download programs in memory of the target system and verify that the programs execute properly in the target system. Now the critical questions are: Is hardware working? If it is not working, how does one troubleshoot the system and debug software in conjunction with hardware? When troubleshooting, embedded systems present unique challenges to answering these questions because such systems usually lack keyboards, screens, and storage devices that are available in PCs. Without a keyboard and display (in most cases) there is no easy way to send an input signal and check the expected output.

B.1 TROUBLESHOOTING TOOLS: IN-CIRCUIT EMULATOR (ICE) AND IN-CIRCUIT DEBUGGER (ICD)

To troubleshoot hardware of a target system, traditional tools such as volt meters and oscilloscopes are inadequate; we need additional tools. In industry, tools commonly used to troubleshoot embedded systems are the in-circuit emulator (ICE) and the in-circuit debugger (ICD). Microchip and other vendors provide these tools. The ICE is much more expensive than the ICD; however, the ICD is a cost-efficient alternative to the ICE. In the following sections, we briefly describe both of these tools and focus on the Microchip ICD 2 that is supported by MPLAB IDE software tools.

B.1.1 In-Circuit Emulator (ICE)

An ICE is a complete diagnostic system that enables the user to test the hardware and software of a microprocessor- or microcontroller-based prototypes such as embedded systems in the development process. The ICE includes resources such as a processor, memory, and I/O that can be borrowed by a target system through a probe (a cable with appropriate pin connections). For example, if we build a part of our prototype target system that includes a 40-pin socket for an MCU, power supply, and a clock, and insert the probe from the ICE into the socket in place of the MCU, we can test our programs using the resources of the ICE. As we begin to add components such as memory to our prototype system, our prototype will be less and less dependent on the ICE. This is similar to a fetus growing in a womb connected by an umbilical cord. Finally, when all components such as memory, I/O, and programs are tested, we can replace the probe by the MCU in our prototype embedded system. The ICE provides the following capabilities in the development process of our embedded system:

1. **Downloading.** Software can be transferred between the ICE, a software development system such as a PC, and the prototype target system.
2. **Resource Sharing.** The ICE allows sharing of any resources (memory and I/O) with the prototype system. Memory and I/Os of the ICE can be mapped (assigned addresses) anywhere to avoid conflict with the memory and I/O addresses of the prototype system.
3. **Debugging Tools.** The ICE software provides the following debugging tools: (a) setting breakpoints, (b) tracing code execution, (c) displaying and modifying register contents, and (d) assembling and disassembling code in-line.

B.1.2 In-Circuit Debugger (ICD)

An ICD is a debugging tool that enables the user to test hardware and software by using the resources of the prototype target embedded system. The ICD does not have independent resources similar to the ICE that can be shared between the ICD and the prototype target.

The ICD requires exclusive use of some hardware and software resources of the target system, communicates serially with the target system through specially assigned pins, and probes the code execution. The limitations of the ICD are: (1) the target system should be partially working, (2) it needs some resources from the target system, and (3) an access should be built in the target system. However, the ICD has the following advantages over the ICE in addition to cost effectiveness. The ICD:

1. can be connected to the embedded system after the production cycle because it can work with the microcontroller that is installed and operational in the system.
2. can re-program the firmware (system software in Flash) in the target system without any additional connections or equipment.

B.2 MPLAB® ICD 2

This section focuses on the specific ICD called MPLAB® ICD 2 designed by Microchip. Some of the features of the ICD 2 listed by Microchip are as follows:

- USB (Full Speed 2 M bits/s) & RS-232 interface to host PC
- Reading/Writing memory space and EEDATA areas of target microcontroller

- Real time background debugging
- Various power connection options for ICD 2 and target systems
- Diagnostic LEDs (Power, Busy, Error)
- Built in over-voltage/short circuit monitor
- Firmware upgradeable from PC
- Supports low voltage to 2.0 volts (2.0 to 6.0 range)
- Programs configuration bits

The first four features are discussed in the following sections.

This in-circuit debugger is used for two purposes:

1. Programming the code and data (reading/writing program memory and EEPROM) into the target.
2. Troubleshooting hardware and software (real time background debugging).

The software necessary to program and troubleshoot the target system is integrated in the MPLAB. The ICD 2 has three connections: (1) power supply, (2) six-conductor modular interface that connects to the target system, and (3) USB or RS-232 (EIA-232) cable that connects to the PC. A block diagram of these three connections is shown in Figure B-1. The ICD 2 is connected to the target PIC18F MCU (such as PICDEM™ 2 Plus) with the six-conductor modular interface cable as shown in Figure B-2. Pin 6 is not used.

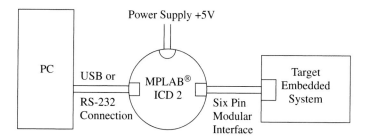

FIGURE B-1   MPLAB® ICD 2 Connections to PC and Target Embedded System

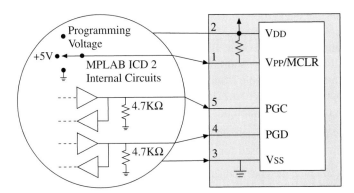

FIGURE B-2   MPLAB® ICD 2 Connections for Programming and Debugging

B.2.1 Applying Power (Refer to ICD 2 User's Guide)

There are a number of configurations for powering ICD 2 and the target system. In this description, we assume that the target system is powered by its own power supply.

1. When using the USB connection to the PC, ICD 2 can be powered from the PC.
2. When using the RS-232 connection to the PC, the ICD 2 must have its own power supply.
3. When the ICD 2 has its own power supply (+ 5 V), it can provide a limited amount of current, up to 200 mA, to a small target board (not recommended and not discussed here—See the user's guide for Power sequence).
4. ICD 2 cannot be powered from the target board.
5. Power should be applied to ICD 2 before applying power to the target.

B.2.2 Setting Up the Environment for Programming and Debugging

1. Connect the USB cable from the PC to MPLAB ICD2. This will power the ICD2 or apply power separately to the ICD 2. Do not power the target yet.
2. Open up MPLAB IDE v X.XX software on your PC.
3. Select the Programmer menu (or Debugger menu). Select Programmer → MPLAB ICD 2. (You may ignore any output error messages until a target system is connected and powered up.)
4. Programmer → MPLAB ICD 2 Setup Wizard > A Welcome screen opens up. → Next
5. Communication: Select Communication Method → USB (or COMX and Baud for RS-232) → Next
6. Power: Select target power source ⊙ Target has own power supply → Next
7. Connection: ⊙ MPLAB IDE connected to the MPLAB ICD 2 → Next
8. Download: ⊙ MPLAB ICD 2 automatically downloads the required operating system → Next
9. Summary: → Finish. (See Figure B-3 – Summary Screen). The summary screen shows MPLAB ICD 2 settings which include serial port (USB or COM Port), power source of the target, automatic connection to ICD 2 and automatic downloading of the operating system.

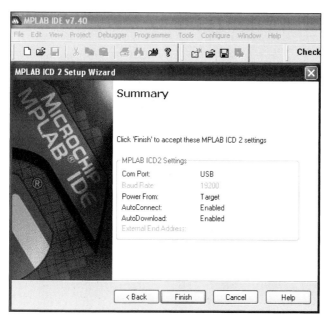

FIGURE B-3   Summary of MPLAB ICD 2 Setup

10. Connect the six-conductor modular cable to a target system (such as PICDEM™ 2 Plus) and apply power to the target system.
11. Select Programmer → Settings → Status (See Figure B-4a after self-test). It shows the status of the settings. → Power (See Figure B-4b for Power connections). It shows the power connections of the target system and ICD 2.

FIGURE B-4a   Status of MPLAB ICD 2 Settings   **FIGURE B-4b**   Power Connections

12. → Program (See Figure B-4c for programming options). (a) For the PIC18FXX2/
    XXX2 devices, memory is erased each time the chip is programmed. Therefore, "Erase
    all before program" option does not have any effect, (b) Check 'Program' option in
    Select Memories section and enter memory addresses for full range from 0x00 to
    0x7DBF (PICDEM 2 Plus). This End address is adjusted to leave memory space for
    the MPLAB ICD 2 Debug Executive.

13. Select Programmer → Connect (See Figure B-4d for the output message). The output
    message should show that ICD 2 is ready. You may ignore the initial error messages
    while setting up the ICD 2.

The above steps can also be accomplished by selecting the Debugger menu → Select Tools
→ MPLAB ICD 2.

FIGURE B-4c   Programming Options

FIGURE B-4d   Output Messages

B.2.3 Setting Configuration Bits

After the selection of a microcontroller, the MPLAB® IDE initializes many options by setting up configuration bits. To set or change configuration bits to be programmed into the device:

- Select Configure → Configuration Bits. The screen shows various options and their settings. By clicking on the text in the "Settings" column, various options can be enabled or disabled.

The following options are recommended for PICDEM 2 Plus:

- Oscillator: EC-OS2 as RA6
- OSC Switch: Disabled
- Power Up Timer: Enabled
- Brown Out Detect and Watchdog Timer: Disabled
- CCP2MUX: RC1
- Low Voltage Programming and Stack Overflow Reset: Disabled
- Background Debug: Enabled
- All other configuration bits are disabled.

B.2.4 Programming and Debugging

The steps for using ICD 2 for programming and debugging are similar in many ways to using the MPLAB simulator. Before we can program (store binary code) in flash memory, we need to write a program, and we have written, assembled, debugged, and executed many programs throughout the text. The steps for writing programs are recalled here for easy reference.

1. Open up MPLAB IDE and write a program (source code) using the MPLAB editor, and save it under a file name.
2. Set up a project and workspace, add the source file created in Step 1, and build (assemble) the project.
3. Use the Debugger to select the MPLAB SIM as a tool. Debug and execute the source code using the MPLAB simulator.
4. Set up the Programmer as outlined in Section B2.2 and program the memory of the target system (such as PICDEM™ 2 Plus) using the Hex file created when the project is built in Step 2 above. Use the Program command.
5. Set up the ICD 2. Setup breakpoints, select registers to monitor in the Watch window, and debug and execute the program in the target system.

The screens that display while the ICD 2 is running are similar to those in the MPLAB simulator.

APPENDIX **C**

OPERATIONAL AMPLIFIERS AND SIGNAL CONDITIONING CIRCUITS

To interface an output of a sensor to an A/D converter module of a microcontroller, it may be necessary to shift and/or scale voltage levels of the output to take full advantage of the dynamic range of the reference voltages of the converter. This process of shifting and scaling voltages is called Signal Conditioning. For example, if the reference voltage range of an A/D converter is 0 to 5 V and the sensor output voltage is 0 to 1 V, the sensor voltage is amplified with the gain of 5. This is called scaling. If the sensor voltage range is -2 V to +3 V, the low end of the sensor voltage is adjusted to 0 volts. This is called shifting. An operational amplifier (known as op amp) is a commonly used device to design signal conditioning circuits. This appendix focuses on how op amp circuits work and how they are used in scaling (amplifying) an output voltage of a sensor (a circuit for shifting a voltage is not discussed here).

C.1 OPERATIONAL AMPLIFIER (OP AMP)

An operational amplifier, usually referred to as an "op amp" is a high-gain voltage amplifier with two inputs (as shown in Figure C-1a), a single output, and, generally, biased with two power supply voltages: +V and -V. It amplifies the difference between the two input voltages—V_1 connected to (+) terminal, called the non-inverting input, and V_2 connected to (−) terminal, called an inverting input. The input voltages V_1 and V_2 are also referred to as Vin+ and Vin−, respectively, or just Vin. The op amp, when used as an amplifier, always employs a feedback circuit (discussed in the following section). The ideal characteristics of an op amp (Figure C-1b) are:

1. Infinite voltage gain ($A_V = \infty$)
2. Infinite input impedance (Rin = ∞)
3. Zero output impedance (Ro = 0)
4. Infinite bandwidth (BW = ∞)

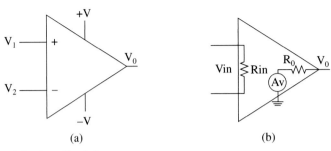

(a)   (b)

(a) Op Amp Symbol

(b) Op Amp Equivalent Circuit

FIGURE C-1

Based on these ideal characteristics, we will analyze the op amp circuits in the following sections.

C.1.1 Inverting Amplifier

Figure C-2 shows an inverting amplifier with an input Vin to the inverting terminal connected through the resistor R_1, and the non-inverting terminal is grounded. The output signal V_o is fed back to the input through the resistor R_f. The voltage gain Av, defined as V_o/Vin of the op amp is calculated with the following reasoning.

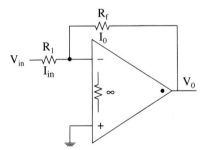

FIGURE C-2   Inverting Amplifier Circuit

1. Because the op amp has infinite gain, it needs a very small signal (almost zero) as an input. If the non-inverting terminal is grounded (zero volts), the inverting terminal must be almost at zero volts, known as virtual ground.
2. The input voltage Vin = $R_1 I_{in}$
3. Because the op amp has an infinite impedance, no current can flow towards the op amp; it must flow through the resistor R_f. Therefore, $I_{in} = Io$ and Vo = - R_f Io (this is a negative voltage because of the direction of the current flow).
4. Gain Av = V_o/Vin = - $R_f I_0/R_1 I_{in}$ = - R_f/R_1 (because $I_0 = I_{in}$).

In this circuit, the voltage gain is negative and equal to the ratio of two resistors $= -R_f/R_1$. This is called an inverting amplifier because the polarity of the output voltage is opposite of the input voltage.

C.1.2 Non-inverting Amplifier

Figure C-3 shows a non-inverting amplifier with an input Vin to the non-inverting terminal (+). The portion of the output voltage V_f is fed back to the inverting terminal (−). The voltage gain Av (V_o/Vin) of the non-inverting op amp is as follows:

1. Voltage $V_f = V_0 (R_1)/(R_1 + R_f)$ – Voltage divider
2. Because the op amp has an infinite gain, the input signal needed to amplify is very small. Therefore, Vin $\approx V_f = V_0 (R_1) / (R_f + R_1)$

$$Av = \frac{V_0}{Vin} = \frac{R_f + R_1}{R_1} = \frac{R_f}{R_1} + 1$$

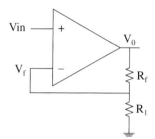

FIGURE C-3   Non-inverting Amplifier Circuit

In this circuit, the voltage gain is equal to the ratio of two resistors $= R_f/R_1 + 1$. This is called a non-inverting amplifier because the polarity of the output voltage is the same as of the input voltage.

C.1.3 Op Amp as a Comparator

Figure C-4 shows the op amp used as a comparator. It compares two input signals, and when one is slightly higher than the other, the output switches from positive to negative power supply voltage or vice versa.

When V_1 is slightly higher than V_2, the op amp with infinite gain drives the output to the saturation voltage limited by the value of the power supply. V_1 is connected to the non-inverting terminal; therefore, the output voltage goes to the positive power supply.

When V_2 is slightly higher than V_1, the op amp with infinite gain drives the output to the negative power supply because V_2 is connected to the inverting terminal.

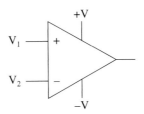

FIGURE C-4   OpAmp As a Comparator

EXAMPLE C.1

Calculate the output voltage in Figure C-5 if the input from the temperature sensor varies from 0 V to 1 V (this is the same problem as in Figure 14.4).

Solution

The Op Amp1 is an inverting amplifier with the voltage gain = – Rf/R1 = – 10 k/ 10 k = – 1

The Op Amp2 is also an inverting amplifier with the voltage gain = – 50 k / 10 k = – 5

The total voltage gain in the circuit = (– 1) x (– 5) = + 5

Therefore, the output voltage Vo for the input voltage 0 V = 0 V

The output voltage for the input voltage 1V = + 5 V

Therefore, the voltage range of the output = 0 to + 5 V

C.2 SIGNAL CONDITIONING CIRCUIT (SCC)

In Figure C-5, the output of the temperature sensor has a range 0 to 1 V. In Figure 14-4 (Chapter 14), in the range of the reference voltages of the A/D converter is 0 to + 5 V; thus the output of the A/D converter is calibrated for 0 to 5 V. If the output of the temperature sensor is directly connected to the A/D converter, the digital output will be limited to 1 V input from the sensor. To take the advantage of the full range, we need to scale (amplify) the input voltage to 5 V.

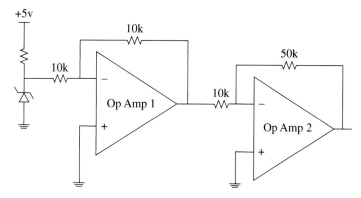

FIGURE C-5 Op Amp Scaling Circuit

In Figure C-5, the sensor output is connected to the unity gain amplifier (Op Amp1), which inverts the polarity of the signal. The Op Amp2 inverts and amplifies the signal with the gain of 5. Thus, the sensor voltage range is scaled to match the range of the reference voltages. In some situations, the minimum sensor voltage output is negative. In such situations, the minimum voltage needs to be shifted to 0V. This process of scaling and shifting the output of a sensor to match the range of reference voltages of an A/D converter is known as conditioning the signal and the circuits as signal conditioning circuits (SCC).

APPENDIX **D**

NUMBER SYSTEMS AND
HEX ARITHMETIC

Computers communicate and operate in binary digits (0 and 1); on the other hand, human beings generally use the decimal system with 10 digits (0–9). Other number systems are also used, such as octal with 8 digits (0–7), and the hexadecimal (Hex) system with digits 0–15. In the hexadecimal system, digits 10 through 15 are designated as A through F, respectively, to avoid confusion with the decimal numbers 10–15.

A positional scheme is used to represent a number in any of the number systems. This means that each digit will have its value according to its position in a number. The total number of digits in a given number system is also referred to as the base. For example, the binary system has base 2, the decimal system has base 10, and the hexadecimal system has base 16.

D.1 NUMBER CONVERSION

A number in any base system can he represented in a generalized format as follows:

$$N = A_nB^n + A_{n-1}B^{n-1} + \ldots + A_1B^1 + A_0B^0$$

N = number, B = base, A = any digit in that base

For example, the decimal number 245_{10} can be represented in various number systems as follows:

Decimal:	$(2\ 4\ 5)_{10}$	$= 2 \times 10^2 + 4 \times 10^1 + 5 \times 10^0$	$= 245_{10}$

Octal:	$(3\ 6\ 5)_8$	$= 3 \times 8^2 + 6 \times 8^1 + 5 \times 8^0$	
		$= 192 + 48 + 5$	$= 245_{10}$

Hexadecimal:	$(F5)_{16}$	$= 15 \times 16^1 + 5 \times 16^0$	
		$= 240 + 5$	$= 245_{10}$

Binary: $(11110101)_2$ $= 1 \times 2^7 + 1 \times 2^6 + 1 \times 2^5 + 1 \times 2^4 + 0 \times 2^3 + 1 \times 2^2 + 0 \times 2^1 + 1 \times 2^0$

$$= 128 + 64 + 32 + 16 + 0 + 4 + 0 + 1$$

$$= 245_{10}$$

The previous example shows the equivalent values of the decimal number 245 into three other number systems and also how to convert a given number in any system into its decimal equivalent. We can also make up new number systems such as base 4, base 5, or base 20. Let us devise a number system with base 4 that includes digits from 0 to 3, name it Quatro, and convert the numbers from base 4 to decimal as in the above examples:

Quatro: $(312)_4$ $= 3 \times 4^2 + 1 \times 4^1 + 2 \times 4^0$

$$= 48 + 4 + 2 = 54_{10}$$

CONVERSION TABLE: DECIMAL, HEXADECIMAL, BINARY, AND OCTAL

Decimal	Hex	Binary	Octal	Quatro
0	0	0000	00	00
1	1	0001	01	01
2	2	0010	02	02
3	3	0011	03	03
4	4	0100	04	10
5	5	0101	05	11
6	6	0110	06	12
7	7	0111	07	13
8	8	1000	10	20
9	9	1001	11	21
10	A	1010	12	22
11	B	1011	13	23
12	C	1100	14	30
13	D	1101	15	31
14	E	1110	16	32
15	F	1111	17	33

D.1.1 How to Convert between Binary, Hex, Octal, and Quatro Numbers

There is a special relationship between these number systems: each one is powers of 2. The Hex number system has 16 digits (2^4) and requires 4 binary digits to convert from binary to Hex; the octal has 8 digits (2^3) and requires 3 binary digits, and the quatro has 4 digits (2^3) and needs 2 binary digits. The following two examples illustrate how to convert between these number systems.

EXAMPLE D.1

Convert the binary number 1 0 0 1 10 1 0 into its Hex, octal, and quatro equivalent numbers.

Solution:

Binary to Hexadecimal

Step 1: Starting from the right (LSB), arrange the binary digits in groups of four: 1 0 0 1-1 0 1 0.

Step 2: Convert each group into its equivalent Hex number = 9 A.

Binary to Octal

Step 1: Starting from the right (LSB), arrange the binary digits in groups of three:

1 0 - 0 1 1 - 0 1 0.

Step 2: Convert each group into its equivalent octal number: 2 3 2.

Binary to Quatro

Step 1: Starting from the right (LSB), arrange the binary digits in groups of two:

1 0 - 0 1 - 1 0 -1 0.

Step 2: Convert each group into its equivalent octal number: 2 1 2 2.

EXAMPLE D.2

Convert the Hex number 3 F 8 into its binary equivalent number.

Solution:

Hexadecimal to Binary

Step 1: Convert each Hex digit into its equivalent binary digits.

3   F   8 = 0 0 1 1 – 1 1 1 1 – 1 0 0 0

D.2 CONVERSION FROM DECIMAL NUMBERS INTO OTHER NUMBER SYSTEMS USING REPEATED DIVISION

To covert a decimal number into any other number system, the decimal number is divided by the base of the number system and the remainder of the first division becomes the least significant digit of the new number. The division is repeated until the quotient becomes 0, and the remainders represent the digits of the equivalent number in the new number system as illustrated in the following examples.

EXAMPLE D.3

Convert the decimal number 245_{10} in octal and Hex number systems.

Solution:

To convert 245_{10} into the octal number, divide the number by the base 8 until the quotient is 0.

$$
\begin{array}{ccc}
0 & 3 & 30 \\
8\,\overline{\smash{)}\,3} & 8\,\overline{\smash{)}\,30} & 8\,\overline{\smash{)}\,245} \\
\underline{-0} & \underline{-24} & \underline{-240} \\
R = 3 & R = 6 & R = 5
\end{array}
$$

Answer: $(245)_{10} = (365)_8$

To convert 245_{10} in the Hex number, divide the number by the base 16 until the quotient is 0.

$$
\begin{array}{cc}
0 & 15 \\
16\,\overline{\smash{)}\,15} & 16\,\overline{\smash{)}\,245} \\
\underline{-0} & \underline{-240} \\
R = 15 & R = 5
\end{array}
$$

Answer: $(245)_{10} = (F5)_{16}$

EXAMPLE D.4

Convert the decimal number 29_{10} in binary, octal and Hex number systems.

Solution:

To convert 29_{10} into the binary number, divide the number by the base 2 until the quotient is 0.

$$
\begin{array}{ccccc}
0 & 1 & 3 & 7 & 14 \\
2\,\overline{\smash{)}\,1} & 2\,\overline{\smash{)}\,3} & 2\,\overline{\smash{)}\,7} & 2\,\overline{\smash{)}\,14} & 2\,\overline{\smash{)}\,29} \\
\underline{-0} & \underline{-2} & \underline{-6} & \underline{-14} & \underline{-28} \\
1 & 1 & 1 & 0 & 1
\end{array}
$$

Answer: $(29)_{10} = 1\ 1\ 1\ 0\ 1$

continued

To convert 29_{10} into the octal number, divide the number by the base 8 until the quotient is 0.

$$
\begin{array}{c|c}
 & 0 \\ \hline
8 & 3 \\
 & -0 \\ \hline
 & 3
\end{array}
\qquad
\begin{array}{c|c}
 & 3 \\ \hline
8 & 29 \\
 & -24 \\ \hline
 & 5
\end{array}
$$

Answer: $(29)_{10} = (35)_8$

To convert 29_{10} into the hex number, divide the number by the base 16 until the quotient is 0.

$$
\begin{array}{c|c}
 & 0 \\ \hline
16 & 1 \\
 & -0 \\ \hline
 & 1
\end{array}
\qquad
\begin{array}{c|c}
 & 1 \\ \hline
16 & 29 \\
 & -16 \\ \hline
 & 13
\end{array}
$$

Answer: $(29)_{10} = (1D)_{16}$

D.3 2'S COMPLEMENT AND ARITHMETIC OPERATIONS

The processors perform the subtraction of two binary numbers using the 2's complement method. In digital logic circuits, it is easier to design a circuit to add numbers than to design a circuit to subtract numbers. The 2's complement of a binary number is equivalent to its negative number; thus, by adding the complement of the subtrahend (the number to be subtracted) to the minuend (the number to be subtracted from), a subtraction can be performed. The method of 2's complement is explained below with examples from the decimal number system.

Decimal Subtraction

EXAMPLE D.5

Subtract the following two decimal numbers using the borrow method and the 10's complement method: (52 – 23).

Borrow Method

Minuend: 5   2
          −
Subtrahend: 2   3

continued

Step 1: To subtract 3 from 2, 10 must be borrowed from the second place of the minuend. Therefore, $(12 - 3) = 9$

Step 2: The second place: $4 - 2 = 2$
The result is: 2 9

10's Complement Method

In the 10's complement method, we find the 10's complement of the subtrahend by subtracting each digit from 0 by borrowing from the next digit. In our example, the subtrahend is 23. First we subtract 3 by borrowing 10 from the next 0 in the tenth place and next we subtract 2 and borrow 1 by borrowing 10 again from the next position, even if there is no digit.

Step 1: $00 - 23 = 77$ with a borrow from the third position or this is equivalent of subtracting 23 from 100 $(100 - 23 = 77)$, which is also known as 100's complement. This can also be stated as 77 is a complement or negative of 23.

Step 2: Add the complement of the subtrahend to the minuend. $77 + 52 = 129$

Step 3: The result is 29 with a carry. Some processors complement (or drop) the carry, making it zero; PIC processors show that the carry is set.

    The processor follows the same procedure to perform a subtraction. To perform a subtraction of two binary numbers, the processor takes 2's complement of the subtrahend and adds to the minuend.

Procedure to Find 2's Complement of a Binary Number

To find the 2's complement of a binary number, we can subtract the number from 0's and find the complement. Or we can follow the next two steps.

Step 1: Find the 1's complement. This amounts to replacing 0 by 1 and 1 by 0.

Step 2: To find the 2's complement, add 1 to the 1's complement. This is equivalent of subtracting the number from 0's.

EXAMPLE D.6

Find the 2's complement of the binary number: 0 0 0 1 1 1 0 0 ($1C_H$ or 28_{10}).

Solution:

Step 1: Find the 1's complement, meaning replace 0 with 1 and 1 with 0: 1 1 1 0 0 0 1 1

continued

Step 2: Add 1 to 1's complement.

$$1\ 1\ 1\ 0\ 0\ 0\ 1\ 1$$

$$+\quad 0\ 0\ 0\ 0\ 0\ 0\ 0\ 1$$

CY   1 1

$$\overline{\rule{3cm}{0pt}}$$

$$1\ 1\ 1\ 0\ \ 0\ 1\ 0\ 0\ = (\text{E}\ 4_\text{H})$$

By examining the result of the example, the following rules can be stated to find the 2's complement of a binary number, instead of the above procedure of the 1's complement.

Rule 1: Start at the LSB of a given number, and check all the bits to the left. Keep all the bits as they are up to and including the least significant 1.

Rule 2: After the first 1, replace all 0's with 1's and 1's with 0's.

These rules *can* be applied to the *given* binary number ($1C_\text{H}$) as illustrated below:

Binary Number: (0 0 0 1 1) (1 0 0) ← Start Here

↑   ↑

Replace 0 with 1   Keep as they are
and 1 with 0

2's complement: 1 1 1 0 0 1 0 0 = E 4$_\text{H}$

Subtraction Using 2's Complement and Borrow Method

A Binary subtraction can be performed by using the 2's complement method, and if the result is negative, it is expressed in terms of 2's complement.

EXAMPLE D.7

Subtract $1C_\text{H}$ (0001 1100) from 42_H (0100 0010).

Solution:

2's Complement Method

continued

Step 1: Take 2's Complement of $1C_H$ =   1 1 1 0 0 1 0 0   (See Example D.6)

                            +

Step 2: Add minuend 42_H =   0 1 0 0 0 0 1 0

                   CY   = 1]   1

              Result: = 1]   0 0 1 0 0 1 1 0

Borrow Method: 4 2   =   0 1 0 0 0 0 1 0

              -    1 C   =   0 0 0 1 1 1 0 0

                 BR   =   1 1 1

           Result   =   0 0 1 0 0 1 1 0

The results in both methods are the same except that the 2's complement method generates a carry.

Subtraction in Hex:

We can perform the same subtraction ($42_H – 1C_H$) using Hex numbers rather than binary numbers as follows:

Minuend:	4	2
Subtrahend:	- 1	C
Borrow:	- 1	
Result	2	6

Step 1: To subtract C_H (12_{10}) from 2, we need to borrow from the next position. When we borrow 1 in Hex, that is 16 (not 10 as in decimal). Therefore, $(16 + 2 – 12) = 6$

Step 2: In the second position, we need to subtract 1 of the subtrahend and borrow 1 from 4 of the minuend. The result is 2.

We can arrive at the same answer by various methods. But the important points to remember are:

1. The processor executes the subtraction by using the 2's complement method.
2. The only difference between the borrow method and the 2's complement method is the borrow. The PIC processor preserves the borrow; other processors (such as Intel and Motorola) complement the borrow to match the result of the borrow method.
3. The negative numbers (integers) are represented in 2's complement.

American Standard Code for Information Interchange: ASCII Codes

Graphic or Control		ASCII (Hexadecimal)
NUL	Null	00
SOH	Start of Heading	01
STX	Start of Text	02
ETX	End of Text	03
EOT	End of Transmission	04
ENQ	Enquiry	05
ACK	Acknowledge	06
BEL	Bell	07
BS	Backspace	08
HT	Horizontal Tabulation	09
LF	Line Feed	0A
VT	Vertical Tabulation	0B
FF	Form Feed	0C
CR	Carriage Return	0D
SO	Shift Out	0E
SI	Shift In	0F
DLE	Data Link Escape	10
DC1	Device Control 1	11
DC2	Device Control 2	12
DC3	Device Control 3	13
DC4	Device Control 4	14
NAK	Negative Acknowledge	15
SYN	Synchronous Idle	16
ETB	End of Transmission Block	17
CAN	Cancel	18
EM	End of Medium	19
SUB	Substitute	1A
ESC	Escape	1B

Graphic or Control		ASCII (Hexadecimal)
FS	File Separator	1C
GS	Group Separator	1D
RS	Record Separator	1E
US	Unit Separator	1F
SP	Space	20
!		21
"		22
#		23
$		24
%		25

Graphic or Control	ASCII (Hexadecimal)	Graphic or Control	ASCII (Hexadecimal)
&	26	S	53
'	27	T	54
(	28	U	55
)	29	V	56
*	2A	W	57
+	2B	X	58
,	2C	Y	59
−	2D	Z	5A
.	2E	[	5B
/	2F	\	5C
0	30	]	5D
1	31	^	5E
2	32	—	5F
3	33	`	60
4	34	a	61
5	35	b	62
6	36	c	63
7	37	d	64
8	38	e	65
9	39	f	66
:	3A	g	67
;	3B	h	68
<	3C	i	69
=	3D	j	6A
>	3E	k	6B
?	3F	l	6C
@	40	m	6D
A	41	n	6E
B	42	o	6F
C	43	p	70
D	44	q	71
E	45	r	72
F	46	s	73
G	47	t	74
H	48	u	75
I	49	v	76
J	4A	w	77
K	4B	x	78
L	4C	y	79
M	4D	z	7A
N	4E	{	7B
O	4F			7C
P	50	}	7D
Q	51	~	7E
R	52	DEL Delete	7F

APPENDIX  **F**

PIC18 SIMULATOR IDE

INTRODUCTION TO PIC18F ASSEMBLER AND SIMULATOR

This textbook includes a CD with a simulator program, PIC18 Simulator IDE, that is designed, developed, and written by Vladimir Soso of Oshonsoft[1] In this appendix, the terms PIC18 and PIC18F are used synonymously. The CD also includes illustrative programs from the textbook that can be assembled and executed using the simulator. The simulator program is included in the textbook for an individual user and should not be used for commercial purposes. The user must complete the installation of the program on one's computer system and agree to the licensing terms. The program is available to the user for one semester after the date of installation. For additional information and site license, see the Help menu.

The PIC18 Simulator IDE (Integrated Development Environment) is a group of software programs that can be executed on IBM compatible PCs under the Windows operating system. The execution of a PIC18F program using the Simulator is an alternative to that of the execution of a program on a hardware trainer. The term IDE indicates that the necessary programs to write, assemble, debug, and execute a program in a simulated environment are grouped together (integrated) in such a way that the results of one program are supplied to the next program to perform the next task. The PIC18 Simulator IDE includes various programs that perform the functions of the following generically named programs:

1. Editor
2. Assembler
3. Debugger
4. Simulator
5. Basic Compiler
6. Interactive Editor

An editor enables the user to write mnemonics (text), an assembler looks up the codes of the mnemonics written by the editor and assigns memory addresses, and a simulator enables the user to observe the internal operations of the PIC18F microcontroller at the instruction level on a PC.

[1]*http://www.oshonsoft.com/* Contact Mr. Vladimir Soso at the web site for questions related to the PIC18 Simulator IDE. For future updates, the user should download the program from *http://www.oshonsoft.com/pic18textbookrelease.html.*

This simulator can execute a program in two modes: Step and Run. In the Step mode (also known as Single Step), one instruction is executed at a time and changes in registers, flags, and output ports can be observed. In the Run mode, the entire program is executed, and the final results can be observed. This simulator is capable of running the programs at various speeds selected by the user.

F.1 HOW TO USE THE PIC18F SIMULATOR:

The PIC18 Simulator IDE can be installed on your hard drive by inserting the CD in its drive. It is a self-extracting program, which means that as soon as you insert the CD, the installation procedure will be displayed on the screen. Follow the instruction, and the simulator will be installed on your C: drive as C:\Program Files \ PIC18 Simulator IDE. If your system does not open up the installation screen when you insert the CD, you can install the simulator by accessing the CD using the My Computer icon or Windows Explorer. As a part of the installation process, the program will list the path that can be accessed through the Start button. The following two symbols are used in this Appendix to describe the steps in using the simualtor: (1) Click: → and (2) Double Click: → →. To open the simulator program, the steps are as follows:

1. Click on Start → Program → PIC18 Simulator IDE → PIC18 Simulator IDE.
2. If you want to create an icon for the simulator on your desktop, drag the PIC18 Simulator IDE on to your desktop instead of clicking it as in Step 1. Now you can → → on the icon PIC18 Simulator IDE.

Simulator Screen: When you → → on the PIC18 Simulator icon, the PIC18F programming model showing various registers appears on the screen. The model also shows the title and menu bars at the top similar to that of any application program (such as Microsoft Word).

The menu bar displays six tasks: File, Simulation, Rate, Tools, Options, and Help. If you click on any of these words, you will see the various functions you can perform as listed below.

File:	Enables you to load an existing file, save a file, and clear memory.
Simulation:	Provides three options: Start, Stop, and Step (single-step execution). The default starting address of the simulation is 0000_H. To begin simulation at any other address, the address in the program counter (PC) must be changed. See Step 1 in Section F1.3 to change the address.
Rate:	Enables you to execute simulation with various speeds beginning with the Single Step.
Tools:	Presently, this provides twenty three different options or tools, also referred to as commands, in the Help menu. The commonly used tools are: (1) Pro gram Memory Editor, (2) Microcontroller View, (3) Assembler, (4) Breakpoint Manager, (5) 8 x LED Board, (6) Keypad Matrix, (7) LCDModule, (8) 7-Segment LED Display, and (9) Watch

Variables. These will be described as necessary with an illustrative program in the next section. These tools (commands) are also described in the Help Topics listed under Help.

Options: There are fifteen options, and they are described in the Help menu. One of the options that must be used the first time is Select Microcontroller.

Help: In this menu, Help Topics includes brief descriptions of various options in the simulator.

F.1.1 Building a Simulated Microcontroller System on the Screen of Your PC

As discussed in Chapter 1, a microcontroller includes three components on a chip: (1) MPU, (2) Memory, and (3) I/O ports. To design a simple system, we need to connect a peripheral device such as LEDs, and write instructions and store them in memory. We can build such a system and represent its internal components on the screen by using the PIC18 Simulator IDE (Figure F-1). The steps are as follows:

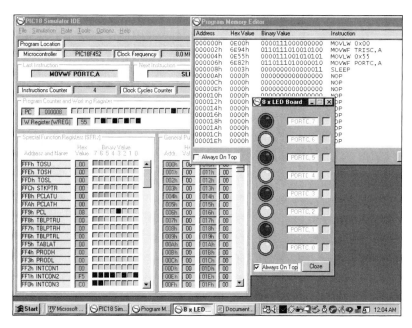

FIGURE F-1   Simulated Microcontroller System

Step 1: Double click on the PIC18 Simulator icon on your desktop. The PIC18F processor with its internal registers is displayed on the screen; it is titled as PIC18 Simulator IDE. Drag the selected PIC18F processor to the left of the screen by using the top bar. Select the microcontroller you are using by selecting Options → and select microcontroller. Select the microcontroller from the list.

Step 2: Select Tool → Program Memory Editor. A 32K flash memory chip with the addresses from 000000 to 007FFF$_H$ is displayed on the screen; it is titled as Program Memory Editor, and it shows addresses in the word format (16 bits) and their contents in Hex and 16-bit binary values. Drag the memory chip to the right of the processor.

Step 3: Select Tool → 8 x LED Board. Drag the LED board and place it under the program memory. The LED board has eight LEDs, and each one can be connected to a pin on any I/O port by selecting a port and a pin. For example, to connect the top LED to PORTC and pin 7, click on the white space to the right of the LED, and the connection dialog opens up. Click PORTC and Pin 7 and → Select. The top LED will be connected to Pin 7 of PORTC, and with appropriate instructions you can turn on that LED.

F.1.2 How to Enter a Program Manually in Memory

Refer to Section 2.6 (Chapter 2) in the text: Illustration – Displaying a Byte at an I/O port. The following program initializes PORTC as an output port and turns on the alternate LED on the 8 x LED board of the simulator. The mnemonics and the Hex code are as follows; the Hex code was obtained by looking up the instruction set manually.

Instructions		Hex Values	Comments
MOVLW	00	0E 00	; Load 0s in W register
MOVWF	TRISC	6E 94	; Place all 0s in TRIS register to set up PORTC as output
MOVLW	0x55[2], 0	0E 55	; Byte to turn on alternate LEDs
MOVWF	PORTC	6E 82	; Output 55$_H$ to turn on alternate LEDs
SLEEP		03 00	; Go into power down mode

Assuming the simulated computer is built on the screen as described in Section F.1.1, Hex code can be entered in memory by clicking on a memory address.

Step 1: To enter Hex code in memory starting from memory location 000000, → on the address 000000. The Memory Editor opens the dialog box and asks you to enter the Hex code.

Enter the code of the first instruction (0E 00 of the example above) and press the Enter key or click on OK. The Memory Editor removes the dialog box, converts the Hex code in binary, and translates it in mnemonics. It displays the Hex code, binary code, and mnemonics in line with the memory address 000000.

Step 2: To enter the next code, → on memory address 000002. You will see again the dialog box. Enter the next code (6E 94 and press Enter). The Hex code, binary code, and mnemonics are displayed against the memory address 000002. Continue to enter the remaining Hex codes. Verify that the program is entered in memory locations from 000000$_H$ to 000008$_H$ (See Figure F-1 – Contents of the Memory Editor from memory locations 000000 to 000008$_H$).

---

[2] In assembly language programming, Hex byte is written as 0xByte or with the postscript H.

F.1.3 Executing (Running) the Program using the Simulator

Step 1: Go to PIC18F Simulator IDE on the screen and → Rate → Step by Step→ Select Simulation → Start. Check the address in the PC register. (It should be 000000. If the address is not 000000, → PC and the dialog box will let you enter the address where you want to begin the execution). **Until you start the simulation, you cannot change the address in the program counter.** As soon as you click on Start, the mnemonics of the first instruction MOVLW 00 will be displayed under the box Next Instruction, indicating that this instruction will be
executed next.

Step 2: To observe the results, adjust the view of the Special Function Registers (SFRs) so that the registers TRISC (F94$_H$) and PORTC (F82$_H$) are visible. Continue to → Step and observe each instruction and the register contents. The simulator always shows the last instruction executed and the next instruction to be executed.

Result of the Simulation:

The first two instructions initialize PORTC as an output port; the contents of TRISC register will be 00. The next instruction loads 55$_H$ in WREG, and the instruction MOVWF PORTC displays 0 1 0 1 0 1 0 1 on the LED port by turning on the alternate LEDs. As each instruction is executed, the program counter PC shows the address of the next instruction to be executed.

F.1.4 Writing a Program Using the Assembler and Executing the Program

Now we will write the same program assembled manually in Section F.1.2. Here it is assembled at the memory location 000020$_H$ (instead of the memory location 000000) using the assembler in the PIC18 IDE. To write this program using the assembler, we need to define the following four pseudo codes or directives (discussed in Chapter 4), which are instructions to the assembler.

ORG	specifies the starting location of memory where the code should be stored
EQU	defines the value of the label (variable) being used
DB	stores a data byte
END	indicates the end of the program to the assembler

In this program, we will introduce only two directives: ORG and END.

Using the Assembler:

Step 1: Open PIC18F Simulator → Tools → Assembler. This opens up a split blank screen divided into two halves— upper and lower. On the menu bar, you get four options: File, Edit, Tools, and Options.

Step 2: Select Files → New. This is an Editor similar to Notepad in Windows that allows you to enter text under a file name. Write the program instructions including the assembler directives ; this is known as the source code, listed below. Do not type anything in column 1 of the screen. Use the Tab key before typing each instruction.

Step 3: Save the program as a file. The Editor adds the default extension .asm to the file. In this illustration, the file is listed as Illust2-6 Displaying Byte.asm. Select Tools → Assemble. This command will assemble the program and list the errors in the lower half of the screen.

Step 4: Correct the errors. Select Tools → Assemble. Repeat the assembly processes until there are no errors. Save the file. At the end of the assembly process, two more files are generated: the List file with the extension .lst, which is used for documentation, and the binary object file with the extension .obj, which is used for execution.

Step 5: Select Tools → Load. This command loads the program Hex code in the simulator memory starting from memory locations 0020_H.

Step 6: Close the Editor.

Step 7: To see where the program is loaded in memory, open the Memory Editor (→ Tools in PIC18F Simulator → Memory Editor). Observe that the Hex codes of the program are stored in memory locations staring from 0020_H.

Step 8: To execute (run) the program using the simulator, follow the steps given in Section F.1.3. Caution: check the starting address in the Program Counter (PC); it will be 000000. Change the address to 0020_H. To change the address → Rate → Step by Step → Simulation → Start → PC. Enter 0020 or just 20 in the dialog box and press Enter.

Listing of Source Program: The program you wrote using the assembler is called the Source Program, and it has a default extension asm. In this example, the Source Program is listed as follows and stored under the file name Illust2-6 Displaying Byte.asm.

Illust2-6 Displaying Byte.asm

```
ORG      0x20      ;Begin assembly at 0000H
MOVLW    00        ;Load 00 into WREG
MOVWF    TRISC     ;Set up PORTC as output port
MOVLW    0x55      ;Load byte 55H to turn on alternate LEDs
MOVWF    PORTC     ;Display 55H at PORTC
SLEEP              ;End of program, power down
END                ;End of assembly
```

F.1.5 Writing a Program Using the Interactive Assembler

The interactive assembler is very useful in the beginning when you are not familiar with the instruction set. It is a graphical method of writing a program by selecting instructions. This interactive editor enables you to write an assembly language program without an error. We will use the same instructions from Section F.1.2: Displaying a Byte at an I/O Port.

Step 1: Open the PIC18F Simulator and → Tools → Interactive Assembler Editor. The screen shows lines on the left with labels starting from L001. The right-hand side of the screen shows various functions such as Delete Line, Insert Line, and Add Lines as shown in Figure F-2.

Step 2: Go to Select Group. This group shows six categories of instructions such as Byte-oriented File Register Operations, Control Operations, and Literal Operations.

Step 3: The first instruction of our program is MOVLW 00. This is a literal operation; loading an 8-bit number in WREG. Follow the next sequence of clicks.

Click Literal Operations → MOVLW k → 0 → 0 → Accept → Append. After the completion of this sequence, you will see the instruction MOVLW 00 on line L001.

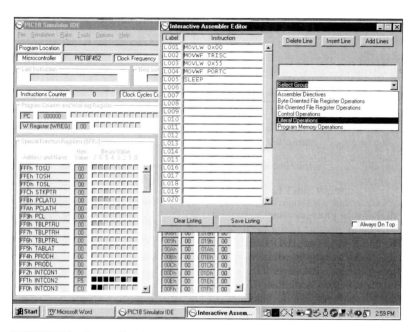

FIGURE F-2   Screen Shot: Interactive Assembler

Step 4: Write all five instructions by selecting appropriate instructions (see Figure F-2).

Step 5: Save the program as a file with a title with the .asm extension and close the Interactive Assembler Editor.

Step 6: Now the program must be assembled. To assemble and load the program → Tools → Assembler → Assemble and Load and close the assembler. Follow the steps in Section F.1.3 to run the program.

F.2 BASIC COMPILER

If you are interested in writing programs using BASIC language instead of the PIC18F assembly language, the IDE includes the BASIC compiler that can compile a program. Once the program is compiled, you can follow the same steps as above to execute and/or simulate the program.

F.3 ADDITIONAL EXAMPLES

F.3.1 Displaying Waveforms on a Simulated Oscilloscope in PIC18 IDE

EXAMPLE F.1

Illustrative Program: Generating Waveforms (Section 5.5.3). This program generates 10 kHz square wave at PORTC, Bit0.

```
    Title "Generating 10 kHz Square Wave"
    List p=18F452, f =inhx32
    #include <p18F452.inc>      ;This is a header file that defines SFRs
                                ; and other parameters


REG1      EQU 0x01             ;Address of Data Register1
REG10     EQU 0x10             ;Address of Data Register10


          ORG     0x20
START:    MOVLW   B'11111110'  ;Load number to set up  Bit0 as an output
          MOVWF   TRISC        ;Initialize Bit0 as an output pin
          MOVWF   REG1         ;Save Bit pattern in REG1


ONOFF:    MOVFF   REG1,PORTC   ;(8 Clk)    Turn on/off Bit0
          MOVLW   D'166'       ;(4 Clk)    Load decimal count in W
          MOVWF   REG10        ;(4 Clk)    Set up REG10 as a counter


LOOP1:    DECF    REG10,1      ;(4 Clk)    Decrement REG10
          BNZ     LOOP1        ;(8/4 Clk)  Go back to LOOP1 if REG 10 =/ 0


          COMF    REG1,1       ;(4 Clk)    Complement Bit Pattern
          BRA     ONOFF        ;(8 Clk)    Go back to change LED
          END
```

Step 1: Write and load the program in the program memory using the PIC18 IDE assembler as outlined in Section F.1.4.

Step 2: Open the Program Memory Editor to display the code and the mnemonics.

Step 3: To display the square wave generated by the program on the simulated oscilloscope, → Tools → Oscilloscope. The four channel oscilloscope is displayed as shown in Figure F-3. The scope shows two functions: Settings and Mode.

Step 4: Click Settings → Turn On/Off Channel 1 → Select Pin. Select PORTC and Bit0. You can also change frequency. The Mode function enables you to change the interval of a wave form.

continued

Step 5: To execute the program starting from the memory location 000020_H, the PC (Program Counter) should have the address 000020_H.

Step 6: To place an address in PC, the simulation should be started in the Step mode. Perform the following steps. Click Rate → Step. Select Simulation and → Start. Click PC. Write the address 000020 in the dialog box. Now go back and change the simulation rate to Ultimate (without refresh). The program generates the square wave at PORTC, Bit0 as shown in Figure F-3.

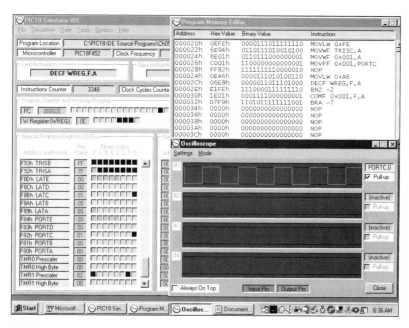

FIGURE F-3   Screen Shot: Simulation of Generating Square Wave

F.3.2 Displaying Digits on a Simulated Seven-Segment LED Panel of PIC18 IDE

EXAMPLE F.2

Connect seven-segment LEDs in the common anode configuration and write instructions to display 2006.

PROGRAM

```
ORG      00
MOVLW    00         ;Byte to enable ports as outputs
MOVWF    TRISA      ;Enable all ports as outputs
MOVWF    TRISB
MOVWF    TRISC
MOVWF    TRISD
MOVLW    0x82       ;Common anode code for digit 6
MOVWF    PORTA
MOVLW    0xC0       ;Common anode code for zero
MOVWF    PORTB
MOVWF    PORTC
MOVLW    0xA4       ;Common anode code for digit 2
MOVWF    PORTD
SLEEP
END
```

Step 1: Write and load the program in the program memory using the PIC18 IDE assembler as outlined in Section F.1.4.

Step 2: To display the digits on the simulated 7-Segment Panel, → Tools → 7-Segment Panel. The four seven-segment LEDs are displayed on the screen. To connect the seven-segment LED on the right to PORTA, follow the instructions given in Step 3.

Step 3: → Setup. An additional display is opened showing a seven-segment LED, eight elements "a" through "h," and eight spaces for eight bits. Click one of the bits (rectangular space showing PORTA, 0). Select PORTA and Bit0 to connect Bit0 to segment "a" and → Select Pin. Continue the same steps until all elements are connected to respective bits. Click Inverted levels, which turns on LED segment with logic 0.

Step 4: Connect the remaining three seven-segment LED to PORTB, C, and D, respectively.

Step 5: To run the program, perform the following steps. Click Rate → Step. Select Simulation and → Start. Click Step and continue to → Step and observe the execution of each instruction. Based on the sequence of the instructions in the program, the rightmost LED is turned showing digit "6." As you continue to single-step, 2006 is displayed as shown in Figure F-4.

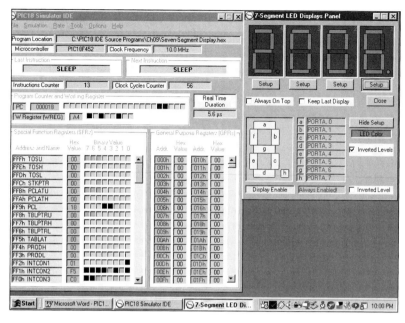

FIGURE F-4   Displaying at Seven-segment LEDs

INDEX